EVOLUTION AND HUMAN BEHAVIOUR

Other books by John Cartwright:

From Phlogiston to Oxygen: A Case Study in Ideas and Evidence from the Chemical Revolution, ASE, 2000
Del Flogisto al Oxígeno, Fundación Canaria Orotava, de Historia de la Ciencia, 2000
Evolutionary Explanations of Human Behaviour, Routledge, 2001
Literature and Science, ABC Clio, 2005 (with B. Baker)

Evolution and Human Behaviour

Darwinian perspectives on human nature

SECOND EDITION

John Cartwright

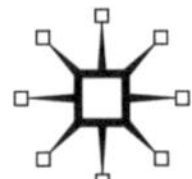

First published 2008 by
PALGRAVE MACMILLAN
Houndmills, Basingstoke, Hampshire RG21 6XS and
175 Fifth Avenue, New York, N.Y. 10010
Companies and representatives throughout the world

PALGRAVE MACMILLAN is the global academic imprint of the Palgrave Macmillan division of St. Martin's Press, LLC and of Palgrave Macmillan Ltd. Macmillan® is a registered trademark in the United States, United Kingdom and other countries. Palgrave is a registered trademark in the European Union and other countries.

ISBN-13: 978–0–333–98632–5
ISBN-10: 0–333–98632–6

This book is printed on paper suitable for recycling and made from fully managed and sustained forest sources. Logging, pulping and manufacturing processes are expected to conform to the environmental regulations of the country of origin.

A catalogue record for this book is available from the British Library.

A catalog record for this book is available from the Library of Congress.

10 9 8 7 6 5 4 3 2 1
17 16 15 14 13 12 11 10 09 08

Printed in China

To Margaret

Brief Contents

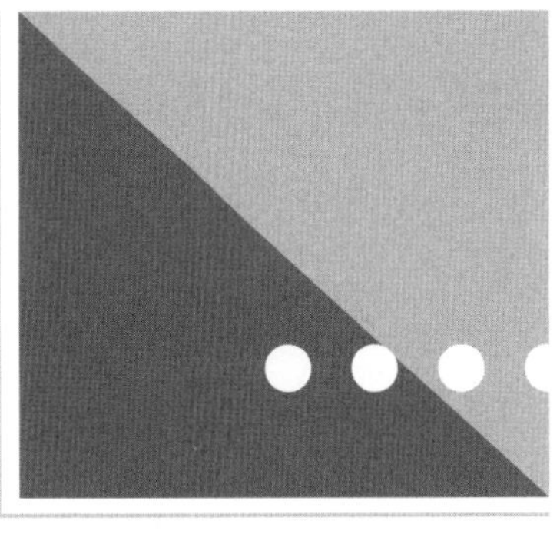

Full Contents

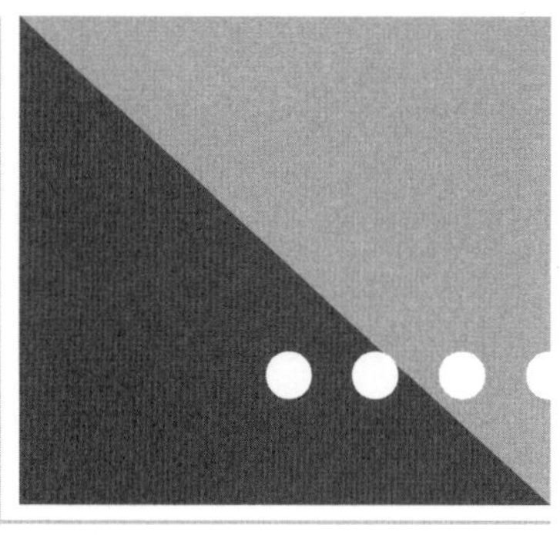

List of Figures

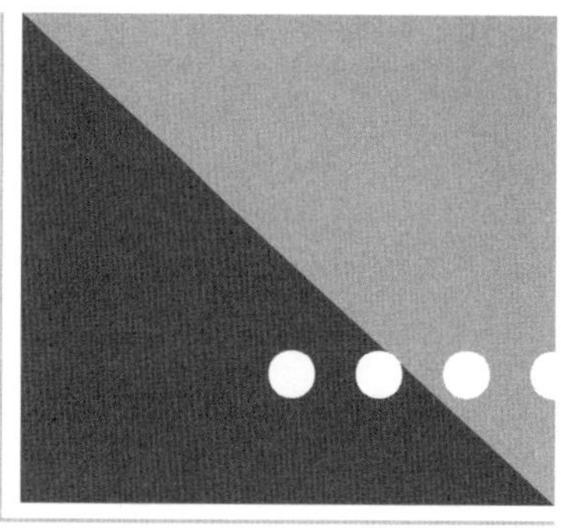

List of Tables

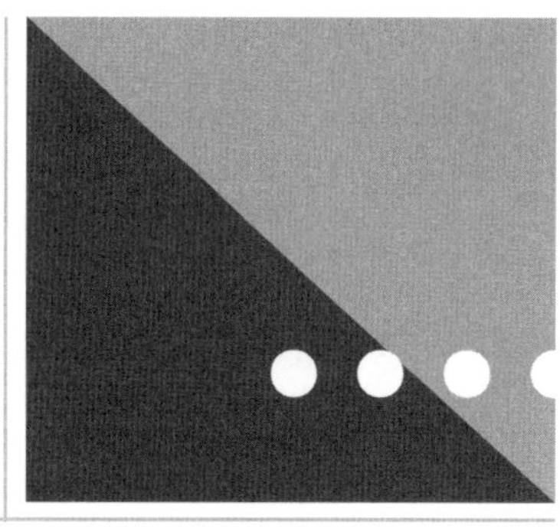

List of Boxes

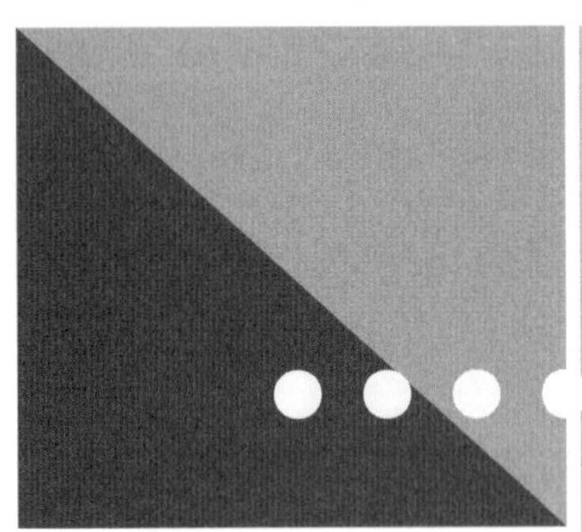

Preface to the Second Edition

This volume represents a considerable expansion over the first edition in terms of both length and range. I have divided the material into seven parts that I think group the concerns and achievements of this field into coherent wholes. I have retained the idea that the scientific content should be sandwiched between introductory and concluding chapters dealing with the historical, philosophical and sociological ramifications of this subject. This is a strongly held conviction of mine: that science has a history and a philosophy worthy of study and does not take place in social isolation.

Part I contains revised material from the first edition and deals with the history of the subject and its fundamental theoretical principles and insights. Many books on evolutionary psychology take the facts of human evolution for granted, but I think it is important to understand in some detail the selective forces that have shaped our journey from Old World primates some seven million years ago to the globally dominant species of today, and so Part II deals with human origins and evolution. It also considers that most puzzling and important question: What factors led to the massive growth in brain size in the hominin lineage? This question is crucial since, as evolutionary psychologists are fond of pointing out, we will have a better understanding of how the brain works when we know the purposes for which it was designed.

In Part III, I have added a complete new chapter on emotions and have included much new material on cognition and reasoning. One of the most successful applications of evolutionary psychology is the understanding of human mating behaviour and mate choice. These topics, as in the previous edition, are given due prominence and in this volume are given a section (Part V) of their own. I am always surprised that most evolutionary psychology texts pay little attention to the Westermarck effect. It seems to me to be crucial to understand why we choose not to mate with some individuals (most notably our kin) and the Westermarck effect promises to explain this in a thoroughly Darwinian fashion. It also serves as a good case study of the interplay between biology and cultural mores. I have, therefore, devoted Chapter 13 to this topic.

If evolutionary psychology really is the long-awaited unifying paradigm that psychology needs, then it should be able to account for both species-typical behaviour and occasions when the mind ceases to function normally. To this end, I have included two chapters in Part VI on the evolutionary approach to mental disorders. This field is in its infancy, but it seems to me that evolutionary psychology is now mature enough to hold its own alongside the variety of other approaches (for example psychodynamic, humanistic and behavioural) that wrestle with the difficult problem of explaining mental illness.

It was with some trepidation that I ventured to write a chapter on ethics, but if, as part of the paradigm of naturalism, evolutionary psychology offers to provide a scientific account of the mind in all its manifestations, then it should be able to illuminate the nature of moral reasoning and help to clarify the source of our strong moral passions. After all, it is not scientifically credible that the origin of our moral convictions should lie outside the plane of human nature.

More generally, in this work I have not detached psychology from its neighbouring disciplines and have tried to convey to the reader the web of complementarity that exists between genetics, evolutionary biology, anthropology, behavioural ecology and the study of the human mind. Overall, I have been mindful of the sentiments and ambitions advanced by E. O. Wilson in his influential work *Consilience* (1998): that we should strive for unity and consistency in our knowledge. I have tried, therefore, to identify the clear bridges and signposts between disciplines that enable the student to walk (admittedly with effort and application) between the different branches of learning without encountering immovable obstacles or irreconcilable foundational principles. Such Enlightenment goals seem to me to remain as urgent as ever.

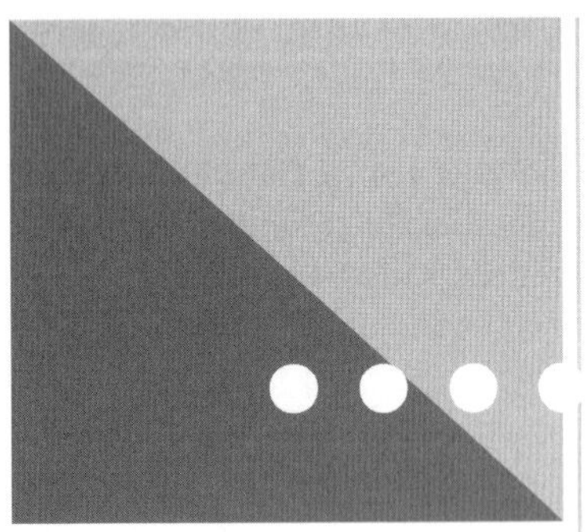

Acknowledgements

I owe a huge debt of gratitude, of course, to fellow scientists and researchers identified throughout the text who have produced the results and ideas on which this book is based. I am also grateful to Anna van Boxel and her team at Palgrave Macmillan for advice and primarily for accepting my argument that to do justice to the subject matter a book such as this needed to be at least twice as long as the originally contracted length. I hope the finished product justifies her faith in the project. I would also like to thank the three anonymous reviewers who offered encouraging feedback on the manuscript and made helpful suggestions for improvement.

Thanks are also due to two departments at the University of Chester: the Department of Biological Sciences for allowing me the flexibility to produce such a work as this, and Graphic Services (especially Angela Bell and Gary Martin) for assisting with several diagrams in this book.

I must also acknowledge the invaluable help offered by Linda, Julie, Maggie and Rose at Aardvark Editorial Ltd. Their diligent and remarkably penetrating editing removed a large number of typos, inconsistencies and infelicitous turns of phrase that otherwise would have been inflicted on a bemused reader. I take responsibility, of course, for any errors that remain.

The author and publisher would like to thank the following organisations and individuals for their help and permission in obtaining the following images and diagrams: Academic Press for Figure 1.2 from Sternglanz et al. (1977) 'Adult preferences for infantile facial features', *Animal Behaviour*, **25**: 108–15; Science Photo Library for Figures 1.2, 1.5 and 2.5 (© A. Barrington Brown); the Editorial Board of *Evolutionary Psychology* for Figure 1.6, adapted from Cornwell, R. E., Palmer, C. et al. (2005) 'Introductory psychology texts as a view of sociobiology/evolutionary psychology's role in psychology.' *Evolutionary Psychology*, **3**: 355–74; Still Pictures for Figure 5.4 (© Mark Carwardine); Chapman and Hall for Figure 3.15, adapted from Voland, E. and Engel, C. (1989) Women's reproduction and longevity in a premodern population, in E. Rasa, C. Vogel and E. Voland (eds) *The Sociobiology of Sexual and Reproductive Strategies*, London, Chapman & Hall; Cambridge University Press for Figure 5.5, adapted from Friday, A. E. (1992) Human evolution: the evidence from DNA sequencing, in S. Jones, R. Martin and D. Pilbeam (eds) *The Cambridge Encyclopedia of Human Evolution*, Cambridge, Cambridge University Press; Springer Publishers for Figure 6.1, adapted from Van Dongen, P. A. M. (1998) Brain size in vertebrates, in R. Nieuwenhuys (ed.) *The Central Nervous System of Vertebrates*, Berlin, Springer, vol. 3: 2099–134; Oxford University Press for Figure 6.3, adapted from Young, J. Z. (1981) *The Life of Vertebrates*, Oxford, Oxford University Press; Cambridge University Press for Figure 6.4, adapted from Deacon, T. W. (1992) The human brain, in S. Jones, R. Martin and D. Pilbeam (eds) *The Cambridge Encyclopedia of Human Evolution*, Cambridge, Cambridge University Press; Oxford University Press and D. Bygott for Figure 6.8, taken from Byrne, R. W. (1995) *The Thinking Ape*, Oxford, Oxford University Press; Cambridge University Press for Figure 6.10, adapted from Dunbar, R. I. M. (1993) 'Coeveolution of neocortical size, group size and language in humans,' *Behavioural and Brain Sciences*, **16**: 681–735; Oxford University Press for Figure 6.11, adapted from Byrne, R. (1995) *The Thinking Ape*, Oxford, Oxford University Press (Figure 14.3, p. 220); Dr Lindsay Murray for a photograph of a common chimp and Dr Allison Fletcher for a photograph of a gorilla, both in Box 6.1; W. H. Freeman for Figure 6.13, based on Passingham, R. (1988) *The Human Primate*, Oxford, W. H. Freeman; Cambridge University Press for Figure 6.15, adapted from Dunbar, R. I. M. (1993) 'Coeveolution of neocortical size, group size and language in humans,' *Behavioural and Brain Sciences*, **16**: 681–735; Adrian Pingstone and *Wikipedia* for the Andelson Chequered square shown in Box 7.1; the American Psychological Association for Figure 7.1, modified from Geary, D. C. (1998) *Male, Female: The Evolution of Human Sex Differences*, Washington, DC (Figure 6.4, p. 180); Oxford University Press for Figure 7.3, modified from Tooby, J. and Cosmides, L. (1992) The psychological foundations of culture, in J. H. Barkow, L. Cosmides and J. Tooby (eds) *The Adapted Mind*, Oxford, Oxford University Press; Oxford University Press for the figures shown in Box 7.7, reproduced from Silverman, I. and Eals, M. (1992) Sex differences in spatial abilities: evolutionary theory and data, in J. H. Barkow, L. Cosmides and J. Tooby (eds) *The Adapted Mind*, Oxford, Oxford University Press (Figure 14.3, p. 538); *Scientific American* and Patrica Wynne for Figure 9.6, adapted from Wilkinson, G. S. (1990) 'Foodsharing in vampire bats,' *Scientific American*, **262**: 76–82; *Wikipedia* Commons for Figure 9.11 A ration party of the Royal Irish rifles; Aldine de Gruyter for Figures 10.5, 10.6 and 10.8, adapted from Daly, M. and Wilson, M. (1988a) *Homicide*, New York, Aldine De Gruyter; Still Pictures for Figure 10.9 (© Shehzad Noorani) and 11.2 (© Mark Edwards); the Library of Congress Prints and Photographic Division for Figure 11.3; Blackwell Scientific for Figures 11.4 and 11.5, both adapted from Harvey, P. H. and Bradbury, J. W. (1991) Sexual selection, in J. R. Krebs and N. B. Davies (eds) *Behavioural Ecology*, Oxford, Blackwell Scientific; Academic Press for Figure 11.6, adapted from Birkhead, T. R. and Moller, A. P. (1992) *Sperm Competition in Birds: Evolutionary Causes and Consequences*, London, Academic Press; Cambridge University Press for Figure 11.7, adapted and redrawn from Short, R. V. and Balban, E. (eds) (1994) *The Differences Between the Sexes*, Cambridge, Cambridge University Press; the American Psychological Association and David Schmitt for Figure 11.9, adapted from Schmitt, D. (2003) 'Universal sex differences in the desire for sexual variety: tests from 52 nations, 6 continents and 13 islands,' *Journal of Personality and Social Psychology*, **85**(1): 85–104 (Figure 1, p. 92); Cambridge University Press and David Schmitt for Figure 11.10, adapted from Schmitt, D. (2005) 'Sociosexuality from Argentina to Zimbabwe: a 48-nation study of sex, culture, and strategies of human mating,' *Behavioural and Brain Sciences*, **28**: 247–311 (Figure 9.2, p. 278); Elsevier for Figure 12.3 data, from Greenless, I. A. and McGrew, W. C. (1994) 'Sex and age differences in preferences and tactics of mate attraction: analysis of published advertisements,' *Ethology and Sociobiology*, **15**: 59–72; and Figure 12.5 data, from de Sousa Campos et al. (2002) 'Sex differences in mate selection strategies: content analyses and responses to personal advertisements in Brazil,' *Evolution and Human Behaviour*, **23**(5): 395–406; *New Scientist* for Figure 12.4, adapted from Dunbar, R. I. M. (1995a) 'Are you lonesome tonight,' *New Scientist*, **145** (1964): 12–16; the American Psychological Association and Devendra Singh for Figure 12.6, modified from Singh, D. (1993) 'Adaptive significance of female attractiveness,' *Journal of Personality and Social Psychology*, **65**: 293–307, and Figure 12.8, modified from Singh, D. (1995a) 'Female judgement of male attractiveness and desirability for relationships: role of waist-to-hip ratios and financial status,' *Journal of Personality and Social Psychology*, **69**(6): 1089–101; Elsevier and Martin Tovee for Figure 12.12, adapted from Tovee, M. J., Swami, V. et al. (2006) 'Changing perceptions of attractiveness as observers are exposed to different culture,' *Evolution and Human Behaviour*, **27**(6): 443–57 (Figure 1, p. 448); Elsevier for Figure 12.15, from Penton-Voak, I. S. and Perrett, D. I. (2000) 'Female preference for male faces changes cyclically: further evidence,' *Evolution and Human Behaviour*, **21**(1): 39–49; Elsevier for Figure 12.16, redrawn from Penton-Voak, I. S. and Perrett, D. I. (2000) 'Female preference for male faces changes cyclically: further evidence,' *Evolution and Human Behaviour*, **21**(1): 39–49; Elsevier and Victor Johnston for Figure 12.17, from Johnston, V. S., Hagel, R. et al. (2001) 'Male facial attractiveness: evidence for hormonemediated adaptive design,' *Evolution and Human Behaviour*, **22**: 251–69; The Library of Congress Prints and Photographic Division for Figure 13.1; Abo Akademis Bildsamlingar for Figure 13.2; Stanford University Press and Arthur Wolf for Figure 13.3, from Wolf, A. (2004) Introduction, in A. P. Wolf and W. H. Durham (eds) *Inbreeding, Incest and the Incest Taboo*, Stanford, CA, Stanford University Press (Figure 4.1, p. 80); Stanford University Press for Figure 13.4 from Pusey, A. (2004) Inbreeding avoidance in primates, in A. P. Wolf and W. H. Durham (eds) *Inbreeding, Incest and the Incest Taboo*, Stanford, CA, Stanford University Press (Figure 3.1, p. 69); Oslo Kommunes Kunstsamlinger, Munch–museet for Figure 14.1, lithograph by E. Munch; the

American Psychological Association for Figure 14.4, adapted from Ohman, A. and Mineka, S. (2001) 'Fears, phobias, and preparedness: toward an evolved module of fear and fear learning,' *Psychological Review,* **108**(3): 483–522, and Figure 15.1, taken from Hagen, E. H. (1999) 'The functions of post-partum depression,' *Evolution and Human Behaviour,* **20**: 325–59 (Figure 1 [data from Field, T., Sandburg, S., Garcia, R. et al. (1985) 'Pregnancy problems, postpartum depression, and early mother–infant interactions,' *Developmental Psychology,* **21**(1): 1152–6]); Elsevier for Figure 15.2, adapted from Lalumière, M. L., Harris, G. T. et al. (2001) 'Psychopathy and developmental instability,' *Evolution and Human Behaviour,* **22**: 75–92; Elsevier for Figure 16.3, taken from Bloom, G. and Sherman, P. W. (2005) 'Dairying barriers affect the distribution of lactose malabsorption,' *Evolution and Human Behaviour,* **26**(4): 301–13 (Figure 2, p. 306); Edinburgh University Press for Figure 16.4, adapted from Miller, G. F. (1999a) Sexual selection for cultural displays, in R. Dunbar, C. Knight and C. Power (eds) *The Evolution of Culture*, Edinburgh, Edinburgh University Press; the Library of Congress Prints and Photographs Division for Figure 18.2.

Every effort has been made to trace all the copyright holders but if any have been inadvertently overlooked the publishers will be pleased to make the necessary arrangements at the first opportunity.

Fundamentals of the Evolutionary Approach

Historical Introduction: Evolution and Theories of Mind and Behaviour, Darwin and After

The reason why psychologists have wandered down so many garden paths is not that their subject is resistant to the scientific method, but that it has been inadequately informed by selectionist thought. Had Freud better understood Darwin, for example, the world would have been spared such fantastic dead-end notions as Oedipal desires and death instincts.
(Daly, 1997, p. 2)

History, if viewed as a repository for more than anecdote or chronology, could produce a decisive transformation in the image of science.
(Khun, 1962, p. 1)

Before Thomas Kuhn wrote those opening words of his *magnum opus*, *The Structure of Scientific Revolutions* (1962), there was a tendency among some historians of science (and especially professional scientists) to portray the growth of science as a steady linear progression from ignorance towards enlightenment, with ideas getting better and more robust as the discipline progressed. The history of psychology illustrates just how naive this image is, and there is much truth behind Daly's remark that psychologists, especially in the 20th century, did indeed wander down many unproductive garden paths. Borrowing Kuhn's terminology, we might say that for most of its history, psychology has lacked a unifying **paradigm** – a set of procedures, assumptions, methodologies and background theories that all its practitioners can agree upon. There are those who would argue that evolutionary psychology has the potential to supply this missing paradigm, and this book is partly an attempt to explore that claim.

Darwin (Figure 1.1) published his two greatest books on evolutionary theory, *On the Origin of Species by Natural Selection*, and *The Descent of Man and Selection in Relation to Sex*, in 1859 and 1871 respectively, and he was convinced that a revolution in psychology would shortly follow. But for the first three-quarters of the 20th century, while biology became more securely based on deepening evolutionary foundations, psychology failed lamentably to exploit the potential of Darwinian thought. There were some exceptions – William James being the most notable – but many psychologists either ignored Darwinism or, more damagingly, misunderstood the message that it held. Psychology was the poorer as a result.

FIGURE 1.1 Charles Darwin (1809–82) from a photograph taken by Elliot and Fry, London (c.1875)

SOURCE: Darwin, F. (1902) *The Life of Charles Darwin.* London, John Murray.

The reasons for this missed opportunity and the background to more recent developments in animal ecology, sociobiology and evolutionary psychology that look to have finally, and one can only hope irreversibly, reinstated lines of communication are the subjects of this chapter. The first half deals with efforts to construct a science of animal behaviour and examines the early hopes, expressed by Darwin and his followers, that the study of animal minds would throw light on human psychology. The second half of the chapter reviews attempts to apply evolutionary ideas directly to human behaviour in the years following Darwin and up to the early 1970s. Early efforts in this direction were weakened by an incomplete understanding of heredity and genetics. When, by the 1930s, a better understanding of inheritance came along, it was too late: powerful ideologies had corrupted the whole enterprise. One theme also explored in this chapter is the influence of the social context on scientific ideas. Although science has its own internal logic and momentum, there are always social forces at work shaping any scientific undertaking. The history of evolutionary theorising about man must necessarily, therefore, take into account the social and political factors that have clouded the study of human nature.

Darwinism began in 1859 when Darwin, in his fiftieth year, finally published his masterwork *On the Origin of Species by Means of Natural Selection*. The book, originally intended as an abstract of a much larger volume, contained concepts and insights that had occurred to Darwin at least 15 years earlier, yet he had wavered and delayed before publishing. The larger volume never appeared, and Darwin was forced to rush out his *Origin* following a remarkable series of events that began in June of the previous year. It is 1858, therefore, that serves as a convenient starting point.

1.1 The origin of species

On 18 June 1858, Darwin received a letter from a young naturalist called Alfred Russel Wallace, then working on the island of Ternate in the Malay Archipelago. When Darwin read its contents, he felt his world fall apart. In the letter was a scientific paper in the form of a long essay entitled 'On the tendency of varieties to depart indefinitely from the original type'. Wallace, innocent of the irony, wondered whether Darwin thought the paper important and 'hoped the idea would be as new to him as it was to me, and that it would supply the missing factor to explain the origin of species' (Wallace, 1905, p. 361). The ideas were far from new to Darwin: they had been an obsession of his for half a lifetime. Wallace had independently arrived at the same conclusions that Darwin had reached at least 14 years earlier, and the demonstration of which Darwin saw as his life's work. Darwin knew that the essay must be published and, in a miserable state, exacerbated by his own illness and fever in the family, wrote for advice to his geologist friend and scientific colleague Sir Charles Lyell, commenting that he 'never saw a more striking coincidence' and lamenting that 'all my originality, whatever it may amount to, will be smashed' (Darwin, 1858).

Fortunately for Darwin, powerful friends arranged a compromise that would recognise the importance of Wallace's ideas and simultaneously acknowledge Darwin's previous work on the same subject. A joint paper, by Wallace and Darwin, was to be read out before the next gathering of the Linnean Society on 1 July 1858. The reading was greeted by a muted response. The president walked out, later complaining that the whole year had not 'been marked by any of those striking discoveries which at once revolutionise, so to speak [our] department of science' (Desmond and Moore, 1991, p. 470). At Down House, Darwin remained in an abject state, coping with a mysterious physical illness that plagued him for the rest of his life and nursing a nagging fear that it might seem as if he had stolen the credit from Wallace. He was also grieving: his young son Charles Waring had died a few days earlier. As the Linnean meeting proceeded, Darwin stayed away and attended the funeral with his wife Emma. By the end of the day, the theory of evolution by natural **selection** had received its first public announcement, and Darwin had buried his child.

After the Linnean meeting, Darwin set to work on what he thought would be an abstract of the great volume he was working on. The abstract grew to a full-length book, and his publisher, Murray, eventually persuaded Darwin to drop the term 'abstract' from the title. After various corrections, the title was pruned to *On the Origin of Species by Means of Natural Selection*, and Murray planned a print run of 1,250 copies.

Darwin, amid fits of vomiting, finished correcting the proofs on 1 October 1859. He then retired for treatment to the Ilkley Hydropathic Hotel in Yorkshire. In November, Darwin sent advance copies to his friends and colleagues, confessing to Wallace his fears that 'God knows what the public will think' (Darwin, 1859a). Many of Darwin's anxieties were unfounded. When the book went on sale to the trade on 22 November, it was already sold out. It was an instant sensation, and a second edition was planned for January 1860. Thereafter, man's place in nature was changed, and changed utterly.

1.1.1 New foundations

In the *Origin*, Darwin was decidedly coy about the application of his ideas to humans, but the implications were clear enough and, in the years following, both Darwin and Huxley began the process of dissecting and exposing the evolutionary ancestry and descent of man. It was towards the end of the *Origin*, however, that Darwin made a bold forecast for the future of psychology:

> In the distant future I see open fields for far more important researches. Psychology will be based on a new foundation, that of the necessary acquirement of each mental power and capacity by gradation. Light will be thrown on the origin of man and his history. (Darwin, 1859b, p. 458)

On the origin of man, Darwin was right, and light continues to be thrown with each new fossil discovery. On psychology, the new foundation that Darwin foresaw has been slow in coming. Over the past 20 years, however, there have been signs that a robust evolutionary foundation is being laid that promises to sustain a thoroughgoing Darwinian approach to understanding human nature. The bulk of this book is about those foundations.

1.2 The study of animal behaviour

A number of disciplines have laid claim to providing an understanding of animal behaviour, including **ethology**, comparative psychology, behavioural ecology and, emerging in the 1970s, sociobiology. The problem for the historian is that these terms were not always precisely defined and the disciplines frequently overlapped. It is appropriate, therefore, to consider the origins of comparative psychology and ethology together.

1.2.1 Comparative psychology and ethology: the 19th-century origins

Comparative psychology

For Darwin, it was clear and indisputable that all life on earth had evolved from lowly origins. Behaviour, morphology and physiology had all been shaped by the twin forces of natural and sexual selection. In this sense, Darwin subscribed to what can be called 'psychoneural **monism**' – the idea that mind and body are not separate entities. Scientific **naturalism**, in accounting for life and its origins, had seemed finally to square the circle. In respect of man, Wallace agreed with Darwin that in bodily structure, humans had descended from an ancestor common to man and the anthropoid apes, but he objected to the view that natural selection could also explain man's mental faculties. Darwin was, however, determined to push through his programme. He stressed the essential unity between animal and human minds, noting that 'there is no fundamental difference between man and the higher animals in their mental faculties' (Darwin, 1871, p. 446).

It was Darwin's *The Expression of the Emotions in Man and Animals*, published in 1872, that more than any other book acted as a spur to the study of both ethology and comparative psychology. In this work, Darwin, in the tradition of the day, described the behaviour of animals using terms that were extrapolated from the mental life of humans. For Darwin, no human mental function was unique: the way to understand human minds was by invoking processes found in the minds of other animals. Darwin was quite clear that instincts were also adaptations: if they varied a little between individuals, and he thought they did, then natural selection had something to act upon. Although he observed the behaviour of his pets, animals at the zoo and his children, Darwin made few experiments of his own on animal behaviour and for the most part relied upon anecdotal information given to him by naturalists, zoo-keepers and other correspondents. With hindsight, the anecdotal approach appears fatally flawed. Darwin's lasting contribution to the ethological approach was, however, to provide an evolutionary framework to the study of behaviour, and to highlight the value of observing animals in their natural settings.

There were many who sympathised with the misgivings of Wallace and, to defend Darwin's approach, there grew up what has been called the 'anecdotal movement' (Dewsbury, 1984). Anecdotalists such as George John Romanes were so called because they relied upon reports and unsystematic observations of animal behaviour to provide an empirical base for their writings. To demonstrate the continuity between animal and human minds, it was thought necessary to show that animals could reason, show complex forms of social behaviour and display human-like emotions. Romanes is important to the history of animal psychology in the sense that it was the reaction against these ideas that was to shape future developments. In his book *Mental Evolution in Man* (1888), Romanes advanced the doctrine of levels of development that could be expressed on a numerical scale. In this system, humans were born at level 16, reaching the level of insects at 10 weeks and that of dogs and the great apes at about 15 months. Romanes also suggested that human emotions were common to other animals according to the complexity involved. Fish, for example, he thought capable of experiencing jealousy and anger, birds pride and resentment, and apes shame and remorse (Romanes, 1887, 1888). But Romanes was as much a Lamarckian as a Darwinist in his thinking and, despite his background in physiology, he conducted few, if any, controlled experiments on animals. Like Herbert Spencer, he thought that habits could become fixed as instincts. In the end he resorted to a sort of Lamarckian progressivism in which animal minds were viewed as evolving towards a human type of intelligence.

In 1894, the year that Romanes died, two books were published that were pivotal in defining the way ahead. The first was Wilhelm Wundt's *Lectures on Human and Animal Psychology*, in which Wundt (1832–1920) criticised the anecdotal method of Romanes. The second book, perhaps more significant, was *An Introduction to Comparative Psychology* (1894) by Conwy Lloyd Morgan (1852–1936). Morgan was a pioneer of comparative psychology and did much to establish the scientific credentials of the emerging discipline and place it on an evolutionary basis. He was one of the first comparative psychologists to employ controlled experiments. He observed the behavioural development of young chicks, for example, and recorded the effect of environmental changes on their behaviour. By such experiments, he came to the conclusion that the anthropomorphic ideas of the Darwin–Romanes school of anecdotalism were fatally flawed. He became the first scientist to be elected to the Royal Society of Great Britain for work in psychology. In his book, Morgan rejected the **Lamarckian inheritance** of acquired traits and outlined his famous canon:

> In no case may we interpret an action as the outcome of the exercise of a higher psychical faculty, if it can be interpreted as the outcome of the exercise of one which stands lower in the psychological scale. (Morgan, 1894, p. 53)

Morgan's canon is of course a sort of law of parsimony in the style of what is called 'Occam's razor'; that is, where two explanations are possible, we choose the simplest and resist entertaining unnecessary hypotheses. In this instance, scientists were urged to avoid interpreting animal behaviour in terms of the thoughts and feelings experienced by humans. The premises underlying Morgan's canon were surprisingly modern in their conception and firmly based on evolutionary logic. Morgan thought that relatively few innate traits, coupled with an inherent learning process (such as the ability to establish links between cause and effects), could yield quite complex forms of behaviour. Therefore, he reasoned, why assume the evolution of complex and costly behavioural processes when simpler and cheaper ones would suffice (Plotkin, 2004).

Many of the early comparative psychologists were concerned with learning. Important and pioneering work in trial and error learning was conducted by the American Edward Lee Thorndike (1874–1949). A typical experimental arrangement set up by Thorndike was that an animal would receive a reward for a type of behaviour that was initially discovered accidentally. The time taken to repeat the behaviour was taken as an indication of learning. Predictably, behaviours that met with favourable consequences for the animal were learnt quickly. This learning process was termed **operant conditioning**, and the acceleration of learning through positive reward became known as the 'law of effect'. After the First World War, Thorndike became almost exclusively concerned with human psychology. In rejecting the anecdotal approach of Romanes and moving towards laboratory studies of caged animals, Thorndike and others were beginning a trend that was to dominate comparative psychology for the next 50 years.

Ethology

The word **ethology** is derived from the Greek word *ethos* meaning character or trait. In Britain, there was a long and, by the late 18th century, popular tradition of natural history involving the observation and documentation of the behaviour of animals. However, the science of ethology probably began in France, with its early pioneers including Jean Baptiste Lamarck (1744–1829), Etienne Geoffroy Saint-Hilaire (1772–1844) and Alfred Giard (1846–1908). In contemporary biology, the name of Lamarck is still well known as that of someone who provided the only serious rival to the mechanism of natural selection as a way of accounting for the preservation of favoured traits. In 1809, the year in which Darwin was born, Lamarck published *Philosophie Zoologique*, in which he advanced the view that species could change over time to new species (transmutationism). One mechanism for such changes that Lamarck suggested was the idea that organisms could, through their own efforts, modify their form, and these modifications could be passed to the next generation. Although this was a small part of Lamarck's entire theory, the notion that acquired characters can be inherited has become indelibly associated with his name (see Chapter 2). Lamarck made few friends among French scientists. He acquired a reputation for publishing inaccurate weather forecasts, which did his career no good; more tellingly, his views on biology were denigrated by the French anatomist George Cuvier (1769–1837). Lamarck died blind and poor in 1829.

In the late 19th century, the concerns of comparative psychology and ethology overlapped, and the careers of some scientists straddled what were only later to become divergent disciplines. Lloyd Morgan, for example, is often cited as being one of the founding fathers of both ethology and comparative psychology. It was during the 20th century that differences in training, methodologies and even fundamental assumptions about the nature of animals led to a schism.

1.2.2 Ethology and comparative psychology in the 20th century

Ethology 1900–70

One of the giants of 20th-century ethology was the Austrian Konrad Lorenz (Figure 1.2). Lorenz originally trained as a doctor but was influenced by the work of Oscar Heinroth at the Berlin Zoological Gardens on the behaviour of birds. Heinroth exploited the analogy between animals and humans in both directions. Animals could be understood using concepts drawn from the mental life of humans, and this understanding could then be reapplied to understand the human condition. Lorenz frequently expressed his debt to Heinroth's approach.

FIGURE 1.2 Konrad Lorenz (1903–89)
The Austrian zoologist is here shown being followed by a group of ducklings as if he were their mother. This is an instinct Lorenz called 'imprinting'.

By conducting experiments at his home on the

outskirts of Vienna, Lorenz observed numerous features of animal behaviour that have become associated with his name. In one classic study, he noted how a newly hatched goose chick will 'imprint' itself on the first moving object it sees. In some cases, this was Lorenz himself, and chicks would follow him about, presumably mistaking him for their mother. Lorenz stressed the importance of comparing the behaviour of one species with another related one and argued for the importance of understanding evolutionary relationships between species. In this respect, Lorenz unashamedly drew parallels between the behaviour of humans and other animals. In his most popular work, *King Solomon's Ring*, for example, he suggested that the

> war dance of the male fighting fish ... has exactly the same meaning as the dual of words of the Homeric heroes, or of our Alpine farmers, which, even today, often precedes the traditional Sunday brawl in the village inn. (Lorenz, 1953, p. 46)

One of Lorenz's early concepts was that of the **fixed action pattern**, which referred to a pattern of behaviour that could be triggered by some external stimulus. Using a term that was later to prove so troublesome for ethology, Lorenz regarded these action patterns as 'instincts' forged by natural selection and common to each member of the species. Fixed action patterns have the following characteristics:

- Their form is constant, that is, the same sequence of actions and the same muscles are used
- They require no learning
- They are characteristic of a species
- They cannot be unlearnt
- They are released by a stimulus.

Evidence of a fixed action pattern that is often cited is the observation that a female greylag goose *(Anser anser)* will retrieve an egg that has rolled outside her nest by rolling it back using the underside of her bill. Lorenz noticed that this action continued even when the egg was experimentally removed once the behaviour had begun. Once it started, the behaviour had to finish, whether it was effective or not. The stimuli that trigger fixed action patterns became known as 'sign stimuli' or, if they were emitted by members of the same species, 'releasers'. An interesting example is to be found in the behaviour of the European robin *(Erithacus rubecula)*, documented by the British ornithologist David Lack in the 1940s. Lack showed that the releaser for male aggression in this species is the patch of red found on the breast of the bird. A male robin will attack another male that it finds in its territory, but it will also attack a stuffed dead robin and even a tuft of red feathers (Lack, 1943).

Once the essence of the stimulus has been identified, it becomes possible in some cases artificially to exaggerate its characteristics and create supernormal stimuli. If a female oyster catcher *(Haemotopus ostralegus)*, for example, is presented with a choice of egg during incubation, it will choose the larger. Even if an artificial egg twice the size of its own is introduced, the oyster catcher still prefers the larger one, even though common sense (to an outsider) would indicate that it is unlikely to be an egg actually laid by the bird.

Box 1.1 Fixed action patterns

One of the few areas in which the concept of fixed action patterns has been successfully transferred from ethology to human psychology concerns the response of adults to babies and the early reactions of newborn babies. Newborn humans show a number of potential fixed action patterns, such as the grasping motion of hands and feet if something touches the palm or sole of the foot. There is also a programmed search for the nipple, which consists of a sideways movement of the head if the lips are touched. Up until about two months of age, even eye-sized spots painted on cardboard will elicit a smiling reaction from a child. Lorenz suggested that the facial features of human babies, such as a large forehead, large eyes, small chin and bulging cheeks, serve as social releasers that act upon **innate releasing mechanisms** to activate nurturant and affectionate behaviour. This intriguing idea could in part explain why humans find baby-like faces, be they on humans, young animals or even toys such as teddy bears, alluring. There does appear to be considerable experimental support for Lorenz's idea (Eibl-Eibesfeldt, 1989; Archer, 1992). Sternglanz et al. (1977) investigated the effect of varying the features of a baby's face on its perceived attractiveness as ajudged by American college students. By altering various parameters of line drawings, the overall conclusion was, as Lorenz suggested, a marked preference for faces with large eyes, large foreheads and small chins. A composite drawing combining all the features with high attractiveness ratings is shown in Figure 1.3.

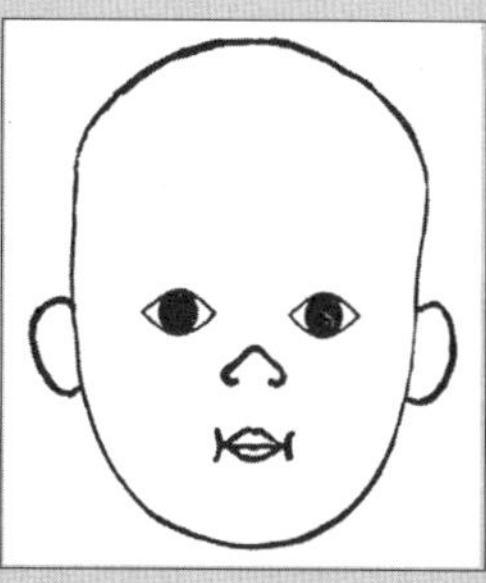

Figure 1.3 Composite drawing of the ideal infantile face

SOURCE: Redrawn after Sternglanz, S. H., Gray, J. L. and Murakami, M. (1977) 'Adult preferences for infantile facial features: an ethological approach.' *Animal Behaviour* **25**: 108–15, Figure 7.

Lorenz had little interest in the individual variation of instincts displayed by different members of the same species, a subject now of great interest to behavioural ecologists. Burkhardt (1983) suggests that this neglect of intraspecific variation in behaviour was partly a reflection of the fact that Lorenz wished to distance himself from animal psychologists and their work on captive animals. Lorenz distrusted inferences from laboratory and domesticated animals on aesthetic grounds and his concern that captive animals showed too much variability in the behaviours they had learnt. This variability was, to Lorenz, a hindrance.

Lorenz believed that animal instincts could be used to reconstruct the all-important evolutionary phylogeny of individual species. This approach can be illustrated by the gift-giving behaviour of species of flies in the family Empididae. In one species, *Hilara sartor*, the male presents the female with a present of an empty silken bag, which the female attempts to unravel, while the male copulates with her. This seemingly pointless behaviour is best understood by comparing it with behaviours displayed by related species. One problem for males in this family is that the female is likely to eat an approaching male. In the species *Hilara quadrivittata*, the male avoids this by presenting a food parcel wrapped in a silken balloon. As the female unravels and eats her gift, the male is able to mate with a reduced risk of being eaten. It seems that *Hilara sartor* has evolved one step further and the male has dispensed with the gift.

It was one of Lorenz's students, Nikolaas Tinbergen (1907–88), who finally completed the establishment of ethology as a serious scientific discipline. Tinbergen joined Lorenz in 1939 and helped to develop methods for studying behaviour in the wild. In 1949, he moved to Oxford and led a research group dedicated to the study of animal behaviour. Tinbergen studied how fixed action patterns interact to give a chain of behavioural reactions. In his classic study of the stickleback, Tinbergen (1952) showed how, during the courtship ritual, males and females progress through a series of actions in which each component of female behaviour is triggered by the preceding behaviour of the male, and vice versa, in a cascade of events. The culmination of the sequence is the synchronisation of gamete release and fertilisation.

Both Tinbergen and Lorenz developed models to conceptualise the patterns of behaviour they observed. Lorenz interpreted his observations as consistent with a psychohydraulic model, sometimes called, somewhat disparagingly, the 'flush toilet model'. If behaviour is interpreted as the outflow of water from a cistern, the force on the release valve can be interpreted as the trigger. The model was more sophisticated than suggested by its comparison with a domestic flush toilet, but its essential feature was the accumulation of 'action-specific energy' in a manner analogous to the accumulation of a fluid in a cistern. Freud employed similar hydraulic metaphors in his thinking about drives and repression. Despite their obvious shortcomings as accurate analogues of mental mechanisms, they are still commonplace in everyday speech. To 'explode with rage' or 'let off steam' are both echoes of the type of models utilised by Lorenz and Freud.

Tinbergen developed an alternative model that, while retaining the concept of accumulating energy that drives behaviour, suggested a hierarchical organisation of instincts that are activated in turn. The models of Lorenz and Tinbergen met with much subsequent criticism. It proved difficult, for example, to correlate the features of the model with the growing body of information from neurobiology about real structures in the brain.

One of Tinbergen's most lasting contributions was a clarification of the types of question that animal behaviourists should ask. In 1963, in a paper called 'On the aims and methods of ethology', Tinbergen suggested that there were four 'whys' of animal behaviour:

1. What are the mechanisms that cause the behaviour? (causation)
2. How does the behaviour come to develop in the individual? (development or **ontogeny**)
3. How has the behaviour evolved? (evolution)
4. What is the **function** or survival value of the behaviour? (function).

It has been noted that a useful mnemonic for these is 'ABCDEF: Animal Behaviour, Cause, Development, Evolution, Function' (Tinbergen, 1963).

To appreciate the application of these questions, it may be useful to consider an example. In many areas of the northern hemisphere, birds fly south as winter approaches. One such species is the wheatear *(Oenanthe oenanthe)*. Wheatears migrate to Africa in the winter even though some groups have moved from their European breeding grounds and have established new populations in Asia and Canada. We could ask of this behaviour: What triggers flocks to take to the air? How do they 'know' when the time arrives to depart and which way to fly? These questions address the **proximate causes** of the behaviour and relate to Tinbergen's first question of causation. The answer lies in specifying the physiological mechanisms that are activated by environmental cues, possibly day length, temperature, angle of the sun and so forth.

Probing further, we could ask how the ability to fly over such vast distances in a species-typical manner is acquired by an individual. Do animals know instinctively which way to fly and how far to travel, or do they have to learn some components from a parent or an older bird? These questions belong to Tinbergen's second category dealing with the development and ontogeny of behaviour.

We could also ask about the evolution of the behaviour to its present form. Is the behaviour found in related species? If so, has it been acquired by descent from a common ancestor? A particular question concerns why even Canadian and Asian wheatears travel to Africa. If the aim is simply to move south, those in Canada and Asia could save themselves thousands of miles of travel. Is the move to Africa by the new populations outside Europe a 'hangover' from when the wheatear only lived in Europe? These questions refer to the evolutionary origin of the behaviour as raised in the third 'why' question above.

Finally, in relation to the movement to Africa, we could ask questions about **ultimate causation**. Why do birds make such arduous and perilous journeys? How does flying to Africa increase the survival chances of those making the journey? It must carry some advantage over not moving, otherwise a mutant that appeared and did not fly away would leave more survivors, and non-migration would gradually become the norm. The type of answer given to this last functional 'why' of Tinbergen would presumably show that the benefits of travelling in terms of food supply and then securing a mate outweigh the drawbacks in terms of risks and energy expenditure. Ultimately, we would have to show that migrating is a better option than staying in one place as a means of leaving offspring. We will then have demonstrated the function or adaptive significance of the behaviour.

Taking an overview of Tinbergen's four 'whys' as a broad generalisation, psychologists (dealing with humans or animals) have tended to be interested in the 'whys' of proximate causation and ontogeny, and less interested in questions about evolution and adaptive significance. The growth of sociobiology and evolutionary psychology is a direct attempt to reverse this trend and supply a unifying paradigm for the behavioural sciences based on the understanding of ultimate causation and evolutionary function (Barkow et al., 1992).

The work of Lorenz and Tinbergen is usually classified as being central to the tradition of classical ethology. Classical ethology was forced to adjust its ground as a result of telling criticisms that appeared in the 1950s and 60s from comparative psychologists and, to be fair, discoveries by the ethologists themselves. Lehrman (1953) in particular was a forceful critic of the use of the term 'innate' in ethology. The attempt to classify behaviour as either innate or learnt was soon seen to be too simplistic (Archer, 1992). The deprivation experiments suggested by Lorenz, in which an individual is reared in isolation from other individuals and hence sources of learning, simply isolate an individual from its social environment rather than from temperature, light and nutrition. Isolation experiments beg the question: From what has the animal been isolated? Peking ducklings *(Anas platyrhynchos)*, for example, are able to recognise the call of their species if it is heard during embryonic development while the chick is still in the shell (Gottlieb, 1971). Moreover, individuals create their own environment as a result of their actions. Aggressive or assertive people create a different environment from shy people, resulting in a different type of feedback. It came to be realised that all behaviour must be the result of both influences.

Important work by Thorpe (1961) on song development in chaffinches demonstrated the mutual interdependence between heredity and the environment. Thorpe demonstrated that whereas the ability of a chaffinch to sing was to some degree 'innate', the precise song pattern depended upon exposure to the song of adults at critical times during the development of the young bird. The form the song took also depended on the ability of the chaffinch to hear its own song. Chaffinches would not, however, learn the songs of other birds, even if exposed from birth. Furthermore, once song development occurred, a chaffinch would not learn other variants. Work such as this showed that the interaction between innate templates of behaviour and the environment is more complex than hitherto thought.

Eventually, the classical ethologists had to accept that behaviour that they had often labelled as innate could be modified by experience. It did not follow from this of course that all behaviour was learnt and unconstrained by genetic factors, as some of the behaviourists seemed to imply. The history of comparative psychology is also one in which fundamental assumptions had to be revised.

Comparative psychology 1900–70

An early exponent of the methods that came to be associated with comparative psychology was Ivan Petrovich Pavlov (1849–1936). Pavlov was the son of a priest who began his career studying medicine, and was awarded the Nobel prize in 1904 for his work on digestion. Pavlov showed that if the sound of a bell accompanied the presentation of food to a dog, the dog would learn to associate the sound with food. Eventually, the dog would salivate at the sound of the bell even in the absence of food. Pavlov thereby produced the first demonstration of what later became known as 'classical conditioning'. By focusing on the observable reactions of animals without presupposing what went on in their minds, Pavlov stressed the objectivity and rigour of these methods in contrast to a psychology that dwelt upon putative inner experiences. Pavlov's work on conditioning became known to Western psychology around 1906 and, although his more ambitious claims for the establishment of a new brain science went unheeded, his methodology proved highly influential.

The focus on the observable reactions of animals and humans under controlled conditions came to be regarded as the hallmark of comparative psychology and what later became known as **behaviourism**. It is easy to overestimate the influence of behaviourism on 20th-century psychology. Smith (1997) notes that it served the polemical interests of cognitive psychologists in the 1960s to represent psychology between 1910 and 1960 as a behavioural monolith from which they were seeking liberation (see also section 1.4.4). As a broad generalisation, however, a pattern was emerging. By the middle of the 20th century, the European approach to animal behaviour and extrapolations to humans was dominated by ethology, while in the USA the experimental approach using laboratory animals was pre-eminent.

One figure who more than any other seemed to symbolise the behaviourist approach in its early days was John Broadus Watson (1878–1958). In his later years, Watson came to be reviled by ethologists as the architect and archetype of an alien approach to the study of behaviour. By adopting a positivist approach to knowledge, Watson claimed that psychology would be retarded in its development unless it ditched its concern with unobservable entities such as minds and feelings. Similarly, both animal and human psychology must abandon any reference to consciousness. A psychology that deals with inner mental events is clinging, he claimed, to a form of religion that has no place in a scientific age. For Watson, the brain was a sort of relay station that connected stimuli to responses.

Watson issued his manifesto in a series of lectures delivered in 1913 at Columbia University. Before the First World War, the reaction was lukewarm: some welcomed his objective approach but warned against its excesses. Others feared that a sole concern with behavioural phenomena as opposed to human consciousness would reduce psychology to a subset of biology. It was after the war that behaviourism became more deeply embedded in American scientific culture. The war had demonstrated the value of objective tests applied in the classification of military personnel. By 1930, behaviourism had become the dominant viewpoint in experimental psychology. In stressing the importance of environmental conditioning, Watson's whole approach was profoundly anti-evolutionary and anti-hereditarian. He denied that such qualities as talent, temperament and mental constitution were inherited. Perhaps his most famous remark on the effect of environmental conditioning, and one of the most trenchant and extreme statements of **environmentalism** in the literature, is his claim for the social conditioning of children:

> Give me a dozen healthy infants, well-formed, and in my own specified world to bring them up, and I will guarantee to take any one at random and train him to become any type of specialist I might select – doctor, lawyer, artist, merchant-chief and yes, even beggar-man and thief, regardless of his talents, penchants, tendencies, abilities, vocations and race of ancestors. (Watson, 1930. p. 104)

In this respect behaviourism fitted in well with American culture. It appeared to be a practical science that promised to deliver socially valuable answers to such things as how to raise children and turn them into effective citizens. It also arrived on the scene during the First World War when anti-German sentiment made the notion of a distinctly American (as opposed to German) science of psychology more appealing.

Behaviourism sought philosophical credibility by allying itself with a philosophy of science known as logical **positivism** and articulated by a group of philosophers known as the Vienna Circle. The logical positivists argued that statements are only meaningful, and hence part of the purview of science, if they can be operationally defined. A statement about the world then only becomes meaningful if it can be verified. The aims of this approach were to outlaw religious and metaphysical claims to knowledge. It was in this approach to epistemology that American behavioural psychology, with its emphasis on empirical, quantifiable and verifiable observations, found a natural ally.

The decline of behaviourism in the 1960s coincided with the downfall of logical positivism as a reliable philosophy of science. Philosophers such as Popper and historians such as Kuhn showed that the verifiability criterion of meaning espoused by the Vienna Circle was untenable both in theory and as a realistic description of the way in which science actually worked. The irony of the linkage between behaviourism and positivism is neatly summed up by Smith (1997, p. 669):

> It appeared as if the behaviourist enterprise had emptied psychology of its content in order to pursue an image of science that was itself a mirage.

One movement in animal psychology related to behaviourism was Skinner's 'operant psychology'. Skinner studied at Harvard and was particularly influenced by the work of Watson and Pavlov. Skinner's programme adopted a number of key principles. One was that he believed that science should be placed on a firm foundation of the linkages between empirical observations rather than speculative theory. For Skinner, and in this respect he resembles Watson, theoretical entities such as pleasure, pain, hunger and love were meaningless and should be expunged from laboratory science. Another essential feature of Skinner's work was that he thought that all behaviour could be resolved and reduced to a basic principle of reinforcement. One typical schedule of reinforcement devised by Skinner was to reward pigeons in a box with grain. By rewarding some forms of behaviour and not others, he was able to make the rewarded behaviour more probable, an approach that became known as operant conditioning.

While behaviourists in America were feverishly attempting to jettison excess metaphysical baggage from psychology, in Europe Freud was weaving a psychology replete with rich and colourful complexes, emotions and subconscious forces. Skinner was both an admirer and critic of Freud. For Skinner, Freud's great discovery was that human behaviour was subject to unconscious forces. This accorded well with the view of behaviourism that conscious reason was not in the driving seat of human behaviour. Freud's mistake was that he encumbered his theory with unnecessary mental machinery, such as the ego, the superego, the id and so on. For Skinner, such entities were not observable and hence were not justified in scientific enquiry.

By 1960, a large number of psychologists in America had been trained in Skinner's methods. Their output was influential in some areas, such as the inculcation of desirable habits in the development of children, but few behaviourists were willing to go as far as Skinner in suggesting that organisms were empty boxes. Skinner faced his most difficult hurdle when, in his book *Verbal Behaviour* (1957), he attempted to interpret language development in terms of operant conditioning. There were already signs that many behaviourists were realising that language threatened to be the 'Waterloo' of behaviourism. Skinner marched on, however, and argued that there was nothing special about language, denying any fundamental difference in verbal behaviour between humans and the lower animals. Skinner had now pushed behaviourism too far, and its weaknesses were fatally exposed.

In 1959, a linguist called Noam Chomsky, then relatively obscure, reviewed Skinner's *Verbal Behaviour* and, in showing that behaviourism failed lamentably when tackling language, also undermined some of its basic pretexts. Chomsky (1959) argued that behaviourism could hardly begin to account for language acquisition. He showed that Skinner's attempts to apply the language of stimulus and response to verbal behaviour lapsed into vagueness and finally hopeless confusion. For Chomsky, behaviourism could not be improved or modified: it was fundamentally flawed and had to go. In their efforts to avoid unobservable mentalist concepts, the behaviourists had indeed transformed psychological terminology; so that memory became learning, perception became discrimination and language became verbal behaviour. As Chomsky remarked, defining psychology in this way was like defining physics as the science of meter readings. Chomsky's review, and his own positive programme for linguistics stressing the creativity of language and its foundation on inherent, deep-seated mental structures, sparked off a revolt against behaviourism that tumbled it into terminal decline. This roughly coincided with increasing reports from animal researchers that animals trained according to operant conditioning methods would occasionally revert to behaviours that seemed to be instinctive.

As Chomsky's work on linguistics and Garcia's work in comparative psychology (see section 1.4.4) were undermining the premises of behaviourism, another development in mathematics was taking place that was to further stimulate evolutionary biology. This was the branch of mathematics called 'game theory' devised by John von Neumann (1903–57) who published *Theory of Games and Economic Behaviour* in 1944 in collaboration with Oskar Morgenstern. Game theory examines the problem of how to choose a strategy in a situation of uncertainty, the uncertainty that arises from not knowing what the other players will do. Game theory was further developed by John Nash who introduced the concept of the Nash equilibrium – a state where no player can do any better by changing his strategy unilaterally. Nash showed in his dissertation of 1950 that Nash equilibria must exist for all finite games with any number of players. Game theory was applied to biology by John Maynard Smith who analysed how the strategies employed by animals in contests contributed towards their **fitness**. An important concept to emerge was that of the evolutionary stable strategy (see also Chapter 9).

1.2.3 Interactions between comparative psychology and ethology

Between 1950 and 1970, there were a variety of approaches to the study of animal behaviour, although ethology and comparative psychology were the dominant players. The exact boundaries of these disciplines and the extent of their interactions were often more complex than is conveyed by a simple 'warfare model'. There was, however, a frank exchange of critical comments and an atmosphere of intellectual rivalry. The ethologists often identified the whole of comparative psychology with the extreme environmentalism of Watson's later years. They saw comparative psychology as a discipline obsessed with what rats do in mazes, bereft of a unifying theory and ignorant of the adaptive and evolutionary basis for behaviour. In return, psychologists saw ethology as lacking in scientific rigour, employing dubious concepts such as innateness and burdened by unsubstantiated models.

An outsider looking into this mêlée would also note that Lorenz and Tinbergen achieved a rapport with the reading public by writing fluent and popular works. In contrast, comparative psychology appeared to be more introspective and plagued with self-doubt. As late as 1984, the comparative psychologist Dewsbury (1984, p. 6) lamented that 'we seem to suffer an identity crisis'. Correspondingly, some of the harshest portrayals of comparative psychology came from the psychologists themselves. It seemed to be a discipline set on a course for self-destruction. In a classic paper entitled 'The snark was a boojum', Beach (1950) demonstrated how, between 1911 and 1945, the output of research on animal behaviour from the perspective of comparative psychology had increased while the number of species actually studied had dwindled. By 1948, the vast majority of articles in this field were concerned with one single species, the Norway rat, and were dominated by reports of learning and conditioning experiments. It was as if, as Lorenz complained, the word 'comparative' was a sham.

The fondness for the Norway rat stood in stark contrast to the much wider diversity of organisms studied by the ethologists. There were other characteristic differences too. In 1965, McGill published a table that captured in broad terms the contrast between ethology and comparative psychology (Table 1.1).

Further surveys involving analyses of the content of journals, in the style of Beach, were conducted at regular intervals after 1950. In the 1970s, several observers concluded that the earlier scenario outlined by Beach, of a subject blinkered by its own concern with a few species, was still very much in evidence (Scott, 1973; Lown, 1975). To be fair to comparative psychology, this narrowness of focus was a conscious decision and did not overly concern its practitioners. They were searching for universal laws of behaviour that, once established for any one species, could be applied to others. However, the day that such laws

Table 1.1 Differences between comparative psychology and ethology

	Comparative psychology	Ethology
Geographical focus	North America	Europe
Background training of researchers	Psychology	Zoology
Animal subjects studied	Norway rat and pigeons	Variety of species
Emphasis	Learning: the development of behaviour as moulded by environmental stimuli	Instinct: behaviour as the expression of evolution-derived characters
Methods	Laboratory work. Statistical analysis of the effect of different variables	Careful observation of natural behaviour. Experiments in the field
Attitude to animal subjects	Objective and dispassionate	Close familiarity, even emotional attachment

SOURCE: Adapted and modified from McGill, T. E. (1965) *Readings in Animal Behaviour*. New York, Rinehart & Winston, p. 2.

could be transferred to illuminate human behaviour seemed to recede as fast as it was chased.

By the late 1970s, comparative psychology was in a sorry state. The field seemed to suffer a crisis of confidence, faced the ignominy of being identified (however unjustly) with a discredited behaviourism, and the epithet 'comparative' sounded alarmingly hollow. The neglect of an adaptive approach to behaviour, as well as an underappreciation of phylogenic differences in learning and intelligence, was increasingly seen as a damaging shortcoming. In 1984, Dewsbury wrote a sustained defence of the discipline that challenged a number of misconceptions and tried to give it a respectable history (Dewsbury, 1984), but even then it was clear that the study of behaviour was moving in radically new directions. The attempt to patch up the image of psychology contrasted sharply with the hybrid vigour demonstrated by the merger of ethology and ecology to yield behavioural ecology and sociobiology.

A rapprochement was eventually established between ethology and comparative psychology, and each side learnt valuable lessons from the other. Two people in particular who sought to bridge the disciplines and effect a reconciliation were the American psychologist Donald Dewsbury and the Cambridge ethologist Robert Hinde. In 1982, Hinde took a sanguine view and concluded that, despite a few remaining differences, the two approaches had broadened in their concerns and 'on the whole the distinction between the two groups barely exists' (Hinde, 1982, p. 187). Dewsbury was also, in 1990, calling for reconciliation between causative and functional (adaptive) approaches to behaviour. By this time, behavioural ecology, and sociobiology in particular, were well on the way to exploiting new insights into the adaptive basis of behaviour generated by concepts such as parental investment, reciprocal altruism, kin selection theory and a revival of Darwin's own theory of sexual selection. Dewsbury (1990) in fact complained that the concentration on the functional approach threatened to imbalance the study of behaviour.

Before examining the remarkable efflorescence of ideas associated with this movement, we will examine the background of applying evolutionary ideas to human behaviour, for it is precisely in this area that behavioural ecology and sociobiology succeeded in bridging the human–animal divide where comparative psychology failed.

1.3 Evolution and theories of human behaviour: Darwin and after

1.3.1 Herbert Spencer (1820–1903)

Herbert Spencer (Figure 1.4), philosopher and critic, was a contemporary of Darwin and someone equally keen to push forward the programme of scientific naturalism. Even before Darwin published his *Origin*, Spencer was attempting to build a philosophy of knowledge on the unifying principle of evolution. In his essay of 1852, 'A theory of population deduced from the general law of animal fertility', he introduced the famous phrase 'survival of the fittest'. It was this phrase that Darwin borrowed from Spencer and used in his fifth edition of the *Origin*.

FIGURE 1.4 Herbert Spencer (1820–1903)
A contemporary of Darwin who coined the phrase 'survival of the fittest'.

Spencer's curiosity in evolution was first aroused when he stumbled across fossils in his first career as a civil engineer. In 1840, he read Lyell's *Principles of*

Geology (1830–33), a work that contained a discussion of Lamarck's evolutionary ideas, and thereafter became attracted to Lamarckian notions of the evolutionary process. Spencer's thinking on evolution applied to the human psyche came to fruition in 1855 when he published his *Principles of Psychology*. The strain of completing this work led him to suffer a nervous breakdown that left him incapacitated for about 18 months; modern readers who wish to plough through Spencer's writings might be warned of a similar risk. In this book, Spencer reasoned that Lamarckian processes could have led to modern human faculties. The love of liberty, for example, may have originated in the fear that animals show when restrained. This then evolved to a political commitment whereby individuals seek liberty for themselves and others as a principle.

What is more interesting is that Spencer advanced a view of the human mind, a sort of evolutionary Kantism, that he thought would provide a solution to the age-old controversy between the disciples of Kant and Locke. For Locke, the human mind is essentially structured by experience: we are born with a blank slate, a tabula rasa, upon which experience scribes. In this empiricist conception of the mind, there are few inherent specific structures or mechanisms. There is a congruence between our mind, our knowledge and the world itself because experience has shaped for us our perceptual categories and modes of perception. Kant, however, held that the human mind is provided at birth with a priori categories or inherent specialised mechanisms (such as Euclidean geometry and notions of space and time) that structure the world for us. This philosophical divide between the views of Locke and Kant was revived in the mid-19th century by debates between J. S. Mill and William Whewell, the former advocating an inductive view of knowledge in the manner of Locke, and the latter espousing a more Kantian position.

Spencer's proposed solution to this age-old conundrum was to suggest that Mill was right in proposing that experience shapes our mental operations but wrong in suggesting that each individual has to start the process from scratch when he or she is born. Like Whewell, Spencer believed that the mind is born ready equipped with perceptual categories and dispositions, but with the crucial addition that these Kantian categories were themselves the consequences of mental habits acquired through inheritance. Our mind structures our experience, but the structures used have been laid down during the evolution of the species. In broad outlines, and if we replace Lamarckian inheritance by natural selection, this is a view that is regarded as substantially correct and one that is perhaps Spencer's greatest legacy. It was this view of human knowledge, sometimes called 'critical realism', that was employed by ethologists such as Lorenz and Eibl-Eibesfeldt later in the 20th century (see also Chapter 7). Unfortunately for Spencer's reputation, and probably as a result of the failure of his views in other areas and his association with a discredited social Darwinism, it is Darwin rather than Spencer who usually receives credit for cutting this particular Gordian knot.

Darwin himself took a curious view of Spencer. In a letter to Lankester in 1870, he suggests that he will be looked upon as the greatest philosopher in England, 'perhaps equal to any that have lived' (Darwin, 1887, p. 120). Elsewhere Darwin is less impressed and complained to Lyell that Spencer's essay on population was 'dreadful hypothetical rubbish' (Darwin, 1860). To Romanes in 1874, he confessed that 'Mr Spencer's terms of equilibration and so on always bother me and make everything less clear' (quoted in Cronin, 1991, p. 374). History has sided with the latter two views rather than the former.

1.3.2 Evolution in America: Morgan, Baldwin and James

Lloyd Morgan is most frequently remembered today for his canon, designed, as noted earlier, to avoid the traps of anthropomorphising. Another of his achievements was the formulation of a novel theory of the interaction between mental and physical evolution. One of the reasons Morgan rejected Lamarckism was precisely because he thought his own theory rendered it superfluous. While on a lecture tour of America in 1896, he gave a lecture in which he outlined his new theory. By a remarkable coincidence, the man who spoke next on the platform, the evolutionary psychologist James Mark Baldwin (1861–1934), delivered exactly the same idea. The theory became known as 'organic selection' but is usually known today as the 'Baldwin effect' and describes how a learnt adaptation could become fixed in the genome.

Suppose a sudden change in environmental conditions causes stress for a particular group of animals and thus exerts a selective pressure. Some who are unable to adapt their behaviours in their own lives will perish, while those who are flexible enough to accommodate to the changed conditions may stand a better chance of surviving. Those individuals that are saved may over time throw up variations that are favourable in their new environment, thus allowing natural selection to fix these variants in the **gene pool**. By this means, what began at the level of phenotype as a non-inherited behavioural reaction to new conditions could over time become fixed by natural selection.

As an illustration, suppose that a change in conditions means that food for some animals is to be found lower down in the soil and that, to reach it, animals now have to dig deeper. Those that have the ability to learn that food lies at a deeper level, or are perhaps less inclined to give up at shallow depths, will survive food shortages. This could then give natural selection enough time to select out genuine genetic differences that enhance food-digging capabilities, such as large claws or an innate tendency to dig deep for food. In short, **phenotypic plasticity** can give rise to genetic adaptability.

It now seems clear that James Baldwin deserves more recognition in the history of psychology than he has been afforded. The distinguished psychobiologist Henry Plotkin (2004, p. 70) rates Baldwin highly, claiming that 'Baldwin is perhaps the most significant person in the history of the relationship between evolutionary theory and psychology.

Baldwin was particularly far-sighted in his view of how children learn. Until the latter part of the 19th century the common view of children was that they were miniature adults. Empiricists and associationists (the name for an early type of behaviourism common in the late 18th century) thought that the difference between the mind of a child and an adult lay primarily in the number and complexity of the associations that had formed. Similarly, although from an opposite perspective, the nativists and rationalists (for example followers of Kant) thought that children, like adults, possessed a priori concepts of such things as space, time and number. Baldwin was to challenge all this. Initially, it was his experience of the maturing development of his own children (daughters born in 1888 and 1891) that convinced him that both these position must be incorrect. Looking at his children, he observed specific developmental stages and concluded that the mind is a product of both an evolved phylogeny and ontogentic influences. Baldwin thought that a child develops its understanding of the world by acting upon it and being, in turn, acted upon. He proposed that the child explores the world and assimilates phenomena into its existing knowledge structures and frameworks, but these structures are periodically forced to change to accommodate new and anomalous information. In this suggestion that a child's understanding of the world develops by a circular process of action, assimilation and accommodation, Baldwin even anticipated the views Piaget advanced some 40 years later. More specifically, he proposed that a child's consciousness passes through three stages: the differentiation of people and objects; the differentiation of self and others; and finally 'ejective consciousness', whereby the child appreciates that others also have mental states different from its own. On this last point, Baldwin was foreshadowing the importance of the theory of mind (see Chapter 15) some 80 years before it became fashionable.

Baldwin employed Darwinian reasoning in two senses. He believed that this developmental process was a product of innate dispositions, but he also thought that the selection of ideas to match reality in an individual was itself a process of Darwinian selection. Baldwin went on to propose that children pay particular attention to the behaviour of other people as they mature. This, of course, makes good evolutionary sense, since in most cases it is better for the maturing child to imitate the behaviour of older, presumably successful group members and parents, than to engage in the costly business of trial and error learning all for itself.

In 1908, having been caught in a police raid on a brothel in Baltimore, Baldwin resigned his job at Johns Hopkins University and his editorship of *Psychology Review*. Soon after, he emigrated first to Mexico and then to France. The unfortunate circumstances of his departure led to a decline in the influence of Baldwin's ideas and smoothed the way for the rise of Watson's brand of behaviourism. Significantly, Watson inherited Baldwin's job at Johns Hopkins and also took over the editorship of *Psychology Review*.

Modern biologists have usually steered well clear of the Baldwin effect and have been reluctant to exploit its explanatory potential. The reason no doubt is that, prima facie, it smacks of Lamarckism – although in reality it is consistent with modern Darwinism. Richards (1987) and Dennett (1995) give sympathetic discussions of the Baldwin effect, and Deacon (1997) uses the Baldwin effect in his attempt to account for the origins of human language. It is likely that the Baldwin effect will be observed in gene–culture coevolutionary models where cultural change sets up new selective pressures and drives genetic change. This is examined briefly in Chapter 16.

Both Morgan and Baldwin had high expectations of the power of evolutionary thinking to explain the development and adaptation of individuals within social groups and even the progress of society itself. So too did the philosopher and psychologist William James (1842–1910), the elder brother of the novelist Henry James. In 1875, James taught a course in psychology at Harvard using Spencer's *Principles* as his text. Later he wrote his own textbook, *Principles of Psychology*, published in 1890. The book, originally intended only as a brief introduction to the subject, took him 11 years to write and became a major landmark in the discipline. Like Darwin and Spencer before him, James looked to animals for the instinctive roots of human behaviour and morality. Parental affection and altruism, for example, were to be seen as evolved traits. Morality as a whole was unashamedly a product of heredity.

In his psychology, James made frequent use of the concept of instinct, ascribing to humans such inherent traits as rivalry, anger, a hunting instinct and so on. He also speculated that some pathological conditions in humans could be remnants of animal instincts possessed by our ancestors that resurface in the human psyche following injury or disease. Many animals, rodents for example, prefer to remain close to cover and only reluctantly rush across open ground. James saw in such behaviour a distant basis for agoraphobia in humans. In all, James listed over 30 classes of instinct possessed by humans – enough to satisfy the keenest ethologist – and regarded humans as being more richly endowed with instincts than any other mammal. Such instincts he saw as being elicited by environmental stimuli. A given stimulus could spark off the emergence of a jumble of potential instinctive reactions, but humans were not mere automata since reason could intervene to select the most appropriate response to the situation.

James's singular contribution was to apply the idea of natural selection to ideas themselves. James raised the philosophical question of how we are to choose the best ideas in a world where humans project a range of competing theories, all claiming to have a correspondence with the truth. To answer this, James advanced an evolutionary epistemology that became known as 'pragmatism'. In this scheme, truth is the idea that works, or, more specifically, the one that survives best in the intellectual environment into which it is cast.

James's thinking, particularly his use of the concept of human instincts, was enormously influential in American psychology in the first few decades of the 20th century. The historian Hamilton Cravens estimated that, between 1900 and 1920, over 600 books and articles were published in Britain and America that employed the idea of human instincts (Degler, 1991).

1.3.3 Galton and the rise of the eugenics movement

William James died in 1911, the same year as saw the demise of Darwin's cousin Francis Galton (1822–1911). Both of these thinkers left an influential legacy of ideas that were to have a major impact on subsequent attempts to apply evolutionary theory to human behaviour. The influence of James was, as noted above, initially highly productive, but it was the association of the ideas of heredity and instinct with the more insidious notions of Galton and the eugenicists that were to have a more lasting, and ultimately damaging, effect on the evolutionary paradigm.

The idea that natural science could throw light on pressing social problems was a typical product of Victorian positivism and scientism. It was this overweening ambition of scientists such as Galton that led to an entanglement of evolution with naive and dangerous political thought. Francis Galton noted that our very humanity served to blunt the edge of natural selection. Care for the sick and the needy led to the procreation of the less fit and could in time, Galton feared, lead to the deterioration of the national character. Whereas Darwin was content to let things be, the remedy advocated by Galton was to propose that the state should intervene to modify human mating choices. Those with heritable disorders or the constitutionally feeble should be discouraged from breeding, while the better sort of person should be positively encouraged. Such a programme Galton called **eugenics** – a term he coined in 1883 from Greek roots meaning 'well born'.

The fallout from the eugenics programme was for the most part detrimental and led to tragic consequences. Between 1907 and 1930, 30 states in America passed laws allowing for the compulsory sterilisation of criminals and mental defectives. By the early 1930s, about 12,000 sterilisations had been carried out in the USA. In America, the eugenicists were also particularly active in lobbying for the restriction of immigrants from southern and eastern European countries, whom they regarded as intellectually inferior. Consequently, in 1924, the US Congress passed the infamous Immigration Restriction Act. The way in which eugenic notions, and the idea that there are heritable mental differences between human groups, infected attitudes to race and immigration, and informed IQ testing and sterilisation practices, has been documented on many occasions (see especially Gould, 1981). Some of the issues raised will be considered again in Chapter 18, where it will be shown that eugenics has no logical linkage to the evolutionary programme.

By the 1930s, however, the damage was done. In the mind of the public and many scientists, the evolutionary approach to human nature had become entwined with an odious set of political beliefs. There thus began in the middle of the 20th century an almost total eclipse of serious efforts in the natural sciences to study human behaviour from an evolutionary perspective, a neglect that lasted for several decades, illustrating James's own observation that ideas are selected and thrive if they suit a given social and intellectual milieu. In this respect, it is also the social context that helps to explain the move away from hereditarian and evolutionary thinking in anthropology and the social sciences that led to the triumph of cultural over biological explanations of human nature.

1.4 The triumph of culture

1.4.1 Franz Boas

Darwin, in common with many anthropologists of his day, subscribed to the notion that differences in the level of culture between peoples were based upon biological inequalities. It was reasoning such as this that allowed racism to enter evolutionary thought. The efforts of Franz Boas (1858–1942) helped to turn the tide away from racist ideas in anthropology, and in so doing, Boas effectively enshrined culture as the central concept to be employed in explaining the social behaviour of man. In his youth, Boas worked in Berlin at the Royal Ethnographic Museum, an institution with a tradition of espousing cultural rather than biological explanations for human differences. In 1888, he joined the faculty of Clark University in America, later being appointed professor of anthropology at Columbia University, where he exerted an immense influence over American anthropology.

In 1911, Boas published two works that had a revolutionary impact on the social sciences. One was a book, *The Mind of Primitive Man*, the other a report entitled 'Changes in the bodily form of descendants of immigrants'. The first work was an assembly of previously published essays and conveyed the central message that the mind of primitive man (people in traditional or tribal cultures) did not differ in mental capability from that of civilised people. In effect, Boas was denying any significant innate differences between indigenous 'savages' and civilised peoples. Boas pointed to history and culture rather than biology as explanations for social differences in behaviour and customs. In place of innate racial differences, Boas substituted a single human nature common to all humans but shaped by culture.

In his report on bodily changes in the form of American immigrants, Boas published a series of remarkable findings that surprised even himself. He looked at whether the cephalic index, the ratio in humans of head length to head width, remained constant after racial groups had emigrated to America. The significance of this unusual statistic was that the cephalic index was widely used in the late 19th and early 20th centuries as a way of classifying human types. It was thought that races could be reliably classified along a spectrum ranging from the wide-headed (brachycephalic) southern Europeans to the long-headed (dolichocephalic) Nordic types. It was generally assumed that this measure was immune from environmental influences and remained stable from generation to generation. In studying the head shapes of immigrants to the USA and their subsequent offspring, Boas found that the head shapes of the children appeared to change if the mother had been in America for 10 years or so before conception. The round heads of the European Jews and the long heads of southern Italians converged towards a common type. So surprising were the results that Boas made further enquiries to rule out illegitimacy. For Boas, such results were proof concrete that culture and environment exert a strong pressure on the most basic of human features. If humans were plastic in their morphology, reasoned Boas, their mental plasticity would also be assured.

As Plotkin (2004) observes, what Boas had established is still open to dispute. In 2002 and 2003, two sets of reports appeared from investigators who had independently re-analysed Boas's original data. One group (published in the *Proceedings of the National Academy of Science*) concluded that differences between the offspring of European and American racial types were insignificant and that exposure to the American environment had no effect on the cranial index of children (Sparks and Jantz, 2002). The other study, published in the *American Anthropologist*, concluded that Boas was largely correct (Gravlee et al., 2003). So even 91 years after Boas's original conclusions, the empirical data is capable of supporting various interpretations and ideological factors may still play a role in guiding interpretation.

Boas's emphasis on the importance of culture did not imply that he was an extreme environmentalist like his contemporary Watson. Boas was primarily concerned with demonstrating the importance of culture and history as explanatory causes for the differences in achievement between (rather than specifically within) racial groups. Boas grew up in a liberal household, but two key influences on him were Rudolf Virchow, a German anthropologist who refused to accept that humans had evolved from other animals, and Theodore Waitz, a committed Lamarckian. Consequently, Boas was suspicious of Darwin all his life and instead clung to the view that Lamarckian inheritance held out the promise of human improvement through environmental manipulation. Boas accepted that heredity played a major role in shaping the traits that a child possessed. His core message was that variations within any racial group are so large that they call into question the concept of race as a useful idea at all, a position largely endorsed by most modern geneticists.

It was the work of Boas and his followers that persuaded many social scientists to abandon Darwinian approaches to human social behaviour. However, relinquishing Darwin was, for many, not such a great sacrifice. Lamarckism always seemed more attractive anyway since it added a hopeful significance to the process of social reform. The notion that striving brought about heritable change also seemed more in keeping with human dignity. Moreover, Lamarckism was able to persist into the early years of the 20th century since biologists were not unanimous in its rejection. When Lamarck was finally ousted from biology in the 1920s, the purposeless process of Darwinian natural selection seemed markedly at odds with the powerful intentionality that emanated from human nature. Better therefore to fly into the arms of a comforting environmentalism than to suffer the cold embrace of natural selection. It is ironic that just as scientists such as R. A. Fisher, J. B. S. Haldane and Sewall Wright were, in the early 1930s, publishing their great summary works, which finally demonstrated the consistency between Darwinian natural selection and Mendelian genetics, most social scientists were moving rapidly in the opposite direction.

Social scientists began to erect what Tooby and Cosmides (1992) were later to call the standard social science model of human nature. The model, which in its various versions dominated psychology and the social sciences from about the 1930s to the 1970s (and is still uncritically accepted in some quarters), contains a number of components. First, it stresses the insignificance of intergroup variations in genetic endowment. In other words, people at birth are by and large everywhere the same. Second, since adult human behaviour varies widely across and within cultures, it must be culture itself that supplies the form to the adult mind, disposes it to think and behave in culturally specific ways, and shapes adult behaviour. On the first point, evolutionary psychologists and sociobiologists are largely in agreement with the social scientists. It is on the second point that the standard model receives its greatest challenge from evolutionary theory.

Having abandoned Darwin, the social sciences found that it was the unity of human nature rather than its diversity that was remarkable. There remained of course one striking feature of the human condition that posed a challenge to environmentalism, and that was sex. Physical differences between the sexes were undeniable, as was the fact that, in virtually every society, males and females differed in their behaviour and their social roles. Darwin, with his theory of sexual selection (see Chapter 3), offered a biological approach that promised to be fruitful. More powerful forces than scientific rationality were, however, at work, and this was the next bastion of biology that came under attack from the social sciences in the 1920s and 30s.

One figure in the vanguard was Boas's student Margaret Mead (1901–78). In her twenties, Mead studied the life of the islanders of Samoa. Her work culminated in a popular and influential book called *Coming of Age in Samoa*, published in 1928. This was followed in 1935 by *Sex and Temperament in Three Primitive Societies*. The central message of both books was that sex roles and sex differences in behaviour were not biologically ordained. Mead concluded that sex roles were interchangeable, so any differences between the personalities of each sex observed in any culture must be socially produced. Human nature, as she put it, is 'almost unbelievably malleable'.

A number of problems were subsequently noted with Mead's interpretation of her data, and this is considered in Chapter 18. What is noteworthy here is the enormous influence that Mead's work had at an academic and popular level. The links between biology and human nature were now being severed as fast as they were once made. It was, however, the revulsion at eugenic ideas put into practice that was decisive.

1.4.2 The revolt against eugenics

Despite the promise offered by the eugenics movement to provide a scientific programme for social improvement, many sociologists had never swallowed the arguments. Experimentally, it was difficult to demonstrate that mental traits, good or bad, could be inherited. The family genealogies that Galton constructed showing how genius and talent ran in families could always be explained by nurture as well as or instead of nature. Moreover, it was not obvious that the most successful in society possessed the best genes. Boas, despite his insistence on cultural explanations for racial differences, accepted that personal qualities could be inherited from parents but questioned whether anyone would agree on what qualities were desirable and worth fixing through eugenics. He also believed, probably rightly, that individuals would not accept the infringement by the state of something so fundamental as the right to procreate. For social scientists, there was the additional drawback that many pressing social problems, such as inequalities in wealth distribution, poverty, prostitution and so on, did not seem particularly amenable to eugenic solutions. The Great Depression in America in the early 1930s, in which the wealthy, the intelligent and the poor alike were all affected by more powerful factors than their genetic endowment, seemed to confirm to many the irrelevance of the purported link between intelligence and success.

Following the Second World War, repugnance in the face of Nazi atrocities – many of them committed with eugenic principles in mind – led to a statement in 1951 by the United Nations Educational, Scientific and Cultural Organisation (UNESCO), that there existed no biological justification for the prohibition of mixed-race marriages. The statement, endorsed by a number of prominent geneticists, also asserted that scientific knowledge provided no basis for believing that human groups differed in their innate capacity for intellectual and emotional development. This of course was not the same as saying that evolutionary theory had nothing to say about human behaviour, but the impression conveyed was that environmentalism had been given official and political sanction.

With whatever justification, an association began to form between eugenic thinking, right-wing ideologies and an evolutionary approach to human nature. Terms such as 'biological determinism' and 'biology is destiny' were applied, however fallaciously, to attempts to identify a biological basis for human nature and then roundly attacked. Compared with environmentalism, biological approaches to human nature seemed tainted by association and looked distinctly reactionary in their implications.

1.4.3 Behaviourism as an alternative resting place

Thus, in the middle years of the 20th century, the insecure foundations and dubious political connotations of the eugenics programme, the rigidity of the concept of instinct, and the paucity of experimental data to substantiate a belief in human instincts gave pause to those who might otherwise have sought a basis to human psychology in biology. There was also some guilt among psychologists over the way in which hereditarian notions about IQ had been used to justify some racist immigration policies.

It is just possible that psychology would not have so readily abandoned the evolutionary paradigm if it had had nowhere else to go, but there was an alternative in the form of environmentalism, or more specifically behaviourism. In 1917, Watson was already claiming that a child is born with just three basic emotions: fear, rage and joy. For some psychologists, this was a positive relief from the dozens of instincts described by James and others, and the various traits claimed by the eugenicists to have a genetic basis. By 1925, Watson was denying the usefulness of the concept of instinct at all in human psychology. Although we should be wary of regarding Watson as typical of behaviourism as a whole, it is fair to say that most psychologists were not too sorry to abandon the theory of human instincts.

There were also methodological considerations. In the 1920s, there was a move away from what were seen to be deductive methods (that is, the deduction, or more strictly the induction, of human characters by extrapolation from animals and our supposed evolutionary past) towards an experimental approach. Behaviourism fitted neatly with the new image of psychology as a discipline based on rigorous experimental methods free from disputatious distractions, which psychologists were keen to cultivate.

In the 1970s, two psychologists, Wispe and Thompson (1976), looking back over the hegemony that behaviourism once exerted over psychology, concluded that the whole idea of instincts never fitted comfortably with the American outlook. In America, where even the humble farmer's son could become president, the burgeoning science of psychology more naturally warmed to a position that did not suggest biological predestination. In contrast to 'biologism', behaviourism (in which anybody, with some exaggeration, could be anything) seemed liberal and liberating. Such sociological considerations may indeed provide a handle on the drift away from biological theories of human nature in the 1930s.

1.4.4 The 'cognitive revolution' and other challenges to behaviourism

As we noted in section 1.2.2, Chomsky's criticism of Skinner's views on language was a factor in the decline of behaviourism. More generally, the rapid decline in the dominance of behaviourist approaches came from the emergence of cognitive science and the so-called 'cognitive revolution' in psychology. It would be prudent, however, to treat the claim that cognitive psychology completely ousted and supplanted a defunct behaviourism with some caution. For a start, psychology (1920–60) was never a unified behaviourist discipline: there were many in Europe, such as gestalt psychologists and followers of Piaget, who carried out non-behaviourist research. Second, the growth of cognitivist approaches retained some of the fundamental assumptions inherent in behaviourism. Both, for example, were interested to explore in materialistic terms how the human machine behaved. In this sense, cognitive psychology extended behaviourism by suggesting that mental states and inherited structures do exist. But to the extent that there was a profound shift in the direction that many psychologists took in the late 1950s, a number of factors and developments stand out as instrumental in bringing this about: the emerging field of psycholinguistics; the delayed influence of Piaget's work in the field of child development; the growth of computing and information technology (IT); and new findings in animal behaviour studies. In his own personal recollection, the eminent psychologist George Miller (2003) dates the crucial moment of the self-conscious conception of cognitive science as September 1956, when a symposium on information theory was held at the Massachusetts Institute of Technology (MIT). Participants left, he claims, feeling that a new approach was beginning to cohere, linking neuroscience, linguistics, experimental psychology and computing.

The effect of Chomsky's review of Skinner's *Verbal Behaviour* (see section 1.2.2) was to lend support to the rapidly growing field of psycholinguistics. One of the beliefs of this new discipline was the idea that there are certain formalities common to all languages and that these formal similarities could be related to the structure of the mind and brain – hence the encouragement to look at internal cognitive processes.

At about the same time, the impact of the work of the Swiss psychologist Jean Piaget (1896–1980) was finally beginning to be felt in the English-speaking world. Piaget began his work on child development at the Jean-Jacques Rousseau Institute in Geneva in the 1920s, but it was not until the 1950s that it was received with enthusiasm by the Anglophone world. Piaget set his developmental psychology in a thoroughly biological framework. He argued that children have a biologically inherited disposition to interact with the world and construct meaning internally. Thus the way knowledge is obtained is a product of experience and inherent organising tendencies.

The strongest stimulus to the growth of cognitive science, however, was probably the development of computing and IT in the 1950s and 60s. The Second World War had thrown up a whole range of problems related to radar, guided weaponry, code breaking and information storage, and so gave a huge stimulus to the study of these areas. As the large and clumsy electronic valve computers of the 1940s gave way to those exploiting the behaviour of transistors, so many brilliant minds were attracted by the possibilities that this research (heavily funded in the Cold War years within the US and the USSR) offered. As the historian Roger Smith observed, computing quickly became a 'defining technology' – a technology that gave a culture its dominant models and metaphors for thinking about the world. Hence, cognitive psychology was the 'imagination of the computer age applied to knowledge of the mind' (Smith, 1997, p. 837). By this process, just as behaviourism felt obliged to replace mentalist terms (such as memory and feeling) with the language of behaviour, so by the 1960s the language of cognition (such as memory storage, parallel and serial information processing, and filtering) became commonplace in mainstream psychology. This also explains the dramatic impact of Chomsky on linguistics: his idea of a natural formal grammar of the human mind could be compared to the grammars used in programming.

Further challenges to behaviourism came from the study of the behaviour of captive animals. A particularly interesting set of results (since they came from ardent erstwhile behaviourists) emerged from the work of Keller and Marian Breland (1961). In a paper that subsequently became a milestone in the history of psychology, they describe how the successful application of operant conditioning had enabled them to engineer the behaviour of over 38 species and 6,000 individual animals for entertainment and commercial purposes. They went on, however, to relate how in their work they repeatedly met instances where operant conditioning failed altogether. They found that many animals in their care persisted in forms of behaviour that were unrelated to the conditioning process and, moreover, slowed down the reward (usually food) that could be obtained. Even more paradoxically, the stronger the drive (that is, the hungrier the animal), the more they persisted in time-wasting behaviour. There were pigs, for example, that could easily be trained to carry dollars to a depository until a specified number earned them a reward of food. Soon, however, the pigs would drop the coin and root it along the way tossing it into the air. This behaviour set in to such an extent and wasted so much time for the pig that they often did not receive enough food. There were also caged chickens that were trained by the Brelands to pull a loop to send a ball across a miniature baseball field, the reward being food. Once the cage was removed, however, these well-trained chickens chased the ball and pecked it even though this did not bring the food reward.

It now seems obvious that these creatures were expressing instinctive forms of behaviour selected to serve the interests of the individuals of the species in its evolutionary past. But to the researchers of the 1960s raised on the paradigm of behaviourism, it came as something of a shock. The Brelands spoke of the 'utter failure of conditioning theory' (p. 684) and confessed:

> After 14 years of continuous conditioning and observation of thousands of animals, it is our reluctant conclusion that the behaviour of any species cannot

> be adequately understood, predicted, or controlled without knowledge of its instinctive pattern, evolutionary history and ecological niche. (p. 684)

A crucial part of Skinner's behaviourism was that the laws of operant behaviour were supposed to be general purpose; they were assumed to have no special features for dealing with specific contexts, content or domains, but rather were thought to hold good for all stimuli and responses. This was called the 'assumption of the equipotentiality of stimuli'. It was John Garcia and his colleagues who provided a devastating challenge to this position. They discovered that when rats were given an electric shock after drinking flavoured water, the rats found great difficulty in learning the association and hence were poor at avoiding the water. In contrast, rats could readily learn to avoid the water when it was followed by experimentally induced nausea (see Garcia and Koelling, 1966; Garcia et al., 1966). These findings were clearly uncomfortable to the behaviourists, so much so that when Garcia pointed out the discrepancy between his data and the assumption of general purpose learning mechanisms and interpreted his findings in adaptationist terms (very unusual at that time), his manuscripts were rejected by the American Psychological Association (APA). Eventually, in 1998, he was awarded the APA Distinguished Scientific Contribution Award.

Soon terms like 'preparedness' and 'species-specific defence reactions', revealing a concession to the evolutionary history of a species, began to appear in the literature. By the 1980s it was widely appreciated that the domain-general learning mechanisms of behaviourism were not the only processes to be found in humans and other animals.

1.5 The rise of sociobiology and evolutionary psychology

1.5.1 From sociobiology to evolutionary psychology

In the 1970s and 80s, the contribution of ethology to the natural sciences was recognised by the award of Nobel prizes in 1973 to Konrad Lorenz, Niko Tinbergen and Karl von Frisch for their work on animal behaviour. Then in 1981, Roger Sperry, David Hubel and Torsten Wiesel were awarded the prize for their work in neuroethology. The classical approach to ethology associated with Lorenz was continued in Germany by Irenaus Eibl-Eibesfeldt (1989). In the UK, the Netherlands and Scandinavia, where the influence of Tinbergen was more dominant, a more flexible approach to the study of animal behaviour was emerging. At Cambridge in the 1950s, for example, under the leadership of William Thorpe, ethologists became particularly interested in behavioural mechanisms and ontogeny, while at Oxford, Tinbergen headed a group more concerned with behavioural functions and evolution. In a sense, these two groups divided up Tinbergen's four 'whys' (Durant, 1986).

Meanwhile, aided by some fresh ideas from theoretical biology, a new discipline called 'sociobiology' was emerging that applied evolution to the social behaviour of animals and humans. Sociobiology, like behavioural ecology, is concerned with the functional aspects of behaviour in the sense raised by Tinbergen. It drew its initial inspiration from successful attempts by biologists to account for the troubling problem of altruistic behaviour. Of all the problems faced by Darwin, he considered the emergence of altruism and, more specifically, the existence of sterile castes among the insects as two of the most serious. Darwin provided his own answer in terms of community selection but also came tantalisingly close to the modern perspective when he suggested that if the community were composed of near relatives, the survival value of altruism would be enhanced. Haldane came even closer when, in his book *Causes of Evolution* (1932), he pointed out that altruism could be expected to be selected for by natural selection if it increased the chances of survival of descendants and near relations.

Sociobiologists such as Barash (1982) suggested that the new discipline represented a new paradigm in the approach to animal behaviour. Others such as Hinde (1982, p. 152) concluded that 'sociobiology' was an 'unnecessary new term' since behavioural ecology covered the same ground. Behavioural ecology is an established and uncontroversial epithet for those who study the adaptive significance of the behaviour of animals. It would be a mistake, however, to treat 'sociobiology' and 'behavioural ecology' as interchangeable terms. Behavioural ecologists tend to focus more on non-human animals than humans and have a particular concern with resource issues, game theory and theories of optimality. Sociobiology itself deals with human and non-human animals, although it started out with a particular concern with inclusive fitness, and can be thought of as a hybrid between behavioural ecology, population biology and social ethology. The term 'human behavioural ecology' comes close to that of 'sociobiology'.

A book that served in some ways as a seed crystal for the new approach – at least in the sense that it galvanised its opponents – was *Animal Dispersion in Relation to Social Behaviour*, written by V. C. Wynne-Edwards and published in 1962. In this work, Wynne-Edwards advanced a position that was already present in the literature but had aroused no real opposition, namely that an individual would sacrifice its own (genetic) self-interest for the good of the group. It was attacks on this idea that catalysed the emergence of a more individualistic and gene-centred way of viewing behaviour.

One example of the new approach that countered group selectionist thinking came in 1964 when W. D. Hamilton published two ground-breaking and decisive papers on inclusive fitness theory (see Chapter 2). Then in 1966, G. C. Williams published his influential *Adaptation and Natural Selection*, in which he argued that the operation of natural selection must take place at the level of the individual rather than that of the group, and in a similar vein exposed a number of what he thought were common fallacies in the way in which evolutionary theory was being interpreted. In the 1960s and 70s, the British biologist John Maynard Smith pioneered the application of the mathematical theory of games to situations in which the fitness that an animal gains from its behaviour is related to the behaviour of others in competition with it. Then in the early 1970s, the American biologist Robert Trivers was instrumental in introducing several new ideas concerning reciprocal altruism and parental investment (see Chapters 3 and 9). It is a sad reflection on the specialisation of academic life and the disunified condition of the social sciences that while this revolution was occurring in the life sciences, psychology initially remained aloof from these new ideas.

The book that encapsulated and synthesised these new ideas more than any other was E. O. Wilson's *Sociobiology: The New Synthesis*, published in 1975. The book became a classic for its adherents and a focus of anger for its critics. In this work, Wilson (Figure 1.5) irritated a number of scientists by forecasting that the disciplines of ethology and comparative psychology would ultimately disappear by a cannibalistic movement of neurophysiology from one side and sociobiology and behavioural ecology from the other. Others were alarmed that Wilson extended biological theory into the field of human behaviour. Although only one chapter out of the 27 in Wilson's book was concerned with humans, a fierce debate over the social and political implications of Wilson's approach ensued (see Chapter 17). It was as if Wilson had stormed the citadel that had been fortified from biology by social scientists over the previous 40 years, with predictable repercussions.

One of these took place in February 1978 when Wilson attended a meeting of the American Association for the Advancement of Science held in Washington DC. As he prepared to deliver his paper, about 10 people charged onto the stage accusing Wilson of genocide. One of them poured a jug of water over his head.

FIGURE 1.5 The American biologist Edward Wilson (born 1929)
Wilson did pioneering work on the social insect before publishing his massive and controversial work *Sociobiology* in 1975.

As sociobiology developed its theoretical tools, mainstream theoretical psychology, as epitomised by the journals *Psychological Review* and *Psychological*

Bulletin, kept its distance and seemed reluctant to engage with this fledgling discipline. Meanwhile, some anthropologists were beginning to explore the application of sociobiological concepts to their field. In 1979, there appeared a ground-breaking work edited by Irons and Chagnon: *Evolutionary Biology and Human Social Behaviour: An Anthropological Perspective*, a book containing articles by Richard Alexander, William Irons and Napoleon Chagnon.

It is significant that, following Wilson's book, a number of new journals, such as *Behavioural Ecology and Sociobiology* and *Ethology and Sociobiology*, appeared to cater for this growing area. The latter journal eventually changed its name to *Evolution and Human Behaviour*, reflecting a new diversity of approach to the study of human behaviour. For many, the term 'sociobiology' is redolent of the painful debates that surrounded Wilson's early work. Partly to avoid associations with the vitriolic debates of the 1980s and partly to reflect a change in emphasis and ideas, the term sociobiology is now rarely used.

The legacy of sociobiology now lives on, in modified form, in the discipline called 'evolutionary psychology'. Evolutionary psychologists focus on the adaptive mental mechanisms possessed by humans that were laid down in the distant past – the so-called **environment of evolutionary adaptedness** (EEA). To be fair, sociobiology too had emphasised the importance of adaptive mechanisms forged in the geological period known as the Pleistocene. There are those who insist that evolutionary psychology and sociobiology are really the same thing. Robert Wright, in his *The Moral Animal: Evolutionary Psychology and Everyday Life* (1994), for example, suggests that 'sociobiology' was simply dropped as a name (for political reasons) although its concepts carried on under new labels:

> Whatever happened to sociobiology? The answer is that it went underground, where it has been eating away at the foundations of academic orthodoxy. (Wright, 1994, p. 7)

The rise of evolutionary psychology was facilitated by the cognitive revolution in psychology. As noted above, one factor in promoting this revolution was perhaps the Second World War and the stimulus it gave to research on information processing by machines or human subjects in conjunction with machines. Cognitive psychology is based on the idea that mental events could be conceived as the processing of information within structures in the brain. As Tooby and Cosmides (2004, pp. 14–16), two of the world's leading evolutionary psychologists, observe:

> The brain's evolved function is computational – to use information to adaptively regulate the body and behaviour ... the brain is not just like a computer. It is a computer – that is, a physical system that was designed to process information.

And just as a good computer has lots of applications and subroutines separate from each other (for example word processing, playing music, handling images), so the brain, in this view, consists of a variety of modular-based capabilities. Parts of the brain useful for finding a fertile mate, for example, will not be called upon or be much use in the task of identifying nutritious foods.

1.5.2 The influence of evolutionary psychology (EP)

An interesting question is whether or not EP has had much impact on mainstream psychology. After all, it was the prediction of Darwin and Wilson that the evolutionary approach would eventually transform psychology. Cosmides et al. (1992) were equally sanguine when they stated that EP had the 'potential to draw together all of the disparate branches of psychology into a single organised system of knowledge' (p. 3). At the time of writing, it is probably fair to say that the evolutionary approach has not yet unified psychology in the way its protagonists hoped it would, and mainstream psychology remains in a state of, to put it charitably, 'conceptual pluralism'.

In an effort to examine what influence EP has exerted, Cornwell et al. (2005) examined 262 introductory psychology textbooks spanning the period 1975–2004. The influence of EP was measured using a variety of criteria, including coverage in the books, accuracy of portrayal and the type of topics discussed. Figure 1.6, for example, shows how the attention given to EP has increased over the past 30 years.

The authors also note that this increasing exposure was accompanied by a generally more accurate

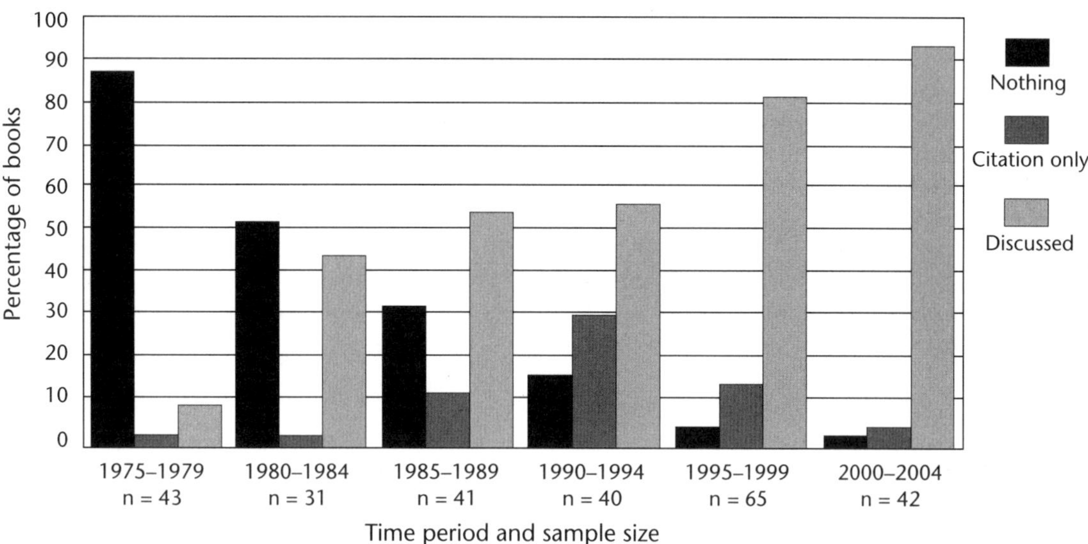

FIGURE 1.6 Coverage of EP in introductory psychology texts 1975–2004

SOURCE: Data from Cornwell, R. E., Palmer, C. et al. (2005) 'Introductory psychology texts as a view of sociobiology/evolutionary psychology's role in psychology.' *Evolutionary Psychology* **3**: 355–74.

treatment, a more positive evaluation and fewer undefended criticisms. One troubling feature, however, was that many textbooks had narrowed the coverage of EP to mating strategies.

As will be explained in Chapter 2, this book takes a catholic approach to the project of 'Darwinising man'. It adopts as a fundamental premise the fact that we evolved from ape-like ancestors and that our early environment has shaped both our bodies and minds. The first species of the genus *Homo* evolved into being about two and a half million years ago, *Homo sapiens* making an entry about 200,000 years ago. If we date the origin of civilization to the Neolithic revolution (the origins of agriculture), at about 10,000 years ago, this represents only 5 per cent of the time since the origins of *Homo sapiens* and only 0.5 per cent of that since the first appearance of the genus *Homo*. It follows from all this that we largely carry genes hard won by our ancestors in the Palaeolithic period (Old Stone Age), genes that are not always optimally suited for modern life. Nevertheless, there are features of modern life, such as making friends, finding a mate and raising children, which we experience in ways not totally dissimilar from our predecessors. Moreover, human behaviour is not completely 'hard wired': our genetically based behavioural mechanisms are moulded by development and modified by learning. Natural selection has also almost certainly endowed us with conditional developmental strategies whereby particular behaviour patterns are triggered by, and are thence adaptive to, specific contexts.

In view of the rather sordid history of some aspects of evolutionary theorising about man, it is necessary, even at the risk of repetition elsewhere, to make some statements about how the Darwinian paradigm now stands on some of the issues raised in this chapter. Nearly all sociobiologists and evolutionary psychologists now assert the psychic unity of mankind; this is done not out of political correctness – a poor foundation for knowledge anyway – but simply because the biological evidence points in that direction. It follows that racial and between-group differences in behaviour are largely attributable to the influence of culture, or at least the interaction between commonly held genes and their fate in different cultures. In this sense, Darwinians agree with Boas: race is not a predictor of inherent potential and capabilities, and there is no racial hierarchy worth taking seriously. Where Darwinians profoundly disagree with the 'nurturists' is the view of the latter that our evolutionary past has no bearing on our present condition; human behaviour and the human mind are moulded and conditioned solely or largely by culture; and, crucially, culture bears no relation to our genetic ancestry and hence can only be explained in terms of more culture.

In contradistinction to this, Darwinism asserts the existence of human universals upon which the essential unity of mankind is predicated. Some cultures may amplify some of these universals and suppress others, and culture itself may in some as yet unclear way reflect the universals lurking in the human gene pool, but the crucial point is that these universals represent phylogenetic adaptations: they have adaptive significance. This is not to suggest of course that there are 'genes for' specific social acts; genes describe proteins rather than behaviour. Human behavioural genetics is rapidly advancing but still in its infancy. It will be genes in concert with other genes and environmental influences (meaning both the cellular and extracellular environments) that are seen to shape the neural hardware of the brain that forms the ultimate basis of human universals.

But to Darwinise man, we need to begin with what Darwin actually said and how subsequent scientific work has modified or confirmed his conclusions. This is the subject of the next chapter.

Summary

- Darwin and some of his immediate followers sought an explanation of animal and human minds based on the principle of psychoneural monism: the idea that the mental and physical lives of animals belong to the same sphere of explanation and that both have been subject to the force of natural selection. The essential continuity between animal and human minds was also assumed. This approach stimulated the application of concepts drawn from the emotional life of humans to account for animal behaviour. At its worst, this tradition relied too heavily on anecdotal information and was prone to anthropomorphising animals.
- In the 20th century, two distinct approaches to animal behaviour emerged: ethology and comparative psychology. The ethological tradition in Europe was based around the work of such pioneers as Heinroth, Lorenz and Tinbergen, entailing the study of a variety of animal species in their natural habitat using a broad evolutionary framework. Comparative psychology took root particularly in America and built upon the work of Thorndike. Watson and Skinner were extreme representatives of this tradition. With the passage of time, comparative psychology became increasingly concerned with the behaviour of a few species in laboratory conditions, to the neglect of an evolutionary perspective.
- Biology became entangled with political movements claiming to seek the improvement of society, or the race, or the national character. Galton was one of the chief founders and exponents of eugenics, the idea that the state should occupy a role in directing the breeding of individuals. Galton and his followers recommended that the able should be encouraged to procreate and the unfit or undesirable discouraged. Eugenic ideas were put into practice in the USA and Germany, with repressive and hideous consequences.
- The theory of human instincts begun by Darwin was continued in America by James and in Europe by Heinroth, Lorenz and other ethologists. The experimental problem of demonstrating the action and existence of human instincts, as well as the reactionary associations of biological theories of human nature, led most social scientists and anthropologists in the middle years of the 20th century to reject biological explanations of human behaviour and assert the primacy of culture.
- In the 1960s and 70s, a number of fundamental papers and books set in motion the sociobiological approach to animal and human behaviour. Sociobiology is

predicated on the view that animals will behave so as to maximise the spread of their genes. Behaviour is therefore examined largely with functional (in the sense of Tinbergen) questions in mind. When applied solely to humans, the approach is also sometimes called evolutionary psychology, some regarding this movement as a new paradigm. This new movement seeks to revive the Darwinian project of demonstrating the evolutionary basis of human behaviour and of many facets of human culture.

- The cognitive revolution in psychology precipitated the demise of behaviourism and allowed evolutionary psychology to flourish.
- Evolutionary psychology is slowly making headway into mainstream psychology and increasingly features in introductory textbooks. Its promise to transform psychology and provide it with a much needed unifying base still seems some way off.

Key Words

- **Behaviourism**
- **Environmentalism**
- **Environment of evolutionary adaptedness (EEA)**
- **Ethology**
- **Eugenics**
- **Fitness**
- **Fixed action pattern**
- **Function**
- **Gene pool**
- **Innate releasing mechanism**
- **Lamarckian inheritance**
- **Monism**
- **Morgan's canon**
- **Naturalism**
- **Ontogeny**
- **Operant conditioning**
- **Paradigm**
- **Phenotypic plasticity**
- **Positivism**
- **Proximate cause**
- **Selection**
- **Ultimate causation**

Further reading

Degler, C. N. (1991) *In Search of Human Nature: The Decline and Revival of Darwinism in American Social Thought*. Oxford, Oxford University Press.

True to its title, this book examines the period 1900–88. A penetrating sociological analysis of the fate of Darwinian ideas in America.

Plotkin, H. (2004) *Evolutionary Thought in Psychology: A Brief History*. Oxford, Blackwell.

An excellent, readable and balanced discussion of evolutionary thinking in psychology leading up to evolutionary psychology.

Richards, R. J. (1987) *Darwin and the Emergence of Evolutionary Theories of Mind and Behaviour*. Chicago, University of Chicago Press.

A detailed and thorough work. Covers the ideas of Darwin, Spencer, Romanes, Morgan and James. Lays particular emphasis on the evolutionary origins of morality. Contains the author's contemporary defence of using evolution as a basis for ethics.

Smith, R. (1997) *The Fontana History of the Human Sciences*. London, Fontana.

A panoramic survey of the human sciences from the scientific revolution to the late 20th century. Many chapters are relevant to the study of evolution.

Thorpe, W. (1979) *The Origins and Rise of Ethology*. New York, Praeger.

An inside account of ethology from someone who helped to shape the discipline. Plenty of anecdotal information.

Darwinism, Inclusive Fitness and the Selfish Gene

If I were to give an award for the single best idea anyone has ever had, I'd give it to Darwin, ahead of Newton and Einstein and everyone else. In a single stroke, the idea of evolution by natural selection unifies the realm of life, meaning and purpose with the realm of space and time, cause and effect, mechanism and natural law. But it is not just a wonderful scientific idea. It is a dangerous idea.
(Dennett, 1995, p. 2)

This chapter outlines the ideas central to Darwinism and examines some of the difficulties that Darwin faced. Many of the problems that confronted Darwinism in the 19th century have been largely resolved by work over the past 75 years. A successful theory, however, has not only to confront empirical evidence, but also to show that it can do so better than alternative accounts. With this in mind, we will contrast Darwinism with Lamarckism as alternative ways of explaining how organisms become adapted to their environments. The chapter also looks at how Darwin's ideas on fitness were extended into the concept of inclusive fitness, a development that helped to solve many of the problems that puzzled Darwin. The chapter concludes with an examination of the gene-centred view of evolution and natural selection. Our understanding of the genetic basis of behaviour and the whole operation of natural selection is also greatly assisted by a clarification of what can properly be considered to be the unit of natural selection, and by the crucial distinction that must be made between the replicators (genes) and the vehicles of these replicators (bodies). The concept of the selfish gene has been much maligned, but properly understood it does not suggest that all individuals must behave selfishly in the pejorative sense. This chapter will demonstrate how altruism can possibly arise in a world of selfish replicators.

2.1 The mechanism of Darwinian evolution

The essence of Darwinism can be summarised as a series of statements about the nature of living things and their reproductive tendencies:

- Individuals can be grouped together into species on the basis of characteristics such as shape, anatomy, physiology, behaviour and so on. These groupings are not entirely artificial: members of the same species, if reproducing sexually, can by definition breed with each other to produce fertile offspring.
- Within a species, individuals are not all identical. They will differ in physical and behavioural characteristics.
- Some of these differences are inherited from the previous generation and may be passed to the next.
- Variation is enriched by the occurrence of spontaneous but random novelty. A feature may appear that was not present in previous generations, or may be present to a different degree.
- Resources required by organisms to thrive and reproduce are not infinite. Competition must inevitably arise, and some organisms will leave fewer offspring than others.
- Some variations will confer an advantage on their

possessors in terms of access to these resources and hence in terms of leaving offspring.

- Those variants that leave more offspring will tend to be preserved and gradually become the norm. If the departure from the original ancestor is sufficiently radical, new species may form, and natural selection will have brought about evolutionary change.
- As a consequence of natural selection, organisms will become adapted to their environments in the broadest sense of being well suited to the essential processes of life such as obtaining food, avoiding predation, finding mates, competing with rivals for limited resources and so on.

We can conceive of the essential elements of this process as a sort of 'Darwinian wheel of life' (Figure 2.1), where organisms (or rather their genes) are propelled around a relentless cycle of reproduction, variation and differential survival.

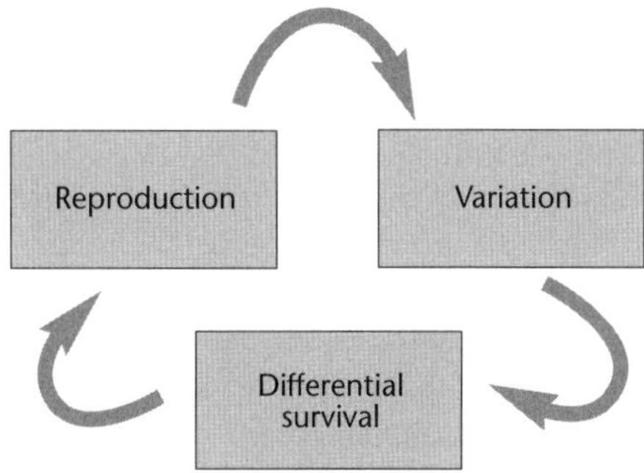

FIGURE 2.1 A Darwinian wheel of life
The birth and death of thousands of individual organisms result in a gradual modification of their genes through differential survival. The result is the gradual formation of new species and, crucially for Darwinian psychology, the emergence of adaptations.

One of the crucial points to appreciate about Darwinian thinking is that evolution is not goal directed. Organisms are not getting better in any absolute sense; there is no end towards which organisms aspire. Creatures exist because their ancestors left copies (albeit imperfect ones) of themselves. One of the great triumphs of Darwinism was that it offered an explanation of the structure and behaviour of living things without recourse to any sense of purpose or teleology. **Teleology** (Greek *telos*, end) is the doctrine that things happen for a purpose or are designed with some express end. It was a cast of mind widespread in the early 19th century that natural phenomena and living things were designed with some motive or purpose in mind. Even by the mid-century, the critic Ruskin could still write that the structure of mountains was calculated for the delight of man and that in their contours 'the well-being of man has been chiefly consulted'. The teleological way of thinking was systematised and popularised by Archdeacon William Paley. In his classic *Natural Theology* (1802) – a work Darwin knew well and once even admired – he advanced the famous watchmaker analogy. Just as a watch could not arise by chance but immediately suggests a designer and maker, so too the intricate organisation of living things suggests a cosmic designer. The whole of creation could then be seen as God's 'Book of Works', a book, moreover, that provided ample evidence for a benign Creator. It followed that creatures appeared to be fitted to their mode of life because they were designed that way by the supreme artificer we call God.

Purpose, teleology and the whole concept of providential design were all to be swept away by Darwin's 'dangerous idea'. For Darwin, there was no grand plan, no evidence that life forms were placed on earth by a Creator, no ultimate purpose or inevitable progress towards some goal. The watchmaker is blind.

2.1.1 The ghosts of Lamarckism

One person to suggest the possibility of the transmutation of species before Darwin was the French thinker Jean Baptiste Lamarck (1744–1829). His views are now virtually totally discredited, but they once served as the only serious alternative to Darwinism as a way of explaining the adaptive nature of evolutionary change. Lamarck argued that the characteristics (**phenotype**) acquired by an individual in its lifetime could be passed on to subsequent generations through the germ line (**genotype**). This view is usually parodied by the suggestion that the large muscles of a blacksmith, acquired through use during his lifetime, will result in slightly larger than average muscles in his son (and presumably daughter). Darwin is known to have rejected **Lamarckism**, but what he was really rejecting was the additional notion within Lamarckism that creatures possessed an inherent tendency to strive towards greater complexity. It was this teleological idea of purpose, rather than Lamarck's mechanism of adaptation, that Darwin wisely attacked. Indeed, it comes as a surprise and shock to many to learn that Darwin accepted the possibility of 'the effects of use and disuse' as a mechanism for influencing the characteristics of the next generation.

Lamarckism has at least two failings. Suppose we suggest that the hind legs of a rabbit developed their strength by the continued action of rabbits running away from predators. The first problem, as noted with the blacksmith analogy, is that, as an empirical fact ascertainable through experimentation, acquired characteristics are not passed on to offspring. The other problem is more serious, relating to the difficulty of explaining why exercise should increase the strength of muscles in the first place. Logically, exercise could reduce muscle strength, leave it unchanged or increase it. We know that it tends to increase it, but Lamarckism needs to explain this. To a Darwinian, it is straightforward: animals possess physiological mechanisms whereby exercise promotes strength because they help to increase their survival chances. Lamarckism is hard pushed to explain the direction of the muscular response.

The ghost of Lamarck was virtually (but not quite) laid to rest by August Weismann (1839–1914). Weismann distinguished between the germ line of a creature and its body, or 'soma'. Characteristics acquired by an individual affected the somatic cells (all the cells of the body other than the sperm or eggs) but not the germ line (information in the sperm or eggs). Weismann's essential insight was that information could flow along the direction of the germ line and from the germ line to the somatic cells, but not from somatic cells to the germ line (Figure 2.2).

In establishing his case, Weismann (largely between 1875 and 1880) cut off the tails of mice over a number of generations and showed that there was no evidence that this mutilation was inherited. This is in fact a rather unfair experiment – to both Lamarck and the mice – since, in strict Lamarckian terms, the mice were not striving in any way to reduce their tail length. It turns out that Weismann performed the experiment as a counterblast to those who argued at the time that it was well known that if a dog's tail was docked, its pups often lacked tails (Maynard Smith, 1982). For an equally forceful experiment, consider the fact that many Jewish male babies are still circumcised despite the fact that they are the products of a long line of circumcised male ancestors.

2.1.2 The central dogma in a modern form

The **central dogma** in its modern form can be understood in terms of the flow of information within individuals and between generations carried by macromolecules, shown in Figure 2.3. There are a small number of cases in which this picture is an oversimplification, but none of these seriously challenges the general application of this schema (Maynard Smith, 1989).

We must still, however, deal with the question of why a two-way flow of information between genotype and phenotype never evolved. After all, male germ cells or **gametes** in the form of spermatozoa are made continuously by the body of most mammals at an incredible rate (adult human males are producing sperm at the rate of about 3,000 per second), so why couldn't information about the current state of the body be fed back to the germ line in the same way perhaps as engineers use their experience of the behaviour of the prototype to alter the blueprint for the next improved version? The answer seems to be that most phenotypic changes (with the exception of learnt ones) are not useful or adaptive. They result typically from disease, injury or ageing. A hereditary mechanism that enabled parents to transmit such changes would not be favoured by natural selection (Maynard Smith, 1989). As is so often the case, alternatives to Darwinism are unlikely, for good Darwinian reasons.

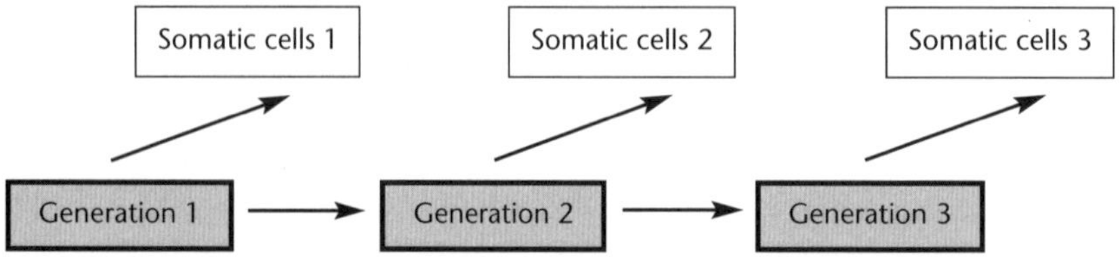

Somatic cells 1		Somatic cells 2		Somatic cells 3
↗		↗		↗
Generation 1	→	Generation 2	→	Generation 3

FIGURE 2.2 The germ line and information flow, according to Weismann

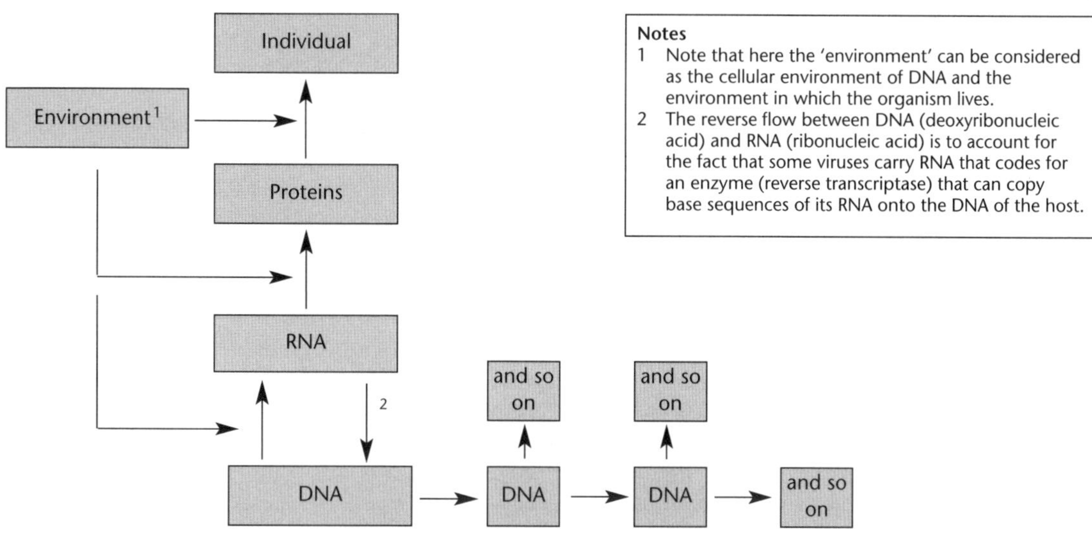

FIGURE 2.3 The central dogma of the molecular basis of inheritance

2.1.3 Darwin's difficulties

There were four notable areas in which Darwin's theory faced serious problems:

1. the mechanism of inheritance
2. the means by which novelty might be introduced into the germ line
3. the existence of altruistic behaviour
4. the fact that some features of animals (for example the tail of a peacock) ostensibly seem to place the animal at a disadvantage in the struggle for existence.

Concerning the first two problems, Darwin knew nothing of course of the schema shown in Figure 2.3. His own theory of inheritance (sometimes called the theory of 'pangenesis', involving 'gemmules' circulating around the body) is so wide of the mark that it is not worth considering. We will show shortly how inheritance and novelty can be explained in terms of molecular genetics. With respect to the last two problems, Darwin made more headway, but a more thorough examination of these problems will be covered in section 2.5 and Chapter 3 respectively.

2.2 Some basic principles of genetics

2.2.1 The genetic code

Earlier we saw how Weismann argued that inheritance is best thought of as a flow of information along the germ line. With this in mind, there are four basic questions that we need to answer in order to appreciate the genetic basis of this information in the context of evolution:

1. How is genetic information stored and preserved?
2. How is this information used to build organisms?
3. How is the information passed on to new organisms in the act of reproduction?
4. How do novelty and change enter the information and thus provide the raw material for natural selection to act upon?

The storage and preservation of information: the language of the genes

The language of genes is written on an extremely long molecule called deoxyribonucleic acid, or **DNA**. The molecule consists of two strands, each strand having a backbone of alternating ribose sugar and phosphate groups, and each sugar group having one of four bases attached to it. This sequence of bases is the genetic code, which prescribes the development of each individual. Since each **base** can only chemically bond with one of the other three (its complementary partner), the sequence of bases on one strand uniquely defines the sequence on the other strand. The base pairs are said to be complementary. There are four types of bases: cytosine (C), thymine (T), adenine (A) and guanine (G). The complementary pairing of the bases is as follows:

A	T	(Two hydrogen bonds)
G	C	(Three hydrogen bonds)

It is estimated that the human **genome** contains about 3 × 109 base pairs, of which, at the time of writing (2007) about 92 per cent have been mapped by the human genome project. Each strand of the DNA molecule is wound into a helical shape, two strands thus giving us the double helix (Figure 2.4). It is the sequence of bases on any one strand that contains the information necessary for the development and functioning of each cell. Since each base determines the base on the opposite strand, this information is therefore contained within just four characters. At first sight, it may appear unlikely that just four characters could carry the information necessary to define the development of complex organisms – even the English alphabet has 26 letters. We must remember, however, that molecules are small and, relatively speaking, the DNA molecule is long. The enormous storage capacity achieved by modern computers is achieved using just two characters: a zero (0) and a one (1), corresponding to the states of transistors in the microcircuitry.

If the sequence of nucleotides in the entire human genome were represented on the page of a book by symbols (so that a sequence would, for example, look like AGTCGAATTGCC ...), on this scale a gene would spread over about three pages, an average chromosome would take up about 50 books of the size of this one, and the entire genome (present in just one of your cells) would spread over about 1,000 books.

The preservation during growth and reproduction of the information contained in the base sequences of DNA raises two related questions. First, when a cell divides during the growth of an organism, how is information passed to the daughter cells in order that

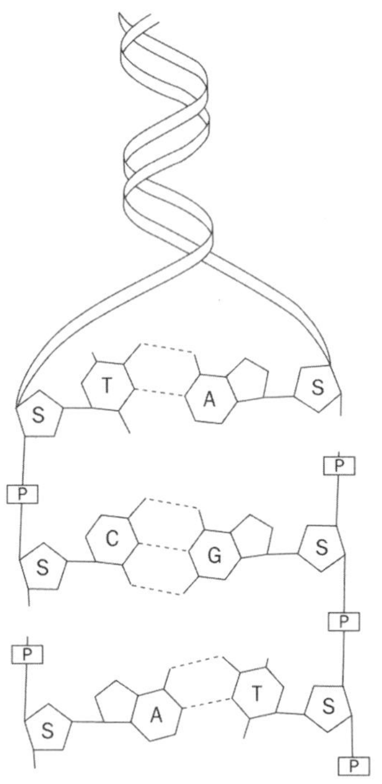

S = sugar; P = phosphate; A, C, T, G = bases

FIGURE 2.4 DNA as a double helix of complementary polynucleotide chains

they develop and perform appropriately? Second, when organisms reproduce either sexually or asexually, how is information passed from parent to offspring? The process, at least in the early stages, is essentially the same. An examination of the structure of DNA in Figure 2.4 shows that if the two strands were divided and each retained its sequence of bases, each strand could serve as a template to create another double helix. This potential was immediately obvious to Watson and Crick (Figure 2.5) when they established the structure of DNA in 1953. Their own suggestion in their paper submitted to *Nature* has become a classic of understatement:

> It has not escaped our notice that the specific pairing we have postulated immediately suggests a possible copying mechanism for the genetic material. (Watson and Crick, 1953, p. 737)

FIGURE 2.5 James Watson (b. 1928), on the left, and Francis Crick (b. 1916), on the right, shown with their model of DNA at the Cavendish Laboratory, Cambridge, England in 1953
Their discovery of the helical structure of DNA ranks as one of the most important scientific findings of the 20th century.

How genetic information is used to build organisms: the translation of the language

Following the rediscovery of Mendel's work by de Vries in 1901, and subsequent work by de Vries and Morgan, it became generally accepted that heritable information was carried on units called **genes**. Even as long ago as 1909, it was suspected that genes influenced the phenotype through the production of enzymes. Archibald Garrod, an English physician, proposed that inherited diseases, so-called inborn errors of metabolism, were caused by defective enzymes, which reflected a defective encoding of the genetic information. Later work supported Garrod's insights and led to the 'one gene, one enzyme' hypothesis. In this view, each unit of information (the gene) specified one enzyme. All enzymes belong to the class of molecules called proteins, and this 'one gene, one enzyme hypothesis' was later modified to 'one gene, one protein'. When it was realised that some proteins consisted of several different chains coded for by different genes and assembled after each chain was produced chemically, the view was modified again to the idea of 'one gene, one polypeptide'.

A gene can thus be regarded as a stretch of DNA that in some way contains the information necessary for the synthesis of a polypeptide. There are about 25,000–30,000 genes per human cell – a number much smaller than the estimated value of 100,000 only a few years ago. Not all the DNA in a cell codes for proteins, and there are some rather mysterious sections that seem to do nothing at all, sometimes called, to the great irritation of geneticists, 'junk DNA' or more respectfully 'introns'. It is estimated that about 97 per cent of human DNA is of the non-coding or 'junk' sort, leaving only 3 per cent with a function that we currently understand. Each of us (excluding identical twins) is genetically unique. But if we examined our DNA closely over most of its length, the variation is small – about 1 base pair difference in somewhere between 100 and 1,000 base pairs. But, fortunately for DNA testing, there are some non-coding regions where we differ greatly and these are the regions used in forensic work and for paternity testing.

Proteins are made up of long chains of chemical units called **amino acids**. There are thousands of different kinds of proteins but only 20 types of amino acid common to all organisms. We have seen how the information along the DNA is in the form of a four-character language: A, C, T, G. Now, if each base coded simply for one type of amino acid, we would only have four possible acids that could be encoded. If a pair of bases such as AT or CG coded for one amino acid, there would still be only 16 (4^2) permutations possible. But if triplets of bases such as AAA, CCG, GCA and so on coded for each amino acid, there are 64 (4^3) possibilities. This is the minimum number of base combinations that are required to encode the 20 or so amino acids used to build organisms. Moreover, this triplet code would allow spare information that could convey instructions such as 'start', 'stop' and so on. It turns out that the triplet code is the one used by DNA.

Proteins are not assembled directly from the DNA template. The fine detail of the biochemistry is beyond the scope of this book, but, in brief, the information along the DNA is first transcribed to a very similar long molecule called messenger ribonucleic acid (mRNA) in the nucleus of the cell. The **RNA (ribonucleic acid)** then carries this information to the cytoplasm, where it is translated into polypeptides according to the triplet language just described.

Deciphering the triplet code began in 1961 when the **codon** TTT was established as coding for the amino acid phenylalanine. The remarkable fact about the 64 codons that have now been deciphered is that they have the same meaning in virtually all organisms. It is as if all living things share one universal language. Hence in laboratories, bacterial cells can translate genetic messages from human cells and vice versa. This consistency of the genetic vocabulary is what makes genetic engineering possible, and it also implies that the code must have been established early in evolution. By examining the similarity of amino acid sequences in proteins in different species, we gain some insight into similarities between their DNA and hence, by making a few assumptions, into how closely related they are in evolutionary time. Haemoglobin, for example, is a molecule found in monkeys, chickens and frogs. The precise arrangement of haemoglobin amino acids in a typical monkey differs by only 5 per cent (8 being different in a chain of 125) from that in humans, whereas that of a chicken differs by about 35 per cent. Studies such as these fit almost perfectly with what one would expect concerning the ancestry of creatures from fossil evidence and morphological similarities. They have led to the oft-quoted remark that we differ from common chimps by only 1.6 per cent, which all sounds rather depressing. We can be assured, however, that differences in phenotype are not a simple linear function of DNA differences.

In summary, the genotype of an organism is determined by the sequence of codons (base triplets) on its DNA. This sequence in turn commands the types of protein that will be synthesised. The assembly of proteins into cellular structures and their action in metabolic pathways, in combination with environmental influences during development, determines to a large degree the form, functions and behavioural patterns that together comprise the phenotype. Now, there is obviously a huge leap between this well-established mechanism of how genes make proteins and the final assembly of an organism in all its multicellular complexity. Embryonic development is a vast subject and one still not properly understood.

How the information is passed on in reproduction: the flow of information through the germ line

The idea that all the information we need to build an organism is found in the DNA has now entered popular culture. Novels and films have explored the idea that an individual can be cloned from a small tissue sample, an idea relying upon the fact that every diploid cell in the human body contains a full complement of DNA. These cells would be invisible to the naked eye, yet each one contains about 3 metres of DNA. If the cell were enlarged to the size of a full stop on this page, the DNA would stretch for 150 metres, although it would be too thin to be visible.

The DNA of each cell nucleus is usually bound to groups of proteins and exists as long thin diffuse fibres that are difficult to see. As a cell prepares to divide, however, the fibres coil into visible structures called **chromosomes**. It is convenient to consider the reproduction of cells in terms of the fate of chromosomes. Chromosomes consist of DNA coiled around protein bodies and then coiled on itself twice again.

We have already seen how the structure of DNA lends itself to replication: the DNA molecule can act as a template for the synthesis of more identical DNA either for cell division within an organism or to form the basis of a new organism. Cell division that contributes to growth and repair in an organism is called **mitosis**, and each new cell is simply a replicate of the original. By the time you have finished reading this sentence, several thousand of your cells will have divided by mitosis.

In order to transmit DNA in the process of sexual reproduction, the cells responsible divide in a different way, called **meiosis**. It is worth examining the process of meiosis in some detail, since it helps us to understand why sexual reproduction should exist at all as well as throwing some light on the essential differences between males and females.

To understand the process of meiosis, and hence the implications for sexual reproduction, we need to tackle the arrangement of chromosomes in cells. The position of a particular gene along a length of DNA, and hence on the chromosome, is called the **locus** of that gene. All human beings belong to the same species and thus have some obvious similarities that must have a genetic basis. In features such as hair type, eye colour and so on, there are also differences, and it follows that there must be different forms of the genes that determine these characters. These different forms are called allelomorphs, or more commonly **alleles**. Now, most cells in a typical animal contain chromosomes in matching pairs. In human cells, there are 46 chromosomes made up of 23 pairs, each member of a pair being very similar except for the two that are called **sex chromosomes**. The sex chromosomes are called X or Y, which refers to their appearance under an optical microscope. It could be said then that humans have 22 matching pairs (of **autosomes**) and then either one X and one Y chromosome (XY) if they are males, or one X and another X chromosome (XX) if they are female.

A B C

a B c

FIGURE 2.6 Complementary pair of chromosomes showing homozygosity for allele B and heterozygosity at A and C

Figure 2.6 shows a pair of chromosomes aligned side by side. The loci of the genes are given by the letters Aa, BB and Cc. In the case of the allele B, the two forms of the gene on each member of the chromosome pair are the same, so the genotype is said to be **homozygous**. If that allele coded for spots on the fur of an animal, the phenotype of the organism would be spotted. In the case of allele A, it is not as simple. Both genes refer to the same trait but exist in different forms; the genotype at this locus is thus **heterozygous**. The final outcome that is expressed depends a lot on the type of gene. If this were the locus for eye colour, a coding for blue eyes and A for brown, the phenotype would show brown eyes. We say that brown is the **dominant allele** and blue the recessive. In other cases, the outcome is intermediate between the homozygous condition for each allele.

Why chromosomes exist in pairs when it seems that one set would do relates to the phenomenon of sexual reproduction. Of the 46 chromosomes in each of your cells, 23 are provided by your mother and 23 by your father. You share 50 per cent of your genome with your biological mother and 50 per cent with your biological father. When cell division occurs in your body to replace damaged cells, or simply as part of growth, each new cell has the same 46 chromosomes as the one from which it grew. When animals produce sex cells or gametes by the process of meiosis, however, the procedure is different. In each human sperm and egg, there are only 23 chromosomes, that is, half the number in normal 'somatic' cells. Fertilisation brings about the **recombination** of these to 46 in 23 pairs, and so the life of a new organism begins.

Figure 2.7 shows a simplified account of meiosis for a simple organism with only one pair of chromosomes (noting that even fruit flies have four). The loci for eye colour and fur colour are shown. Figure 2.7 focuses on meiosis leading to the production of sperm (**spermatogenesis**), but the same steps occur in

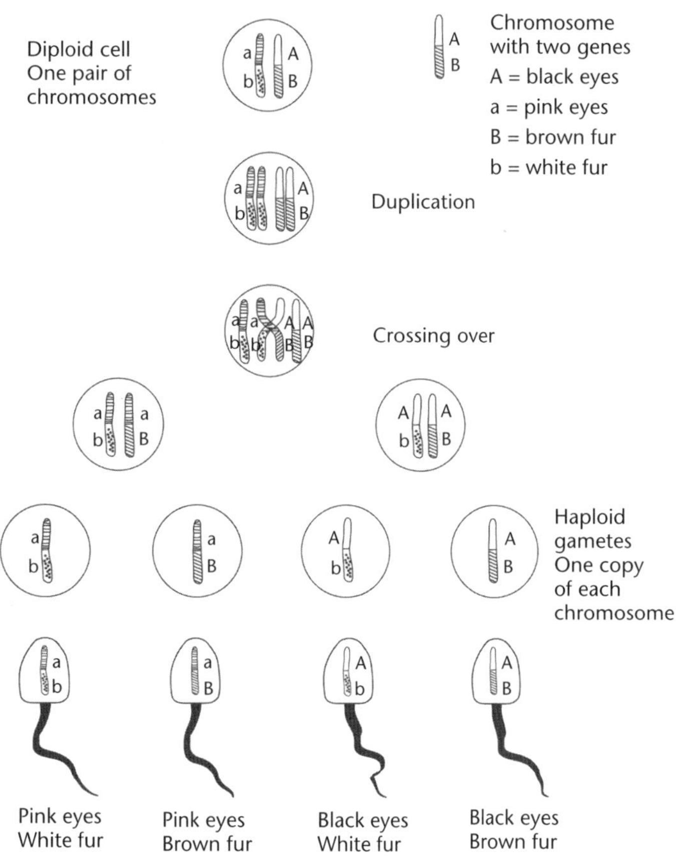

FIGURE 2.7 Simplified picture of meiosis and spermatogenesis

the formation of the eggs in the female (**oogenesis**). Figure 2.8 shows the fusion of two gametes (sperm and egg) to produce a fertile zygote.

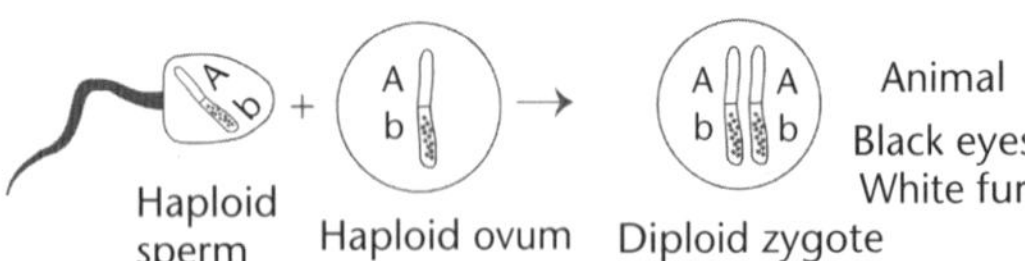

FIGURE 2.8 Fertilisation of an ovum

Notice that, in Figure 2.7, one pair of chromosomes with just two genes on each gives rise to four different gametes. The number of possible gametes in human spermatogenesis or oogenesis from 23 pairs of chromosomes with 30,000 genes is vast.

In Figure 2.7 we can also see the process of **crossing over**. During meiosis, chunks of DNA from each complementary chromosome are swapped. The effect of this process is extremely important and goes to the heart of why the gene must be considered to be the unit of natural selection (see below), as well as perhaps to why sex exists at all. The consequences of crossing over can be understood by comparing gametes with ordinary cells. Looking down an imaginary high-powered microscope at a pair of complementary chromosomes inside an ordinary cell, it would in principle be possible to see one as coming from your mother and one as originating from your father. Looking along the single chromosome inside one of your **haploid** gametes, we would see a patchwork of genes from your mother alternating with genes from your father. Meiosis has thus stirred up the genes from your father and mother, reconfigured them and presented a whole new array to the world.

How novelty enters the information: mutations and meiosis

Modern genetics provides an answer to the problem that so troubled Darwin of how spontaneous novelty arises and allows natural selection to take effect. Much of the apparent novelty that arises in any one generation comes from the shuffling of genes during meiosis and sexual reproduction, as discussed above. Fundamental changes must, however, be caused by changes in the base sequence of the DNA. We now know that chemical and physical agents such as high-energy radiation can have a mutagenic effect on DNA and cause alterations in its structure. Mutagenesis can also occur spontaneously by errors in replication. It seems likely that sexual reproduction may have begun as a way of reducing the number of these spontaneous errors.

One example of the debilitating effects of mutations is the condition known as phenylketonuria (PKU). People suffering from this disease are unable to produce the enzyme phenyl hydroxylase (PAH). This enzyme is essential in the conversion of the amino acid phenylalanine (obtained from the diet) into tyrosine. As a result, the phenylalanine accumulates in the body and is eventually converted into phenylketones which can be detected in the urine of those with the disorder. If children with this disease are left untreated, they become increasingly prone to hyperactivity, seizures and eventually mental retardation. The incidence of PKU is about 1 in 15,000 births, but this figure varied widely among different population groups.

The PAH gene is found on chromosome 12. The precise mutations that can disrupt the successful synthesis of PAH are numerous and over 400 different sites have been documented. PKU is an autosomal recessive disorder. This means that the defective gene (and functional versions) are found on an autosome – that is one of the 22 chromosome pairs that are not sex chromosomes (X or Y). It is recessive in the sense that to develop PKU a child must inherit two defective copies of the gene (one from each parent) and so be homozygous for the recessive defective allele. If each parent carries just one copy of the defective allele they are said to be carriers and by Mendelian genetics any child they produce will have a 1 in 4 chance of being homozygous and so developing this disease.

Why the allele responsible for PKU persists and has not been eliminated from the gene pool by natural selection is a subject of ongoing research. One possibility is that the heterozygous condition (that is, that of carriers) confers some advantage in protecting against toxins from moulds that once grew on stored grains and other foodstuffs in damp climates.

Box 2.1 gives another example of the severe practical consequences of a genetic mutation.

Box 2.1 Queen Victoria's gene: haemophilia and the breaking of nations

A particularly fascinating example of how a small change in one gene can have profound effects concerns the inheritance of **haemophilia** in Queen Victoria's family.

Of the 23 pairs of chromosomes in human cells, all are homologous except for the pair of sex chromosomes. In females, this pair is referred to as XX (from the shape they appear under an optical microscope), and that in males XY. When male cells divide to produce haploid gametes, it follows that half will carry

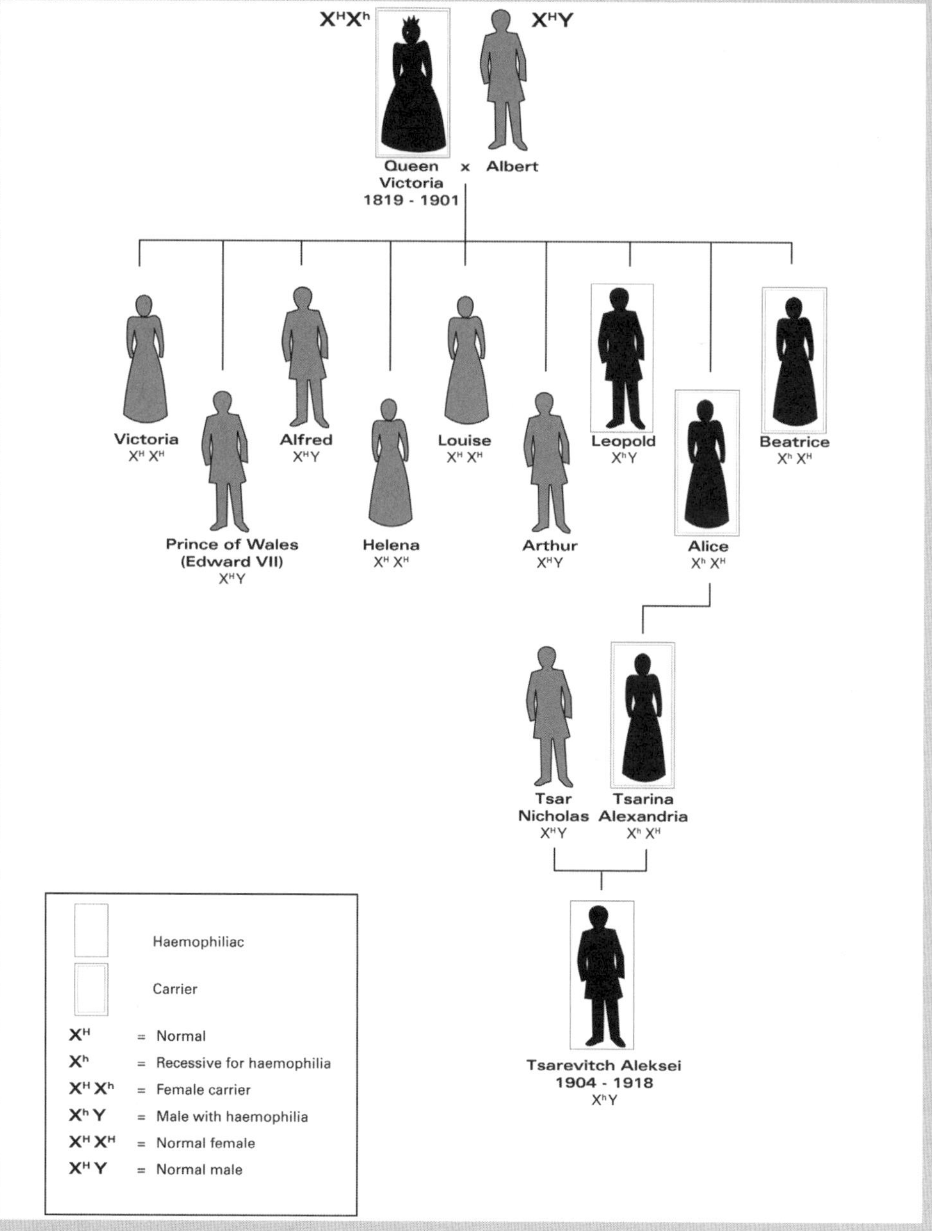

FIGURE 2.9 Haemophilia in the monarchies of Europe

an X chromosome and half a Y chromosome. Females produce eggs carrying only X chromosomes. When a Y male gamete fertilises a female egg, a zygote (XY) is produced that will grow into a male. When an X gamete from the male meets an X gamete from the female, an XX zygote is produced, which grows into a female. Hence the sex of a child is determined by sperm from the male, and on average an equal number of males and females is produced.

It seems that Queen Victoria must have been a carrier on her X chromosome of a defective allele leading to haemophilia. Haemophilia is caused by an allele producing a defective version of a protein involved in blood clotting. We can represent Queen Victoria as XHXh, meaning that she 'carried' the mutant allele (h) on one chromosome but had the normal allele on the other (H). Consequently, Victoria herself had no symptoms of the disease but passed the defective gene down the generations.

One way of consolidating power and strengthening alliances is through marriage, and this tendency led to the defective gene being passed through the courts of Spain and Russia. Potts (1995) charted the fate of this gene. He argues that, in the case of the Russian monarchy, Tsar Nicholas's desire to keep secret the fact that his only son and male heir, Aleksei, suffered from haemophilia was a factor in the downfall of the Romanov regime. The Serbian peasant Rasputin seemed able to exert a hypnotic affect over the Tsarevitch Aleksei and enabled his internal bleeding to subside (restriction of movement enabling some healing to occur). When the February revolution began, the new government offered to make Aleksei a constitutional monarch. Instead, the Tsar, reluctant to allow his haemophilic son to succeed to the throne, abdicated. In 1918, the family was tracked down by revolutionaries and shot in Siberia. In July 1998, the bones of the Romanovs were finally given a state funeral in Moscow. They were identified as the authentic remains by DNA fingerprinting.

A complete account of the way in which Queen Victoria acquired the defective gene has not yet been written. There is no sign of haemophilia in her mother's family since it fails to appear in the numerous descendants of her first marriage. It is possible that a mutation occurred in Queen Victoria herself or in her father, the Duke of Kent. Strangely, the Duke's own genetic defect (porphyria) was not passed on to Victoria or her family. Potts raises the intriguing possibility that because of the pressure on Queen Victoria's mother (Mary Louise Victoria, Duchess of Kent) to produce an heir, she may have engaged in 'extrapair copulation'.

2.2.2 From genes to behaviour: some warnings

Our current knowledge of the molecular basis of inheritance, gene expression and the inheritance of genetic disorders is considerable. The conditions discussed above – haemophilia and sickle-cell anaemia – are, however, physiological rather than behavioural conditions, and it is important to point out that a simple one-to-one correspondence between genes and behaviour is hard to find. Part of the reason for this is that most behavioural characteristics are polygenic, that is, they are the result of the expression of many genes rather than just one (as in the case of sickle-cell anaemia).

In this whole area, there remains a great deal of heated discussion about the idea that behaviour is 'innate' or has a genetic basis. Marian Dawkins (1986) suggests that much of the dispute can be related to a lack of precision in terminology. It should in fact be pointed out that when we say 'gene for', we are not implying a simple one-to-one correspondence. The trait in question, be it morphological or behavioural, could be the result of many genes interacting with the environment. It would perhaps be better to say that the behaviour in question has some genetic basis and that genetic differences exist between individuals displaying this trait. A spectrum clearly exists in the genetic 'determination' of behaviour. We can identify at least four positions on this spectrum:

1. *Unlearnt patterns of behaviour invariant with respect to environment*
   Clearly, all genes need an environment in which to thrive and replicate. If *Teleogryllus* crickets are deprived of oxygen (an environmental commodity), they will die and not sing. In a wide band of normal environmental conditions, however, they sing a song that is so characteristic of the species that even if individuals are reared in isolation, have never heard the song or indeed are subjected during development to a cacophony of non-song-like sounds, they still grow up to sing the species-typical song of the cricket. Male sticklebacks act aggressively towards the red bellies of other males when confronted with them for the first time, even if they have never seen another fish or even their own reflection. Both these behaviour patterns are clearly something with which the creatures are born; they must have some genetic foundation. An environment is needed only in the trivial sense of providing a platform for the actions. The singing behaviour is 'hard wired'.

2. *Unlearnt but modified by learning*
   Laughing gull *(Larus atricilla)* chicks have an innate tendency to peck at their parents' bills to persuade them to regurgitate food. Very young chicks will even peck at long, thin red knitting needles. As they mature, however, they become more discriminating and gradually learn to recognise the finer details of a gull's bill.

3. *Selectively learnt or selectively expressed*
   To sing properly in a chaffinch-like way, young chaffinches must learn the song from other birds, but they are not simple mimics. If exposed to a wide variety of song sounds, they will tend to pick out those sounds characteristic of their species. One cannot make a chaffinch sing or call like a duck. It is clear that even simple creatures such as scorpion flies or bush crickets have a range of behavioural strategies that are contingently expressed in relation to environmental conditions. Different environments trigger different responses. In these cases, genes have coded for the mental machinery that solves a problem in a particular environment. Language acquisition is probably another example of learning that has a genetic basis. We have an innate disposition to learn a language, but we need cues from the environment to trigger the process and shape the outcome.

4. *Learnt behaviour*
   Perhaps the classic case of this is the ability of tits to open the foil tops of milk bottles to extract the cream. It is not something with which they are born, and milk bottles have not formed part of their 'environment of evolutionary adaptiveness', yet the behaviour has spread, each tit learning it from another. The disposition to peck and explore and the shape of beak are all obviously laid down by genes, but this particular behavioural phenotype has to be learnt from scratch.

2.2.3 Heritability

Despite these qualifications and notes of caution, we may begin tentatively to approach human behaviour. A great deal of human behaviour must necessarily fall into the learnt category: there are no genes for playing tennis, watching television or playing the violin. But given that behaviour is under the control of a nervous system, and that every nerve cell has a full set of genes that regulate the chemistry of the cell, it would be surprising if there were to be found no genetic influences on even complex aspects of human behaviour. Studies on identical twins are increasingly confirming this.

Dizygotic or fraternal (non-identical) twins result from the separate fertilisation of two eggs by two sperm, but identical twins are **monozygotic**; that is, a single female cell is fertilised by a single sperm, but as the **zygote** grows, for reasons that are unclear, it separates into two. An inspection of the process of meiosis reveals that, on average, fraternal twins share half their genome, whereas truly zygotic twins have their entire genome in common. It is now possible to use DNA fingerprinting to establish zygosity with some confidence. Research has shown that even though environmental influences have an effect on the behaviour of these people, zygotic twins are more alike both physically and behaviourally than fraternal twins. Moreover, zygotic twins, even when separated and raised in different environments, reveal bizarre similarities (Plomin, 1990).

It is worth repeating here, however, the point made in Chapter 1 concerning **heritability**. Heritability estimates refer to the variation between individuals that can be accounted for by differences in the genome. For the evolutionary psychologist, it is features that have low heritability that are interesting since these point to similarities in anatomy and behaviour that are common to humans and that may be explained in adaptive terms. This is because, through time, natural selection will tend to favour advantageous genes, and a population will become increasingly homogeneous. In other words, the genetic variation will be used up. The crucial point is that heritability is not a property of a gene or a trait but instead reflects the distribution of alleles in a population and the state of the environment at any time. In highly genetically homogeneous populations, such as may be the case for many psychological adaptations common to humans, phenotypic variation will largely be caused by environmental influences, so heritability will be low. Suppose, however, that the environmental influences to which people are exposed become more similar. Differences will then largely be due to genetic differences, so heritability will increase. The interpretation of heritability estimates must be carried out with great caution (see Bailey, 1998).

2.3 The unit of natural selection

2.3.1 The unit of selection: replicators and vehicles

It comes as a surprise to find that, despite a considerable knowledge of the structure of DNA and the processes of transcription and translation, there still remain fundamental questions about what exactly is evolving and being selected. The issue is often called the 'unit of selection' question. Dawkins takes the view that if, for the unit of selection, we are looking for an irreducible entity that persists through time, makes copies of itself and on whose slight changes natural selection can act, the individual organism will not do. The reason is that organisms do not make facsimiles of themselves: in sexual reproduction, offspring often differ markedly from their parents. The process of meiosis and recombination found in sexual reproduction serves as a method of shuffling genes every generation. It is as if a football team is forced to split and swap players with another team before every new match. (In this sense, natural selection favours 'teams of champion players' rather than 'champion teams'.) Moreover, accidental changes to an organism (acquired characteristics) are not inherited by the offspring. We are looking for a unit in which changes that have an effect on reproduction are inherited. Even organisms that reproduce asexually are not the replicators. An aphid that has lost a bit of leg does not reproduce by parthenogenesis to make copies of 'itself' complete with shortened leg. It 'strives' (apparently) to make copies of its genome. Consequently, the unit of selection is neither the group, so-called **group selection**, nor the individual organism but the gene itself (Dawkins, 1976; Hull, 1981).

The question is somewhat clarified by making a distinction between units that reproduce and entities that expose themselves to natural selection. This distinction is important because genes are not exposed directly to selective forces: the environment 'sees' not the genes but only the phenotypic expression of many genes working together. Dawkins introduced the terms 'replicators' and 'vehicles' for this purpose. In this view, organisms become vehicles for the replicators. The properties of these organisms are conferred on them by genes (in concert with environmental influences), and these properties influence the survival of the organism and ultimately the replicators.

In a sense, it does not matter in principle where the properties that are adaptive are manifested: they could be in the individual organism, the group or the population. When bees cluster together in winter to form a tight ball, this has the effect of reducing the surface area to volume ratio of the colony and thus reducing the heat loss, which has obvious survival value. This geometric property is a product of a colony of animals even though the gene(s) responsible are carried by individuals and probably amount to some simple instruction about moving close to their neighbours. In most cases, it is far more probable that these properties are expressed at the level of individual organisms. This is largely because the different genes in any body have the same 'hoped-for route' into future generations, and same-body genes must collaborate to pursue a common purpose. This of course has the effect of keeping at bay any conflict of interests between genes found in one body.

Interestingly, as genes pursue their reproductive 'goals', there does arise the possibility that they will care more for themselves than for the body that harbours them. If DNA can survive and be replicated without phenotypic expression, then so be it. This could explain why many organisms have large amounts of repetitive or 'junk' DNA whose function, if any, is not yet known. This may look non-adaptive and puzzling until we take a gene-centred view. This 'parasitic' DNA (assuming that it carries no benefit for its vehicle and that its replication is a net drain on resources) is highly adaptive for itself (Doolittle and Sapienza, 1980).

The whole issue is, however, still not resolved. If we accept the gene as the unit of selection (which is compelling for the reasons outlined above), we still face the problem of **linkage disequilibrium**. Some genes do not behave as separate units during reproduction but remain linked with other genes. So, how large or small a fragment of genome should count as the fundamental unit? This debate is also likely to continue.

A great deal of science relies upon metaphors, and evolutionary theory is no exception. If we speak of DNA as the 'blueprint for life', this conveys a misleading impression of how individual genes affect the phenotype. On an architect's blueprint, it is possible to estimate how the lines on the paper will translate into the finished building. If we add or remove a few features, we can make a good guess on the effect on the final outcome. But genes are not like this. A better analogy might be a recipe for a cake. From the list of ingredients and instructions for assembly, it is not easy to predict the appearance of the product and even more difficult to forecast the taste: adding and removing ingredients could have profound effects on both. We could say that the level of selection for cakes is the cake itself – its appearance and taste. Cakes are exposed to humans and are judged accordingly. Individual cakes do not survive, but recipes may persist if they are successful. By analogy, genes are selected according to the success of the vehicles that they influence. This gene-centred view of selection helps to explain some aspects of altruism, as we shall see in the next section.

2.4 Kin selection and altruism

To ensure its survival, a gene normally impresses itself on its vehicle in ways that enhance the chance of the vehicle, and ultimately itself, reproducing. It is because of this strong association between genotype and phenotype that the behaviour of the Hymenoptera (the group that includes ants, wasps and bees) appears extremely puzzling. In some of these species, individuals care for the offspring of the queen, defend and clean the colony and, in short, devote their lives to the survival and reproduction of other individuals in the colony while they themselves remain sterile. For this and other reasons, Darwin declared that the insects posed a 'special difficulty, which at first appeared to me insuperable, and actually fatal to my whole theory' (Darwin, 1859b, p. 236). In terms of the language of vehicles and replicators, the problem becomes how to explain the existence of a replicator that instructs an individual to behave altruistically if the individual that carries it does not reproduce to pass it on. In a world where the 'fittest survive', why do some individuals forego their own reproductive interests in favour of others? How could neuter wasps and bees have evolved that leave no offspring but instead slave devotedly to raise the offspring of their queens? Why does the honeybee die when it stings?

It is tempting to respond to this difficulty by reference to what Cronin has called 'greater goodism' – the idea that individuals will serve the greater good at a cost to themselves. The **altruism** of insects described above could thereby be explained by suggesting that such behaviour serves the interests of the hive, the group or even the species. This line of argument is often associated with Wynne-Edwards (1962), who argued that the dispersal of animals in relation to food supply is such that the final population density reached is optimal for the group. In this view, groups possess some mechanism whereby the selfish inclinations of individuals to overgraze are restrained in favour of the longer term interests of the group. Wynne-Edwards' book, *Animal Dispersion in Relation to Social Behaviour*, had the major effect not of converting biologists but of rousing several prominent Darwinians, chief among whom were George Walden and John Maynard

Smith, to the attack. The way out of this conundrum was not through group-level thinking but was provided instead by Hamilton's (1964) theory of **kin selection**.

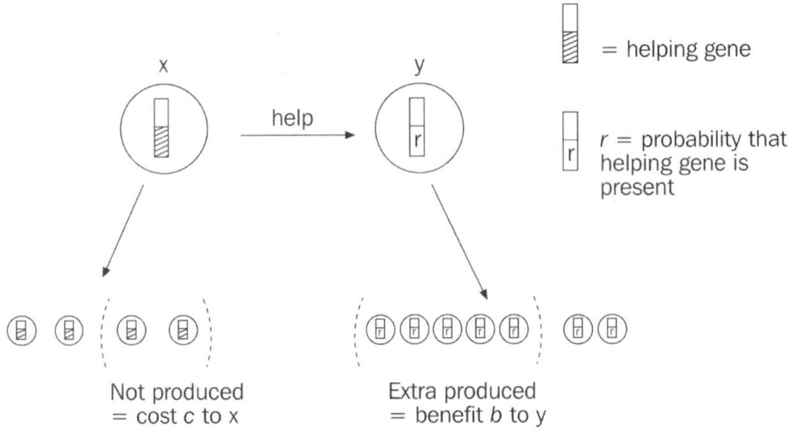

FIGURE 2.10 Conditions for the spread of a helping gene
Notice that the focus is on the helping gene and not just any gene shared in common, since it is precisely the spread of this helping gene that has to be explained.

2.4.1 Hamilton's rule

In 1963, Hamilton laid down the conditions required for gene coding for social or altruistic actions to spread. This theory is also known as the **inclusive fitness** theory. The mathematics of Hamilton's original papers is extremely complex. It was West-Eberhard (1975) who showed how Hamilton's rule could be simplified, and it is the simplified form that is now commonly encountered.

Consider two individuals X and Y, which are related in some way, and that X helps Y. An altruistic act can be defined as one that increases the reproductive success of the beneficiary (Y) at the expense of the donor (X).

Let b = benefit to recipient
c = cost to donor
r = the **coefficient of relatedness** of the recipient to the donor. This is the same as the probability that the gene for helpful behaviour is found in both the recipient and the donor.

The condition for assistance to be given and thus for the gene to spread is: 'help if $rb - c > 0$', which is the same as $rb > c$. Figure 2.10 shows an example of this. It should be clear that although the reproductive success of the helping gene in X is reduced, this is more than compensated for by the potential increased success of the gene appearing in the offspring of Y.

2.4.2 Coefficient of relatedness

The situation presented in Figure 2.10 is highly simplified. Figure 2.11 shows a more detailed example of two **diploid** individuals I and J, with alleles aA and mM respectively, that mate sexually to produce four types of offspring: am, aM, Am and AM.

The coefficient of relatedness can be understood in a number of ways. The important thing to remember is that we are concerned with how closely individuals are related to each other rather than how similar they are. There are three ways of conceptualising this:

1. r is a measure of the probability over and above the average probability (which is determined by the gene's average population frequency) that a gene in one individual is shared by another
2. r is the probability that a random gene selected from I is identical by descent to one present in J
3. r is the proportion of genes present in one individual that are identical by descent with those present in another.

The terms 'identical by descent' and 'probability above average' are there to take into account the fact that a gene may be present in two individuals because it is common in the population (that is, it has a high average probability) or because they are siblings and the gene came from the father or mother.

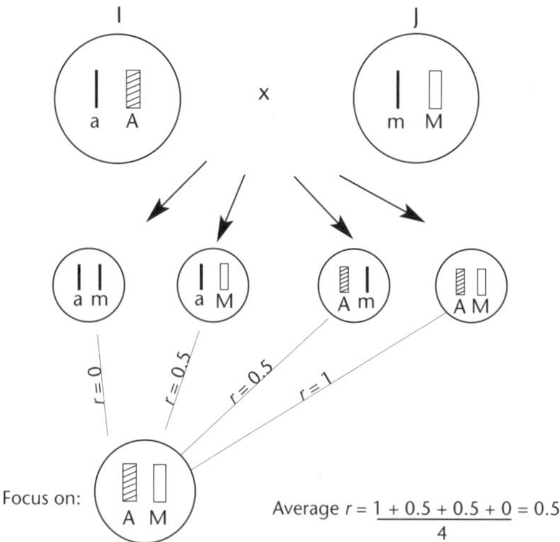

r = coefficient of relatedness, which can be thought of as the probability that a gene sampled at random in one individual is also present in another. In this case, r between siblings is 0.5.

FIGURE 2.11 Coefficient of relatedness between siblings

Figure 2.11 shows how two unrelated parents produce siblings with an average value of $r = 0.5$. It follows that it may pay siblings to help each other if the gene for helping thereby increases in frequency. This is, in effect, what the distinguished biologist Haldane was suggesting when he said (reputedly in the Orange Tree Pub on the Euston Road in London) that he would lay down his life for at least two brothers or eight cousins. Table 2.1 shows the r values associated with diploidy.

Table 2.1 Coefficients of relatedness r between kin pairs in humans

r values: diploid (e.g. humans)	
Parent <--------> offspring	0.5
Full siblings (same parents)	0.5
Half-siblings	0.25
Identical twins (zygotic)	1.0
Grandparent <---> grandchild	0.25

2.4.3 Application of Hamilton's rule and kin selection

We are now, post Hamilton, in a position to answer Darwin's problem about the **eusocial** insects and their extreme form of self-sacrifice. The fact that worker bees have barbed stings that, when used to attack predators, remain embedded in the victim, causing the death of the worker, is understandable when we realise that other members of the nest are close relatives. This extreme form of self-sacrifice is maintained as a behavioural and morphological trait in the face of natural selection because the survival of the genes responsible is ensured by their presence in the rest of the colony. In measuring the reproductive fitness of an organism, we can now distinguish between 'Darwinian fitness', which is a measure of numbers of direct offspring, and inclusive fitness, which measures the successful transmission of genes in direct offspring and relatives.

Cooperative breeding, as illustrated by the Hymenoptera, is just one example of altruistic behaviour. More generally, numerous studies on cooperative breeding illustrate the application of kin selection theory (see Metcalf and Whitt, 1977; Emlen, 1995), but altruism can take many other forms: sharing food, grooming, raising alarm calls and so on. For animals that live in social groups, the raising of an alarm call by one individual at the approach of a predator is often considered to be altruistic behaviour since the call itself could enable the predator to locate the caller. It would presumably pay the first observer of a predator to skulk silently away, leaving others to their fate. Numerous studies have been carried out on such calls, most reporting that animals are much more likely to give alarm calls when the beneficiaries are close relatives. In experiments reported by Hoogland (1983), it was found that black-tailed prairie dogs *(Cynomys ludovicianus)* were much more likely to give an alarm call at the sight of a predator if close relatives were present in the social group.

Although human altruism is examined more thoroughly in Chapter 9, it is worth noting here that humans are more inclined to behave altruistically towards kin than unrelated individuals. Most people choose to live near their relatives, sizeable gifts are exchanged between relatives, wills are nearly always drawn up to favour relatives in proportion to their genetic relatedness and so on. In an intriguing direct experimental investigation of this effect, Dunbar (1996a) asked volunteers to sit with their back against the wall in a skiing posture but with no other means of support. After a while, it becomes painful to maintain this position. The subjects were given 75 pence for every 20 seconds they could maintain this posture, with the proviso that the money must go to one of three categories of people: themselves, relatives of varying degrees of genetic relatedness or a major children's charity. The results were unambiguous – subjects tended to endure more discomfort and thus earn more money for themselves or close relatives than distant relatives or a charity.

Human kin make powerful claims on our emotional life. It is significant that movements that extol the 'brotherhood of man' or group solidarity, such as the world's great religions or organised labour, often resort to kin-laden language. A trade union leader may typically use the term 'brothers and sisters' when addressing fellow members. The problem with groups, however, especially modern ones, is that they are vulnerable to free riders – individuals who claim membership but who are not genetically related to anyone in the group or who will extract favours from the group without any repayment in kind. What would be desirable in these situations is some sort of cultural badge, a mark of belonging to the group that indicates commitment. It could be that language dialect serves as some sort of badge of membership; as George Bernard Shaw suggested, no sooner does one Englishman open his mouth to speak than another despises him. Groups break down into smaller groups, and there begins an 'in-group' and 'out-group' morality. Dunbar has explored the suggestion that language dialects may serve a marker function. He found, using computer simulations, that as long as dialects change slightly over a few generations, cheaters (that is, outsiders who drift into the group to extract benefits and then drift out) find it difficult to gain a foothold. Dunbar (1996a, p. 169) thinks it likely that 'dialects arose as an attempt to control the depredations of those who would exploit people's natural co-operativeness'. This raises the further interesting possibility that the diversification of dialects into different languages is related to the need for group cohesiveness.

2.5 Kin recognition and discrimination

For kin selection to operate, it requires animals to be able to recognise or at least discriminate between kin and non-kin. The evidence for kin recognition or 'reading r values' is necessarily indirect since it is an internal process, but if we observe animals treating kin differently from non-kin, this **kin discrimination** could be used as evidence for kin recognition.

There are probably two basic reasons why it is in the interests of an animal to recognise kin. First, kin selection requires that acts of altruism be directed according to Hamilton's formula $rb > c$. This requires some assessment of costs, benefits and r values. Second, it is important for sexually reproducing organisms not to mate with close relatives, otherwise deleterious gene combinations may result. It follows that evidence for kin discrimination does not automatically imply the existence of an altruistic gene. We will outline some kin recognition mechanisms and evidence in their favour and then return to this second and important point.

There are probably at least four mechanisms that are available to animal species: location, familiarity, phenotype matching and recognition alleles ('green beards'). In the case of *location*, for animals living in family-based groups, for example in burrows or groups of nests, there is a good chance that neighbours will be kin. In these circumstances, a simple mechanism such as 'treat anyone at home as kin' may suffice. Cuckoos of course exploit this.

Familiarity is also a rough and ready cue of kinship. Work by Holmes and Sherman (1982) showed that non-sibling ground squirrels raised together were no more aggressive to each other than were siblings reared together, both these groups being far less aggressive than non-sibling young squirrels reared apart. The mechanisms here may be one of association rather than location. If you have spent some of your early life in close proximity with another individual, the chances are that you are related.

Phenotype matching is more complicated. Some animals seem capable of assessing the similarity of genotype using some characteristic of the phenotype, possibly odour. Many creatures probably carry such labels that allow kin to establish their closeness. The sniffing behaviour on greeting, exhibited by many rodents, seems to be designed as a kin recognition system.

It is important for many animals to avoid inbreeding since close relatives may be homozygous for deleterious **recessive alleles**. It is estimated that each human probably carries between three and five lethal recessive alleles. Mating with close relatives increases the likelihood that the chromosome will be homozygous at the loci for these recessive and defective alleles. This could be the genetic basis for the strong taboos in incest found in numerous human societies. Olfactory cues have some significance in human mating. Work by Claus Wedekind and his colleagues in Switzerland has shown that human females actually prefer the smell of males who are different from them in terms of the **major histocompatibility complex** (MHC) region of their genome. This region is deeply involved in self-recognition and the immune response. Differences in this region between individuals can be tested by measuring the antigens produced in their body fluids. It is important for a female to choose a mate who will differ in the MHC region since this provides a cue for genetic relatedness: close relatives will be similar in this region. Moreover, differences in the MHC between a woman and her partner may allow females to produce offspring with a more flexible response to parasites. Evidence has been provided that male odours are related to an individual's MHC. In controlled conditions, women were more likely to find the odours produced by males pleasant if they differed in their MHC. Significantly, the effect was reversed if the women subjects were taking oral contraceptives (Wedekind et al., 1995). The avoidance of inbreeding is discussed in depth in Chapter 13.

The potential importance of *recognition alleles* is illustrated by the **green beard effect**. Two factors ensure a high value for *r*, which is needed if altruism is to spread when costs are significant: kinship and recognition. There is nothing intrinsically special about the location of the altruistic gene in kin; it is merely that kin have a reliable probability that the gene for helping will be present. But it is just a probability. In Figure 2.11 above, if the individual 'AM' were to help 'am', which it would if the gene 'A' simply instructed it to help siblings, its actions would be wasted.

The 'green beard' effect was named after a thought experiment of Richard Dawkins. The idea was in fact first proposed by Hamilton in 1964 and given its memorable title by Dawkins (1976). Dawkins considered a gene for helping that would also cause its vehicles to sprout a green beard. This would be an ideal way to focus altruistic efforts. Kin would be ignored if they did not possess the green beard, but anyone with a green beard would be helped without regard to how closely related they were. It is often pointed out that a gene for helping is unlikely to be able also to command the ability to recognise a label and the ability to produce a label, but if two or more genes are closely linked such that they tend to occur together (linkage disequilibrium), this does become a theoretical possibility. The reason why green beard effects are not common probably results from meiotic crossover and recombination (Haig, 1997).

Strictly speaking, a green beard effect is not necessarily kin recognition since the altruistic gene would behave favourably towards any other recognisable 'green beard' irrespective of its kinship. Some evidence for green beard effects is observed in the congregating behaviour of the larvae of the sea squirt *(Botryllus schlosseri)* as studied by Grosberg and Quinn (1986) and the behaviour of red fire ants *(Solenopsis invicta)* as documented by Keller and Ross (1998). The real importance of green beards is, however, a thought experiment. If you understand why altruism towards green beards should spread faster than that simply towards relatives, you have understood the force of Hamilton's equations and are a long way down the road to a gene-centred view of natural selection.

2.6 Levels of altruism

2.6.1 Selfish genes and compassionate vehicles

So far, we have implicitly adopted a working definition of altruism as 'an act that enhances the reproductive fitness of the recipient at some expense to that of the donor', but we did not specify whether the donor was a gene or a vehicle. If we consider our definition at the level of the gene, we are looking for a gene that helps another non-identical gene at some cost to itself. In kin selection or kin altruism, the gene responsible is merely increasing its own multiplication rate; it is not assisting another gene. The fact that the gene is present in another vehicle is irrelevant. It becomes clear that there can be no truly altruistic genes. A gene that helped others at expense to itself would quickly become extinct when confronted with a variant that simply helped itself. Following this line of thought, it becomes clear that kin altruism is true altruism only at the level of vehicles. The genes responsible may be increasing their propagation in another vehicle, but, by definition, the host vehicle (since we are at this level) is making a sacrifice in favour of another host vehicle.

So despite the reductive approach employed in much of this chapter, we can affirm that altruism does exist. Genes, of course, are not 'selfish' in a conscious or pejorative sense: the term is used merely as a metaphor to succinctly capture their apparent behaviour. But genes, driven only by the remorseless logic of survival, competition and multiplication rates, can direct the growth of vehicles such as human bodies with a complex emotional life, displaying empathy and sympathy for others. What started as a system for ensuring gene survival by recognising and helping relatives has led to human subjects capable of compassion for others, whether they are related or not. This is a subject further explored in Chapters 9 and 16.

2.6.2 The stupidity of genes

At this point it is also worth pointing out that genes as well as being metaphorically selfish are also metaphorically 'stupid'. Kin-directed altruism began as a way of perpetuating genes that caused their vehicles to behave altruistically. Imagine two organisms (loveless A and loving B) long ago each watching a sibling drown. There is a 1 in 10 chance, say, that going to the aid of the drowning sibling will cause both the victim and the rescuer to drown. Loveless A (that is, an organism not endowed with altruistic genes) stands by and watches his brother or sister drown, grateful that he didn't risk his own life needlessly. Loving B (endowed with genes for altruism) jumps in and saves her brother or sister. Sometimes B will drown but the chance of the altruism-directing genes surviving is much higher than the chance of them failing (there is a 50 per cent chance that the sibling of B will contain the altruism genes and only a 10 per cent chance that they will both die). Over time, it is easy to see that the loving behaviour of B will do better than the selfish behaviour of A. The crucial point is that A types will die out and B types will become the norm. This means of course that these altruism-directing genes will become the norm in the gene pool – all organisms of the species will have them. This is where the gene 'stupidity' resides: we still have systems that incline us to treat our kin more favourably than non-kin even though the very genes that benefited from such actions are now possessed by all. If we did want our altruism to be based on hyperrationality of gene frequencies, then it would make more sense to give gifts of money to suffering strangers of reproductive age than our own children, since copies of the loving genes inside us will do much better by the act of helping strangers who otherwise might not reproduce. From a gene-centred point of view, the persistence of biasing our loving care to close relatives is now irrational; from a human-centred point of view, that's the way we are.

Summary

- The theory of evolution provides a naturalistic account of the variety, forms and behaviour of living organisms. In constructing his theory, Darwin jettisoned the idea of purpose in nature (teleology) and the notion that organisms conform to some abstract and pre-existing blueprint or archetype.
- Evolution occurs through the differential reproductive success of genes. To understand the mechanism of natural selection, we need to distinguish between genotype and phenotype. The genotype consists of the genes that carry the information needed to build organisms. The phenotype is the result of the interaction between genes and the environment within an individual. Whereas envir-

onmental factors may strongly influence the way in which genes are expressed within an organism, the outcome of this interaction cannot be communicated to the genotype. Characteristics acquired in the lifetime of an individual are thus not inherited by the offspring.

- Darwin was unable satisfactorily to explain the mechanism of inheritance and how novel and spontaneous differences between offspring and parents (which form the raw material for evolution) could arise. Modern genetic theory supplies answers to these problems.
- The existence of altruistic behaviour posed a problem for Darwin that he did not satisfactorily resolve.
- The information that instructs the behaviour of each cell in any organism and that passes to offspring along the germ line is contained in the sequence of base pairs found on DNA molecules. DNA is found in the nucleus of cells, packed with proteins in the form of chromosomes. In all mammals, the somatic cells contain complementary (homologous) pairs of chromosomes (diploidy), one copy in each pair being inherited from each parent. Reproductive cells (gametes) contain only one copy of each chromosome (haploidy). Fertilisation occurs when a sperm meets an egg to produce a zygote. A zygote has the potential to develop into a new individual.
- A phenotype is the result of environmental factors acting upon the genotype. The determination of behaviour may be interpreted along a spectrum of weak to strong genetic control.
- Although environmental stresses and selection pressures act upon groups and individuals, it is logically more consistent to consider the gene as the fundamental unit of natural selection.
- Kin recognition is important both to ensure that altruism is effectively directed and also to avoid inbreeding.
- Altruism can be understood biologically in terms of kin selection. In this process, genes direct individuals to help other individuals that are related and thus share genes in common. Kin selection offers insights into the biological basis of altruism in humans.

Key Words

- **Allele**
- **Altruism**
- **Amino acid**
- **Autosome**
- **Base**
- **Central dogma**
- **Chromosome**
- **Codon**
- **Coefficient of relatedness**
- **Crossing over**
- **Diploid**
- **Dizygotic**
- **DNA**
- **Dominant allele**
- **Eusocial**
- **Gamete**
- **Gene**
- **Genome**
- **Genotype**
- **Green beard effect**
- **Group selection**
- **Haemophilia**
- **Haploid**
- **Heritability**
- **Heterozygous**
- **Homozygous**
- **Inclusive fitness**
- **Kin discrimination**
- **Kin selection**
- **Lamarckism**

- **Linkage disequilibrium**
- **Locus**
- **Major histocompatibility complex (MHC)**
- **Meiosis**
- **Mitosis**
- **Monozygotic**
- **Oogenesis**
- **Phenotype**
- **Recessive allele**
- **Recombination**
- **Ribonucleic acid (RNA)**
- **Sex chromosomes**
- **Spermatogenesis**
- **Teleology**
- **Zygote**

Further reading

Cronin, H. (1991) *The Ant and the Peacock*. Cambridge, Cambridge University Press.

Excellent historical account of the theories of kin selection (the ant) and sexual selection (the peacock). Closely argued and packed with references. A book for the serious historian of ideas.

Dawkins, R. (1976) *The Selfish Gene*. Oxford, Oxford University Press.

Now a classic, this is probably the best account of gene-centred thinking ever written. A more recent edition was published in 1989.

Dawkins, R. (1982) *The Extended Phenotype*. Oxford, W. H. Freeman.

Shows how the effects of genes reach outside their vehicles.

Dugatin, L. A. (1997) *Co-operation Among Animals*. Oxford, Oxford University Press.

A thorough work, with numerous empirical examples. Dugatin accepts a role for a model of group selection as proposed by D. S. Wilson. An excellent book for the review of altruism among non-human animals, but contains no discussion of humans.

Ridley, M. (1993) *Evolution*. Oxford, Blackwell Scientific.

A good overview of the whole theory of evolution.

Sex, Sexual Selection and Life History Theory

And nothing gainst Time's scythe can make defence,
Save breed to brave him, when he takes thee hence.
(Shakespeare, Sonnet 12)

Each generation is a filter, a sieve; good genes tend to fall through the sieve into the next generation; bad genes tend to end up in bodies that die young or without reproducing.
(Dawkins, 1995, p. 3)

For individuals of sexually reproducing species, finding a mate is imperative. It is through mating, essentially the fusion of gametes, that genes secure their passage to the next generation; without it, the 'immortal replicators' are no longer immortal. It is hardly surprising then that sex is an enormously powerful driving force in the lives of animals and is attended to with a sometimes irrational and desperate urgency. At a fundamental level, sex is basically simple – a sperm meets an egg – but it is in the varied forms of behaviour leading to this event that complexity is to be found and needs to be understood. In order to understand human sexuality, we need to raise some basic questions concerning the causes, consequences and manifestations of sexual activity in animals as a whole. This chapter begins this task by looking at some current theories of the origin and maintenance of sexual reproduction. It also addresses some fundamental questions, such as why female gametes (eggs) are usually at least 100 times larger than male gametes (sperm) – a phenomenon known as **anisogamy** – or why the male to female ratio remains so close to 1:1, albeit with some slight but significant variations.

It was once thought convenient to classify sexual behaviour in terms of mating systems, and the terminology of such systems is introduced here. It will be argued, however, that a better approach is to focus on the strategies of individuals rather than the putative behaviour of whole groups. This individualistic approach will reveal that sex is as much about conflict as about cooperation, each sex employing strategies that best serve its own interests.

3.1 Why sex?

For about 100 years after the publication of Darwin's *On the Origin of Species* (1859b), the existence and function of sex was not really seen as a problem. Sex was viewed as a cooperative venture between two individuals to produce variable offspring. Variation was required to secure an adaptive fit to a changing environment and constant variation was needed to ensure that species did not become too specialised and face extinction if the environment changed.

From a modern, gene-centred perspective, these arguments now appear fatally flawed. Variation and selection cannot act for the good of the species; genes only care for themselves. So from a gene's eye view, what advantage does sex carry?

3.1.1 Why do males exist?

The question 'why sex?' resolves itself into the question 'why do males exist?' – a question that has probably occurred to many women from time to time. All organisms need to reproduce, but some manage this asexually: females simply make copies of themselves by a sort of cloning process. This form of reproduction is known as **parthenogenesis** ('virgin birth'), and although it is not found among mammals and birds, it is not uncommon in fish, lizards, frogs and plants. Males are a problem because, in the absence of male care (which is very common), a mutation that made a sexually reproducing organism switch to parthenogenesis (which some organisms can do anyway) should be favoured since it would produce more copies of itself and rapidly spread throughout the population. Put another way, with a given set of environmental limitations, females should be able to produce twice as many grandchildren by **asexual reproduction** compared with sexual reproduction. Sexual reproduction also has other problems compared to parthenogenesis. They include:

- Time and effort is spent attracting, defending and copulating with mates. Such effort could have been directed into reproduction.
- Individuals may be vulnerable to predation during mating, especially during intercourse or courtship displays.
- Risk of damage during physical act of mating.
- Risk of disease transmission from one individual to another.
- Recom`bination of genes that follows sex may throw up a homozygous condition for a dangerous recessive allele.
- Sex introduces same-sex competition. Where polygamous mating is common, an individual may not find a mate at all.
- Sex breaks up what may have been a highly successful combination of genes. If it 'isn't broke', sex still 'fixes it'.

Almost as if to mock the doubts of biologists concerning the functions of sex, the natural world teems with sexual activity. Intriguingly, asexual species seem to be of fairly recent origin; they comprise the 'twigs' of the phylogenetic tree rather than its trunk or main branches. Some asexual species still betray their sexual ancestry. In the case of the Jamaican whiptail lizard, for example, the female will lay a fertilised egg only when physically 'groped' by a male. The male provides nothing in the way of genetic material, but its physical presence seems to trigger self-fertilisation. In some frog species, the male provides sperm for the activation of the development of the female's eggs, but again no genetic material is transmitted – as Sigmund (1993) observed, a case for the male of 'love's labours lost'. This behaviour is probably of fairly recent evolutionary origin or else the males would have caught on and such time wasting would be selected against – another caution against always interpreting animal behaviour as optimal. Nearer to home, everyone's back garden probably contains a few dandelions. The gaudy yellow flowers at first sight appear to be made like any other flower to attract pollinators, but dandelions are entirely self-fertilising; their flowers are leftovers from their sexual past when cross-pollination did occur.

There are, however, four major theories advanced to explain the persistence of sexual modes of reproduction in the face of all the apparent disadvantages. These are discussed below.

The lottery principle

The American biologist George Williams (1975) was one of the first to suggest that sex introduced genetic variety in order to enable genes to survive in changing or novel environments. He used the lottery analogy: breeding asexually is like buying many tickets for a national lottery but giving them all the same number; sexual reproduction is like making do with fewer tickets but having different numbers. The essential idea behind the lottery principle is that since sex introduces variability, organisms have a better chance of producing offspring that survive if they produce a range of types rather than more of the same. On the positive side for this theory, it may help to explain why creatures such as aphids, which can breed both sexually and asexually, choose to multiply asexually when environmental conditions are stable but switch to sexual reproduction when facing an uncertain future. In the steady months of summer, aphids multiply at a fast rate on rose bushes by parthenogenesis, but as winter approaches, they have bouts of sex to produce numerous and variable cysts that survive the winter and wait for the return of warmer conditions.

Parasites also provide an illustration of this principle. When a host is first invaded, parasites typically reproduce asexually to fill the host as rapidly as possible. When this niche is filled, new offspring have to leave and infect other hosts. At this stage, the parasite typically switches to sexual reproduction to take advantage of the fact that sex produces variation that may be useful for success in the next round of infecting unknown hosts – some of which may be resistant to the genotype of the parent parasites. In short, sex precedes dispersal.

The tangled bank hypothesis, or spatial heterogeneity

The lottery principle idea of Williams was developed to form the 'tangled bank' theory of Michael Ghiselin (1974). This term is taken from the last paragraph of Darwin's *Origin*, where he referred to a wide assortment of creatures all competing for light and food on a tangled bank. According to this theory, in environments where there is an intense competition for space, light and other resources, a premium is placed on diversification. From a gene-centred point of view, a gene will have an interest in teaming up with a wide variety of other genes in the hope that at least one such combination will do well in a competitive environment. Although once popular, the tangled bank theory now seems to face many problems, and former adherents are falling away. The theory would predict a greater interest in sex among animals that produce lots of small offspring (so-called r selection) that compete with each other. In fact, sex is invariably associated with organisms that produce a few large offspring (K selection), whereas organisms producing smaller offspring frequently engage in parthenogenesis.

The Red Queen hypothesis

The Red Queen hypothesis, which now offers one of the most promising explanations of sex, was first suggested by Leigh Van Valen. Van Valen (1973) discovered from his study on marine fossils that the probability of a family of marine organisms becoming extinct at any one time bears no relation to how long it has already survived. It is a sobering thought that the struggle for existence never gets any easier: however well adapted an animal may become, it still has the same chance of extinction as a newly formed species. Van Valen was reminded of the Red Queen in *Alice in Wonderland*, who ran fast with Alice only to stand still.

The application of this theory to the problem of the maintenance of sex is captured by the phrase 'genetics arms race'. A typical animal must constantly run the genetic gauntlet of being able to chase its prey, run away from predators and resist infection by parasites. Parasite infection in particular means that parasite and host are locked in a deadly 'evolutionary embrace' (Ridley, 1993). Each reproduces sexually in the desperate hope that some combination will gain a tactical advantage in attack or defence. William Hamilton summed this up in a memorable fashion when he compared sexual species to 'guilds of genotypes committed to free fair exchange of biochemical technology for parasite exclusion' (quoted in Trivers, 1985, p. 324).

The Red Queen hypothesis also gains support from the comparative approach to sexual reproduction developed by Graham Bell in Montreal. Bell (1982) found that sex is most commonly practised in environments that are stable and not subject to sudden change. The lottery principle suggests that sex is favoured by a variable environment, yet an inspection of the global distribution of sex shows that where environments are stable but biotic interactions are intense, such as in the tropics, sexual reproduction is rife. In contrast, in areas where the environment is subject to sudden change, such as high latitudes or small bodies of water, it seems that the best way to fill up a niche that has suddenly appeared is by asexual reproduction. If your food supply is already dead, it cannot run away, so the best policy if you are an organism feeding on dead matter (a decomposer) is to propagate your kind quickly to exploit the food resource and forego the time-wasting business of sex. In the world of the Red Queen, organisms have to run fast to stay still. A female always reproducing asexually is 'a sitting duck for exploiters from parasitic species' (Sigmund, 1993, p. 153).

Further support for the parasite exclusion theory comes from the fact that genes that code for the immune response – the major histocompatibility complex (MHC) – are incredibly variable. This is consistent with the idea that variability is needed to keep an advantage over parasites. As we will see in

Chapters 12 and 13, human females can be choosy about their prospective partners in relation to their MHC genes, genes that are different from their own being preferred.

The DNA repair hypothesis

Why are babies born young? The question at first sight appears to be a rather stupid one; surely babies are young by definition? But the question we are really asking is how, despite the ageing of somatic cells in, for example, the skin and nervous tissue of the parents, the cells of the newborn have their clocks set back to zero. Somatic cells die, but the germ line appears to be potentially immortal. Bernstein et al. (1989, p. 4) lay claim to a solution to this problem:

> We argue that the lack of ageing of the germ line results mainly from repair of the genetic material by meiotic recombination during the formation of germ cells. Thus our basic hypothesis is that the primary function of sex is to repair the genetic material of the germ line.

As we have already noted in Chapter 2, the primary features of sex from a gene's point of view are meiotic recombination followed by the fusion of a gamete with that from an unrelated individual (this latter phenomenon is known as 'outcrossing'). Bernstein et al. interpret both these events as responses to the need for repair.

DNA faces two types of disruption. It can be damaged in situ by ionising radiation or mutagenic chemicals, or a **mutation** can occur through errors of replication, which are best thought of as change rather than damage. Damage to the DNA can take a number of forms, repair mechanisms often being suited to each type. Single-strand damage can be made good by enzymes using the template provided by the other strand, but double-strand damage is more serious: the cell may die or possibly make use of the spare copy in haploid cells. During crossing over in meiosis (see Figure 2.7), the chromosomes line up and the spare copy is used to repair double-strand breaks.

Most mutations are deleterious, but fortunately they are recessive and their effects consequently swamped by viable alleles on the complementary chromosome. As cell division proceeds, however, the burden of mutation steadily increases, and there will come a time when a genome becomes homozygous for a dangerous recessive allele. This is an example of an effect called 'Muller's ratchet': as time passes, mutations accumulate in an irreversible fashion like the clicks of a ratchet. With the outcrossing brought about by sex, these mutations can be masked in the heterozygous state.

In asexual reproduction, any mutation in one generation must necessarily be passed to the next. Ridley (1993) likened this to photocopying: as a document is copied, and copies made from the copies and so on, the quality gradually deteriorates. In accumulating mutations at a steady rate, asexual organisms face the prospect that they may eventually not be viable. In sexually reproducing species, on the other hand, some individuals will have a few mutations while some will have many. This arises from meiosis and outcrossing. Sexual reproduction involves the shuffling of alleles; some individuals will be 'unlucky' and have a greater share than average of deleterious mutations in their genome, and some will be 'lucky', with a smaller share. The unlucky ones will be selected out. In the long term this has the effect of constantly weeding out harmful mutations through the death of those that bear them (Crow, 1997). Eyre-Walker and Keightley (1999) have reported a mutation rate in humans of about 1.6 deleterious mutations per person per 25 years. This would have devastating consequences if it were not for sexual reproduction.

In summary, we have four major types of theory to account for the origin and maintenance of sex:

1. sex produces variable offspring to thrive as environments change through time
2. sex produces variation to exploit subtle spatial variations in environmental conditions
3. sex enables organisms to remain competitive in a world where other organisms are poised to take advantage of any weakness
4. sex serves to keep at bay the effects of damage wreaked daily on our DNA and thus weed out deleterious mutations.

There is perhaps no one single explanation for the maintenance of sex in the face of severe cost. Genes that promote sexual reproduction could flourish for a variety of reasons. In this respect, we should note that the models are not mutually exclusive: all rely upon sex to maintain genetic variability.

3.2 Sex and anisogamy

Individuals in sexually reproducing species exist in two forms: males and females. The question is, how do we define 'maleness' and 'femaleness'? In most higher animals, the distinction is pretty clear. Even if males and females are morphologically different, we could say that males inject sperm into females. However, to cover cases of external fertilisation, as practised by many fish species, we need a better definition than this. A more comprehensive definition would be that males produce small mobile gametes (sperm) that seek out the larger, less mobile gametes (eggs) produced by the female.

Yet the ancestral state of life on earth must have been that of primitive, single-celled asexual organisms. Now a further problem confronts us: since the first sexually reproducing organisms probably produced gametes from males or females of equal size **(isogamy)**, how have we arrived at the situation where, for virtually all cases of sexual reproduction, the size of the gametes from males and females is vastly different? Figure 3.1 shows how great the discrepancy is.

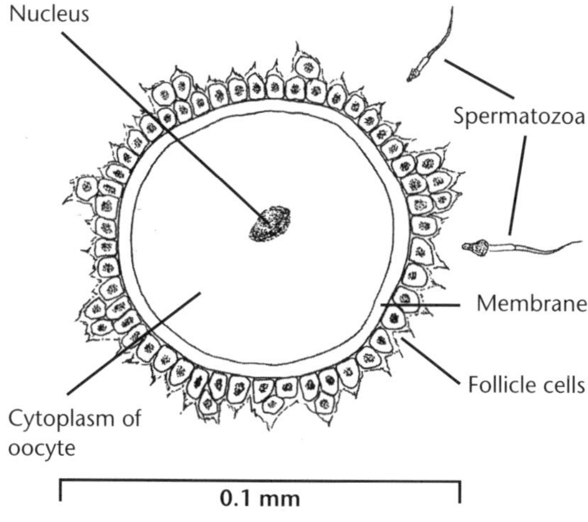

FIGURE 3.1 Relative dimensions of an egg from a human female and a sperm from a male

Parker et al. (1972) suggest one probable scenario. Their argument is essentially that an ancestral state of equally sized gametes quickly breaks down into two strategies: providers and seekers. Parker et al. were also able to show that these two strategies are stable, in the sense that they can resist invasion from other strategies. Once the archetypal female hit upon the strategy of provisioning her gamete (egg) with nutrients to give it a head start after fertilisation, the archetypal males responded by producing lots of sperm (smaller gametes) in an effort to be the one to successfully fertilise this valuable prize. Males and females became locked into their separate strategies.

3.3 Describing mating behaviour: systems and strategies

There is as yet no universally accepted precise system to classify patterns of mating behaviour. This largely results from the fact that a variety of criteria can be employed to define a mating system. Commonly used criteria usually fall into two groups: mating exclusivity and pair bond characteristics. In the former, a count is made of the number of individuals of one sex with which an individual of the other sex mates. In the latter, it is the formation and duration of the social 'pair bond' formed between individuals for cooperative breeding that is described.

Both sets of criteria have their problems. In the case of mating exclusivity, the crucial act of copulation is not always easy for fieldworkers to observe. With humans, we can issue questionnaires and hope for an honest response, but with other animals, copulation may be underground, in mid-air, at night or generally difficult to see. Such are the problems here that copulatory activity has often been inferred from the more visible non-copulatory social forms of behaviour such as parental care and nest cohabitation. It is obvious that this approach is open to errors of interpretation. Bird species such as the dunnock *(Prunella modularis)* that were once classified as monogamous on the basis of shared nest building and parental care of offspring have turned out to be more varied in their mating habits (Davies, 1992). Consequently, terms such as **extrapair copulations** and 'sneak copulations' have entered the repertoire of behavioural terminology.

Table 3.1 Characteristics of mating systems

System	Mating exclusivity and/or pair bond character	Examples
Monogamy	Mating with one partner. Some animal behaviourists now make the distinction between social monogamy (living with one member of the other sex) and genetic monogamy (DNA evidence that the monogamous father really is the parent). Many species that were once assumed to be socially and genetically monogamous (e.g. swans) have been shown to engage in extrapair copulations and are therefore not genetically monogamous	About 90 per cent of all bird species are socially monogamous, but a much smaller percentage (possibly as low as 10 per cent) are genetically monogamous. Monogamy is rare in mammals (about 3 per cent or all species) About 19 per cent of preindustrial human societies are monogamous
Polygamy	One sex copulates with more than one member of the other sex. There are two types	
Polygyny	Where a male mates with several females. Females mate with only one male	Most mammalian species are polygynous. Examples include lions, gorillas and elephant seals. About 80 per cent of preindustrial human societies allow a man to behave polygynously (that is, take more than one wife)
Polyandry	One female mates with several males. Males mate with only one female	Very rare in human societies, only a few cultures documented. Found in birds (e.g. the Galapagos hawk, the northern jacana), some primates (e.g. marmosets), and insects (e.g. field crickets, *Gryllus bimaculatus*)
Polygynandry or 'promiscuity'	No stable pair bonds, males will mate with several females and females with several males. Promiscuity is a rough shorthand for this behaviour but avoided by most biologists since it carries moralistic overtones	Chimpanzees. Occasionally found among human groups (e.g. communes) Anthropologists use the term 'group marriage'

A more individually focused approach to the study of mating reveals as much conflict as cooperation between the sexes, and the so-called 'pair bond' for many species could be seen with equal validity as a sort of 'grudging truce'. Bearing in mind these reservations, Table 3.1 shows a simplified classificatory scheme combining both sets of criteria.

The fact that naturalists have found it difficult to devise hard and fast definitions for mating systems need not concern us too much. The attempt to match a species with a particular mating system faces a more fundamental array of problems than mere observational difficulty.

3.3.1 Problems with the concept of mating systems

The 'systems' approach to understanding mating is not entirely satisfactory for a number of reasons:

- *The label is sex specific*; in the case of polygyny, for example, males and females are behaving differently although they are of the same species. In the case of elephant seals, some males are highly polygynous and successful ones may mate with dozens of females in a season, whereas females are monogamous and mate with only one male.
- *Species in themselves do not behave as a single entity*; it is the behaviour of individuals that is the raw material for evolution. Natural selection can only act upon individuals, ruthlessly exterminating those that are unsuccessful and allowing those that are better adapted to pass their genes to the next generation. To understand behaviour, we need to consider how that behaviour helps or retards the chances of individuals reproducing.
- Within any species, individuals even of one sex may differ and utilise different strategies. Where there is a genetic basis to this, it is sometimes known as a *polymorphism* (many shaped). Humans

are polymorphic for a number of characteristics such as blood groups, or presence or absence of freckles. Hence, in the gene pool of a species, there may be genes that predispose some individuals more towards monogamy and others towards polygyny.

Some animal behaviourists prefer to describe the social arrangement that facilitates mating in terms of the number and sex of the individuals. Table 3.2 shows how this works. Hence we would say that the mating behaviour of gorillas is uni-male, multi-female, since they live in polygynous groups where one male has access to a 'harem' of females. The mating behaviour of chimps would be described as multi-male, multi-female, since they live in mixed groups where both males and females may have several sexual partners. Table 3.2 shows alternative formulations for the traditional mating systems approach. Hence, polyandry becomes uni-female, multi-male, since one female shares several males.

Table 3.2 Four basic mating systems

	Uni-male	Multi-male
Uni-female	Monogamy	Polyandry
Multi-female	Polygyny	Promiscuity/polygynandry

3.4 The sex ratio: Fisher and after

3.4.1 Why so many males?

Let us recall some of the facts of human anisogamy, which are in many respects typical of mammals as a whole. Each ejaculate of the human male contains about 280×10^6 sperm, enough, if they were all viable and suitably distributed, to fertilise the entire female population of the USA. Moreover, they are produced at the phenomenal rate of about 3,000 each second (Baker and Bellis, 1995). In contrast, the human female only produces about 400 eggs over her entire reproductive lifetime of 30–40 years. Now, the ejaculate of the male is of course not evenly distributed, and a male must impregnate the same woman many times to have a good chance of fathering a child. Even so, the longer period of fertility experienced by the male, the fact that females are incapable of ovulating when bearing a child or breast-feeding, and the heavy demands of child-bearing that fall unevenly on females all imply that a single male could, in principle and in practice, fertilise many women.

The obvious question that follows from this is why nature has bothered to produce so many men. It would seem that a species would do better in terms of increasing its number by skewing the **sex ratio** in favour of women, thereby producing fewer men. Men who remained would then be destined to mate polygynously with more women. Yet unfailingly, the ratio of males to females at birth for all mammals is remarkably close to 1:1.

The statistics of polygynous mating seem ever more wasteful. In cases where a few males fertilise the majority of females, such as in leking and polygynous species, given a 1:1 sex ratio at birth, it follows that some males are not successful at all. In evolutionary terms, it seems as if their lives have been pointless and, for the parents that produced them, a wasted expenditure of paternal effort. It was Fisher who pointed a way out of this conundrum.

3.4.2 Fisher's argument

A superficial answer to the question of why roughly even numbers of human males and females are born is that every gamete (oocyte) produced by the female contains an X chromosome but that gametes produced by the male contains either a Y or X chromosome, these two types being produced in equal numbers. Consequently, there is an equal probability of an XX and an XY fusion, and it follows that boys (XY) are just as numerous as girls (XX). This is, in fact, the mechanism used for all mammals and birds (except that in birds the females are XY and the males XX).

This is of course only part of the answer. The X/Y chromosome system provides a proximate mechanism for sex determination, but we know that this is subject to some variation. In humans, it is estimated that, three months after conception, the ratio of males to females is about 1.2:1 and that because of the higher *in utero* mortality of male embryos, the ratio falls to 1.06:1 at birth. It evens out at 1:1 at age 15–20. What we are looking for of course is an ultimate evolutionary argument that explains the adaptive significance of the proximate mechanism. The argument that is now widely accepted was first provided by Fisher in his *The Genetical Theory of Natural Selection* (1930). Fisher's reasoning can be expressed verbally in terms of negative feedback. First we must rid ourselves of the species-level thinking that lies behind the view that species would be better off with fewer males. Species might be better off, but selection cannot operate on species. Selection acts on genes carried by individuals, and what might seem wasteful at a group level might be eminently sensible at an individual level.

Consider the fate of a mutant gene that appeared and caused an imbalance of the sex ratio in favour of females. This could take the form of a gene influencing the probability of fertilisation or survival of the XY zygote in a positive way. Or a gene that influenced the number of X and Y gametes produced by the male. Let us further suppose that, for some reason, this gene gained a foothold and shifted the ratio of males to females to 1:2. Consider now the position of parents making a 'decision' (in the sense of the selection of possibilities over evolutionary time) of what sex of offspring to produce. In terms of the number of grandchildren, sons are more profitable than daughters since, in relative terms, a son will, on average, fertilise two females every time a daughter is fertilised once. More grandchildren will be produced down the male line than down the female line. It therefore pays to produce sons rather than daughters. In genetic terms, the arrival of a gene that now shifts the sex ratio of offspring in favour of males will flourish.

The argument also works the other way round. In a population dominated by a larger number of males, it is more productive of grandchildren to produce a female since she will almost certainly bear offspring, whereas a male (given that there is already a surplus) may not.

The logic of the argument also works for polygamous mating. Suppose only one male in ten is successful and fertilises ten females. It still pays would-be parents to produce an equal number of males and females even though nine out of ten males may never produce offspring, because the one male in ten that is successful will leave many offspring and the gamble is worth it. This one successful male will have ten times the fertility of each female.

Fisher's argument has logic in its favour, but what of empirical confirmation? Interesting support for Fisher's theory has come from the work of Conover (1990) on a species of fish called the silverside *(Menidia menidia)*. The silverside is a small fish common in the Atlantic that has its sex partly determined by the temperature of the water at birth. A low temperature yields females and a high temperature males. This mechanism in itself probably has adaptive significance. A low temperature indicates that it is early in the season, and it is known that an increase in size boosts the reproductive performance of females more than males. Parents should prefer females to males early in the season since this gives the opportunity for growth to increase fertility. Conover kept batches of fish in tanks at various constant temperatures. At first, the sex ratio drifted away from 1:1 in the way expected from the temperature effects, but after a few generations, it returned to 1:1 in a way expected from the negative feedback effects of Fisher's argument.

3.5 Sexual selection

3.5.1 Natural selection and sexual selection compared

Darwin's idea of natural selection was that animals should end up with physical and behavioural characteristics that allow them to perform well in the ordinary processes of life such as competing with their rivals, finding food, avoiding predators and finding a mate. The life and death of thousands of our ancestors should have ensured that by now our characteristics are finely tuned to growth, survival and reproduction, and so most features of plants and animals should, therefore, have some adaptive function in the struggle for existence. Nature should allow no extravagance or waste. So what about, for example, the spectacular train of the peacock? It does not help a peacock fly any faster or better. Neither is it used to fight rivals or deter predators – in fact, the main predator of peafowl, the tiger, seems particularly adept at pulling down peacocks by their tails. It would seem to be an irrelevance, a magnificent one to be sure, but nevertheless an encumbrance that should have been eliminated by natural selection long before now. Nor is the peacock's tail an exception: many species of animals are characterised by one sex (usually the male) possessing some colourful adornment that serves no apparent function (or even seems dysfunctional), while the other sex, like the peahen, seems much more sensibly designed. Such features seem, at first glance, to challenge the power of natural selection to explain the behaviour of animals.

When males and females differ like this in some physical characteristic, they are said to be sexually dimorphic (two shapes). **Sexual dimorphism** is found to varying degrees in the animal kingdom. Humans are moderately dimorphic: men are, on average, taller than women and more muscular, and grow more facial hair. Now some of these differences could, in principle, be due to natural selection: males and females may exploit different food resources and female mammals are generally adapted to provide more care to offspring than males. If ancestral men hunted while women stayed at home nurturing children (albeit a rather oversimplified picture), then height and muscularity would have benefited males more than females. Yet, however ingeniously we work to apply the principle of natural selection, we are still drawn back by the startling spectacle of the peacock's tail.

It was Darwin himself who provided the answer to this seeming paradox. In his *Descent of Man and Selection in Relation to Sex* (1871), he gave the explanation that is still accepted (with refinements) today. Darwin realised that the force of natural selection must be complemented by the force of **sexual selection**: individuals posses features that make them attractive to members of the opposite sex, or help them to compete with members of the same sex for access to mates (Figure 3.2). In essence, the train of the peacock has been shaped for the delectation of the peahen. For reasons that are still debated, peahens are excited by males with fancy tails. Natural and sexual selection now form the twin pillars of the Darwinian programme that seeks to demonstrate how the features of organisms have evolved (sometimes called the 'adaptationist paradigm'). Natural selection gives rise to adaptations favoured by the biotic environment, while sexual selection results in adaptations selected by the social environment.

FIGURE 3.2 Sexual selection as found in the humming bird species *Sparthura underwoodi*
Darwin used this illustration in his second edition of *The Descent of Man and Selection in Relation to Sex* (1874). The male is the one with the long and highly ornamented tail. Darwin realised that the tail grew like this to please the female and so help the male gain a sexual partner.

3.5.2 Intersexual and intrasexual selection

We should really distinguish between two types of sexual selection: 'intra' and 'inter'. Even among species where polygamy is the norm, the sex ratio (males to females) usually remains close to 1:1; in other words, the number of males and females in a population is roughly the same. So where conditions favour polygyny, males must compete with other males for access to females; this follows from the obvious reason that if each male is intent on mating with two or more females to the exclusion of other males, then there simply are not enough females to go round. This leads to **intrasexual selection** (intra = within). Intrasexual competition can take place prior to mating or, in the case of sperm competition (see below) after copulation has taken place. On the other hand, for many species, a female investing heavily in offspring, or only capable of raising a few offspring in a season or lifetime, needs to make sure she has made the right choice. There will probably be no shortage of males clamouring for her attention but the implications of a wrong choice for the female are graver than for the male, who may, after all, be seeking other partners anyway. Under these conditions, a female can afford to be choosy and pick what she thinks is the best male. This leads to **intersexual selection** (inter = between).

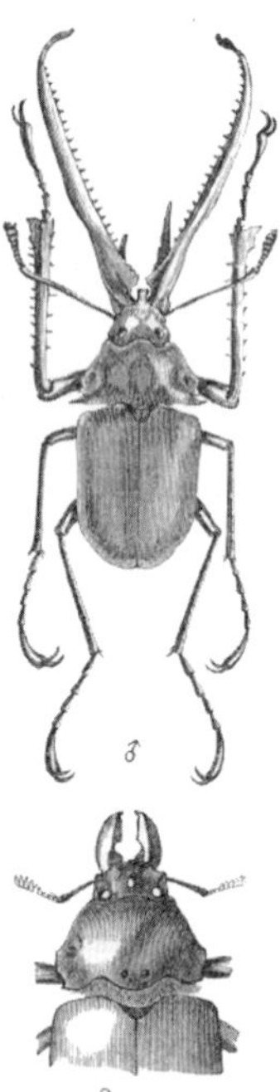

FIGURE 3.3 Sexual dimorphism in the beetle species *Chiasognathus Grantii*

The male (top) has long protruding mandibles, which it uses to fight and intimidate other males. The female (bottom) lacks these characteristics. Darwin concluded that although these devices were selected for fighting, they seemed to be excessive in size even for this purpose and noted that 'the suspicion has crossed my mind that they may in addition serve as an ornament'.

SOURCE: Darwin, C. (1874) *The Descent of Man and Selection in Relation to Sex*. London, John Murray.

This distinction between intra- and intersexual selection helps us to begin to understand some of the more bizarre manifestations of sexual activity in the natural world. The head-butting and twisting of antlers of male deer in the rutting season, while the females look on, is an example of intrasexual selection: males compete with males for the prize of impregnating many females. Many insect species also show evidence of intrasexual selection (see Figure 3.3). The screeching displays of the peacock in front of a demur and far less brightly coloured peahen is an example of intersexual selection: the female chooses the male with the most elaborate and impressive train (tail). Figure 3.4 shows a comparison of inter- and intrasexual selection. We now need to consider how these concepts help us understand human sexual behaviour and how the type of selection that takes place can be understood in relation to the biology of the species.

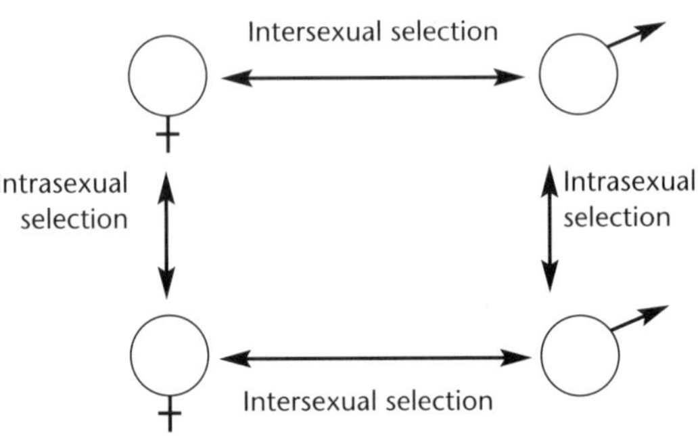

FIGURE 3.4 Inter- and intrasexual selection

3.5.3 Parental investment

As a rough guide, the degree of discrimination exercised by an individual in selecting a partner is related to the degree of commitment and investment that is made by both parties. Male black grouse that provide no paternal care will mate with anything that resembles a female black grouse. Female black grouse, however, that bear the brunt of the consequences of fertilisation in terms of egg production and incubation, are rather more careful in their choices, and select what they take to be the most suitable male from those displaying before them. Among humans, both males and females have a highly developed sense of physical beauty, and this aesthetic sensibility is consistent with a high degree of both maternal and paternal investment. The more investment an individual makes, the more important it becomes to choose its mate carefully – genes that direct for poor choices (for example infertile or unhealthy individuals) are destined to be eliminated from the gene pool. All this decision-making results in the selective force of intersexual selection. It follows that if a high degree of investment by one sex sets up the force of intersexual selection, then the lack of investment by the other sets up competition between that sex for access to the sex that invests most. These ideas were formalised into the concept of **parental investment** by Robert Trivers in 1972.

The concept of parental investment seemed to promise, and to a degree did deliver, a coherent and plausible way of examining the relationship between parental investment, sexual selection and mating behaviour. The sex that invests least will compete over the sex that invests most, and the sex that invests most will have more to lose by a poor match and so will be choosier over its choice of partner. Trivers (1972, p. 139) defined parental investment as:

> any investment by the parent in an individual offspring that increases the offspring's chance of surviving (and hence reproductive success) at the cost of the parent's ability to invest in other offspring.

Using this definition, Trivers concluded that the optimum or ideal number of offspring for each parent would be different. In the case of many mammals, a low-investing male will have the potential to sire more offspring than a single female could produce. A male could, therefore, increase his reproductive success by increasing the number of his copulations. On the other hand, females should prefer quality rather than quantity.

The logic is clear but in practice it has proved difficult to measure such terms as 'increase in offspring's chance of surviving' and 'costs to parents'. Consequently, deciding which sex invests the most is not always easy. One concept that may help to circumvent these difficulties is that of potential reproductive rates.

3.5.4 Potential reproductive rates: humans and other animals

Clutton-Brock and Vincent (1991) have suggested that a fruitful way of understanding the mating behaviour of animals is to focus on the potential offspring production rate of males and females, rather than trying to measure mating effort or parental investment. In this view, it becomes important to identify the sex that is acting as a 'reproductive bottleneck' for the other. Applying these ideas to human mating, it should be clear by now that there are large differences between the potential reproductive rates of men and women. Harems may have been common in ancient civilizations but there are no examples recorded of female rulers guarding a company of male studs or 'toy boys'. Biologically, what would be the point?

It is noteworthy then that the record often claimed for the largest number of children from one parent is 888 for a man and 69 for a woman. The father was Ismail the Bloodthirsty (1646–1727), an emperor of Morocco (Figure 3.5). The mother was a Russian lady who between 1725 and 1765 experienced 27 pregnancies with a high number of twins and triplets. Most people are more astonished by the female record than the male.

The figure of 888 looks extreme compared to most cases of fatherhood but would at first glance seem to be a practical possibility. Ismail died at the age of about 82 and could have enjoyed a period of fertility of at least 55 years. Over this time, he could have had sex with his concubines once or twice daily. The performance claimed for Ismail has recently been questioned, however, by Dorothy Einon of University College London. Einon (1998) analysed the mathematical probability of conception by members of his harem and concluded that 888 children may be an exaggerated figure. The problem for a man with access to a large number of females and intent on breeding is that he is uncertain when his womenfolk are ovulating. The fact that ovulation takes place between 14 and 18 days before the next menstruation was not known until 1920. Before this, men knew that sex led to offspring but not when sex was at its most effective. Einon has argued that the viability of sperm inside the female genital tract is only 3.5 days, giving a potentially fertile period of

3.5 days within each month. More recently, the reputation of Moulay Ismail has been restored somewhat by Gould (2000), who points to errors in Einon's analysis – he lived till about 82 for example, not 55 as Einon thought, and sperm remains viable for 6 days, not 3.5. Gould concludes that Moulay Ismail could have fathered 888 children if he sustained a rate of 1.2 couplings per day over 62 years. An exhausting prospect but not beyond the bounds of possibility.

Figure 3.5 Moulay Ismail Ibn Sharif (Ismail the Bloodthirsty) emperor of Morocco (1646–1727)

SOURCE: Windus, J. (1726) *Reise nach Mequinetz, der Residentz des heutigen Käysers von Fetz und Marocco welche C. Stuart als Groß-Britannischer Gesandter Anno 1721 zur Erledigung der dortigen Gefangenen abgelegt hat.* Hannover, Förster.

It is possible of course that men know subconsciously when ovulation is taking place and are thereby able to better target their reproductive efforts. Males may produce 280 million sperm with each ejaculate, but most are wasted in that they fail to meet an egg. The obvious question is why women have evolved to conceal ovulation. If they behaved like chimps, then a red swelling and obvious odours would announce the fact unambiguously and the social life of humans would be vastly different. This feature of human sexuality illustrates once again that behaviour is not designed to benefit the species but to serve the individual. Although concealed ovulation is still a puzzle in evolutionary biology, it now seems likely that it may have evolved as a tactic by females to elicit more care and attention from males than they would otherwise give.

We should also note that most men in history have not been emperors, and the harems enjoyed by Ismail and his like would not have been a regular feature of our evolutionary past. It is probably true to say, however, that in *Homo sapiens*, the limiting factor in reproduction resides marginally with the female. This by itself would predict some male v. male competition and both intra- and intersexual selection can be expected to have moulded the human psyche.

3.5.5 The operational sex ratio

The potential reproductive rate and the **operational sex ratio** (OSR) are closely related concepts. Although for most mammals there are roughly equal numbers of males and females, not all the males or females may be sexually active and there may be local variations in the ratio of the sexes. This idea is contained in the concept of the operational sex ratio:

$$\text{Operational sex ratio} = \frac{\text{Fertilisable females}}{\text{Sexually active males}}$$

When this ratio is high, the reproductive bottleneck rests with males and females could compete with other females for the available males. When the ratio is low, the situation is reversed and males will vie with other males for the sexual favours of fewer females (Figure 3.6).

At first sight, it would seem that females are always the limiting resource for male fecundity. Consider the following facts. If you are a young fertile male, while you are reading this you are producing sperm at the phenomenal rate of about 3,000 per second. If you are a young fertile female, you are holding onto a lifetime's supply of only 400 eggs. In addition, a man could impregnate a different woman every day for a year, whereas over this same period a woman can become pregnant only once. We need to consider this with caution, however. Imagine a male who mates with 100 different women over the course of a year and a female who mates with 100 different men over the same period. The woman is likely to become pregnant and bear one offspring in the same year. Using the reasoning advanced by Gould and Einon earlier, and the notion that sperm can survive for about six days in the genital tract of the female, if the male avoids the time of menstruation, he has about a 26 per cent chance (6 days out of 23) of impregnating a woman during her fertile period. Only about 70 per cent of the female ovarian cycles will be fertile, and implantation will only take place about 50 per cent of the time. The number of women a man could expect to make pregnant is about 9 (100 × 0.26 × 0.7 × 0.5). In one sense, women are a limiting resource but not to the extremes that might be indicated by the differences in the size or rate of production of gametes.

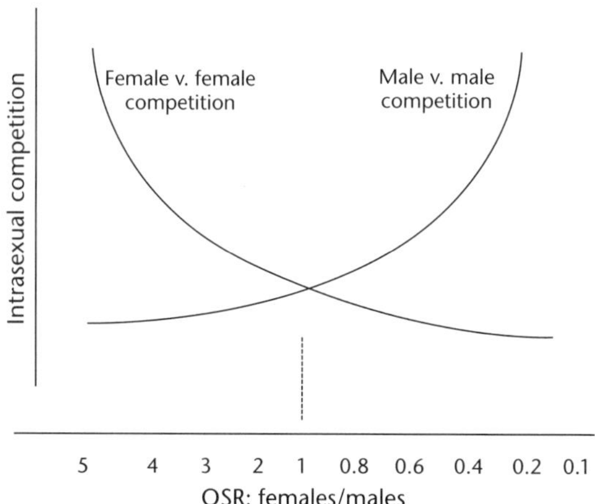

FIGURE 3.6 Intrasexual competition and the OSR

Source: Adapted from Kvarnemo, C. and Ahnesjo, I. (1996) 'The dynamics of operational sex ratios and competition for mates.' *Trends in Evolution and Ecology* **11**: 4–7.

The operational sex ratio (females/males) for a group of humans with a 1:1 sex ratio (that is, equal numbers of men and women in the population) will be less than one if we measure it in terms of males or females that are fertile. It is likely that there will be more sexually fertile males than females. This arises from the fact that men experience a longer period of fertility compared to women. It is counterbalanced somewhat by the higher mortality rates for men than women, but not entirely. The picture is complicated, however, if the population is growing. Under these circumstances, the fact that women tend to prefer to marry slightly older men will mean more young women looking for slightly older partners than there are slightly older men available, since the cohort of marriageable men will be smaller than the number of marriageable women in the expanding cohort below. Guttentag and Secord (1983) have argued that this in itself can be a contributing factor to the development of social mores. In the USA from 1965 to the 1970s, because of the postwar baby boom, there was an oversupply of women compared to the cohort of slightly older men. This would have the effect of decreasing male–male competition and increasing female–female competition. This allowed men to pursue their own reproductive preferences, especially in terms of an increased number of partners, to a greater extent than women could pursue theirs. Guttentag and Secord (1983) suggest that this could be a contributing factor to the liberal sexual mores of those decades, characterised by high divorce rates, lower levels of paternal investment and a relaxed attitude to sex. They stress that sex ratios by themselves are not a sufficient cause for such social changes but may be part of the equation.

Such arguments are extremely difficult to support conclusively. Following the Second World War, other more profound changes occurred in Western cultures such as rising affluence and, crucially, the availability of the contraceptive pill. An interesting study could be made on the changes in social values in France, Britain and Germany following the First World War when the carnage of war would have biased the sex ratio towards females. Here again, however, other changes were taking place such as votes for women, changes in the economic status of women and so forth. Perhaps a more realistic application of sex ratio thinking may be found in the analysis of traditional cultures where social values shift less rapidly. In South America, there are two indigenous Indian groups with different sex ratios. The Hiwi tribe shows a surplus of men, while the Ache people have a sex ratio of females to males of about 1.5 (Hill and Hurtado, 1996). The ecology of the two groups is otherwise similar, but whereas among the Ache people extramarital affairs are common and marriages are unstable, among the Hiwi marital life is more stable. This pattern is what one would expect from the anticipated effect of sex ratios on mating strategies.

In virtually all cultures, however, it is significant that men engage in competitive display tactics and are more likely to take risks than women are. It is also men who tend to pay for sex, this being one way of increasing the supply of the limiting resource (Figure 3.7).

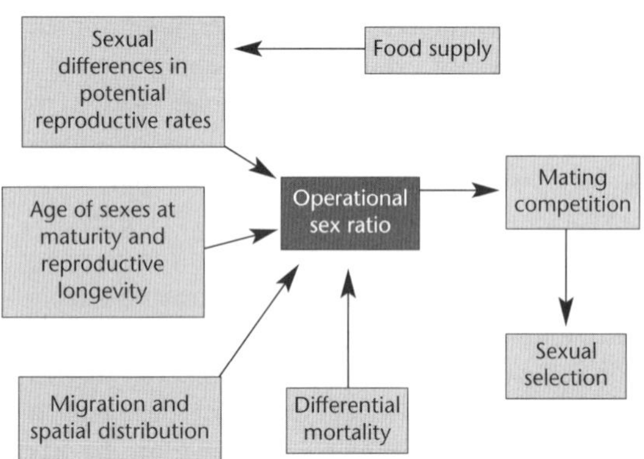

FIGURE 3.8 Influences on the operational sex ratio

FIGURE 3.7 Detail from *The Rake's Progress* by William Hogarth
Here the young rake (a man of loose morals) is shown visiting a brothel. Prostitution is said to be the world's oldest profession. The fact that the vast majority of prostitution consists of women exchanging sex for money from men is understandable from an evolutionary standpoint, since women are the limiting resource for men. Consequently, men will seek ways to improve the supply of the limiting resource.

3.5.6 The operational sex ratio and contingent strategies

We have seen how the operational sex ratio may influence mating competition and hence drive the direction of intersexual selection: the sex in short supply will be competed for by the other sex. We should note, however, that this ratio is not some fixed property of living things, it can vary with other factors such as food supply, differential mortality and time to reach sexual maturity (Figure 3.8).

If the sex ratio can vary, then we might expect the strategies employed by animals to vary accordingly. This brings us to the important point, further explored in a human context in Chapter 12, that sexual strategies may vary adaptively with local conditions.

3.6 Consequences of sexual selection

Just as natural selection has left humans with bodies and brains suited to the processes of finding food, avoiding predators and resisting disease, so sexual selection has left its mark on our bodies and our sexual inclinations. So much so, that once we understand the basic ideas of sexual selection, we can examine humans and make predictions about what type of sexual behaviour patterns may be typical of our species.

3.6.1 Sexual dimorphism in body size

Darwin argued that intrasexual selection was bound to favour the evolution of a variety of special adaptations, such as weapons, defensive organs, sexual differences in size and shape and a whole range of devices to threaten or deter rivals (Figure 3.3 above).

The importance of size is illustrated by a number of seal species. During the breeding season, male northern elephant seals *(Mirounga angustirostris)* rush towards each other and engage in a contest of head-butting (Figure 3.9). Such fighting has led to a strong selection pressure in favour of size, and consequently male seals are several times larger than females. Elephant seals are in fact among the most sexually dimorphic of all animals: a typical male is about three times heavier than a typical female. The mating system of these seals is described as female defence polygyny and a male needs to be large both to fight off contenders to take his place among a harem of females and to defend his position once he is there. As a result of the intense competition, many males die before reaching adulthood without ever having mated.

FIGURE 3.9 **Elephant seals**

Humans show sexual dimorphism in a range of traits (Figure 3.10). Men, for example, have greater upper body strength and more facial and body hair than do women. Men also have deeper voices, later sexual maturity and experience a higher risk of infant mortality than do females. The pattern of fat distribution between men and women is also different. Women tend to deposit fat more on the buttocks and hips than do men. These of course refer to average tendencies. It is likely that many of these are the results of sexual selection.

The fact that human infants need prolonged care would ensure that females were alert to the abilities of males to provide resources. In addition, the fact that a female invests considerably in each offspring would make mistakes (in the form of weak or sickly offspring that are unlikely to reproduce) very expensive. It has been estimated that human females of the Old Stone Age would have only raised successfully to adulthood two or three children. Females would therefore be on the lookout for males who show signs of being genetically fit and healthy and who are able to provide resources. Both these attributes, genetic and material, would ensure that her offspring receive a good start in life.

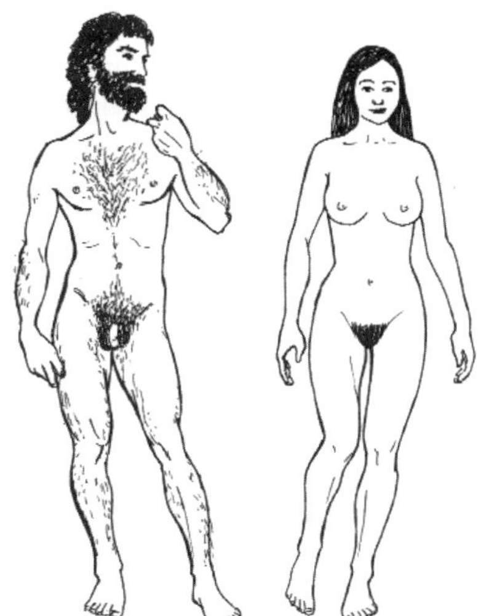

Males compared with females

On average, males have:

- Greater upper body strength
- More facial and bodily hair
- Greater height and mass
- Deeper voices
- Riskier life histories and higher juvenile mortality
- Later sexual maturity
- Earlier death
- Broader and more prominent chins
- Lower levels of fat deposited on buttocks and hips

FIGURE 3.10 **Human sexual dimorphism**

One possible way in which men and women could assess the genetic fitness of a potential mate is through the symmetry of their features. The logic here is that symmetry is an honest signal since it is difficult to fake and physiologically difficult to achieve. Invasion by parasites and vulnerability to environmental stress factors both reduce the symmetry of an organism. Only the fittest genomes are capable of engineering symmetry in the body of an animal. Evidence that humans are sensitive to symmetry in their appraisal of the attractiveness of mates comes from work on facial attractiveness explored later in Chapter 12.

3.6.2 Post-copulatory intrasexual competition: sperm competition

At first glance, it may seem that once copulation has taken place, then intrasexual competition must cease: one male must surely have won. But the natural world has more surprises in store. Some females mate with many males and retain sperm in their reproductive tracts; sperm from two or more males may then compete inside the female to fertilise her egg. The concept of **sperm competition** illuminates many features of male and female anatomy in non-human animals. Male insects have evolved a variety of devices aimed at neutralising or displacing sperm already present in the female. The male damselfly *(Calopteryx maculata)*, for example, has evolved a penis designed to both transfer sperm and, by means of backward pointing hairs on the horn of the penis, remove any sperm already in the female from a rival male.

Many animals have also evolved other tactics to outwit rivals in sperm competition. When a male garter snake mates with a female, it leaves behind a thick sticky mass, technically known as a 'copulatory plug', that effectively seals off the reproductive tract of the female from other would-be suitors. When a male wolf mates with a female, its penis becomes so enlarged that even after ejaculation it remains stuck in the vagina of the female for up to half an hour after impregnation. The male and female remain fixed like this in some apparent discomfort, but the mechanism ensures that the sperm of the successful male gets a head start over any rivals.

We should not think of females as passive in this process of sperm competition. The female may exercise choice over the sperm once it is inside her (Wirtz, 1997). Many female insects store sperm that they use to fertilise their eggs at oviposition (egg laying) as the eggs pass down the female's reproductive tract. It has been suggested that the function of the female orgasm in humans is to assist the take-up of sperm towards the cervix (Baker and Bellis, 1995). Randy Thornhill and his colleagues carried out a study to show that the bodily symmetry of the male is a strong predictor of whether or not a female will experience a copulatory orgasm. Symmetry is thought to be an indicator of genetic fitness and the possession of a good immune system (Thornhill et al., 1996). The orgasm therefore ensures that sperm from exciting and desirable males, who presumably are genetically fit and unlikely to be transmitting a disease, stand a good chance of meeting with the female's egg. In this way, the human female may be extending her choice beyond courtship (Baker and Bellis, 1995).

The more sperm produced, the greater the chance of at least one finding the egg of the female: 50 million sperm are twice as effective as 25 million and so on. In species where sperm competition is rife, we would expect males to increase the number of sperm produced or ejaculated compared to species where sperm competition is less intense. Between species, this prediction has been supported indirectly by measurements on levels of sperm expenditure as measured by testis size. Species facing intense sperm competition have larger testes than those where sperm competition is less pronounced (see Chapter 11).

Baker and Bellis (1995) (then working at Manchester University) provided evidence to support the idea that the number of sperm in the ejaculate of men is adjusted according to the probability of sperm competition taking place. In one study, when couples spent all their time together over a given period, the male was found to ejaculate about 389×10^6 sperm during a subsequent sexual act. When the couples only spent 5 per cent of their time together, men typically ejaculated 712×10^6 sperm. Baker and Bellis interpret this as consistent with the idea that the male increases the number of sperm in the latter case to compete more effectively against rival sperm which may have entered the female if she had been unfaithful. Baker and Bellis (1995) have been successful in generating new ideas in an area of research that faces innumerable experimental and ethical difficulties. They have also been successful in disseminating their ideas, helped partly by a media eager for such theories and partly by the popularisation of their work in such books as *Sperm Wars* (Baker, 1996).

In the 'sperm wars' of post-ejaculate intrasexual competition, males can adopt various tactics: they can produce sperm in large numbers, attempt to displace rival sperm, insert copulatory plugs or produce sperm that actively seek out to destroy rivals. Baker and Bellis (1995) have developed this latter idea into a 'kamikaze sperm hypothesis', claiming that a wide variety of animals including humans produce sperm whose function is to block or destroy rival sperm. Part of the evidence used by Baker and Bellis is the number of deformed sperm found in any ejaculate. They argue that the function of some of these deformed sperm is to seek out and destroy sperm from rival males.

Following the initial publication of the ideas of Baker and Bellis in 1988, considerable debate ensued concerning the existence of these kamikaze sperm. In one study, sperm from different males was mixed in vitro (that is, in glassware in laboratory conditions) and its viability compared to sperm mixed from the same male. If the kamikaze sperm hypothesis is correct, then mixed male sperm should show signs of a lack of function as rival sperm kill each other off. The result, however, was that the performance of mixed male sperm was not noticeably different to that of same male sperm (Moore et al., 1999).

In a careful analysis of the evidence, Harcourt (1991, p. 314) concluded that kamikaze sperm did not in all probability exist, and that 'the function of all mammalian sperm is to fertilize, and that sperm competition in mammals occurs through scramble competition, not contest competition'. Harcourt's conclusion is based on the fact, among others, that male primates in polyandrous species (that is, where a female will mate with more than one male) do not produce more deformed (that is, non-fertilising) sperm than primates in monandrous (one male, one female) species. Yet if the kamikaze hypothesis were to be correct, this is the reversal of what would be expected. Harcourt does agree, however, that the males of many mammalian species produce secretions from accessory glands that serve to coagulate semen and act as a copulatory plug.

Long before sperm competition takes place, however, a male has to be accepted by a female or vice versa. Passing this quality control procedure has also left its mark on the anatomy and behaviour of humans and it is to this process we now turn.

3.6.3 Good genes and honest signals

Darwin had difficulty in explaining in adaptationist terms why females find certain features attractive. A peahen may have forced male peafowl to sport long trains to please her, but why in functional or ultimate terms, should she be pleased by a long rather than a short tail? If, as Darwinism informs us, beauty is in the eye of the genes, what genetic self-interest is served by finding long trains or colourful tails beautiful? If we can crack this problem, then perhaps the basis of human physical beauty can be understood. Answers to this puzzle tend to fall into two camps: the 'good sense' and the 'good taste' schools of thought (Cronin, 1991).

The good taste school of thought stems from the ideas of Fisher who investigated the problem in the 1930s. Consider a male characteristic such as tail length that females once found attractive for sound evolutionary reasons, such as it indicated the species and sex of the male, or it showed the male was healthy enough to grow a good sized tail. Fisher argued that, under some conditions, a 'runaway effect' could result, leading to longer and longer tails. Such conditions could be that sometime in the past an arbitrary (that is, non-functional) drift of preference led a large number of females in a population to prefer long tails. Once this fashion took hold it could become self-reinforcing. Any female that resisted the trend, and mated with a male with a shorter tail, would leave sons with short tails that were unattractive. Females that succumbed to the fashion would leave 'sexy sons' with long tails and daughters with the same preference for long tails. The overall effect is to saddle males with increasingly longer tails, until the sheer expense of producing them outweighs any benefit in attracting females. The argument and the precise conditions under which this mechanism could work are complex, but evolutionary biologists are now convinced that the Fisherian runaway process is a distinct possibility. In this model, although individuals are choosing a set of genes by examining a feature such as a long tail that is expressed by such genes, the genes are not necessarily good for anything else. In a sense, they have become an arbitrary fashion accessory to which individuals find themselves attracted. A more recent formulation and extension of Fisher's idea is the 'sensory bias model' (Kirkpatrick and Ryan, 1991). In this model, an animal's preference for something such as food of a specific colour spills over into other domains. Hence a preference for green fruits or red fruits may drive a mating preference for greenness or redness in males.

The 'good sense' view suggests that an animal estimates the quality of the genotype of a prospective mate through the signals he or she sends out prior to mating. In addition, individuals could also assess each other on the basis of the level of resources a mate is likely to be able to provide – something that in itself could be a reflection of the quality of genes that an animal carries. These ideas are now some of the most promising lines of inquiry in sexual selection theory, with many suggestive applications to human mate choice (Table 3.3).

Table 3.3 Mechanisms of intersexual competition

Category	Mechanism
Good taste (Fisherian runaway process)	Initial female preference becomes self-reinforcing. A runaway effect results that leads to elaborate and often dysfunctional (in terms of natural selection) traits, for example peacock's train
Good sense (genes and/or resources)	One sex uses signals from the other to estimate the quality of the genome on offer. Such signals may indicate desirable characteristics such as resistance to parasites or general metabolic efficiency. One partner may also inspect resources held by the other and the likelihood that such resources will be made available

The **good genes** dimension of good sense would explain why, in polygynous mating systems, females share a mate with many other females, even though there may be plenty of males without partners, and despite the fact that the males of many species contribute nothing in the way of resources or parental care. Females are in effect looking for good genes. The fact that the male is donating them to any willing female is of no concern to her. The important point here is that the female is able to judge the quality of the male's genotype from the **honest signals** he is forced to send. In this respect, size, bodily condition, symmetry and social status are all signals providing the female with information about the potential of her mate. Likewise, some human females find some men sexually attractive, even though they know such men may be unreliable, philandering and untrustworthy.

There are several ways in which the features of an animal may serve as signals of genetic prowess. Males and females can send signals about their health and reproductive status in a variety of ways. Consider the time-honoured principle of fashion that 'if you have it, flaunt it, if you haven't, hide it'. This applies to cosmetics as much as clothes. This leads to a distinction between honest and dishonest signals. Hiding signs of genetic weakness or false advertising are really dishonest signals. Dishonest signals are rare among non-human animals since they are likely to be spotted and eliminated in favour of honest ones. Humans, however, with their clever brains and sophisticated culture are particularly adept at sending both honest and dishonest signals about themselves to others.

A particularly fruitful line of research that has emerged in recent years is costly signalling theory (CST). This theory suggests that honest signals will emerge and be attended to by the receiver if one or two of the following conditions hold:

- The signal must be honestly linked to the quality of the trait it is trying to advertise. When this is the case, imitation by inferior individuals is impossible since the linkage would ensure that the advertisement actually revealed the poor quality of the signaller.
- The signal must be a **handicap**, that is, it must impose a cost on the signaller. In this way, only high-quality individuals can afford the handicap and hence advertise.

Smith and Bleige Bird (2000) are anthropologists in the USA who have applied this theory with some success to turtle hunting among the Meriam islanders of Torres Strait, Australia. The Meriam people live on the barrier reef island of Mer about 100 miles from New Guinea. They have numerous public feasts during which time men engage in competitive dancing, hunting, diving and boat racing. During one type of feast involving funerary rites, large amounts of turtle meat are consumed. To provide this meat, turtles have to be hunted, and turtle hunting seems to qualify as an example of a costly and honest signal since:

- Hunting involves taking a small boat out to sea, locating a turtle and then jumping on its back with a harpoon – a procedure that requires strength, agility and is physically dangerous.
- When the turtles are captured, they are shared communally during a public feast. In fact, the hunters take virtually no share of the meat and have to bear the full cost of the hunt themselves.
- During the feast, people are attentive to who provided the largest turtles and who returned with small ones or none at all.

In short, the ability to return with a large turtle and offer it as a communal gift is an honest advertisement of physical strength and vigour as well as wealth. It is difficult to see how a physically unfit and resource-poor male could return from the hunt with a large turtle. The fact that the hunt is a signal rather than an economic necessity is indicated by the fact that this sort of hunting is too risky to be pursued except during public feasting. In other words, the activity does not make much economic sense except as a device for young men to show off their virtues (Smith and Bleige Bird, 2000).

In some species, the males provide nothing except a few drops of sperm. The females have come to expect nothing except genes, so if they are choosy, it will be for good genes rather than any parental after-sales service. Most human females, however, expect males to bring something to mating in addition to their DNA, and may judge, therefore, the ability of the male to provide resources before and after copulation. In the case of humans, resources could be indicated by the social and financial status of a male and, just as importantly, his willingness to donate them in a caring relationship. This could of course be one of the functions of courtship in human societies. As well as allowing an assessment of character and health, courtship, viewed in this dispassionate light, also enables each sex to 'weigh up' their prospective partner in terms of their commitment to a relationship and the resources they are likely to bring, both genetic and material.

There is little doubt that wealth is sexy. In modern societies, the phenomenon of the 'sugar daddy' is well known. Rich and powerful men seem to be able to attract younger and highly attractive females as their partners. One British chat show host caused great mirth when she asked the young attractive wife of an elderly but wealthy and well-known TV star: 'Tell me Mrs ... what first attracted you to your millionaire husband?' This dimension to human mating will be visited again in the study of advertisements for partners (see Chapter 12).

3.7 Life history theory

3.7.1 Life history variables

To understand the adaptive function of both behavioural and physical attributes, it is often necessary to look across the whole lifespan of an animal. What may seem maladaptive over a narrow time frame, such as a long birth intervals or early death, may make perfect sense when considered across the complete lifespan.

Life history theory (LHT) is based on the idea that organisms must balance competing demands on their time and energy budgets. In this respect, we can think of organisms as capturing energy from the environment through feeding and allocating it to three types of activity (Gadgil and Bossert, 1970):

- *Growth:* organisms must grow to a size that they can fend for themselves and reach a size such that they are sexually mature and large enough to attract a mate. This investment impacts on future reproduction.
- *Repair and maintenance:* all organisms are daily buffeted by shocks and damage from their external and internal environments. Some energy

must therefore be allocated to coping with infections and parasites and repairing damaged tissue. This investment also impacts on future reproduction.

- *Reproduction:* all organisms must produce direct or indirect offspring. This investment, by definition, is concerned with current reproduction.

It can be seen that these three allocations resolve to the problem of a trade-off between future and current reproduction possibilities.

The allocation of energy into these three domains will require compromises and so it follows that activities which may seem maladaptive in the short term (such as delayed sexual maturity) may be perfectly sensible when viewed over the whole lifespan. In this light, even death and senescence begin to make sense. Why, we may ask, has natural selection not engineered an organism that lives forever and produces copies of itself (or spreads it genes by sexual reproduction) at regular intervals for ever? The answer is that such an organism would have to allocate an enormous quota of energy to maintenance and repair at the expense of reproduction. A rival organism that allocated more energy to reproduction early in life, at the expense of later senescence and death, would quickly begin to leave more offspring (children, grandchildren and so on). Such rapidly multiplying shorter lived offspring would quickly fill the ecological niche and become the norm. In this way, natural selection is largely blind to diseases and degenerative processes that take effect after the period of reproduction (Figure 3.11). Williams (1957) even postulated the possible existence of genes that exhibit antagonistic pleiotropy. Pleiotropic genes (genes that code for more than one effect) would become antagonistic if they had a positive effect on fertility in youth and a negative effect later in life. They would be favoured even though they were detrimental to the later life of the organism that had already reproduced. In essence then, post-reproductive death is natural selection's answer to the trade-off conflict between reproduction and repair: bodies are disposable, temporary vehicles.

The relevance of LHT to evolutionary psychology is that, for humans, complex decisions have to be made about the allocation of resources and so it may be supposed we have psychological adaptation to assist us in this task. This differential allocation in relation to the age of the organism does not require the existence of a centrally coordinated set of conscious calculations for its determination. Rather, endocrine systems have evolved to serve this role. In puberty, for example, the release of hormones initiates changes to energy budgets that can manifest effects over several years. In both males and females, growth subsides while energy is allocated to reproductive functions and secondary characteristics. More specifically, in females such hormones lead to fat storage and regular menstrual cycling. In males, **androgens** lead to increased musculature and competitive mating displays. In both sexes, investments to the immune system are reduced. Once humans have reached middle age and have reproduced, the investment in sexual display is less of a priority. Consistent with this is the finding that when men become fathers, testosterone levels subside, enabling effort to be redirected from mating to fathering (Gray et al., 2002).

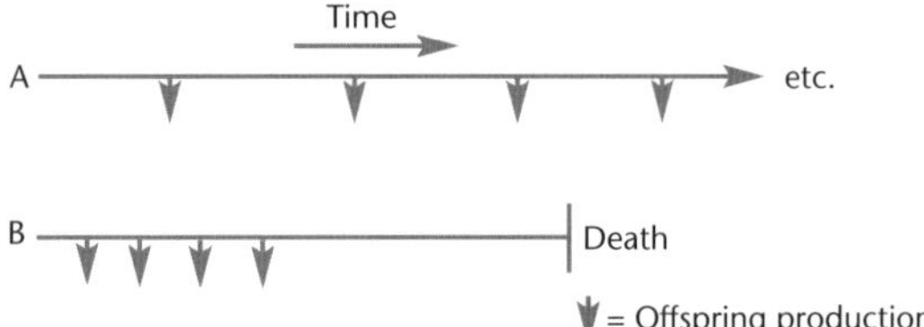

FIGURE 3.11 Alternative growth, reproduction and repair strategies
Organism A delays sexual maturity and opts for high levels of investment in repair and maintenance. Organism B breeds more rapidly but at the expense of an earlier death. Over time, organism B will leave more offspring and, other things being equal, become the norm. In this way, death and senescence are inevitable.

3.7.2 Quantity and quality, mating and parenting

K or r selection?

Within the category 'energy directed to reproduction' (see above), there are still choices (either by the individual or in the phylogenetic line) to be made. Should an individual invest energy in a few high-value offspring or large numbers of lower value offspring? Some species (for example fish, frogs) are said to be **r selected**, that is, they produce large numbers of ener-

Table 3.4 Some features of K and r selected organisms

	r selected	K selected
Climate	Variable and unpredictable	Constant and/or predictable
Population size	Variable over time. Often below carrying capacity of environment. Periodic recolonisation of habitat	Fairly constant and at equilibrium, near to carrying capacity of environment
Lifespan of organisms	Usually short, typically less than one year	Longer than one year
Reproduction	Production of many small offspring. Usually small body size and rapid growth to sexual maturity	Fewer but larger progeny. Delayed reproduction. Slower growth to sexual maturity
Mortality rates	High	Low
Exemplary species	Mouse lemur, frogs, oysters	Gorillas, elephants, humans

getically cheap offspring; others, such as humans and elephants, are said to be **K selected** and produce few but energetically expensive offspring (Table 3.4).

There arises the possibility that complex organisms that make day-to-day decisions about their lives can bias their reproductive allocation to a more r or K selected strategy in the direction of adaptive optima for local conditions.

Birth spacing

The outcome of the compromise between quantity and quality reveals itself in the pattern of birth spacing. A long interval between births, for example, would imply that a woman is choosing to invest more in each offspring. The !Kung San people of Botswana have been the focus of a number of studies to investigate if birth spacing represents an adaptive strategy. Studies by Lee (1979) have shown that the birth interval for !Kung mothers was about four years and that an average woman bears 3.8 offspring in her lifetime. A factor that determines birth interval in this society is the harsh reality that in the dry season !Kung women have to travel several miles to collect their staple diet of mgongo nuts. Children under four years are carried by their mothers on these stressful foraging expeditions. Blurton Jones and Sibly (1978) developed models to see how the backload of women varied with their inter-birth interval (IBI). Their model showed that as the IBI fell from ten years, the backload remained fairly constant, until at four years the predicted load rose sharply. In the model at least, an IBI of below four years would place a high strain on the mother and could jeopardise her health. From this simplified perspective, it looks like four years may be the optimal IBI for !Kung mothers. There is an ongoing debate on this subject, however, and other factors may also be involved (see Pennington, 1992).

Mating and parenting

Organisms also have to decide how to allocate effort between mating and parenting. Often this decision had been made for an individual through the phylogeny of the species. Hence, in about 95 per cent of all mammalian species, females provide all the investment to offspring, whereas males expend more energy on mating through displays and contests. It is also significant, however, that the allocation to mating and parental investments can, within a species, vary with ecological conditions.

Age of menarche and trade-off between growth and reproduction

Within LHT, the age of **menarche** can be viewed as a 'decision' on when to begin reproduction so as to maximize some aspect of reproductive fitness (this could be some function such as the total number of offspring times the probability of survival of each to maturity integrated over the lifetime of the mother). If this is the case, then it might be expected that this age is subject to alterations according to how cues from local contexts might be interpreted as indictors of reproductive opportunities and survival probabilities (see section 11.3.2).

3.7.3 Age-related activation of mental modules

LHT may inform the approach of evolutionary psychology by noting that the costs and benefits of the operation and activation of mental mechanisms may alter with age. So, if men have cuckoldry avoidance and response mechanisms, for example, we need to consider that such mechanisms involve costs: time spent guarding a mate or, if a mate is rejected, looking for a new mate. These costs are in themselves moderated by the risk of false positives. It could be that one's partner was not having an affair after all and that cuckoldry avoidance effort was wasted energy. It is possible then that the intensity of this jealous response might fall as the risks associated with cuckoldry decline. Figure 3.12 lends some support for this notion. It shows that adultery as a reason for divorce becomes progressively less important for men as they age. This may reflect a combination of a reduced cuckoldry guarding response together with the fact that as the age of the female partner increases, then the likelihood and consequences of cuckoldry from a reproductive point of view decline.

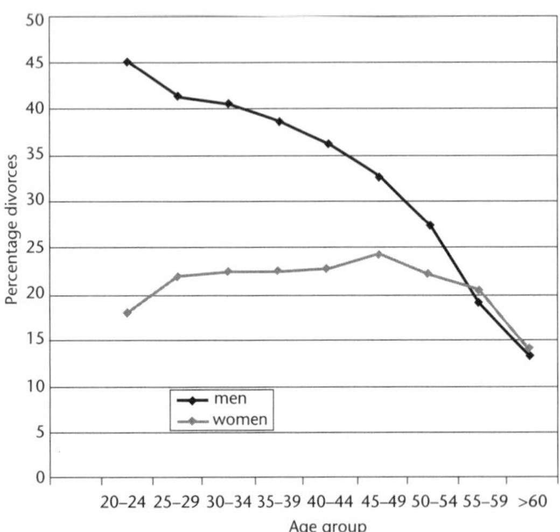

FIGURE 3.12 Divorces as a result of adultery as a percentage of total in each age group plotted against age of petitioning party for population of England and Wales, 1995

SOURCE: Cartwright, J. (2001) *Evolutionary Explanations of Human Behaviour*. Hove, Routledge, p. 69.

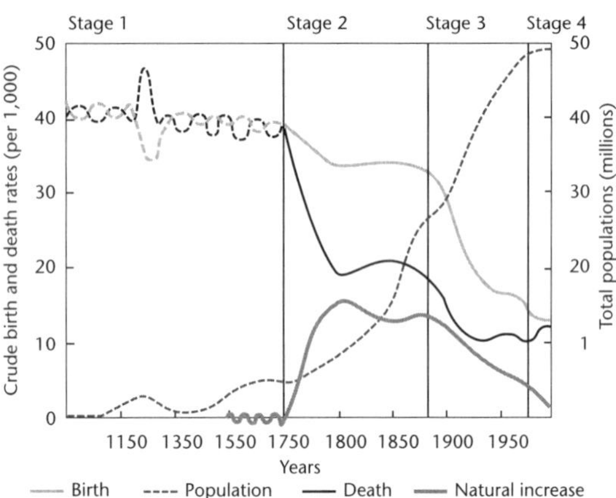

FIGURE 3.13 The demographic transition illustrated by the UK

With economic development, both death rate and birth rate fall.

3.7.4 LHT and the demographic transition

A special challenge for human behavioural ecology (predicated as it is on measuring real fitness gains through offspring numbers) is the explanation of the **demographic transition** (Figure 3.13). How can humans still be regarded as fitness maximisers by choosing to have fewer children, when presumably (since the transition is associated with the growing wealth of a culture) they could afford more? Data on reproductive rates and birth spacing for foraging societies, rural societies and modern societies reveal considerable differences in the life history decisions made by women (Figure 3.14).

A life history theory framing of this problem would start by seeing parents making a decision to increase investment in each child, or even a single child, compared to a lower investment per child in a larger number of children. The rationale for this might be that, in complex societies, children can only grow to be effective and competitive adults, and hence attract mates, if they have undergone extensive training to acquire skills useful in such cultures. If this is the case, then children who have received higher investment (but fewer brothers and sisters as a consequence) should in the long term do better, that is, leave more offspring than children from larger families who received less investment. We might also expect higher status adults to eventually leave more offspring than low earners. But studies exploring such ideas have met with numerous problems. Kaplan et al. (1995), for example, looked at the number of third-generation descendents (that is, grandchildren) of men in Albuquerque, New Mexico, and found it to be highest among men who initially had the most children. Indeed, many studies seem to indicate that whereas higher fertility is associated with lower educational and economic status of offspring, the reduced earning capacity of such offspring does not reduce their fertility. Furthermore, although some studies show that in preindustrial societies there is a positive correlation between status and resources and reproductive success (Barkow, 1989; Voland, 1990), studies on low fertility societies, such as those associated with modernity, reveal no such correlation or even a negative one (Kaplan et al., 1995).

It is here then that there has been a proliferation of ingenious theories to account for these anomalies within life history theory and the human behavioural ecology approach. Within evolutionary psychology, the problem disappears since this whole approach suggests that current actions and lifestyle practices are not necessarily fitness maximizing. We are driven to have sex, which is a natural adaptation, but in figuring out contraception, we have thwarted

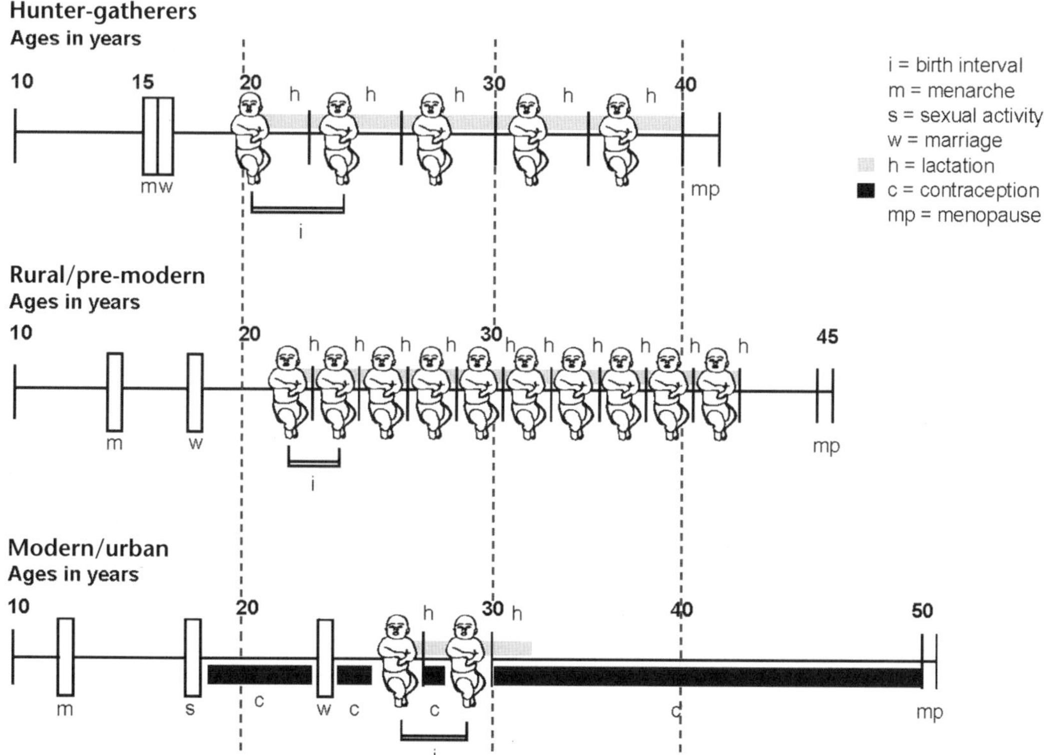

FIGURE 3.14 Birth intervals, reproductive rates and other variables for women in foraging, rural and modern societies

Modern women reach menarche much earlier than their forebears and also spend much less of their overall life lactating. Some of these parameters are almost certainly related to lifestyle and environmental factors such as nutrient supply and healthcare. But clearly modern women are making the decision to delay the age when they first conceive and have fewer children than their ancestors. Trying to identify adaptive reasons for these decisions is a difficult task.

SOURCE: Adapted and redrawn from Zihlman, A. (2000) *The Human Evolution Colouring Book*. New York, HarperCollins, section 6.9.

what were once the inevitable consequences. A variety of suggestions have been forthcoming:

- One obvious solution is to suggest, as Barkow and Burley (1980) have done, that the combination of human intelligence and birth control has enabled women to make a conscious choice to have fewer children than their biological optimum.
- In a similar vein, Perusse (1993) suggested that men's psychology is adapted to securing resources, since in the past this reliably increased mating opportunities and, hence, in the absence of contraception, number of offspring. Today male psychology still dictates wealth pursuit but without the same ensuing reproductive gains.
- Similarly, Irons (1983) and Turke (1989) have suggested that humans are evolved to track cultural success as a proxy for what enhances fitness and the costs of achieving such success in some modern cultures (for example the time demands of a successful career) impact negatively on the number of children.
- Lancaster (1997) has suggested that parents need to increase investment in children at the expense of number of children to enhance the competitiveness of their children in the marriage market.
- Kaplan and Lancaster (1999) have developed a model that emphasises the importance of skill acquisition by offspring. As noted earlier, human life history is such that children mature slowly to acquire the complex skills needed to function in the ecological niche to which we are adapted. According to this model, parental psychology may be such as to detect the relationship between investment in offspring and the return on this investment (that is, the income or proxies for income) in the offspring. As societies modernise, therefore, and increasingly complex skills are required to function successfully, so the resources that parents need to invest in offspring (schooling and so on) increase and can only be met by a lower overall fertility. In response, contraceptive technologies are sought. If this model is accurate, it follows that contraception is the outcome not cause of forces tending to lower fertility.
- Turke (1989) has suggested that it is the development of the nuclear family and the break-up of extended family networks that follow modernisation that have reduced fertility. With modernisation comes social and physical mobility, this in turn breaks down networks of kin support for couples rearing children. In effect, the burden of children on the parents (and especially the mother) rises and it becomes optimal to have fewer of them.

3.7.5 Old age, the menopause and the function of grandparents

The evolutionary purpose of the declining years in humans is still something of a mystery. It may be that the period of life after 60 years is merely a non-functional period of decline that simply takes years to complete. The alternative is that even this period has been shaped by natural selection to yield fitness benefits. As Hill (1993) notes, these two alternatives make different predictions. If old age is simply a time of somatic collapse, then all faculties could be expected to decay roughly synchronously – there is no point in maintaining a healthy heart, for example, if the liver is shutting down and cannot detoxify the blood. On the other hand, if older people have been selected to play a role in, for instance, skill and knowledge transfer, then the decline in mental faculties might be expected to be slower than the rest of the body. There is indeed some evidence that this is the case. Age-related neuropathology in humans is less pronounced than in macaques, for example (Finch and Sapolsky, 1999).

At around the age of 50, the **menopause** sets in for women (the average for developed countries being 50.5 years), whereas male fertility continues beyond this and gradually declines with age rather than experiencing the abrupt cessation seen with female fertility. Now, although this fact can help to explain mate choice preferences, it is a feature that calls for an explanation in itself. Most mammals of both sexes, including chimpanzees and gorillas, experience a gradual decline in fertility with age, totally unlike the menopause of human females. So why the sudden shutdown experienced by human females? An explanation may be found by examining the mortality risks of human childbirth.

At birth, the human infant is enormous compared with the offspring of chimps and gorillas; 7 lb (3.18 kg) baby humans emerge from 100 lb (45.5 kg) mothers compared with 4 lb (1.8 kg) gorilla babies emerging from 200 lb (91 kg) mothers. Consequently, on a relative scale, the risk of death to the mother during childbirth is huge for humans and minute for chimps and gorillas. Because human infants are effectively born premature (see Chapter 6), they remain dependent on parental (especially maternal) care for a long period of time. For a woman who already has children, every extra child, while increasing her reproductive success, involves a gamble with the risk that she may not survive to look after the children. Now, the risks of death in childbirth increase with age, and there comes a point at which the extra unit of reproductive success of another child is exactly balanced by the extra risk to her existing reproductive achievement. Beyond this point (at which an economist might say the marginal benefit equals the marginal cost), it is not worth proceeding. To protect the mother's prior investment in her children, natural selection probably instigated the menopausal shutdown in human fertility. Since childbirth carries no risk to the father, and fathers can always increase their reproductive success with other partners, men did not evolve the menopause (Diamond, 1991). Given the difference in parental investment in the early years of a child's life, we can see that children born to old men but young women have a better chance of survival than those born to old women.

If reproductive shutdown at the menopause occurs to protect a mother's investment, it could be supposed that females should live just long enough after the menopause to protect and raise their children to independence. In fact, women tend to live longer than is strictly necessary. Among other mammals, a mother will spend only about 10 per cent of her life after her last birth, whereas human females can live nearly a third of their lives after their last child. This has led to much speculation about the evolutionary function of grandmothers. But grandmothers aside, it is a reasonable hypothesis to suppose that women who bear children late in life and survive childbirth subsequently live longer. Voland and Engel (1989) claim to have found support for this hypothesis. Modern medical care will probably iron out any effect in modern cultures, so they examined the records of 811 women born between 1700 and 1750 in a rural district of Germany and found that life expectancy did increase with the age of the last child. In the same study, these authors confirmed that childbirth on the whole decreases life expectancy since married childless women tended to live longer than married mothers. However, if women became mothers, their life expectancy was increased significantly by a late age of the last child (Figure 3.15).

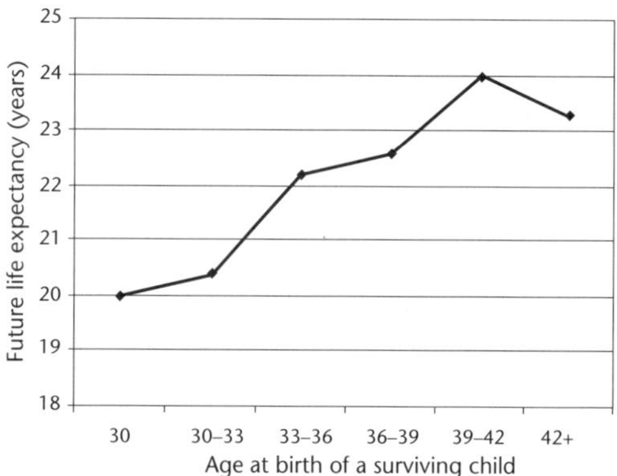

FIGURE 3.15 Future life expectancy of 47-year-old women by their age at the birth of the last surviving child

Pearson's correlation coefficient = 0.06980; P = 0.02787

SOURCE: Data taken from Voland, E. and Engel, C. (1989) Women's reproduction and longevity in a premodern population, in E. Rasa, C. Vogel and E. Voland (eds) *The Sociobiology of Sexual and Reproductive Strategies*. London, Chapman & Hall.

But we still have to explain the fact that although menopause begins around 50, women live another 20 years or so – longer than seems necessary to make sure her last offspring has survived to reach sexual maturity. The conjecture to explain this continued longevity is called the 'grandmother hypothesis' and it suggests that women can still achieve fitness gains late in life by diverting investment to grandchildren. It is a plausible idea but one that is difficult to test. There are other arguments, however, that suggest that post-reproductive women are advancing their genetic fitness. As Hill and Hurtado (1997) point out, if elderly women had no reproductive value, then natural selection would be powerless to resist the accumulation of recessive alleles that would lead to senescence and death. In which case, we would expect women to die sooner than men since men are reproductively active after 50. This is patently not the case, and the fact that women live slightly longer than men suggests that fitness gains are still being gained by elderly women. Since this is not through direct offspring (they are beyond child-bearing) they must be having a positive effect on kin.

Blurton Jones et al. (1999) have also suggested that if the grandmother hypothesis is correct, then women should have shorter inter-birth intervals than predicted from life history theory applied to individual women. The reasoning here is that if grandmothers are helping their family, then a mother can raise more offspring than simply predicted by considering her own time constraints and access to resources. Some support for this comes from Sear et al. (2000) who conducted a study on nearly 2,000 children in Gambia. They found that children with living grandmothers had better nutritional status and higher survival chances (the study used longitudinal data spanning 25 years) than children without. Interestingly, the effect was confined to maternal grandmothers and not paternal grandmothers. This is consistent with the notion of discriminatory grandparental solicitude in relation to paternity certainty (see Chapter 9).

A variant on this approach is that offered by Peccei (1995) who suggested that women in the Pleistocene Epoch probably did not live long enough to have an effect on the welfare of their grandchildren. Instead, Peccei argues for a 'mothering effect' such that natural selection favoured the female menopause to enable mothers to invest effort into the raising of offspring to young adulthood. But a review of the archaeological and ethnographic data by Lancaster and King (1995) suggests that, in traditional societies, a significant number of women do (and probably did in prehistorical times) survive to see and help their grandchildren.

Summary

- Sexual reproduction carries costs and benefits for individual organisms. The formidable cost of sex is probably offset by the genetic variation conferred on offspring: genes for sex may find themselves in new winning combinations. Such variation is invaluable in enabling organisms to compete with others.
- At a superficial level, the mating behaviour of animals can be described in terms of species-characteristic mating systems. A deeper understanding is gained, however, by looking at the strategies pursued by individuals as they strive to maximise their reproductive success.
- Even where there is to be found considerable variance in the reproductive success between males and females, the sex ratio remains remarkably close to 1:1. The best ultimate explanation of this so far is that of Fisher, who suggested that natural selection gives rise to stabilising feedback pressures tending to maintain unity.
- Sexual selection results when individuals compete for mates. Competition within one sex is termed intrasexual selection, and typically gives rise to selection pressures that favour large size, specialised fighting equipment and endurance in struggles.
- Individuals of one sex also compete with each other to satisfy the requirements laid down by the other sex. An individual may require, for example, some demonstration or signal of genetic fitness or the ability to gather and provide resources. Selective pressure resulting from the choosiness of one sex for the other is studied under the heading of intersexual selection. Such pressure often gives rise to elaborate courtship displays or conspicuous features that may indic-

ate resistance to parasites, or may possibly be the result of a positive feedback runaway process.

- The precise form that mating competition takes (such as which sex competes for the other) is related to the relative investments made by each sex and the ratio of fertile males to females. If females, for example, by virtue of their heavy investments in offspring or scarcity, act as reproductive bottlenecks for males, males will compete with males for access to females, and females can be expected to be discriminatory in their choice of mate.
- In cases where a female engages in multiple matings and thus carries the sperm of more than one male in her reproductive tract, competition between sperm from different males may occur. The theory of sperm competition is successful in explaining various aspects of animal sexuality, such as the high number of sperm produced by a male, the frequency of copulation and the existence of copulatory plugs and infertile sperm.
- It is probable that many features of human physiognomy and physique have been sexually selected. In examining females, males can be expected to look for features that indicate youth and fertility (nubility), health and resistance to parasites. Females can be expected to look for strength, wealth, health and status as well as parasite resistance in prospective male partners. Symmetry is an attribute valued by both sexes and may correlate with physiological fitness.
- Life history theory (LHT) is concerned with how organisms make decisions about the allocation of energy into different essential functions over the lifespan. This time-oriented perspective is needed to help make sense of behaviour that can only be appreciated as adaptive over a longer perspective.

Key Words

- Androgens
- Anisogamy
- Asexual reproduction
- Demographic transition
- Extrapair copulation
- Good genes
- Handicap
- Honest signal
- Intersexual selection
- Intrasexual selection
- Isogamy
- K selected
- Life history theory (LHT)
- Menarche
- Menopause
- Monogamy
- Mutation
- Operational sex ratio
- Parasite
- Parental investment
- Parthenogenesis
- Polyandry
- Polygamy
- Polygynandry
- Polygyny
- r selected
- Sex ratio
- Sexual dimorphism
- Sexual selection
- Sperm competition

Further reading

Alcock, J. (2005) *Animal Behaviour: An Evolutionary Approach* (8th edn). Sunderland, MA, Sinauer Associates.

A good general book on evolution and animal behaviour. Contains only one short chapter on humans, but see Chapters 10 and 11 for mating theories.

Andersson, M. (1994) *Sexual Selection.* Princeton, NJ, Princeton University Press.

Extremely thorough book that reviews a wide range of research findings. Tends to concentrate on non-human animals.

Geary, D. C. (1998) *Male, Female: The Evolution of Human Sex Differences.* Washington DC, American Psychological Association.

Geary explains the principles of sexual selection and how these can be used to understand differences between males and females. Good discussion of the evidence for real cognitive differences between males and females.

Gould, J. L. and Gould, C. G. (1989) *Sexual Selection.* New York, Scientific American.

Readable, well structured and well illustrated. Its main drawback is a lack of references in the text to support the evidence. Mostly covers non-human animals.

Rasa, A. E., Vogel, C. and Voland, E. (1989) *The Sociobiology of Sexual and Reproductive Strategies.* London, Chapman & Hall.

A useful series of case studies on humans and other animals.

Ridley, M. (1993) *The Red Queen.* London, Viking.

An enjoyable and well-written account of sexual selection theory and its application to humans.

Short, R. V. and Balaban, E. (1994) *The Differences Between the Sexes.* Cambridge, Cambridge University Press.

A valuable series of specialist chapters by experts. Covers humans and non-humans.

Foundations of Darwinian Psychology

In the distant future I see open fields for far more important researches. Psychology will be based on a new foundation, that of the necessary acquirement of each mental power and capacity by gradation. Light will be thrown on the origin of man and his history.
(Darwin, 1859b)

Although Darwin's new foundation for psychology was a long time in the making, we noted in Chapter 1 how the eventual resurgence of evolutionary thinking in psychology was linked to the cognitive revolution in psychology, coupled with theoretical developments in biology that re-energised the study of animal behaviour to give rise to behavioural ecology and sociobiology. Within the broad church of Darwinian approaches to psychology, however, a number of variants exist, sitting, sometimes uncomfortably, side by side. In this chapter, we will examine the plurality of Darwinian perspectives on human nature and the theoretical and methodological problems the discipline has to face.

4.1 Testing for adaptive significance

The whole of evolutionary psychology (EP) is based on the premise that behaviour is driven by adaptations. To say that a feature or behavioural trait is adaptive is to say that it promotes or once promoted reproductive success. To demonstrate this effectively, we need to show how a feature in question confers some reproductive advantage. This is by no means easy. Giraffes with long necks may have an advantage over rivals with shorter necks in terms of grazing from tall trees, but they also probably had a better view of approaching predators. They may also have the edge in aggressive disputes with rivals of the same sex. It is all too easy to jump to conclusions: a given trait may be advantageous in a number of different ways in a given species or even in different ways in different species. Rabbits may have large ears to detect predators, but the large ears of an African elephant probably have more to do with heat regulation than sound detection.

It is also easy to find adaptations that are not there. Consider for a moment balding in human males; what adaptive function could it have? You may suggest that it helps exposure to sunlight and the synthesis of vitamin D. It could show that the male has high levels of testosterone and is thus virile. It could be an adaptive response to the need to lose heat on the African savannah plains (this, after all, is probably why humans lost their body hair). Bald men especially will be good at devising flattering and functional explanations. Most of them are probably false and amount to what Gould, after Kipling, has called 'Just so stories'. This particular trap is examined below.

4.1.1 Pitfalls of the adaptationist paradigm: 'Just So stories' and Panglossianism

In his *Just So Stories*, Rudyard Kipling (1967) gave an amusing account of how animals came to be as they are. The basic structure of the stories is that when the world was new, animals looked very different from today's types. Something then happened to these ancestral species that left them in the form we see now. The elephant, for example, once had a short nose, but after a tussle with a crocodile, its nose was pulled and stretched into a trunk (Figure 4.1). In evolutionary biology, the 'Just so story' has become a metaphor for an evolutionary account that is easily constructed to explain the evidence but makes few predictions that are open to testing.

FIGURE 4.1 How the elephant acquired a long trunk
There was a time when elephant trunks were short but an ancestral crocodile stretched the trunk of the first elephant.
SOURCE: Kipling, R. (1967) *Just So Stories*. London, Macmillan – now Palgrave Macmillan.

A similar trap is what Gould and Lewontin (1979) have referred to as 'Panglossianism'. In Voltaire's book *Candide*, Dr Pangloss is the eternal optimist who finds this world to be the best of all possible worlds, with everything existing or happening for a purpose – our noses, for example, being made to carry spectacles. In evolutionary thinking, Panglossianism is the attempt to find an adaptive reason for every facet of an animal's morphology, physiology and behaviour. Panglossian explanations are fascinating exercises in the use of the creative imagination. Consider why blood is red. It could help to make wounds visible, it could indicate the difference between fresh and stale meat and so on. Yet blood is red simply as a consequence of its constituent molecules, for example haemoglobin, and has probably never been exposed to any selective force. The evolutionist must be prepared to accept that just as some genetically based traits may no longer be adaptive, so some adaptive features may not be directly genetically based – although learning mechanisms will themselves have a genetic basis (Box 4.1).

Box 4.1 Behaviour that is genetic but not adaptive, or adaptive but not genetic

We must be wary of interpreting the basis of all behaviour as genetic adaptation. There may be non-adaptive or non-genetic explanations for the phenomenon we are investigating. Here are some such alternative explanations.

Genetic but not adaptive

Phylogenetic inertia

Organisms may show signs of an ancestry they are unable completely to escape from even though the features in question are no longer adaptive. It is not optimal, for example, for a hedgehog to curl into a ball as a defence against oncoming traffic. The human skeletal frame is not an optimum form for vertical posture, as anyone with a bad back will confirm. When a moth circles a candle flame, sometimes ending its revolutions by incinerating itself, it is obeying a genetic rule to non-adaptive ends. The rule is one that helps it to navigate by moonlight (or the sun). Moth genes haven't kept up with artificial lights.

Genetic drift

Some genetic polymorphisms may exist in a population as a result of chance mutations that are neither advantageous nor disadvantageous or have not yet had time to be weeded out by natural selection. One special case of **genetic drift** is the **founder effect**. If a new population is formed from a few individuals, then alleles may be fixed in the population that were once only a partial sample of a larger population. The new populations may look different but not for adaptive reasons but simply the effect of the founders being a limited sample from a larger and more diverse gene pool. The fact that the blood group B is virtually absent from North American Indians is probably a result of genetic drift rather than adaptive change.

Adaptive but not genetic

Phenotype plasticity

The phenotype of an organism can often be moulded by external influences during ontogeny to suit the prevailing environmental conditions – phenotypic plasticity. Bone, for example, grows in such a way to resist adequately the pressures applied. The growth of corals and trees is well adapted to the direction of water and air currents. We could say of course that the mechanism to so adapt is genetic and so heritable but the adaptation itself is not.

Learning

Humans in particular have a great capacity to learn from each other, from experience and from their culture. If humans in widely dispersed and different cultures show similar patterns of behaviour that appear well adapted, it may because of similar shared genes but it could also be that they have come to the same conclusions as to how to behave by parallel social learning.

It was Williams who, in 1966, helped to clarify what is meant by an adaptation. An **adaptation** is a characteristic that has arisen through and been shaped by natural and/or sexual selection. It regularly develops in members of the same species because it helped to solve problems of survival and reproduction in the evolutionary ancestry of the organism. Consequently, it can be expected to have a genetic basis ensuring that the adaptation is passed through the generations. Williams (1966) suggested that three criteria in particular should be employed to ascertain whether the feature in question is truly an adaptation – reliability, economy and efficiency. Reliability is satisfied if the feature regularly develops in all members of the species subject to normal environmental conditions. Economy is satisfied if the mechanism of a characteristic solves an adaptive problem without a huge cost to the future success of the organism. Finally, the characteristic must also be a good solution to an adaptive problem; it must perform its function well. If these three criteria are satisfied, it looks increasingly unlikely that the feature could have arisen by chance alone.

In searching for adaptations, we must be wary of the pitfalls of Panglossianism and try to avoid them by making precise predictions about how a feature or behavioural pattern under investigation confers a competitive advantage. Some of the specific procedures that can be used to test hypotheses are considered in the next section.

4.1.2 The testing of hypotheses

There is no such thing as a single scientific method. Different disciplines have different ways of gathering evidence, performing experiments, constructing models and testing hypotheses. One crucial feature common to all sciences, however, is the rigorous interplay between theory and experience. One of the most successful methods that structure this interplay is the so-called 'hypothetico-deductive method'. It was the philosopher Karl Popper (1959) who particularly noted the importance of this technique. The essential idea is that a **hypothesis** is framed to account for a particular phenomenon. The consequences of the hypothesis being correct are deduced and turned into predictions. These predictions are tested by experiment or by the analysis of other evidence and, if they are found not to hold, the original hypothesis from which the predictions were deduced is rejected or at least considerably modified. If a hypothesis successfully predicts an outcome, we can cautiously say that the hypothesis is supported (Figure 4.2).

A useful classification of different methods of testing hypotheses is suggested by Buss (1999), who distinguishes between 'theory-down' and 'observation-driven' approaches. The theory-down approach can be used to derive specific hypotheses from higher level theories. The theory of sperm competition, for example, is one such high-level theory that can be used to derive subsidiary hypotheses. The theory suggests that aspects of the physiology and mating behaviour of males can be understood by the fact that, in some species, sperm from more than one male are likely to be present at the same time in the reproductive tract of a female. From this, we could derive the hypothesis that in conditions where the risk of sperm competition is high, males will tend to produce and/or ejaculate more sperm. This can then be tested either within a species in variable conditions or between species with different mating habits (see Chapters 3 and 11).

The observation-driven strategy is a sort of bottom-up approach and is a useful one, partly because we may notice many patterns before we have a scientific explanation for them. It is crucial to realise that it is not simply forcing the ideas to fit the facts. In the physical sciences, a similar technique is sometimes known as 'retroduction'. As an example, astronomers knew of Kepler's laws of the motion of planets around the sun before a theory could explain them. Newton and others partly used the method of retroduction to decide what a higher level explanation would have to look like in order to generate the known laws. Newton's answer was the inverse square formula for gravitational attraction. In evolutionary theory, the method is sometimes called **reverse engineering**: the features of an organism can be used to infer backwards to the function for which it was designed. Box 4.2 gives an example of this type of reasoning.

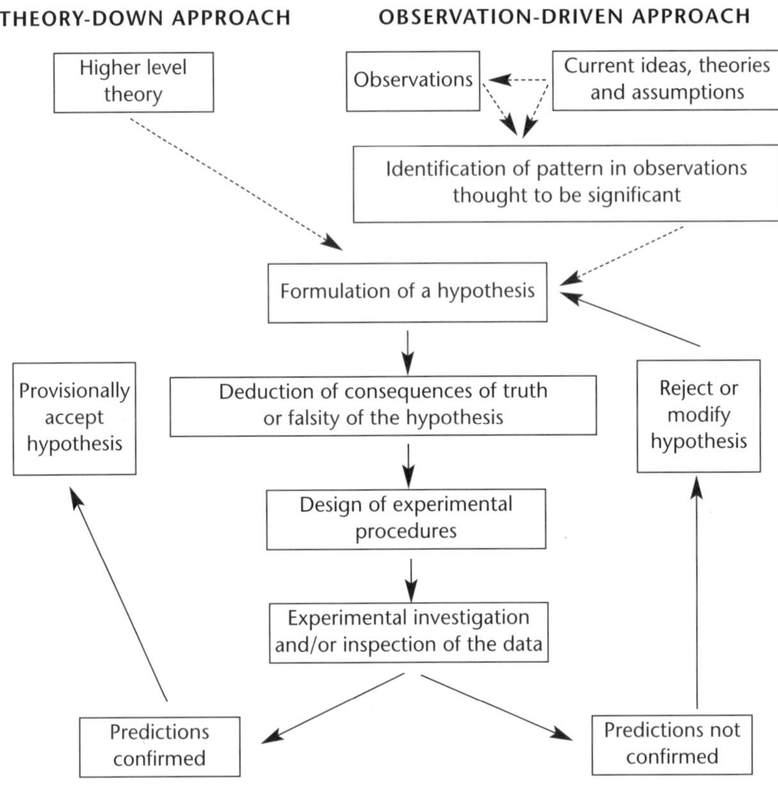

FIGURE 4.2 The hypothetico-deductive method applied to evolutionary hypotheses: an idealised view

Box 4.2 Case study: pregnancy sickness

Explaining the functional basis of pregnancy sickness provides an instructive illustration of the testing of an adaptationist hypothesis. Years of observations on the health of women in all cultures reveals a common pattern: during the first trimester of pregnancy (0–3 months) many women experience the symptoms of 'pregnancy sickness': vomiting, nausea and food aversions.

As we shall see in Chapters 14 and 15, we should not be surprised if many illnesses and disorders turn out to be normal reactions to some life event that may even have some survival value. With this perspective in mind, Profet (1992) hypothesised that pregnancy sickness was an adaptation to reduce the intake of toxin-containing foods. There are several facts that are consistent with this hypothesis. Pregnancy sickness occurs during the first trimester of development when the major organs of the fetus are under development. Moreover, it is known that the effects of drugs (for example thalidomide) and stress from disease (for example rubella) are at their most severe early in pregnancy. The hypothesis is consistent with established data but hypotheses tend to command more respect when they predict and are supported by novel findings (otherwise there remains the suspicion that they were just constructed to fit known facts in a 'just so' fashion – a process known as retroduction). One prediction that would follow if the hypothesis were to be correct is the expectation that women who experience pregnancy sickness (and so avoid foodstuffs containing potential toxins) should have a lower rate of spontaneous abortion (since the fetus is protected) than those who do not. In a meta-review of the literature covering more than 20,000 pregnancies, Flaxman and Sherman (2000) found that the spontaneous abortion and fetal death rate of women who did not experience pregnancy sickness was indeed significantly higher that those that did. Other findings supporting the hypothesis are that women tend to acquire aversions to food known to be potentially high in toxins, such as meat and poultry, alcohol and some vegetables, but rarely to items low in toxins such as cereals and bread (Tierson et al., 1985).

If we find that a prediction made by a hypothesis is found to be the case, we must not mistake this for proof. Popper and others pointed out that to suppose that the hypothesis is proved by a successful prediction is to commit the fallacy of affirming the consequent. The fallacy arises because it is conceivable that a false hypothesis could give rise to a successful prediction. Where we are more certain is in the conclusion that if the prediction is not observed, there must be something wrong with the hypothesis. This distinction was Popper's most valuable contribution to the problem of establishing criteria for what counts as science and what is non-science – the so-called 'demarcation problem'. The essence of science, according to Popper (1959), is that it formulates hypotheses that are in principle falsifiable. Non-sciences (astrology may be an example here) make do with vague general statements that conveniently avoid an open confrontation with facts.

Some critics of modern evolutionary theory have suggested that evolutionary hypotheses are non-scientific since they are ad hoc and, being specific only to the trait in question (like the *Just So Stories*), lack generality to allow testing elsewhere. It has also been suggested that evolutionary theory is not truly predictive (as good science should be) since it refers to past events rather than future occurrences. These points, as a general critique of evolutionary reasoning, are in fact easily dismissed. Evolutionary hypotheses are often of necessity post hoc in that they refer to evolutionary processes that operated long ago when we were not there to observe them, but they are not inevitably ad hoc. In addition, the prediction of future events is not a necessary condition for a discipline to be scientific; if it were, we would be forced to re-evaluate the status of geology, palaeontology and other sciences dealing with the past. It is of course important that scientific theories predict unknown findings, but these could be in the past, present or future. Evolutionary theories, can, for example, make predictions about fossils of intermediary species yet to be discovered. In essence, evolutionary theory is in principle falsifiable, although it has not yet been falsified.

4.1.3 Adaptations and fitness: then and now

The terms 'adaptive' and 'fitness' are troublesome ones. An adaptive character should obviously palpably assist the survival and reproductive chances of its bearer, but this criterion is not always easy to apply. 'Fitness' is a term with its own difficulties. As Badcock (1991) points out, human males would probably be fitter in the sense of living longer and enjoying a reduced susceptibility to disease if they were castrated or had evolved lower levels of testosterone, but this may not be the best way to produce offspring. Work by Westendorp and Kirkwood (1998) has also shown that life expectancy may be increased for couples who do not have children, but again this is not a good **strategy** for increasing Darwinian fitness.

Darwinism is ultimately about the differential survival of genes rather than about the fitness of the gene carriers.

We must also not be tempted to hybridise Darwin with Pangloss and expect every adaptation to be perfect for the task in hand. In some cases, the environment can change more rapidly than natural selection can keep up with, and an adaptation is left high and dry, looking imperfectly designed. Some features may also be caught in an adaptive trough such that a large change would take the organism to a higher (better adapted) peak but small changes would decrease its reproductive fitness. In these cases, the organism will be stuck with a less than perfect adaptive feature.

As we saw in Chapter 3, an adaptation must always represent a trade-off between different survival and reproductive needs. A big body may be helpful in fighting off predators, but big bodies need lots of fuel and time to grow. It is also important to consider how behaviour leads to fitness gains over the whole lifespan of the animal. An animal must devote resources to growth, repair and reproduction. Over a single year, behaviour may not seem to be optimal, whereas over a lifespan a different picture may emerge. This selective allocation of resources is often known as a 'life history strategy' and was considered in Chapter 3. In addition, evolution must start from an existing organism and not a blank sheet. The direction that evolution can take and the solutions it comes up with are therefore also limited by phylogenetic inertia (see Box 4.1 above).

Reverse engineering and adaptive thinking

In EP two crucial forms of thinking are adaptive thinking and reverse engineering, both sometimes lumped together under 'adaptationism' (Griffiths, 2001). Put simply, adaptive thinking claims to infer the solution from the problem, while reverse engineering claims to be able to suggest the problem by examining the solution. As an example of the former, our male ancestors must have faced the problem of paternity certainty and so, by inference, we probably have mental circuitry to incline us to guard our mates and react jealously to situations where our reproductive interests are threatened by a rival. We can then proceed to study the fine detail of this mechanism and how it is activated in a variety of contexts (see Chapters 8 and 10). As an example of reverse engineering, we have lost our bodily hair and so it is a plausible hypothesis to suggest that the environment in which this happened was open grassland savannah where humans hunted by day and needed to keep cool as we expended energy (hence also our sweating mechanism).

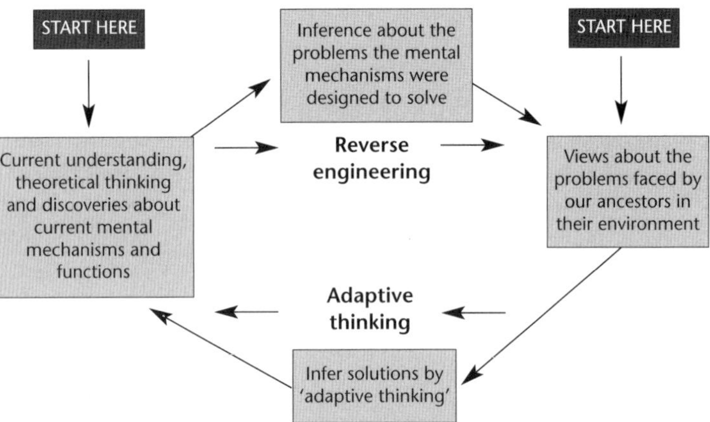

FIGURE 4.3 Reverse engineering and adaptive thinking
Reverse engineering starts from what we know about the human mind and human behaviour and infers backwards what adaptive problems had to be solved. Adaptive thinking starts from a knowledge of what problems our ancestors faced and predicts likely mechanisms that solved them. The trick is not to go round in circles.

But reverse engineering does have its pitfalls. Engineers normally start with a problem (a gorge to be spanned for example) and design an artefact or system to solve it (in this case a bridge). The problem with biology is that we are left with an abundance of solutions (the living world) but with variable information about the problems to which the physiology and behaviour of contemporary organisms are responses. In the case of cognitive phenomena, the system to be explained or reverse engineered might call forth a whole bundle of hypotheses about the nature of the ancestral environment, many of which are difficult to refute because of the paucity of information. One obvious example concerns the growth in the size of the human brain over the past two million years (encephalisation; see Chapter 6). Just what was the problem(s) to which the enormous growth of the hominin brain was a response? In answer to this, there are a variety of competing hypotheses that have proved infuriatingly difficult to judge between.

A danger inherent in adaptive thinking is the temptation in EP to believe that if a particular trait makes adaptive sense, then it must exist. To be sure, adaptive thinking can serve a positive heuristic: it suggests likely areas of research since a hypothesis that makes adaptive sense is worth pursuing. But, as Griffiths (2001) suggests, adaptive hypotheses can have a negative effect if they tend to lead to a neglect of empirical evidence that does not quite fit.

Another problem is that of historicity or contingency. A weakness of naive adaptationism is the belief that natural selection acts like some all-seeing eye that can scan all possible solutions and select the best one. But the hands of natural selection are tied by various contingencies and unpredictable events. The lineage of any organism moves through an adaptive landscape accumulating baggage along the way. As it encounters new problems, its range of solutions are constrained and the outcome may not seem that optimal long after the event. On the other hand, the brain is such a complex organ that it is often easier to make reasonable inferences about the past and the adaptive problems our ancestors faced than to look directly into the brain to establish a functional wiring diagram. Adaptive thinking provides a much needed guidebook in a complex territory.

4.1.4 Evolutionary psychology or Darwinian anthropology?

Suppose we identify some physical or behavioural trait of humans, such as hairlessness (compared at least with other primates) or the specific mating preferences of either sex, and attempt to demonstrate that they have some **adaptive significance** and have been the subject of natural or sexual selection. Among other problems, one serious question to address is 'adaptation to what – current conditions, or conditions in the past?' The problem is the same for non-human animals: a trait that we study now may have been shaped for some adaptive purpose long ago. The environment may have changed so that the adaptive significance of the trait under study is now not at all obvious. Indeed, it may now even appear maladaptive. When human babies are born, they have a strong clutching instinct and will grab fingers and other objects with remarkable strength. This may be a leftover from when grabbing a mother's fur helped to reduce the number of accidents from falling. It is not, however, clear that it helps the newborn in contemporary culture.

The problem is especially acute for humans since, over the past 10,000 years, we have radically transformed the environment in which we live. We now encounter daily conditions and problems that were simply absent during the period when the human genome was laid down. It could be expected that we have adaptations for running, throwing things, weighing up rivals and making babies but not specifically for reading, writing, playing tennis or coping with jet lag. One crucial question is whether the human mind was only designed to cope with specific problems found in the environment of our evolutionary adaptedness (the EEA – a phrase coined by John Bowlby) before the invention of culture (roughly 2,000–40,000 years before the present), or whether those selective conditions gave us minds now flexible enough to give rise to behaviours that still maximise reproductive fitness in current environments. The problem is so serious that it has led to two basic schools of thought in the application of evolutionary theory to human behaviour: the evolutionary psychologists, who argue for the former model of the mind, and the Darwinian anthropologists, sometimes called 'human sociobiologists' or 'biological anthropologists', who argue for the latter. Within the former group, the dominance of the American approach, with its emphasis on domain-specific mental modules (see Chapter 7), initially pioneered by Tooby and Cosmides led to the appellation of this group as 'the Santa Barbara church of psychology' (Laland and Brown, 2002).

Evolutionary psychologists would argue that human behaviour as we observe it today is a product of contemporary environmental influences acting upon ancestrally designed mental hardware. The behaviour that results may not be adaptive in contemporary contexts. We should focus then on elucidating mental mechanisms rather than measuring reproductive behaviour. We should expect to find mind mechanisms that were shaped by the selection pressures acting on our distant ancestors. An analogy is often drawn with the human stomach. We cannot digest everything we put in our mouths; the human stomach is not an all-purpose digester. Similarly, the mind is not a blank slate designed to solve general mental problems because there were no general mental problems in the Pleistocene Epoch, only specific ones concerning hunting, mating, travelling and so on.

A mundane example concerns our food preferences. As humans, we are strongly attracted to salty and fatty foods high in calories and sugars. Our taste buds were probably a fine piece of engineering for the Old Stone Age when such foods were in short supply and when to receive a lot of pleasure from their taste was a useful way to motivate us to search out more. Such tastes are now far from adaptive in an environment in developed countries where fast food, high in salt, fat and processed carbohydrates, can be bought cheaply, with deleterious health consequences such as arteriosclerosis and tooth decay.

At this point, it is worth stressing the difference between behavioural adaptations and cognitive adaptations. A pattern of behaviour may be adaptive without any cognitive component. The greylag goose noted in Chapter 1 that rolls its egg back to the nest is a case in point. The behaviour is highly adaptive – millions of eggs must have been saved by this device – but the fact that the behaviour continues when the egg is removed from beneath the bird's feet shows that is it rather simply constructed, inflexible and not 'thought about'. In contrast, as Tomasello and Call (1997) argue, a cognitive adaptation:

- involves decision-making among a variety of possible courses of action
- is directed towards goals or outcomes
- probably involves some sort of mental representation that goes beyond the information immediately presented to the senses.

Table 4.1 Contrast between the methods and assumptions of Darwinian anthropology and evolutionary psychology

DARWINIAN ANTHROPOLOGY **Human sociobiology, Darwinian social science, human behavioural ecology, human ethology**	EVOLUTIONARY PSYCHOLOGY
Behaviourist approaches	*Cognitivist approaches*
Culture should be viewed as part of a fitness maximisation programme. Humans are flexible opportunists and so **optimality** models (for example foraging and birth intervals) can be used. Game theory can help in investigating decision-making	Fitness maximisation of current behaviour is not a reliable guide to the human mind since current environments differ enormously from ancestral ones. Design (by natural selection) is manifested at the psychological not the behavioural level
Concentration on behavioural outcomes, not beliefs, values, emotions and so on	Studies should look for mental mechanisms that evolved to solve problems of the Pleistocene Epoch, the EEA
Measure lifetime reproductive success of individuals in relation to their environments. Count current babies. Methods typical of those of behavioural ecologists	Need to focus on conditions and selective pressure of ancestral and not contemporary environments
Ancestral adaptations have given rise to domain-general mechanisms	Ancestral adaptations have given rise to domain-specific modules designed to solve specific problems. The mind is like a Swiss army knife, comprising discrete tools or problem-solving algorithms. Such modules may now function in maladaptive ways
Genetic variability can still exist and is particularly influential in mate choice	The evolved mental mechanisms we now possess show little genetic variability; they point to a universal human nature

In a sense, cognitive adaptations are the result of evolutionary processes that have relinquished hard-wired solutions to directing behaviour optimally in favour of judgements made by an individual organism. In such cases, an organism (although not necessarily consciously) has goals, makes decisions and calculations according to context and its own life experiences, and hence chooses an appropriate strategy.

Daly and Wilson (1999) argue that the use of the term 'evolutionary psychology' only for humans introduces an unnecessary species divide. The divide is unwarranted by the history of the subject, since many ideas from animal behaviourists have been incorporated into human behavioural studies and, moreover, the principles that apply to the human animal must also apply to non-human animals. Consequently, they prefer the term 'human evolutionary psychology' when referring only to humans.

For the sake of convenience, we will include human sociobiology, human behavioural ecology and human ethology under the heading of 'Darwinian anthropology'. Darwinian anthropologists argue that ancestral adaptation was not so specific and that we possess 'domain-general' mechanisms that enable individuals to maximise their fitness even in the different environment of today. They suggest that different contemporary environments will give rise to different fitness maximisation strategies. The way to look for adaptations is not to try to find ancient mental mechanisms but to look at current behaviour in relation to local environmental conditions. In this sense, their approach is similar to that of the behavioural ecologists who study non-human animals. The differences between Darwinian anthropology and evolutionary psychology are shown in Table 4.1.

4.1.5 The EEA: the land of lost content

Objections to the approach of evolutionary psychology have focused on the mysterious EEA. Much clearly depends on the EEA, so what was it like, and how far do we go back? Betzig (1998) makes the point that, for the past 65 million years, our ancestors existed as some sort of primate, so do we look to the selective pressure on primates over this period as a clue to the human psyche? We could narrow it down by suggesting that we have spent the past 6 million years as one species or other of hominin (*Homo habilis, Homo erectus* and so on; see Chapter 5), so should we consider those environmental conditions? Archaic *Homo sapiens* and the subspecies *Homo sapiens sapiens* (that is, us) have been around for about 200,000 years, mostly spent in a hunter-gatherer lifestyle – or rather lifestyles, for even among contemporary hunter-gatherers, there are large differences in mating behaviour, paternal investment and diet – so perhaps we should focus on this period.

Over the past two million years, there have been a series of EEAs rather than just one. The EEA has acquired an almost fabled status, but Tooby and Cosmides (1990, p. 386) point out that there was not one single EEA. They argue that it is 'a statistical composite of the adaptation-relevant properties of the ancestral environments encountered by members of ancestral populations, weighted by their frequency and their fitness consequences'. This is a fine definition in theory, but integrating such factors across time for human evolution is extremely difficult.

If we are only allowed to speculate about adaptations to an EEA, and given that establishing the features of an EEA will be difficult enough, we are restricted in how far we can understand human nature with any precision. Moreover, it has been pointed out that, as humans, we know the probable responses of our own species to questionnaires and other measures of thinking, so there arises the temptation to select features from what we imagine to be the EEA to predict correctly outcomes that are already foreseen (see Crawford, 1993). It would be better if palaeontologists and palaeogeographers supplied the conditions of the EEA and an intelligent chimp made the predictions about human behaviour.

Betzig takes the approach that we should look at the contemporary behaviour of humans in all its cultural manifestations through Darwinian spectacles, and that the behaviour of modern humans is still governed by the iron logic of fitness maximisation. We must also be wary of treating the EEA as some sort of 'land of lost content' when the human genome was in total harmony with its surroundings. In reality, no organism is perfectly adapted to its environment. Adaptations are compromises between the different requirements of an animal's life.

Darwinian anthropologists focus on 'adaptiveness' to different environments, implying that humans will tend to maximise their reproductive output in the circumstances in which they find themselves. In contrast, evolutionary psychologists focus on 'adaptation' – discrete mechanisms and traits that an organism carries as a result of past selective pressures. An important point, however, is that behavioural ecologists have shown that non-human animals do display adaptiveness: they have a range of strategies that are evoked in different environmental conditions. It would be a simple or a foolish and very 'unDarwinian' animal that only had a single behavioural strategy to be expressed in all circumstances. With respect to humans, Baker and Bellis (1989) provide evidence that human males adjust the number of sperm in their ejaculate according to the likelihood of sperm competition (see Chapter 11). The whole question of how humans adjust their sexual strategies in relation to context is examined in Chapter 12.

It is also important to note that natural selection can shape how development and learning occur in relation to local environments. It follows that behaviour does not have to be forced into the category of a 'hard-wired mental module'. Natural selection could have shaped our minds to respond to what is fitness enhancing under prevailing conditions, and to behave accordingly. As an example of this, Bobbi Low at the University of Michigan has provided evidence that the training of children is related to the type of society in a way that tends to maximise fitness. Low (1989) found that the more polygynous the society, the more that sons were reared to be aggressive and ambitious. The logic here is that, in polygynous societies, successful males stand to secure more matings. If Low is right, the behaviour of adult males is strongly influenced by their childhood training, but such training is a response to local social and ecological conditions. It shows that humans may have a repertoire of strategies activated by environmental cues.

There are problems, however, in choosing measures of fitness. Consider how a male could maximise his fitness in the late 20th century. A scenario envisaged by Symons (1992, p. 155) is both amusing and instructive:

> In a world in which people actually wanted to maximise inclusive fitness, opportunities to make deposits in sperm banks would be immensely competitive, a subject of endless public scrutiny and debate, with the possibility of reverse embezzlement by male sperm bank officers an ever-present problem.

The answer of course is that natural selection did not provide us with a vague fitness maximisation drive. The genes made sure that fitness maximisation was an unconscious urge; like the heartbeat, it is too valuable to be placed under conscious control. Instead, males and females were provided with sexual drives that could be moderated according to local circumstances. Counting the size of the queue outside a sperm bank would be a fruitless way of assessing a male's fitness – maximising sex drive and counting partners and real sexual opportunities might be better. If sperm banks were set up to allow males to deposit sperm more naturally *(in vivo)*, I suspect that the queues would be longer.

In an important article, Crawford (1993) suggests that we should look for similarities rather than differences between ancient and current environments. It is easy to draw sharp contrasts between ancestral and modern environments by choosing features that have drastically changed, such as population densities, jet travel, computers and so on, but such a comparison is pointless unless we specify the nature of the adaptations that are supposed to be out of place as a result. Moreover, if the world is so fundamentally different, why do humans seem to be thriving? We live surrounded by space-age technology, but the fundamental patterns of life go on: couples meet and have babies, people make friends and enemies, argue and settle arguments, gossip intensely about each other and so on.

It must also be remembered that we have built modern culture around ourselves. We visit or live near to our relatives, houses are designed for the nuclear family, we work in groups with hierarchies – all features probably not far removed from those of the ancestral condition. Crawford advises that we should assume a basic similarity between ancient and contemporary environments with respect to particular adaptations, unless there are signs of stress and malfunction in humans, or the behaviour is rare in the ethnographic record, or unusual reproductive consequences are observed. Polyandry (the sharing of one wife by several men), for example, is rare in human society, and there are strong reasons for suspecting that we are not well adapted to this way of life.

In this book, the approach taken is that there is room for both methodologies (Sherman and Reeve, 1997). The methodological issues are serious, and the interested reader is referred to texts at the end of this chapter, but the human brain is complex enough and powerful enough to accommodate behaviours that are learnt or unlearnt, behaviours that are soft wired and hard wired, behaviours that adjust to local conditions and behaviours that are invariant. In subsequent chapters, we offer significant findings and successful predictions from both perspectives (for a review, see Daly and Wilson, 1999).

4.2 Orders of explanation in evolutionary thinking

One of the great strengths of evolutionary thinking is that it enables us to answer 'why' questions in a scientific and non-metaphysical fashion. Consider the questions posed in Table 4.2. There are at least three types of answer to these questions: teleological, proximate and ultimate.

Table 4.2 Types of explanation in evolutionary thinking

Question	Teleological	Proximate	Ultimate
Why does the fur of stoats *(Mustella erminea)* turn white in the winter?	To become better camouflaged	Hormonally mediated response to day length and ambient temperatures	Advantages once (and still) conferred: differential survival of genes
Why do humans sweat when hot?	To lose heat by evaporative cooling	Response of sweat glands to high temperatures	Advantages once (and still) conferred: differential survival of genes

If we answer that the reason the fur of a stoat turns white in winter is to help with camouflage, we are, strictly speaking, reasoning teleologically. Camouflage is a consequence of the fur turning white; it is an effect of the change of colour, and an effect cannot be a cause. To avoid this, we might resort to identifying prior causes that triggered the change in colour, such as a hormonal response to falling temperatures and reduced daylight. This may be a correct response physiologically but is somewhat unsatisfying; all we have done is to identify a proximate causal mechanism. We have provided the 'how' of the process but have not explained why such processes exist. In the language of Tinbergen, we have identified a causal mechanism (see Chapter 1).

The ultimate causal explanation rests in the third column of Table 4.2: genes that code for a change in coat colour exist because they conferred a survival value on the stoats that possessed them. Natural selection cannot think ahead like the teleologist and plan a set of genes to achieve some purpose. We sweat because a chance mutation in our ancestral genes conferred some fertility advantage on our predecessors; sweating is an adaptive or functional response. One of the remarkable features of Darwinism is that, for the first time in the life sciences, it provided satisfactory answers to 'why' questions. Without Darwin, nothing in life really makes sense. Consider the question 'why are we here?', which carries a miasma of spurious profundity. To a committed Darwinian, we are here because we carry genes that were successful at self-replication. Similar genes that were less successful are not here for us to observe: none of us is descended from sterile ancestors. The nature of genes is to make copies of themselves not with any grand plan or purpose in mind but simply because that is what they do. In a sense, that is non-tautological, we are here because we are here. To paraphrase Wittgenstein, a cloud of metaphysics is thereby condensed into a drop of Darwinism.

If we compare the terms 'proximate' and 'ultimate' to the four 'whys' of Tinbergen considered in Chapter 1 (causation, development, evolution and function), we can note that proximate translates to causal and ultimate to functional. The distinction is important as a means to an intellectual understanding of evolution, but, practically speaking, the four questions of Tinbergen should not be treated as isolated areas of inquiry but instead as concerns that are interdependent. Natural selection has shaped behaviour to serve its present function in ensuring the survival of the genes responsible, but we must also acknowledge that natural selection has determined the way in which the causal mechanisms that initiate the behaviour begin their work. Ontogeny (development) may also be linked to function, in that the precise course of development in an individual may be sensitive to local conditions in order to achieve the best adaptive fit to current circumstances.

Examples of this are discussed in Chapter 3, where it is shown that the mating behaviour shown by an individual is sensitive to variables such as the behaviour of others and the abundance of resources. Individuals may have several strategies for mating that are triggered by different environmental events. Another example concerns the Westermarck effect discussed in Chapter 13, in which the ontogeny of sexual desire is influenced by members of the opposite sex associated with during childhood. If Westermarck was right, we develop to experience no sexual desire for those with whom we grew up in close proximity. The adaptive significance of this, and hence of the incest taboo that proscribes mating between kin, is that sibling mating can produce congenital defects in the newborn. Thus in this case, we see an integration of function, causal mechanism and ontogeny.

One of the most neglected questions of Tinbergen in behavioural studies has often been that of evolutionary history – the very question that Lorenz thought most crucial. Recent studies on phylogeny are redressing this imbalance and promise to throw light on functional questions. The evolution of concealed ovulation in human females, for example, could help to elucidate the function it served and serves. In Chapter 5, we explore this further by examining the evolution of *Homo sapiens* from other hominid species.

Summary

- We should expect the behaviour of animals to be adaptive in the sense that it has been selected by natural and sexual selection to help confer reproductive success on individuals. There are various ways in which the adaptive significance of behaviour can be demonstrated and investigated. Some of these involve experimental manipulation of the natural state, and some involve looking for correlations between behaviour and environmental factors. In such studies, we must constantly be wary of finding convenient but spurious explanations designed post hoc to fit the facts.
- Applying evolutionary thinking to the human mind makes use of reverse engineering and adaptive thinking. Reverse engineering tries to infer what problems our ancestors faced in the light of what we know about how current minds operate. Adaptive thinking starts with knowledge of the problems faced by our ancestors and predicts what adaptations humans should possess.
- A debate exists over the correct way to apply Darwinian reasoning to human behaviour. Darwinian anthropologists, sometimes called human sociobiologists, human behavioural ecologists or human ethologists, suggest that current human behaviour measured in terms of reproductive success shows signs of adaptiveness. Evolutionary psychologists argue that the correct Darwinian approach is to look for adaptations to ancestral environments that can now be identified with discrete problem-solving modules in the brain. It is suggested here that there is room for both interpretations.
- Seeking explanations for human behaviour in terms of adaptations lies at the heart of Darwinian psychology and provides 'ultimate' style explanations for current behaviour. Such explanations are responses to two of the four 'whys' of animal behaviour set out by Tinbergen (see section 1.2.2).

Key Words

- Adaptation
- Adaptive significance
- Founder effect
- Genetic drift
- Hypothesis
- Optimality
- Reverse engineering
- Strategy

Further reading

Barkow, J. H., Cosmides, L. and Tooby, J. (1992) *The Adapted Mind*. Oxford, Oxford University Press.

See Chapters 1 and 2 for discussion of the approach of evolutionary psychology. Discusses some complex methodological issues.

Buss, D. M. (2005) *The Handbook of Evolutionary Psychology*. Hoboken, NJ, John Wiley & Sons.

Numerous chapters by leading authorities on the whole field of evolutionary psychology.

Human Evolution and its Consequences

The Evolution of *Homo Sapiens*

Thus, from the war of nature, from famine and death, the most exalted object which we are capable of conceiving, namely, the production of the higher animals, directly follows. There is a grandeur in this view of life … whilst this planet has gone cycling on according to the fixed law of gravity, from so simple a beginning, endless forms, most beautiful and most wonderful have been, and are being, evolved.
(Darwin, 1859, p. 491)

In this chapter we will be looking at the series of events that led to the evolution of our own species. By investigating how the characteristics that define *Homo sapiens* arose, and the mode of life that led to the adaptations we now possess, we should be in a better position to understand our contemporary behaviour. This whole field is, of necessity, rich in speculation and conjecture, but the broad outlines of the story are reasonably clear. We can be sure that new fossil evidence will continue to be unearthed and there are still probably a few surprises in store.

5.1 Systematics

In this chapter, our primary concern is with one species, *Homo sapiens*, where it came from and what makes it special. It will be helpful to start therefore with some basic concepts in systematics and classification.

Systematics is the study of the diversity of life and the relationship between the different categories into which organisms can be grouped. A **taxon** (plural taxa) is a category of organisms in a classification system. The scheme that is universally accepted for naming a species was introduced by the 18th-century naturalist Carl Linnaeus. This is a hierarchical scheme, in which the name of each species is a binomen (two parts) consisting of the genus plus the specific name. *Homo erectus* and *Homo neanderthalis*, for example, are two species of hominins belonging to the **genus** *Homo*. Above the level of genus lies the taxon of 'family'. We might say therefore (based on morphological similarities) that humans belong to the family Hominidae, which also contains chimps and gorillas. Table 5.1 shows how this hierarchy can be extended upwards. It follows that we start from one species and rise upwards into groupings containing ever more species types. The kingdom Animalia in Table 5.1, for example, contains about one million different species from earthworms to humans. We should not be too fazed by the Latin names in this exercise. The groupings, although we hope they mirror objective relationships, are also human inventions. To reduce the mystery of this, Table 5.1 also

Table 5.1 Traditional taxonomy of humans and a hypothetical classification of a motor vehicle illustrating Linnaean classification

	Human	A motor vehicle
Kingdom	Animalia	Machines
Phylum	Chordata	Self-propelled
Class	Mammalia	Four wheel
Order	Primates	Petrol engine
Infraorder	Anthropoidea	Domestic motor vehicle
Superfamily	Hominoidea	Volvo
Family	Hominidae	Coupé
Genus	*Homo*	S60
Species	*Homo sapiens*	S60 GT

shows how we might classify a motor vehicle. We should not take the analogy too far, although it is tempting to note that just as members of a species can interbreed to produce fertile offspring, so cars on the bottom rung of the classification in Table 5.1 can usually swap parts and still remain functional.

5.1.1 How to classify humans and their relatives

As we try to classify species, the obvious question is on what basis is classification made. When Linnaeus devised his system in the 18th century, he had no notion of evolutionary change and classified organisms on the basis of physical similarities. Darwin himself argued that since species are similar because of genealogical descent, classification should be based on phylogenies (lines of descent). Surprisingly, the debate about proper schemes of classification continues to this day. There are three major schools of thought about how classification should be done: phenetics, cladistics and evolutionary systematics. Phenetics stresses results of adaptations and hence morphological and anatomical similarities. Proponents of this school argue that it is based on real observable features (see Figure 5.1).

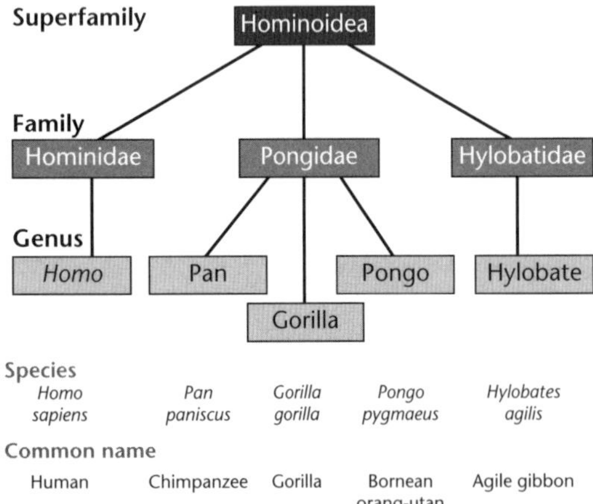

FIGURE 5.1 Traditional (phenetic) classification of apes and humans
It was thought that in view of the differences between humans and chimps, gorillas and orang-utans, only humans now belonged to the family Hominidae.

Cladistics (Greek *klados* = branch), on the other hand, only considers evidence relating to phylogenies of organisms and is concerned to group them according to how they diverged from common ancestors (see Figure 5.2). This may seem a more objective system but the challenge then becomes how to infer what the branching times and sequences were. Evolutionary systematics is a scheme that combines elements of both.

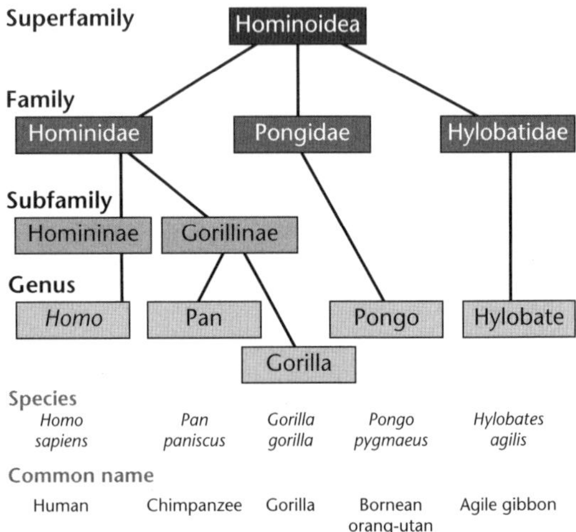

FIGURE 5.2 Classification of humans and the great apes based on phylogenetic information
In this scheme, humans, chimps and gorillas are based in the same family: Hominidae. Humans are then assigned to the subfamily Homininae, and chimps and gorillas to Gorillinae.

The cladistic approach was developed initially by the German systematist Willi Hennig in 1950. Cladistics has some advantages over classical **taxonomy**, in that by concentrating on branching points, it relies less on morphological similarities, which may be open to subjective interpretation and weighting. This leads, however, to tension between different modes of classification. Most traditional classical taxonomists, for example, would argue that, on the basis of the unique features of the human body, mind and behaviour, only humans now belong to the family Hominidae. It is now recognised from molecular studies that humans are closer to gorillas and chimps (the African apes) than either of these are to orang-utans (the Asian ape). Put another way, we last shared a common ancestor with chimps and gorillas much more recently than we did with orang-utans (see Figure 5.5 below).

On this basis then, we would place chimps, gorillas and humans in the family Hominidae; humans would then be called 'hominins'. We currently have the rather unsatisfactory situation in the literature that whereas the term 'hominoids' is taken to mean humans and the great apes (since we belong to the same superfamily), the term 'hominids' can either mean humans and their ancestors (traditional classification) or, following cladistics, the African apes and humans (phylogenetic classification).

In this book, we present the cladistic or phylogenetic system of classification as being the most valuable way of considering the relationship between humans and other species of apes. The term 'hominoids' will be taken to include humans, chimps, gorillas, orang-utans, gibbons and their ancestors to about 25 million years ago. **Hominids** include humans, chimps, gorillas and their ancestors to about 15 million years ago. The term **hominin** will be reserved for humans and their extinct ancestors after the **lineage** broke away from chimps some 7 million years ago. We should note that there is still not universal agreement on this scheme (some would place chimps and humans in the subfamily Homininae; others have even suggested that chimps and humans should be assigned to the genus *Homo*). Roger Lewin (2005) provides an accessible discussion of these issues.

As Table 5.1 shows, humans belong to the order primates and there are about 200 living species of primate left. Their weights range from about 80 gm in the case of the mouse lemur to gorillas at 150 kg.

Some important features of primates that distinguish them from other mammals are:

1. They have the ability to grasp with their hands and feet – the phenomenon of opposable thumbs; humans, for example, can touch each of their fingers with the thumb of the same hand. Human feet are an exception to this.
2. The fingers and toes of higher primates have nails and not claws.
3. Eyes are set at the front of the head giving stereoscopic vision.
4. The olfactory sense is diminished compared to other mammals and so noses tend to be smaller. It looks like a trade-off has been made to reduce the sense of smell and enhance the sense of sight. The reason may be arboreal mobility: you can't smell a branch ahead of you.

Most primates are **precocial**, meaning that the young are born relatively mature and can soon look after themselves. Humans are the exception here and are **altricial**: the young are very immature and require a long period of nurture and protection. This turns out to be a crucial feature in human evolution. This tendency probably increased as the hominins evolved.

5.2 Origins of the hominins

5.2.1 Speciation and earth history

In 1871, Darwin predicted that it was probably a species of ape living in Africa that was the common ancestor of modern humans and today's great apes. Advances in geology, climate reconstruction and genetics, coupled with the recovery of key fossils, lead us to believe that Darwin was almost certainly right. Many puzzles and problems remain but we now have enough information to describe the main outlines of our primate ancestry with some confidence. To put human evolution in an even broader time frame, Table 5.2 shows some key dates in the history of life on earth.

Table 5.2 Some key events in life history

Era	Epoch	Date, years before Christian era (BCE), or before present (BP)	Event
CENOZOIC	Holocene 10,000 BP	1859	Darwin's *Origin of Species*
		c.2,600 BCE	Great Pyramid of Cheops at Giza built
		c.3,000 BCE	Bronze begins to replace stone as a material for tools and weapons in Egypt and Western Asia
		c.4,000 BCE	Agriculture reaches Western Europe
		c.7,000 BCE	Beginning of Neolithic Age. People settle in villages and begin the domestication of plants and animals. Origins of agriculture

Table 5.2 cont'd

Era	Epoch	Date, years before Christian era (BCE), or before present (BP)	Event
CENOZOIC	Pleistocene 1.75 million BP	10,300 BP	End of last ice age
		c.30,000–12,000 BP	Successive waves of migrant *Homo sapiens* enter North America
		c.30,000 BP	Neanderthals become extinct
		c.40,000 BP	Colonisation of Australia by *Homo sapiens*
		c.70,000 BP	Beginning of last ice age
		c.74,000 BP	Large supervolcanic eruption in Toba, Sumatra, Indonesia. Six-year nuclear winter may have led to population of *Homo sapiens* reducing to 2,000 worldwide
		c.90,000 BP	*Homo sapiens* move into Asia
		c.200,000 BP	Emergence of modern *Homo sapiens* in Africa
		c.1 million BP	*Homo erectus* migrates out of Africa
		c.1.4 million BP	Use of fire by *Homo erectus*
	Pliocene 5.3 million BP	*c*.1.8 million BP	Appearance of *Homo erectus* as a species in Africa
		c.2.3 million BP	*Homo habilis* uses stone tools
		c.3.6 million BP	Bipedal hominins (possibly **Australopithecines**) leave footprints in a trail of volcanic ash in Laetoli, Tanzania
	Miocene 23.8 million BP	*c*.7 million BP	The last common ancestor of chimps and humans inhabits Africa
		c.15 million BP	Beginning of the partial isolation of East Africa from the rest of Africa along the line of the Great Rift Valley
		c.35 million BP	Ancestors of New World primates reach South America from their origin in Africa
		c.65 million BP	Likely collision event between asteroid or comet and Earth, leading to the extinction of dinosaurs and new opportunities for mammals
		c.65 million BP	Earliest primates
		c.100 million BP	Common genetic ancestor of mice and humans
		c.125 million BP	Earliest placental mammal: *Eomaia scansoria*. Probably resembled a modern dormouse
		c.300 million BP	Origin of reptiles
		c.480 million BP	Earliest jawed, bony fishes
		c.600 million BP	Earliest multicellular organisms – probably resembled sponges
		c.1,200 million BP	Beginning of sexual reproduction
		c.1,300 million BP	Earliest plants
		c.1,600 million BP	Blue-green photosynthetic algae appear
		c.3,900 million BP	Beginning of life on Earth
		c.4,500 million BP	Formation of the Earth
		c.13,500 million BP	Best current estimate for origin of Universe

5.2.2 Hominin speciation

Humans, chimpanzees and gorillas all originated in Africa and share some features in common, but all unambiguously belong to different species. The concept of a **species** has a fairly clear biological definition: organisms are said to belong to the same species if they can interbreed to produce fertile offspring. Some different species, such as donkeys and horses, are closely related and produce viable offspring, but such offspring are inevitably infertile. But the fact that horses and donkeys are so similar is because they shared a common ancestor fairly recently in geological time; more recently than, say, a horse and a wolf. On the basis of viable offspring, present-day humans all belong to the same species called *Homo sapiens* ('wise humans').

Through time, a single species can give rise to others through a process that Darwin and his followers called transmutation, which today is usually known as **speciation**. Speciation occurs when **populations** are reproductively isolated by geographical barriers such as islands or mountain ranges, isolated temporally by gradually breeding at different times, or separated in some other way. By gradual mutation and genetic drift, the two populations of what was once a single species reach a state in which gene flow between them ceases and cannot be revived by the removal of the barrier (Figure 5.3).

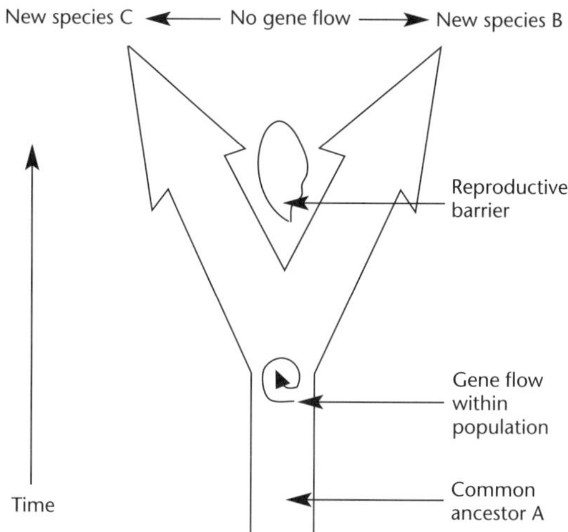

FIGURE 5.3 Diagrammatic illustration of speciation around a reproductive barrier

Where speciation is not quite complete, we often observe the existence of subspecies. In the USA, for example, there are distinct geographical populations of deer mice *(Peromyscus maniculatus)* in which gene flow between the groups is very limited. Hence, we have the subspecies *Peromyscus maniculatus borealis* in the north and *Peromyscus maniculatus sonoriensis* in the south. In time, these may drift further apart to form distinct species. Until recently, there was debate over whether Neanderthal man and early *Homo sapiens* were really distinct species or whether they did, in fact, interbreed. Neanderthal man (with the possible exception of *Homo floresiensis*; see section 5.3.2) was the last hominin species to die out. They lived in Europe as recently as 30,000 years ago and coexisted, at least in time, with *Homo sapiens*. Based on studies of DNA, the consensus now seems to be that they were distinct species and would not have interbred with modern humans.

A key event in the origin of the hominins was the formation of the Great Rift Valley. Twenty million years ago, Africa was fairly level and covered from the west coast to the east coast in tropical forests housing a wide range of species of monkeys and apes. About 15 million years ago, tectonic activity caused a profound disruption to this landscape. Volcanic activity threw up the great lava-based highlands called the Kenyan and Ethiopian domes, and eastern Africa began to separate from the West. As the plates drifted apart, so the rocks collapsed, leading to a truly massive fault line that extends from Mozambique in the south to Ethiopia in the north. In fact, this same fault line extends out of Africa, running northwards through the Red Sea, the Gulf of Aqaba, the Dead Sea, the Jordan Valley, the Sea of Galilee and the Beqaa Valley in Lebanon. The Great Rift Valley created an east–west geographical barrier to the movement of animal populations and so promoted differential speciation each side of the divide. This process of the formation of new species from isolated populations is known as **allopatric speciation**; an example of this on an even more massive scale being the separation of Old World and New World monkeys from a common stock that diverged some 50 million years ago as the continents of Africa and South America separated.

In eastern Africa, the highlands formed by volcanic activity produced a rain shadow to the east of the valley. This caused the formerly continuous forest to shrink and fragment. The new geographical conditions led to the formation of an ecological mosaic of different environments, including hot and arid lowland desert, open savannah grassland and cool and moist uplands. The rapidly eroding uplands also gave rise to sedimentation processes that were effective in forming fossil remains. The whole environment seems to have been a good place for primates to live and evolve and, from the perspective of palaeontologists, a good place for them to die and leave remains. Consequently, our knowledge of early hominins is heavily based on finds in these regions.

The rain shadow effect of the eastern highlands was complemented by a cooling and drying of the climate in the Miocene Epoch (about 25–5 million years BP). It is commonly thought that it was on this border between forest and open savannah that the speciation events occurred, forcing the separation of the lineages leading to the chimps and gorillas and hominins. As the forest cover retreated, so the apes remained in the forest regions and retained an arboreal lifestyle. At the same time, new species of hominoids emerged, adapted to wandering over longer distances in the open. One of the first of these adaptations was probably bipedalism (see below).

The cooling during the Miocene Epoch was followed by a series of periodic cooling episodes associated with ice age effects. Evidence from the measurement of oxygen isotope ratios, and the depth of layers of wind-blown dust on the floors of the oceans around Africa, points to the existence of several substantial global cooling episodes over the past 5 million years. The most severe of these occurred some time between 3.5 and 2.5 million years ago and was associated with the first major build-up of Arctic ice and an increase in the volume of Antarctic ice. The causes of this phenomenon may be related to the rise of the Panamanian Isthmus, which joined North and South America about 2.5 million years ago and so changed the direction of ocean currents. Superimposed on top of these major fluctuations were the so-called 'Milankovitch cycles' (caused by the changing orbit of the earth around the sun), with periods of 100,000, 41,000 and 23,000 years. The result was a complex oscillating pattern of cooling and warming, the extremes of which usually exceeded the habitat tolerances of most species then alive. Fossil evidence, however, shows that the average lifespan of a terrestrial mammalian species is several million years. Since this is much longer than the time it takes for the climate to oscillate, species must have made adaptive responses. Perhaps the commonest is dispersal: species move north in periods of warming and south in periods of cooling.

But major climatic events could also bring about species change. Elizabeth Vrba (1999) has proposed a habitat hypothesis that, put simply, linked climate change to speciation. She argues that the major cooling of three million years ago coincided with the emergence of many new mammalian species, especially in Africa. She also observes that many of the new species have larger body sizes compared to their predecessors. Large bodies of course run at lower metabolic rates and are more capable of surviving in low temperatures.

5.2.3 Bipedalism

Since no other primate walks or runs with an upright gait, bipedalism represents one of the major characteristics that separate us from the other primates. There is some debate about what the other characteristics should be (see Foley, 1997), but candidates proposed include hairlessness, changes in dentition, a rich material culture, language and an enlarged brain. The problem of brain enlargement (encephalisation) is considered in Chapter 6. Bipedalism occurred before the rapid enlargement of the hominin brain and there are a variety of theories to account for its origins and adaptive advantages compared to quadrupedalism or occasional knuckle walking as practised by other primates. The list below gives a flavour of some of the competing theories.

1. *Man the hunter-scavenger*

It is fairly easy to demonstrate that bipedalism is slower and less energy efficient than quadrupedalism when running at high speed (think of yourself in a running race against a dog or a horse). But, taken at a slow pace, such as when tracking an animal, following a herd, or scavenging over a wide area, it can be efficient and allow for great stamina. The problem with this theory is its reliance on the search for meat: hominin diets were mostly vegetarian until about 1.8 million years ago, long after the onset of bipedalism.

2. *Improved predator avoidance*

A bipedal hominin would be able to see further and better detect the approach of predators than quadrupeds; a useful skill in open savannah.

3. *Woman the gatherer and foraging efficiency*

One striking advantage of bipedalism is that it frees the hands of primates to gather and transport plant food – something that was probably practised predominantly by women. Peter Rodman and Henry McHenry (1980) have proposed that as the dietary resources of hominoids in the late Miocene became scarce and more dispersed, so an efficient mode of travel over large distances became advantageous. Lynne Isbell and Truman Young (1996) have more recently added a new twist to this theory. They suggested that the ancestors of chimps and humans responded to a decreasing resource density in two ways. The line that led to chimps adapted by reducing their group size, taking advantage of the fact that a smaller group needs a smaller foraging area; the branch that led to the hominins maintained their group size but evolved a more efficient gait, namely bipedalism, to travel over the large distances necessitated by resource scarcity and the provisioning of a larger group.

4. *Reduced exposure to sunlight*

Peter Wheeler (1991) suggests that standing upright would reduce the harmful exposure to sunlight. For formerly shade-loving organisms, the sun of the open savannah would have proved debilitating. Walking upright would reduce exposure by one-third of that experienced walking on all fours.

Whatever the causes of bipedalism, fossil evidence shows that hominins became progressively bipedal after the branching away from their ancestors shared with the chimps about six million years ago.

5.3 Phylogeny of the Hominoidea

5.3.1 Branching sequences and dates

Early hominins are gone, and there is now only one species left: ourselves, *Homo sapiens*. However, several species of great ape still exist today, and it is instructive to consider how closely we are related to them and when we last shared a common ancestor with each of them. To do this we need a phylogenetic tree detailing the evolution of the primates. There are two basic disciplines that can help us construct such a tree: comparative anatomy and molecular biology. Comparative anatomy examines the similarities between primates in terms of their basic body plan. Since humans more closely resemble chimpanzees than they do ring-tailed lemurs, it seems reasonable to suppose that we are more closely related to chimps than lemurs. By 'more closely related', we mean that we more recently shared a common ancestor. By such methods, humans and the four species of great ape (gorillas, chimpanzees, bonobos and orang-utans) were, by 1943, placed in the same superfamily called Hominoidea (Figure 5.1 above) where they still remain. Morphological evidence, however, tends to be qualitative, and it has proved difficult to push the method further to establish unambiguously a branching order among the hominoids. More recent evidence from molecular biology has provided a clearer and more consistent picture of genetic similarities between the species of this taxon. Since anatomical similarities between the hominoids are the product of genetic similarities, molecular biology, by going straight to the genetic level, provides a more fundamental approach that should be less open to biases in interpretation.

Genetic similarities between species can be ascertained by either examining DNA directly or looking at the structure of proteins that result from DNA expression. There is now a range of techniques that can be used to measure the degree of similarity between proteins or DNA from different species. Similarities between proteins can be estimated using antibody reactions or by direct sequencing of the component amino acids. Although it is a slow process, the base sequences on DNA can be determined and compared between species for regions of the genome that code for the same proteins. When molecular studies began in the 1960s, the exciting news was that virtually every technique agreed in general terms with the broad conclusions from comparative anatomy. In terms of the amino acid sequences in a range of blood proteins, and hence the genes responsible, we are virtually identical to chimpanzees, slightly different from gibbons and different in many amino acids from the Old World monkeys. To construct fully a phylogenetic tree based on molecular differences, however, we need to introduce two assumptions:

1. That quantitative differences between the amino acid sequences of proteins or base sequences in complementary regions of DNA represent relative differences in terms of evolutionary divergence of the species involved.
2. That differences in molecular (protein or DNA) structures can be translated into absolute times of divergence by the introduction of a molecular clock.

Box 5.1 The great apes (Hominidae): our nearest relatives among the primates

Name	Location, ecology and social organisation
Common chimp *(Pan troglodytes)* 	Found in tropical Africa dwelling in forest, woodland and open savannah. Troglodytes means 'cave-dwelling', which reflects an early European misunderstanding. Chimps are arboreal and terrestrial but spend over 50 per cent of their time in trees. Mixed diet of fruit, vegetation and some meat. Multi-male, multi-female groups, 'promiscuous' mating system but with dominant males. Most copulations are opportunistic, with little competition. Female **exogamy,** that is, females leave their native group at puberty. Male–male grooming accounts for nearly 50 per cent of all adult interactions. Tool use common, for example termite sticks to extract termites from nests to eat, stones used as hammers, and munched leaves to act as a sponge to soak up water from otherwise inaccessible places
Bonobo *(Pan paniscus)* 	Found in tropical Africa south of Zaire river in lowland forests. Diet similar to common chimps – fruit, shoots, buds, insects and some mammals. Multi-male, multi-female groups. Females sexually receptive through most of the oestrus cycle. Less overall aggression in groups compared with common chimps. Despite their name (they were once called pigmy chimps), bonobos are only fractionally smaller than common chimps
Gorilla *(Gorilla gorilla)* 	Distributed in Central and West Africa in three subspecies: the eastern lowland gorilla *(Gorilla gorilla graueri),* the western lowland gorilla *(Gorilla gorilla gorilla)* and the mountain gorilla *(Gorilla gorilla beringei).* Ground-dwelling shy vegetarians, utilising over 100 species of plant. Polygynous harems of one dominant male (the so-called 'silver back'), several females and infants. Groups are sometimes raided for females, and invading males can kill infants. Large degree of tolerance within groups. Copulation rate is low: once about every 1–2 years per female

These two assumptions seem reasonable enough. When two species diverge from a common ancestor, mutations begin to accumulate in each species. Since it would be extremely unlikely for the same mutation to occur in both species, differences in DNA sequences accumulate as time passes. Controversy in this field arises when we try to relate the degree of mutational differences to a 'mutation clock'. If we take the simplest assumption of linearity, if species A differs from B by 2 per cent and from species C by 4 per cent, then A and C shared a common ancestor twice as long ago as A and B shared one (Figure 5.4).

It then remains to translate this relative scale into an absolute scale using palaeontological evidence. If fossils show that A and B shared a common ancestor 5 million years ago, it would seem to follow that the ancestor to A, B and C lived 10 million years ago. In principle, it all looks very easy, but it is now appreciated that molecular clocks do not run so regularly, much effort is needed, and uncertainties remain in translating relative differences into absolute time.

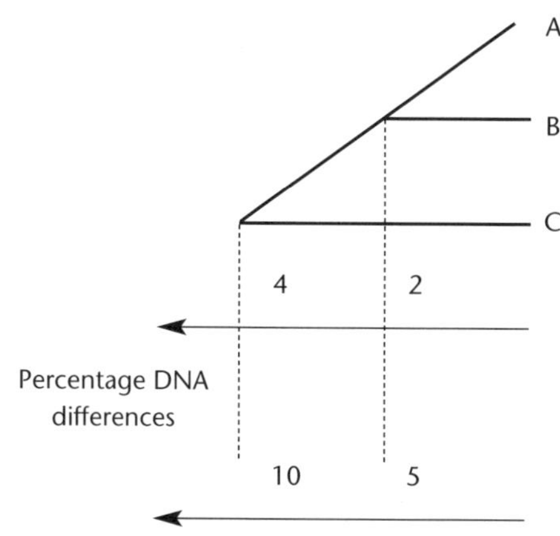

FIGURE 5.4 Hypothetical relationship between three species A, B and C

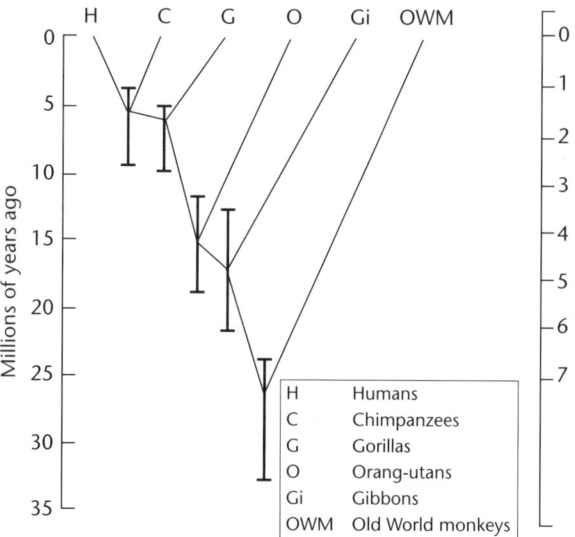

FIGURE 5.5 Evolutionary tree of the primates based on haemoglobin and mitochondrial gene sequences

The pattern is a consensus of recent studies. Bars indicate the range of estimated times for branch points.

SOURCE: Adapted from Friday, A. E. (1992) Human evolution: the evidence from DNA sequencing, in S. Jones, R. Martin and D. Pilbeam (eds) *The Cambridge Encyclopedia of Human Evolution*. Cambridge, Cambridge University Press.

Bearing in mind the promise and difficulties of molecular comparisons, we can now construct a **phylogeny** for the hominoids based on molecular clockwork calibrated using fossil evidence. There are in fact two independent fossil calibrations available for the primates: the divergence between monkeys and apes (about 7 per cent DNA difference), which occurred about 24–34 million years ago, and that between orang-utans and gorillas (with about 4 per cent difference in DNA), which occurred about 12–18 million years ago. Figure 5.5 illustrates a likely branching sequence.

The 1.6 per cent difference

The implications of Figure 5.5 are profound. The first point to note is that chimpanzees, and not gorillas, are our closest relatives. In fact, chimpanzees are more related to us than they are to gorillas. We differ in our DNA from the chimps by just 1.6 per cent, the remaining 98.4 per cent being identical. Our haemoglobin, for example, is the same in every one of the 287 amino acids units as chimpanzee haemoglobin; in terms of haemoglobin, we are chimps. We must beware of making too much of the 1.6 per cent difference. Information coded on DNA is not a linear system; that 1.6 per cent could and has effected profound changes. A small change in DNA can bring about an enormous change in morphology through a change to the timing of the development of any particular feature. One obvious difference between chimps and ourselves is that we have a much larger brain; and even four million years ago, our ancestors had larger brains than modern-day chimps. Also, since DNA is a linear sequence of four bases, any given length will be at least 25 per cent similar to that of any organism (for example a daffodil or a slug).

5.3.2 Early hominins

The first fossilised remains of humans were found in 1856 in the Neander Valley near Dusseldorf. Such specimens quickly became called Neanderthal man, and for a time there was much speculation on whether they were ancestors of *Homo sapiens*. Modern molecular evidence suggests that they were cousins rather than ancestors of modern humans. They split away from the *Homo* lineage about 500,000 years ago and were a distinct species that died out about 35,000 years ago (see Table 5.2).

It was in the 1920s that the first hominin fossils were found in Africa. In 1925, Raymond Dart described the species *Australopithecus africanus* based on a specimen called the Taung child found in a cave at Taung in South Africa. The brain of this creature was small at about 410 cc, but the position of the foramen magnum (the hole in the base of the skull through which the end of the spinal cord enters the brain) suggested bipedal locomotion. It took a further 20 years for Dart's views on the bipedal stance of *A. africanus* to be taken seriously, since the common assumption then was that a large brain preceded bipedalism. It was also widely assumed that Asia not Africa was the cradle of mankind.

An even older hominin was found by Donald Johanson and Tom Gray on 24 November 1974. The

Table 5.3 Data on early hominins

Species	Timespan Ma = million years before present	Mean cranial capacity (cc)	EQ calculated from observed volume/ 0.058 (body mass)$^{0.76}$	Distribution	Other popular names and exemplars	Body size: height (m), mass (kg)		Sexual dimorphism in body mass (males/ females)
						Males	Females	
Australopithecus afarensis	4.0–2.9 Ma	380–450	2.3	Eastern Africa (Ethiopia)	Lucy	1.51 44.6	1.05 29.3	1.52
Australopithecus africanus	3.2–2.5 Ma	457	2.7	Eastern and possibly southern Africa	Taung child, Mrs Ples	1.38 40.8	1.15 30.2	1.35
Homo habilis	2.3–1.6 Ma	552	3.3	Southern and Eastern Africa (Tanzania, Kenya)	Handyman	1.57 37	1.25 31.5	1.17
Homo ergaster	2.0–0.5 Ma	854	3.5	Eastern Africa and then into Asia	Nariokotome boy	63	52	1.21
Homo erectus	1.8 Ma–300,000	1,016	4.1	Africa, Asia, Europe	Java man, Peking man	1.8 63	1.55 52	1.20
Homo neanderthalis	150,000–30,000	1,512	5.7	Europe, Asia, Middle East	Neanderthal man	1.7 73.7	1.6 56.1	1.31
*Homo floresiensis**	?–13,000	380	2.5–4.6	Flores, eastern Indonesia	'Hobbit'	? ?	1.06 16–28.7	?
Homo heidelbergensis	500,000–250,000	1,198	4.6	Africa, Europe	Boxgrove man	1.8 62	? ?	?
Homo sapiens (mean of several human groups)	200,000	1,355	5.4	Africa, Asia, Europe, New World (in that order)	Cro-Magnon man (W. Europe)	1.64 60.9	1.54 50	1.21

SOURCE: Lewin, R. (2005) *Human Evolution: An Illustrated Introduction*. Oxford, Blackwell; Campbell, B. G., Loy, J. D. and Cruz-Uribe, K. (2006) *Humankind Emerging* (9th edn). Boston, MA, Allyn & Bacon; Foley, R. (2002) Hominid evolution, in M. Pagel (ed.) *The Encyclopaedia of Evolution*. Oxford,Oxford University Press.

*see Brown, P., Sutikna, T. et al. (2004) 'A new small-bodied hominin from the Late Pleistocene of Flores, Indonesia.' *Nature* **431**: 1055–61. For EQ data see Martin, R. D. (1981) 'Relative brain size and basal metabolic rate in terrestrial vertebrates.' *Nature* **293**: 57–60.

Beatles song 'Lucy in the sky with diamonds' was playing in their camp during the evening celebrations of the discovery and the name 'Lucy' stuck and she has since become one of the most famous of early hominin specimens. The brain size of *A. afarensis* is estimated at about 400 cc and is often said to be about the same size as that of a modern gorilla or chimp. We should note, however, that early Australopithecines were smaller than modern gorillas and that the ancestors of modern gorillas almost certainly had smaller brains three million years ago when they coexisted with *A. afarensis*. In short, brain expansion had already begun at the time of *A. afarensis*.

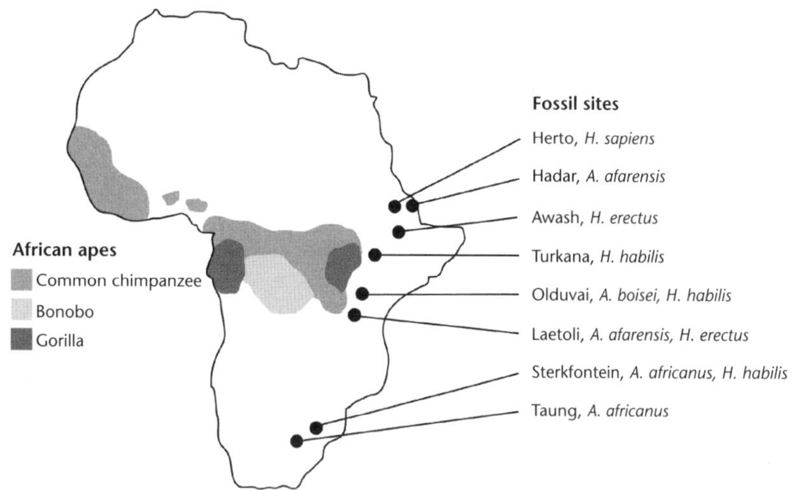

FIGURE 5.6 Current distribution of the African apes and find spots of early hominin species

Since the 1920s, numerous fossil remains have been unearthed and have enabled a whole raft of conjectures about the hominin lineage to be constructed. Figure 5.6 shows the current distribution of the African apes together with some find spots of early *Homo* species. Tables 5.3 and 5.4 provide some information on these discoveries in summary form, and Figure 5.7 shows a tentative branching sequence. The concept of EQ (a measure of intelligence) shown in Table 5.3 is considered in the next chapter. New discoveries and reinterpretations will certainly add to this picture as the years go by.

By its very nature, the naming of fossils provides ample opportunity for debate and controversy. The problem of identifying the species to which a fossil belongs is compounded by the fact that remains are often fragmentary. In addition, the nature of the evidence precludes the application of the usual criterion for a species: whether or not two organisms can breed to produce viable offspring. As a rough guide, in common with

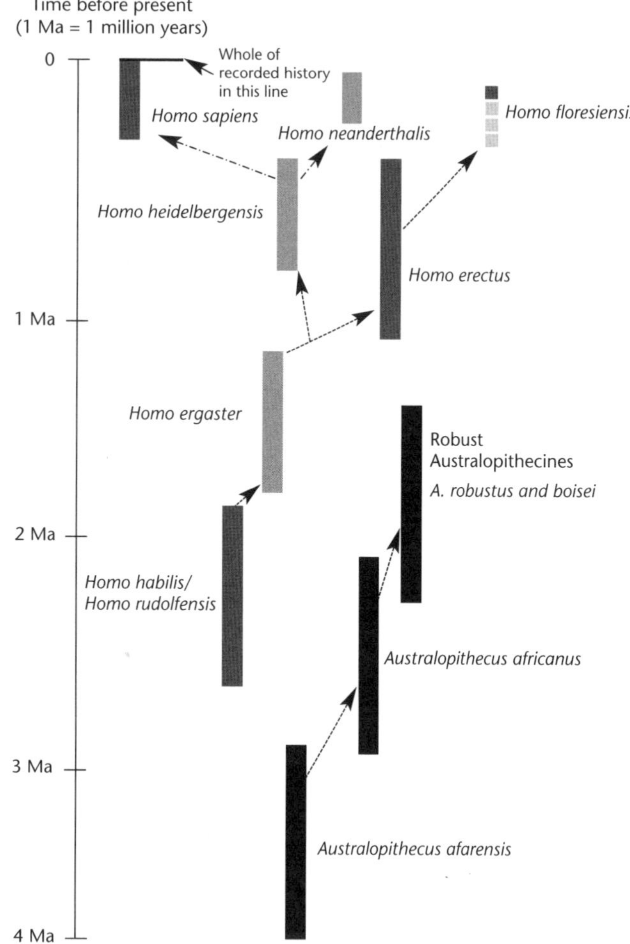

FIGURE 5.7 Time chart of some early hominins
The exact lineages (that is, which species descended from which) remain uncertain and controversial.

Table 5.4 Summary of features of selected hominins

Species	Time range	Key features
Earliest hominins		
Ardipithecus ramidus	5.5–4.4 Ma	The most primitive known hominin. Found in Ethiopia in 1994. Only fragments are known. Possibly bipedal, although it was found in an environment that would have been wet and forested
Australopithecines		
Australopithecus afarensis	3.7–2.9 Ma	Found in Ethiopia and Tanzania. Ape-like brain size but skeleton shows clear evidence of bipedalism. Lucy found at Hadar in 1974 by Donald Johanson. Best known Australopithecine. Footprints also show bipedalism. High degree of sexual dimorphism. Toes and fingers curled, suggesting forest habitation as well
Australopithecus africanus	3.5–2.5 Ma	One of first fossils found to provide evidence of African origin for hominins. First found in South Africa at Taung by Raymond Dart in 1924. Smaller canines than chimps and *A. afarensis*; large back teeth suitable for grinding plant matter. Bipedal but also adaptations for climbing trees. Brain endocasts fail to show a Broca's area, the part of the brain associated with language. *Afarensis* and *africanus* known as gracile Australopithecines because of slight and slender build
Robust Australopithecines		
Australopithecus robustus and *Australopithecus boisei*	2.3–1.0 Ma	Very similar species from South and East Africa. Show enlarged molars adapted to eating plant material (Leakey called one specimen of *A. boisei* 'nutcracker man'). Notice they coexisted with the *Homo* genus before becoming extinct. Some palaeontologists classify these in genus *Paranthropus* instead of *Australopithecus*
Transitional forms		
Homo habilis and *Homo rudolfensis*	2.3–1.7 Ma	Contemporary with Australopithecines but larger brains and smaller teeth. *H. habilis* first described by Leakey in 1964. Found in Tanzania and shores of Lake Turkanya. Thought to be first hominins associated with tool making. Some doubt if these species are very different to Australopithecines, others merge *H. rudolfensis* and *H. habilis* into a single *Homo* species. Brain endocasts show a rudimentary bulge in Broca's area, suggesting onset of language
Archaic *Homo* species		
Homo erectus and *Homo ergaster*	1.9–0.4 Ma	Found mainly in Asia. First discovered in Java in 1891. Very widespread, fossils found in South Africa, Tanzania, Kenya, Italy, Germany, India and China. Oldest fossils come from East Africa, species seems to have dispersed out of Africa and across the Old World as soon as species emerges, resulting in geographically variable local populations. Some favour term *Homo erectus* for Asian specimens and *Homo ergaster* for African and Georgian fossils. Some argue that distribution of *erectus* and archaeological evidence suggest that *H. erectus* preyed on other animals. Tool use known from worked lavas and quartzites from Oiduvai Gorge at around 2.6 Ma. The **Acheulean** (also spelt Acheulian) hand axe shows advance in sophistication and suggestion of a mental template. First hominins found outside Africa. First use of fire and some evidence of systematic hunting. In this species, dietary dependence on meat increases. Meat obtained from scavenging and possibly hunting. Brain size and pelvic measurements show that infants needed prolonged care

Table 5.4 cont'd

Species	Time range	Key features
Archaic Homo sapiens/Homo heidelbergensis	0.78 Ma–128,000 yrs ago	Archaic *Homo sapiens* was the appellation given to early humans that were thought to form the common stock for modern *Homo sapiens* and Neanderthals (*Homo sapiens neanderthalis*). However, recent evidence suggesting that Neanderthals were a separate species (viz. *Homo neanderthalis*) invalidates this term. Instead, the term *Homo heidelbergensis* is now used for this species. Found in Africa and Europe, showing signs of brain enlargement and strongly associated with hand axe technology
Homo neanderthalis	200,000–30,000 yrs ago	Named after one of the earliest finds of a partial skeleton in the Neander Valley, Germany in 1856. Now more than 275 individuals have been recovered from over 70 sites from Western Europe to Uzbekistan. Large brained, thick boned and large bodied, showing adaptations to a cold glacial climate. Used fire and stone tools and buried dead. Overlapped with modern *Homo sapiens* but whether any cross-breeding took place is debateable; evidence from mtDNA studies suggests not. Last common ancestor with *Homo sapiens* probably about 600,000 years ago
Homo floresiensis	?–15,000 yrs ago	Remarkable (but controversial) recently discovered species, identified from a skeleton of a female discovered in a cave in Liang Bua on the island of Flores in Indonesia. Named in 2004. Probably a dwarfed descendent of Javanese *Homo erectus*. Small in stature (approx 1 m) with a small brain (350 cc). Possibly the smallest species of hominin
Modern humans		
Homo sapiens sapiens (modern humans)	0.15 Ma–present	Modern humans appeared in Africa about 150,000 years ago. The 'Out of Africa' hypothesis (now the most favoured) suggests that this species then spread out of Africa to Europe and Asia between 100,000 and 60,000 years ago. Thought to have entered the New World about 12,000 years ago. Highly encephalised, sophisticated culture and technologies. Adult brain has capacity of about 1,350 cc. On this basis, expected brain size for a child if *sapiens* was typical primate would be 725 cc. In fact, the figure is 385 cc, showing that the human brain grows rapidly after birth

taxonomists in other branches of biology, there are the 'lumpers' who would rather group small variations found in fossils into one species, and the 'splitters' who use variations in new finds (such as slight changes in morphology) to argue for a new species. The splitters, for example, have argued (successfully, it seems) that African *Homo erectus* should be called *Homo ergaster*. They would also like *H. erectus* fossils found in Georgia to be called *Homo georgicus*.

The enigma of *Homo floresiensis*

The discovery in 2003 of fossilised remains of diminutive humans dating to about 18,000 years ago on the island of Flores, near to Java in Indonesia, illustrates how new finds continue to radically reshape our views on hominin evolution (Brown et al., 2004). The discoverers called this new species *Homo floresiensis* and argued that it was a descendent of *Homo erectus*. If this turns out to be correct, *Homo floresiensis* replaces *H. neanderthalis* as the last hominin to go extinct. To the annoyance of the teams that found the fossil, journalists quickly began to use the term 'hobbit' to describe this new species. The species does indeed pose a problem for hominin phylogeny. It was small in stature, some 1.5 metres tall, and had a brain the size of a modern chimpanzee; yet its bones are associated with tools found in the same area. Such are the difficulties with these finds that many have suggested that it is not a new species at all but rather a variant of *H. sapiens* suffering from pathological dwarfism.

5.4 Some important features of hominin evolution

Inferences from the data obtained from the fossil discoveries detailed in Tables 5.3, 5.4 and Figure 5.7 reveal several highly important patterns that have a massive bearing on the distinctive characteristics of humans and their behaviour. In summary, the trends we observe are:

1. Increase in body size from Australopithecine species to the *Homo* genus
2. Increasing bipedalism
3. Rapid increases in brain size in the *Homo* genus
4. Decrease in sexual dimorphism from Australopithecines to the *Homo* genus – this has important implications for the likely sexual groupings of our ancestors
5. From the evidence of stone tools, dentition and animal remains, the transition to the *Homo* genus shows an increased reliance on meat.

Some of these changes are explored below.

5.4.1 Body size

The vast majority of mammals, including most non-human primates, are considerably smaller than humans. It has long been recognised that, as mammals evolve, there is a tendency for them to grow in size. The Eocene (the geological period about 56–34 million years BP) ancestor of the modern horse, for example, was only the size of a small dog about 40 million years ago. This trend, although not universal, is sometimes called Cope's law, and humans, as mammals, have been part of this trend towards increasing size. Estimating the body mass of early hominins is fraught with difficulties, but there is general agreement on the broad picture: early Australopithecines were small, *Australopithecus africanus*, for example, probably weighing between 18 and 43 kg; body weight rose from *Homo habilis* to *Homo erectus*; Neanderthals were larger than modern humans; and the height and weight of *Homo sapiens* has declined slightly over the past 80,000 years (ignoring the effects of culture on diet and health since the Neolithic revolution). As noted earlier, *Homo floresiensis* is not part of this trend. Instead, it could be that this species reduced in size (as island species often do) in the absence of predators.

The causes of this gradual increase in body size are difficult to establish. It may be that a move from forests to a more terrestrial habitat relaxed the constraints on the size of arboreal species, or that in the more open environments that early hominins occupied, predation risks were greater and this selected for increased size (Foley, 1987). Whatever the causes, the ecological and evolutionary consequences were profound. One of the most significant effects relates to the fact that the metabolic rate of an animal rises with body weight in line with well-established physiological principles and according to the following equation:

$$M = KW^{0.75}$$

where: M = metabolic rate (energy used per unit of time)
W = body weight
K = some constant.

The effect of this relationship is that, as the size of hominins grew, so the absolute requirement for the intake of calories grew, but the rate of input of calories per unit of body mass fell. This effect can easily be seen in mice and humans. Mice consume absolutely less food per day than a human, but a mouse will typically eat half its own body mass of food in one day and spend most of the day finding it. Humans eat only about one-twentieth of their body mass in food each day and thankfully have plenty of time left over to read books on evolution. The consequence for hominins of the absolutely larger amount of food that had to be found as size grew was that their home range increased. This was not the only option available of course. Gorillas, in contrast, grew in size but became adapted to eating large quantities of low-calorie foodstuffs, such as leaves. This requires a large body to contain enough gut to digest the plant material, and this is essentially the strategy also pursued by large herbivores such as cattle. Chimps and early humans clearly opted for the high-calorie, mixed diet that required clever foraging to obtain.

Increases in size may in themselves have been a response to ecological factors, but it is important to note that such increases can have major effects on the lifestyle of an animal and, through complex feedback effects, force it to adapt further in other ways. As noted earlier, an increase in size means that animals incur an increase in absolute metabolic costs, costs that could be met by increasing the size of the foraging range. Large animals also have a smaller surface area to volume ratio than small ones, which in tropical climates would lead to problems of overheating and consequently a greater reliance on water. The upright stance of hominids may have been a response to this, since standing upright exposes less surface area to the warming rays of the sun. Body fur loss may also have helped with temperature regulation. Larger animals take longer to mature sexually, so offspring become expensive to produce and require longer periods of care. As a result, the hominins became increasingly K selected (see Chapter 3) and the kin group and larger social groups probably became important for care and protection (see Foley, 1987).

5.4.2 Brain size

The most remarkable feature of the period between the Australopithecines and *Homo habilis*, however, is that brains grew larger than expected from body size increases alone. The causes of this phenomenon are explored in the next chapter, but, whatever the factors responsible, an increase in brain size posed at least two problems for early hominins: how to obtain enough nourishment to support energetically expensive neural tissue, and how to give birth to human babies with large heads. In respect of the first problem, we need to note that our brains are notoriously expensive to run. A chimp devotes 8 per cent of its basal metabolic rate to maintaining a healthy brain, whereas for humans the figure is 22 per cent, even though the human brain represents only about 2 per cent of body mass (incidentally, a human brain consumes about 20 watts of power). The initial increase in brain size about two million years ago seems to correlate with a switch from a largely vegetation-based diet of the Australopithecines to a diet with a higher percentage of meat, as found with *Homo habilis*. Meat can provide high-quality food, and meat eating in turn reinforces cooperation, sharing and sociality.

The second problem was solved by what is, in effect, a premature birth of all human babies. One way to squeeze a large-brained infant through a pelvic canal is to allow the brain to continue to grow after birth. In non-human primates, the rate of brain growth slows relative to body growth after birth. Non-human primate mothers have a relatively easy time, and birth is usually over in a few minutes. Human mothers suffer hours of childbirth pains, and the brain of the infant still continues to grow at prebirth rates for about another 13 months. Measured in terms of brain weight development, a full term for a human pregnancy would, if we were like other primates, be about 21 months – by which time the head of the infant would be too large to pass through the pelvic canal. As in so many other ways, natural selection has forced a compromise between the benefits of bipedalism (requiring a small pelvis) and the risks to mother and child during and after childbirth. Human infants are born, effectively, 12 months premature.

5.4.3 Reduced sexual dimorphism

The premature birth of human infants required a different social system for its support than the unimale groups of our distant Australopithecine ancestors. As brain size grew, so infants became more dependent on parental care. Women would have used strategies to ensure that care was extracted from males. This would lead to the emergence of a more monogamous mating pattern since a single male could not provision many females. It is significant that the body size sexual dimorphism of hominids during the Australopithecine phase was such that males were sometimes 50 per cent larger than females. This dimorphism was probably selected by intrasexual selection (see Chapter 3) as males fought with males for access to multi-female groups. By the time of *Homo sapiens*, this figure had reduced to 10–20 per cent, signalling a move away from polygyny towards monogamy. Women probably ensured male care and provisioning for their offspring by the evolution of concealed ovulation. The continual sexual receptivity of the female and the low probability of conception per act of intercourse ensured that males remained attentive. This is a subject explored further in Chapter 11.

Altriciality

So when did the prolonged care of helpless human infants (altriciality) begin? One argument developed by Robert Martin (1990) relies upon knowledge of the way brains grow after childbirth. We know that in modern apes the infant brain doubles in size between birth and adulthood, but this in itself does not require an extended period of parental care for the young as is found in humans. Now, the pelvic opening of the Turkana boy (a virtually complete skeleton of a nine-year-old *Homo erectus* boy found in the Lake Turkana region of northern Kenya) suggests that *Homo erectus* mothers could give birth to neonates with a brain of about 275 cc or about half the size of the average modern brain at birth. If this brain doubled in an ape-like developmental pattern to adulthood, this would give a brain of about 550 cc, which is much smaller than the actual adult brain of *H. erectus* at 900 cc. The tripling of brain volume between birth and adulthood needed to achieve this is akin to modern human growth patterns and so was probably associated with prolonged infant care. Unfortunately, we do not have any pelvic measurements for *Homo habilis* to see if this conclusion extends to species before *Homo erectus*.

Female exogamy

Another feature of the life of *Homo sapiens* we can piece together is that females probably left their homes and encampments and went to live in the settlements of their male partner. This phenomenon is known as female **exogamy** and is still the common pattern in modern populations. There is evidence from studies on **mitochondrial DNA** (mtDNA) that this has been the condition of human bonding for tens of thousands of years. The useful aspect of mtDNA is that it is only passed on through the female line (see Figure 5.9 below). But molecular studies show that the variation (between people) in mtDNA is similar to that of DNA found on the autosomal (that is, non-sex) chromosomes; so much so that the average local population contains about 80–85 per cent of all the variation in mitochondrial and autosomal DNA. In contrast to this, most local populations contain only about 36 per cent of the full genetic variation found on the Y chromosome (only passed down through the male line). This suggests that women have migrated more than men. Although each individual female's journey to her mate's home would (until recent times) have been a short one, over hundreds of generations, this has had global effects on DNA distribution (Pennisi, 2001).

5.5 The supremacy of *Homo sapiens*

5.5.1 Out of Africa or multiregionalism?

If we turn the clock back to about 150,000 years ago, we would find several hominin species still extant. In Asia we would find *Homo erectus*, in Africa and Southeast Asia *Homo sapiens*, on the island of Flores possibly *Homo floresiensis*, and in Europe *Homo neanderthalis*. But by 30,000 years ago (again with the possible exception of *H. floresiensis*), this taxonomic diversity had vanished, leaving one dominant species, *Homo sapiens*, now spread across East Asia, Southeast Asia, Africa, Europe and poised to move into the New World. To explain this diminished diversity, various models of speciation, replacement and genetic mingling (interbreeding) have been proposed. The debate is involved and complex but can be simplified into two main schools of thought. One, the **Out of Africa hypothesis**, suggests a single origin for *Homo sapiens* in Africa followed by an exodus that replaced other *Homo* species (Stringer and Andrews, 1988). The other, the **multiregional model**, argues for several regional populations of *Homo erectus* evolving simultaneously into *Homo sapiens* in different parts of the globe (Wolpoff et al., 1984). A variation on the multiregional model is the idea that the primary features distinguishing *Homo sapiens* arose in Africa but then spread to other populations of archaic *Homo* species by genetic mixing and population flow. Table 5.5 shows some of the key tenets and predictions of the competing theories.

At present, there is more support for the Out of Africa model than its rival. There are several lines of evidence:

1. There is a marked lack of transitional fossils in Europe and Asia that would be needed to show that evolution from *H. erectus* to *H. sapiens* was taking place. Cave sites in Israel (for example

Table 5.5 Origin of modern humans: competing hypotheses

Multiregional evolution	Out of Africa hypothesis
Basic tenets	*Basic tenets*
1. Migration of *Homo erectus* from Africa into Asia and Europe about two million years ago 2. Local adaptations and gene transfers between regional populations gradually giving rise to *Homo sapiens* with local morphological differences ('races') we see today 3. No clear breach between *Homo erectus* and *Homo sapiens*, part of the same species lineage	1. *Homo erectus* migration out of Africa between 1.8 Ma and 700,000 years ago. But *Homo erectus* then went extinct in all its localities 2. *Homo sapiens* was a species that arose initially in one place: Africa about 200,000 years ago 3. *Homo sapiens* migrated out of Africa 130,000 to 35,000 years ago 4. *Homo sapiens* replaced all other species without interbreeding
Key predictions	*Key predictions*
Transitional fossils between archaic humans *(H. erectus)* and modern humans should be found in all parts of the Old World Continuity of anatomy between archaic humans and modern ones should be observed Modern humans should be found in the Old World over a broadly similar period	Transitional fossils between archaic hominids to modem humans should only be found in Africa There should be no evidence of hybridisation between archaic and observed modern humans Modern humans should appear in some places (for example Africa) earlier than others

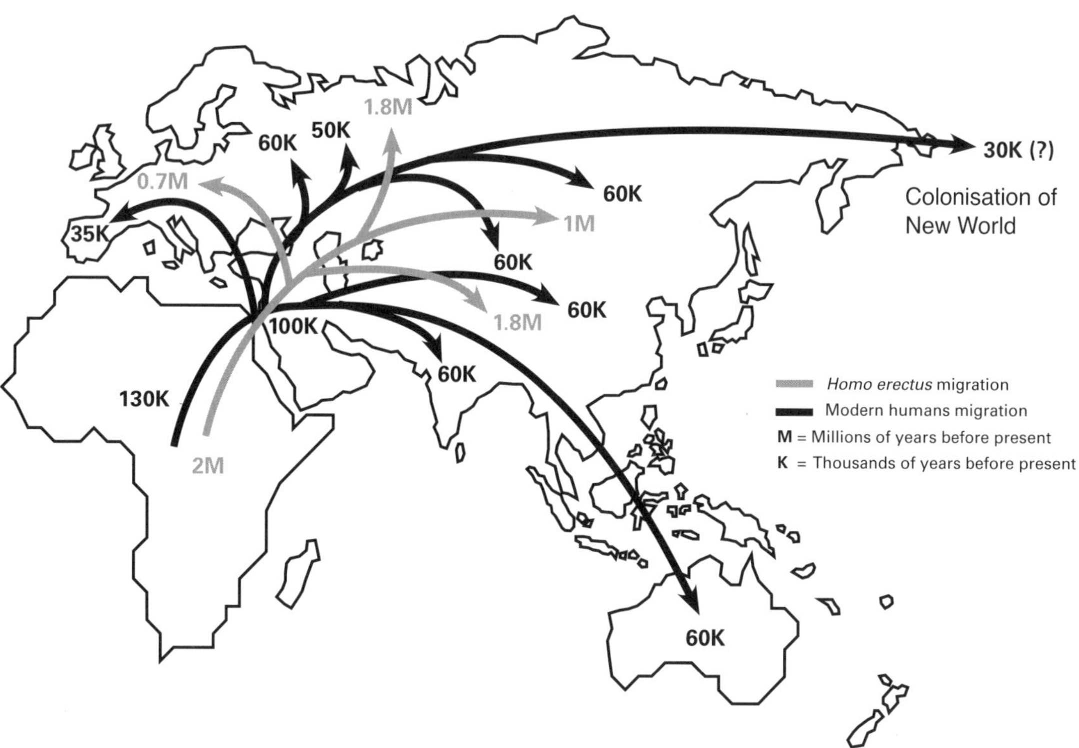

FIGURE 5.8 The diaspora of *Homo erectus* and *Homo sapiens*

Qafzeh and Skhul) dating to 100,000 BP show skeletons of anatomically modern humans and not *H. erectus/H. sapiens* intermediaries.

2. Examination of Neanderthal DNA shows it to be a distinct species and not an intermediary between *H. erectus* and *H. sapiens*. According to the multiregional model, Neanderthal DNA should resemble European DNA more than it does African DNA, since Neanderthals lived in Europe long after *H. erectus* moved out of Africa. Yet Neanderthal DNA is totally distinct from European and African DNA. But to be fair to the multiregional model, this tells us about Neanderthals but not other Old World species.
3. The work of Cann et al. (1987) on mtDNA and the discovery of 'mitochondrial Eve' lend support to the Out of Africa model (see below).
4. Compared to chimpanzees, human DNA is remarkably homogeneous. Yet if *Homo sapiens* evolved simultaneously in numerous areas from *H. erectus*, we would expect much more variation.

Figure 5.8 shows the dispersal of *Homo sapiens* across the globe.

5.5.2 Mitochondrial Eve and Y-chromosome Adam

The mitochondrial Eve hypothesis

The **mitochondrial Eve** concept is one of those scientific ideas that quickly permeates culture but is prone to numerous misunderstandings. To understand the idea and its implications, we need to consider the nature of mtDNA. Mitochondria are components (or organelles) of cells responsible for energy metabolism. There are hundreds of mitochondria in most human cells and mitochondria and mtDNA have some remarkable features and properties that make them very useful to study human evolution. Some of these key features are:

- Mitochondria contain their own genome: a circular molecule of DNA about 16,500 base pairs long (compared to the 3,000 million base pairs in the nuclear genome). Many copies of this genome are found in each mitochondrion
- The DNA of mitochondria codes for 37 genes and accumulates mutations about 10 times faster than nuclear DNA. The DNA is haploid (that is, only one copy)
- MtDNA is only inherited from the maternal line. This is because the mother supplies about 25,000 mitochondria in her ova but the mitochondria of male sperm (relatively few in number that supply the energy for sperm to swim) fail to penetrate the ovum (see Figure 5.9)
- MtDNA does not become mixed by sexual recombination as is the case with autosomal DNA. This has the effect of magnifying diversity and identifying bottlenecks in populations. Lineages and divergence times can be inferred from mutations on 'junk DNA' in the mitochondrial genome since these are not subject to natural selection.

The mitochondrial Eve hypothesis came to the fore in the 1980s in a ground-breaking paper by Cann et al. (1987). By examining the mitochondria of a sample of 147 modern humans, they concluded that the mtDNA of all living people had descended from a single female who had lived in Africa about 200,000 years ago. The name 'mitochondrial Eve' quickly entered public consciousness. There are two

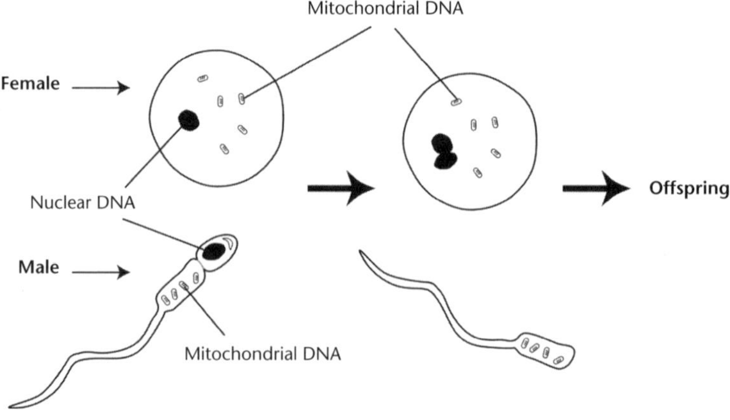

FIGURE 5.9 Mitochondrial DNA is inherited from the mother only

common confusions: first, that mitochondrial Eve was the only woman living at the time; and second, that all existing humans were descended from this individual. To appreciate the true picture, imagine a line linking your mtDNA with your mother. The mtDNA you have in your cells was inherited from your mother; your mother inherited it from her mother and so on backwards into the depths of the past. If the line of each individual is compared, it will be noticed that they start to converge in common female ancestors. This must happen of course since there are more people alive today than there were in the past so we all must be sharing numerous ancestors. If this mtDNA is pursued backwards along the female line of descent, eventually all lines will converge on a single women living about 200,000 years ago. There were of course many other women alive at this time, but Eve happens to be the only one with an unbroken line of matrilineal descent to the present day. Many other women at the time of Eve will also have descendents still alive today through a mixture of male and female lines.

There was some criticism of the original work by Cann et al. (only two of the 147 people sampled, for example, were sub-Saharan Africans). The work was repeated by Max Ingman and colleagues employing various refinements such as using a longer section of mtDNA and more Africans in the sample. They came to a similar conclusion that there was an African mitochondrial Eve that lived about 172,000 (+ or – 50,000) years ago (Ingman et al., 2000).

From recent studies on mtDNA, it has also emerged that variability of mtDNA is low – only about one-tenth of that of chimps and gorillas. From such measurements, Ruvolo et al. (1994) concluded that the two most different humans on earth are less different in mtDNA terms than two lowland gorillas in the same forest in West Africa. This low variation in human populations implies that the human lineage faced a bottleneck at some time in the past. In addition, Africans display the greatest variation in mtDNA. This could imply that this is the oldest human population. Although there are still uncertainties about how to interpret the data and construct branching sequences, the balance of evidence points to the existence of a mitochondrial Eve who lived in Africa about 200,000 years ago before the exodus of *Homo sapiens* from that continent to others.

Y-chromosomal Adam

Human males possess one Y chromosome that they inherit from their father. Most of this chromosome does not undergo sexual recombination and is passed down intact (apart from periodic mutation and insertion events) along the male line. The Y chromosome therefore acts a little like the male counterpart of mtDNA; and just as there was a mitochondrial Eve, so there existed a Y-chromosomal Adam. Over 90 per cent of the DNA on the Y chromosome is so-called junk DNA and so is resistant to the effects of natural selection. The sequence of DNA here can be used to identify lines of descent and, by making some reasonable assumptions about mutation rates, the dates of key events in its history. Such studies show that all living men can trace back their ancestry to a single male living about 60,000 years ago (Underhill et al., 2000). As with mitochondrial Eve, Y-chromosomal Adam was not the only man alive at the time. It is interesting to note that mitochondrial Eve and Y-chromosome Adam were separated by about 80,000 years. The more recent date for the genetic Adam reflects the greater variance in reproductive success of males compared to females. In other words, the number of offspring from a fertile woman would vary much less than those from men: some men may be very successful with hundreds of offspring, while others may leave no offspring at all. Consequently the Y-chromosome line converges more quickly to a single source as we follow it backwards in time. Y-chromosome studies lend enormous support to the Out of Africa model.

In conclusion then, the fossil and genetic evidence so far points to the likelihood of *Homo sapiens* emerging as a result of speciation events in Africa about 200,000 years ago. The main issue for the future will probably be whether *Homo sapiens* replaced other species in the Old World completely, leaving them to go extinct, or whether there was some genetic admixture through interbreeding. This may be a difficult question to answer if the level of interbreeding was low but not zero. Of all the changes in the anatomy of the hominins, the most startling and revolutionary has to be the enlargement of the brain. This subject is explored in the next chapter.

Summary

- According to phylogenetic classification, humans belong to the subfamily Homininae (hence they are hominins), the family Hominidae (hence they are also, like chimps and gorillas, hominids) and the superfamily Hominoidea (hence, like chimps, gorillas, orang-utans and gibbons, they are hominoids). Hominins are humans and all their ancestors leading back to the last common ancestor shared with chimpanzees.
- The hominins originated in East Africa and became adapted to an open savannah environment. A whole range of hominin species once existed but *Homo sapiens* is the last remaining species.
- The evolution of the hominins in the direction of *Homo sapiens* is associated with reduced sexual dimorphism, increased brain size and increased altriciality.
- The most popular theory of the global dispersion of *Homo sapiens* is the Out of Africa hypothesis, which proposes that *Homo sapiens* originated in Africa about 200,000 years ago and gradually spread across the Old and then the New Worlds.

Key Words

- **Acheulean**
- **Allopatric speciation**
- **Altricial**
- **Archaic *Homo sapiens***
- ***Ardipithecus ramidus***
- **Australopithecines**
- **Cladistics**
- **Exogamy**
- **Genus**
- **Hominid**
- **Hominin**
- **Lineage**
- **Mitochondrial DNA**
- **Mitochondrial Eve**
- **Multiregional model**
- **Order**
- **Out of Africa hypothesis**
- **Phylogeny**
- **Phylum**
- **Population**
- **Precocial**
- **Speciation**
- **Species**
- **Taxon**
- **Taxonomy**

Further reading

Campbell, B. G., Loy, J. D. and Cruz-Uribe, K. (2006) *Humankind Emerging* (9th edn). Boston, MA, Allyn & Bacon.

Lewin, R. and Foley, R. (2004) *Principles of Human Evolution* (2nd edn). Oxford, Blackwell Science.

Both these books provide authoritative and accessible acounts of the latest ideas on human origins and evolution.

Encephalisation and the Emergence of the Human Mind

Plac'd on this isthmus of a middle state,
A being darkly wise, and rudely great:
With too much knowledge for the Sceptic side,
With too much weakness for the Stoic's pride,
He hangs between; in doubt to act or rest,
In doubt to deem himself a God or beast;
In doubt his Mind or body to prefer.
Born but to die, and reas'ning but to err.
(Pope, *Essay on Man*)

6.1 The sizes of animal brains

6.1.1 What makes humans so special?

Is there an area of the brain unique to humans? In the middle of the 19th century (*c.*1858), Richard Owen, an anatomist who vigorously opposed the application of Darwinism to humans, thought that there was. He claimed that humans have a special structure in the brain called the 'hippocampus minor' that is not found in apes. This, he argued, was clear evidence that we could not have descended from the apes; here was the seat of human distinctiveness. His hopes for a special status for humans were, however, short-lived: in 1863, Darwin's 'bulldog' champion, Thomas Henry Huxley, rushed to the fray a few years after Owen announced his thoughts and conclusively demonstrated that apes possessed the same structure that Owen had identified. The debate was parodied in popular culture. In Charles Kingsley's *Water Babies*, published in 1863, for example, there is much talk of 'hippopotamus majors'.

Since the time of Owen and Huxley, there have been numerous attempts to establish which features of the human brain, if any, confer upon humans their unique qualities. It is tempting to think that we simply have bigger brains than other mammals, but even a cursory examination of the evidence rules this out. Elephants have brains four times the size of our own, and there are species of whale with brains five times larger than the average human brain. We should expect this of course – larger bodies need larger brains to operate them (Table 6.1). The next step would be to compare the relative size of brains among mammals (that is, the ratio of brain mass to body mass). The results are unedifying: we are now outclassed by such modest primates as the mouse lemur (*Microcebus murinus*), which has a relative brain size of 3 per cent compared with 2 per cent for humans, leaving us roughly on par with bats and squirrels (Figure 6.1).

Table 6.1 Absolute brain weights of selected organisms

Organism	Brain weight (g)
Whales	2,600–9,000
African elephant	4,200
Man	1,250–1,450
Bottlenose dolphin	1,350
Horse	510
Gorilla	430–570
Chimpanzee	330–430
Dog	64
Cat	25
Rat	2
Mouse	0.3

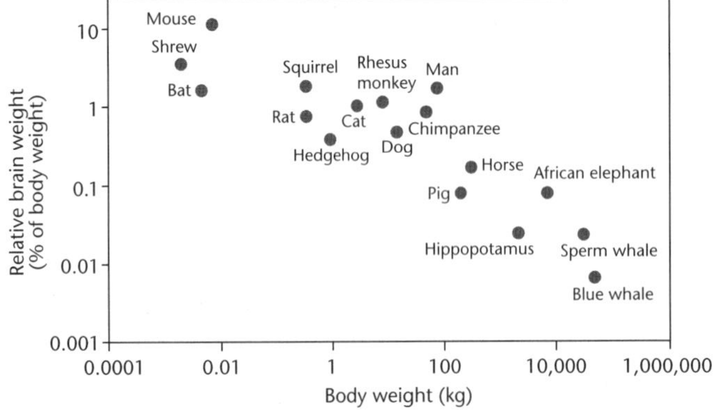

FIGURE 6.1 Brain size as a percentage of body size for selected mammalian species plotted on log–log coordinates
As can be seen, some small mammals such as mice and shrews have larger brains in relative terms (up to 10 per cent of body weight) than humans, with a brain representing 2 per cent of body weight.

SOURCE: Data from Van Dongen, P. A. M. (1998) Brain size in vertebrates, in R. Nieuwenhuys (ed.) *The Central Nervous System of Vertebrates*. Berlin, Springer, vol. 3: 2099–134.

6.1.2 Allometry

If we wish to find some basis for our special status, we can find some reassurance in the phenomenon of **allometry**. As an organism increases in size, there is no reason to expect the dimensions of its parts, such as limbs or internal organs, to increase in proportion to mass or volume. If we simply magnified a mouse to the size of an elephant, its legs would still be thinner in proportion to its body than those of an elephant. This happens in primates too: the bones of large primates are thicker, relatively speaking, than the bones of smaller primates and there are sound mechanical reasons for this. In fact, there is a fairly predictable relationship between brain and body size in mammals:

Brain size = C (body size)^{k}
Or brain size = C(W)^{k}

where C and k are constants and W is the body weight (in grammes) of the organism (equation 1).

The constant C represents the brain weight of a hypothetical adult animal weighing 1 g. The constant k indicates how the brain scales with increasing body size and seems to depend upon the taxonomic group in question. Much of the pioneering work in developing these equations was carried out by Jerison (1973), who concluded that, for the entire class of mammals, k = 0.67 and C = 0.12. There is much discussion about the precise values for these constants, and even within primate groups k varies from 0.66 to 0.88. Martin (1981) revised Jerison's work and concluded that k = 0.76 and C = 0.058.

If we plot a graph of brain size against body weight for mammals on linear scales, a curve results, showing that brain size grows more slowly than body size (Figure 6.2).

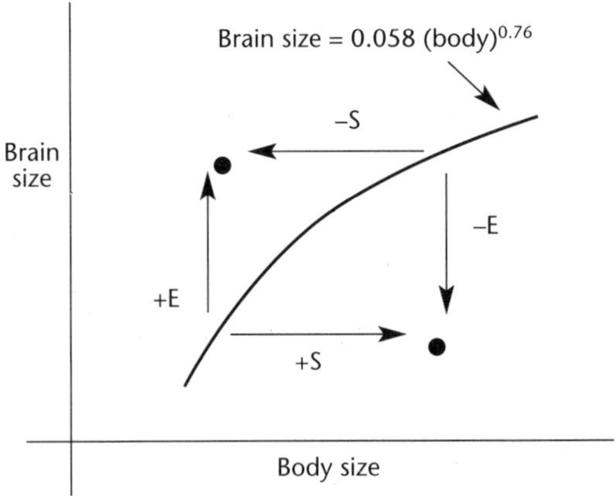

FIGURE 6.2 Growth of brain size in relation to body size for mammals
An animal occupying a point above the line is said to be encephalised, that is, it has a brain larger than expected for an animal of its body mass. This could be the result of a relative growth in brain size (positive **encephalisation**, +E) or a diminution of body size relative to brain mass (negative somatisation, –S). A point to the right indicates that the organism has a smaller brain than expected; this could be due to positive somatisation (+S) or negative encephalisation (–E) (see Deacon, 1997).

If we take log values of both sides of equation 1 with the constants for mammals inserted, then:

Predicted weight = 0.058 (W)$^{0.76}$
log (brain size) = 0.76 log (W) + log (0.058)

Thus, a plot of the logs of brain and body size, or a plot on a logarithmic scale, should give a straight line of slope 0.76.

Figure 6.3 starts to give an indication of what makes humans so special: we lie well above the allometric line seen for other mammals. If we insert a value of 65 kg as a typical body mass for humans into equation 1, our brains should weigh about 264 g. The real figure is in fact nearly 1,300 g. Our brains are about five times larger than expected for a mammal of our size and about three times larger than that expected for a primate of our size.

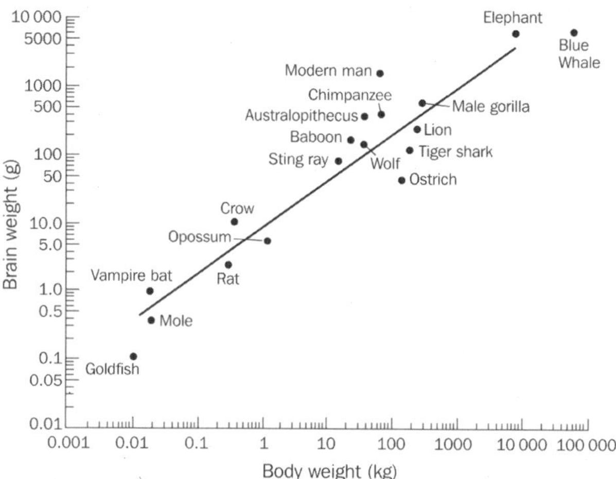

FIGURE 6.3 Logarithmic plot of brain size against body size

Organisms above the line have brains larger than expected from their body sizes. The line of best fit is drawn as: brain size = 0.058 (body size)$^{0.76}$.

SOURCE: Adapted from Young, J. Z. (1981) *The Life of Vertebrates*. Oxford, Oxford University Press.

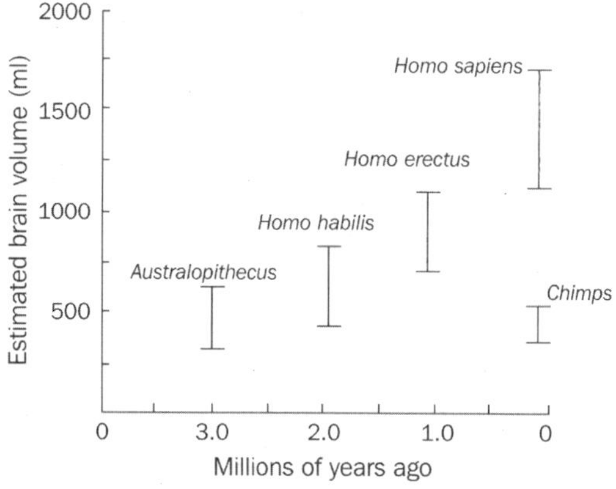

FIGURE 6.4 Growth in brain size during human evolution

SOURCE: Adapted from Deacon, T. W. (1992) The human brain, in S. Jones, R. Martin and D. Pilbeam (eds) *The Cambridge Encyclopedia of Human Evolution*. Cambridge, Cambridge University Press.

6.1.3 Ancestral brains and encephalisation quotients

A reasonable estimation of the size of the brains of our early ancestors can be obtained by taking endocasts of the cranial cavity of fossil skulls (see section 6.2.3). There is some debate over how to interpret the fine detail of these casts (such as evidence for folding), but there is consensus on the general trend: about two million years ago, the brains of hominins underwent a rapid expansion (Figure 6.4). Australopithecines possessed brains of a size to be expected from typical primates of their stature, but *Homo sapiens* now have brains about three times larger than a primate of equivalent body build. The departure of brain size from the allometric line is known as the **encephalisation quotient** (EQ).

The encephalisation quotient (EQ) is defined as:

$$EQ = \frac{\text{Actual brain weight}}{\text{Brain weight predicted from allometric line}}$$

The interpretation of encephalisation remains controversial. Intelligence is likely to be far too complex to have a simple relationship with the EQ. This is illustrated by what Deacon (1997) has called the 'Chihuahua fallacy'. Small dogs such as the Chihuahua and Pekinese are highly encephalised, that is, they lie to the left of an allometric line of brain weight against body weight for carnivores. The reason is that they have been deliberately bred for smallness in body size, but since brain size is far less variable, the breeding programme that led to small dogs has left them with relatively larger brains. In humans, the condition of dwarfism also yields a high EQ. The best way to explain the highly encephalised condition of Chihuahuas is by the concept of negative somatisation (see Figure 6.2 note above). The important point to realise is that human dwarves or Chihuahuas are not noticeably more intelligent than their normally sized counterparts.

But reviewing the evidence, Deacon concludes that the high EQs of humans are not the result of negative somatisation. In fact, the fossil record shows that hominid body size has been increasing over the

Table 6.2 Body weights, brain weights and EQs for selected apes and hominids

Species	Typical body weight (g)	Typical brain weight (g)	EQ
Pongo pygmaeus (orang-utan)	53,000	413	1.83
Gorilla gorilla (gorilla)	126,500	506	1.16
Pan troglodytes (common chimp)	36,350	410	2.42
Homo habilis	40,500	631	3.43
Homo erectus	58,600	826	3.39
Homo sapiens	65,000	1,250	4.74

SOURCE: Data from Boaz, N. T. and Almquist, A. J. (1997) *Biological Anthropology*. Englewood Cliffs, NJ, Prentice Hall; Lewin, R. (2005) *Human Evolution*. Oxford, Blackwell.

past four million years and human brain size has grown rapidly in relation to increases in body size (see Figure 6.5). Instead, it seems that in humans and other primates, body size growth during development is slower than for other mammals. After birth, the human brain continues to grow larger than would be expected for a primate, and body size growth in adulthood stops earlier than expected. This leaves humans with a brain larger than would be expected on the basis of our body size (Deacon, 1997).

The calculation of EQs does lead to some odd results. For example, imagine two species, one with a body weight of 5 g and the other 50 kg, but both with brain volumes 50 per cent higher than expected from allometric scaling. The EQ of both is the same at 1.5, yet it seems unlikely that they are cognitively equal since the large brain has vastly more neural tissue.

It seems clear that EQ is a better measure of intelligence than absolute brain size. But it is probably not a perfect measure of intelligence; for criticisms of the EQ concept, see Holloway (1996) and Roth and Dicke (2005).

Now we have established that there is something unique about the human brain, we need to examine why such a risky organ should have evolved to the proportions that it has. There are two related questions that quickly arise, and answering the first helps with the second. The first is: why did primates evolve larger brains than other mammals? The second is: why, among our own ancestors, did brain size increase well beyond that typical of other primates? Current thinking on these issues is interesting and important, since it points a way towards understanding not only the roots of human cognition, but also the origin and behavioural significance of language. But first, primate brain size.

6.2 Origins of primate intelligence and theories of human brain enlargement

As humans, it is easy to take the value of high intelligence and large brains for granted and assume that they had self-evident survival value. It may seem obvious that natural selection would eventually deliver intelligent creatures

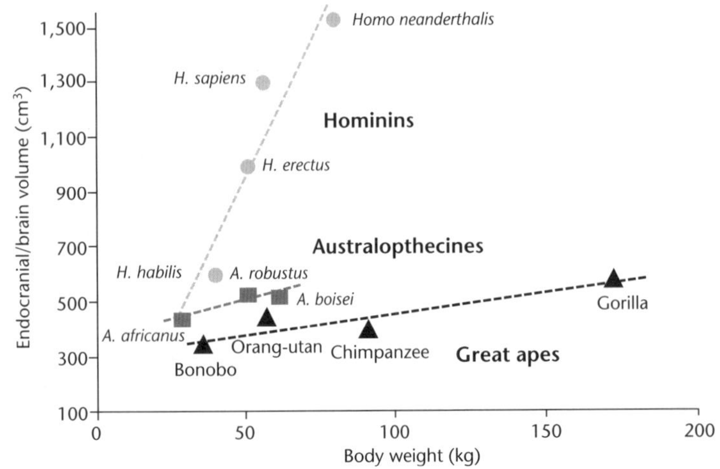

FIGURE 6.5 Brain volume in relation to body weight for the great apes and selected hominins

Brain weight grew much more rapidly than expected in the *Homo* genus.

SOURCE: Data from Jerison, H. J. (1973) *Evolution of the Brain and Intelligence*. New York, Academic Press.

capable of contemplating their own origins. But progress to large brains was never inevitable. Natural selection is not interested in brainpower per se – if small brains are sufficient for the purpose of genetic replication, all the better. There is a curious species of sea-squirt (one of the tunicates or 'dead men's fingers' – an edible invertebrate that the French and Japanese consume avidly) that uses its brain to find a suitable rock to cling to; once found, there is no need for a brain and it is absorbed back into the body (the joke in university circles is that this is what happens to academics once they gain tenure).

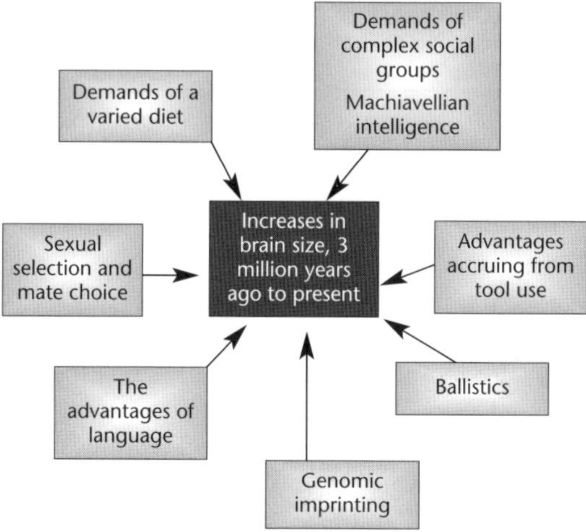

FIGURE 6.6 Some possible evolutionary stimuli on the growth of hominin brains
The chain of cause and effect is likely to be more complex than the one-way influence shown here. Some factors would have interacted with each other and growth in brain size would itself have brought about changes to living conditions.

There are a variety of theories that address the problem of why primates are vastly more intelligent than most mammals. We will examine in detail a number of theories concerning the problem of primate brain size and the extreme manifestation of this in humans (Figure 6.6). But first it is worth noting just how energetically expensive brains are.

6.2.1 The energetic demands of brains

Brains are costly to produce and expensive to run. An adult brain accounts for only about 2 per cent of body mass but consumes about 20 per cent of all the energy ingested in the form of food. The idea that we only use part of our brains is one of those urban myths: brains are made of expensive tissue, natural selection would not continue to fuel brains if there was not some payback. As we rest, the power output of the brain is somewhere between 16 and 20 watts. Unfortunately for thinkers who want to lose weight, this percentage and power output value hardly alters whether we think a lot or little.

We should be aware, however, that the heart, liver, kidney and guts are nearly as expensive. Together with the brain, these organs consume about 70 per cent of the basal metabolism of the human body. Such considerations lie behind the 'expensive tissue hypothesis' proposed by Aiello and Wheeler (1995), in which it is argued that increases in the brain size of hominins must have been balanced by a reduction in the demands of other organs. This argument arises from the fact that the overall metabolic rate for humans is as expected for mammals of our size, and, since our brains are using much more energy than expected, some other organ or organs must have a reduced energy requirement. Table 6.3 shows some calculated expected values.

In hominins, the trade-off may have been gut size for brain size: reducing the size of the gut by utilising higher quality foods would have freed energy to devote to brain growth. The crucial point about the Aiello and Wheeler hypothesis is that whatever the forces driving up encephalisation, the process must have been accompanied by a reduction in gut size and an increase in the quality of foodstuffs (Figure 6.7).

But the real bottleneck in supplying energy to a demanding brain appears to be in periods of prenatal and early childhood growth. Prenatal brains require 60 per cent of basal metabolism and this high figure is maintained for several years after birth (Aiello et al., 2001). This obviously places a huge demand on the energy budgets of mother and infant. It could have been met by changes to social structures and the advent of more support from the father, grandmothering and the sharing of food between unrelated adults.

Table 6.3 Expected and observed masses of human organs (for a 65 kg primate)

Tissue	Observed mass (kg)	Expected mass (kg)	Predicted metabolic increment (that is, changes in metabolic costs due to difference between observed and expected organ masses) (W)
Brain	1.3	0.450	+9.5
Heart	0.3	0.320	−0.6
Kidney	0.3	0.238	+1.4
Liver	1.4	1.563	−2.0
Gastrointestinal tract	1.1	1.881	−9.5
Totals	4.4	4.452	

NOTE: Humans have brains bigger than expected and guts smaller than expected for primates of our size. The increased metabolic demands of the brain could be met by the reduced metabolic demands of the gut.

SOURCE: Data from Aiello, L. C. and Wheeler, P. (1995) 'The expensive-tissue hypothesis: the brain and the digestive system in human primate evolution.' *Current Anthropology* **36**: 199–221; Aiello, L. C., Bates, N. and Joffe, T. (2001) In defence of the expensive tissue hypothesis, in D. Falk and K. R. Gibson (eds) *Evolutionary Anatomy of the Primate Cerebral Cortex*. Cambridge, Cambridge University Press.

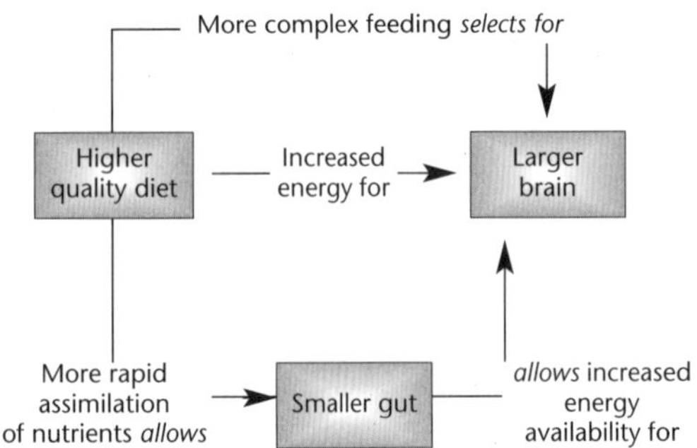

FIGURE 6.7 Relationship between brain size, gut size and energy availability in the evolution of the hominin brain

SOURCE: Adapted from Aiello, L. C. and Wheeler, P. (1995) 'The expensive-tissue hypothesis: the brain and the digestive system in human primate evolution.' *Current Anthropology* **36**: 199–221, Figure 5, p. 207.

The diversion of energy from gut size to brain size looks like an important facet of human evolution. We should remember that there was nothing inevitable about this: the redirected metabolic energy could have been channelled to other organs to increase reproductive output. The fact that it seems to have been channelled to the brain shows that the brain must have been under strong selective pressure at this time.

6.2.2 Environmental factors: lifestyle, food and foraging

Early theories of primate brain enlargement looked at ecological factors as causal agents. In the early twentieth century, for example, Grafton Eliot-Smith (1927) advanced the 'arboreal theory', in which he proposed that moving and feeding on land and in trees required stereoscopic vision and dextrous hands, both of which in turn required a large brain to coordinate and control. But, as Cartmill (1974) pointed out, many arboreal species (for example tree squirrels) manage without stereoscopic ability, have great manual dexterity and yet are not renowned for their high intelligence. Cartmill suggested that stereoscopic vision facilitated insect predation by primates, noting that forward-facing eyes are also found in other predators such as cats.

Most **herbivores**, such as cattle, tend to have specialised guts that enable them to digest and ferment low-calorie food. Such animals need a large gut to contain the bacteria needed to help with breaking down the cellulose, and they need to spend considerable time grazing. For this style of life, a large brain is not necessary (although it may help with predator evasion), and it may be significant that cattle and horses lie below the allometric line shown in Figure 6.3 above. Carnivores, on the other hand, receive a balanced diet with every kill: the meat of another animal is ideally suited to building new flesh. What is needed here is speed, strength and the perceptual apparatus to catch prey. Apart from the great apes, most primates are too small to have sufficient gut to ferment large quantities of cellulose, so resort to a more varied diet. They are, in effect, unspecialised vegetarians, and, at least for some species, obtaining a balanced diet can be fairly intellectually demanding.

Obtaining food can be broken down into a series of stages: travelling to find food, locating food using a perceptual system and extracting food from its source. For primates that rely on high-calorie foods, such as fruits, there may only be a few fruit trees in season in a large patch of forest. Remembering where they are is difficult work, and there is ample evidence that primates employ efficient cognitive maps to remember the location of the fruit and the optimum route between the trees (Garber, 1989). Fruits are easily identified by their colours, but colour vision is common in reptiles and birds (all those showy plumage colours are not there for human delight), so it may be thought that the intellectual demands of colour vision are slight. Certainly, reptiles are not renowned for their high intelligence. But most mammals are colour-blind, and it is therefore extremely likely that primates had to rediscover colour vision at a later date in evolution after it was lost.

When food is found, the work of a primate is not over. Most primates feed upon plant matter and, whereas many plant species are only too glad to allow animals to eat their fruits and spread their seeds, they are equally averse to having their leaves and stems destroyed. Consequently, they have evolved counterattack mechanisms in the form of stings, prickles and poisons to deter foragers. Processing plant material to overcome such deterrents requires some cognitive skill and could have helped to drive up primate intelligence.

Within primates, there are differences in feeding behaviour and habitat and if the ideas linking foraging and intelligence are correct, then it should follow that the primate species most dependent on patchy and dispersed (spatially and temporally) foods should show more cerebral development than those reliant only more uniform resources. Milton (1988) tested these ideas using two contrasting species: howler monkeys *(Alouatta palliata)* and spider monkeys *(Ateles geoffroyi)*. Both these species eat leaves and fruit but howlers are more folivorous (leaf eating) than spider monkeys, while spider monkeys are more frugivorous (fruit eating) than howlers. One consequence of this is that spider monkeys have to deal with a food supply area 25 times as large as that of howler monkeys. When these two species are compared, there are several features that suggest spider monkeys are more intelligent than howler monkeys (Table 6.4). The social behaviour of spider monkeys seems to be more complex and they show a greater range of facial expressions than howler monkeys.

Milton is generally sceptical about the notion that primate intelligence is related to social complexity:

> Data sets on primates, small mammals, bats and marine mammals therefore suggest that diet (and complexities associated with its procurement) show an association with relative brain size. Little support is found for the view that social systems or breeding systems have a similar effect on brain size. (Milton, 1988, p. 298)

There are other possible interpretations of the effect of diet, however. EQ measures rely upon deviation of brain sizes from the allometric line, but it is quite feasible that the selective forces involved also acted on the body sizes of primates. It is possible that large bodies (and hence relatively small brains) are a consequence of folivory, since a diet of mainly leaves requires a large gut and hence a large body to house it. In essence, what may appear as relative negative encephalisation may really be positive somatisation (Figure 6.2 above).

Table 6.4 Comparison of brain sizes of howler and spider monkeys

Species	Diet	Body weight (kg)	Brain weight (g)	EQ
Howler *(Alouatta palliata)*	Leaves and fruit	6.2	50.3	0.66
Spider *(Ateles geoffroyi)*	Mostly fruit	7.6	107	1.2

NOTE: EQs are calculated from formula shown earlier.

SOURCE: Data cited in Milton, K. (1988) Foraging behaviour and the evolution of primate intelligence, in R. W. Byrne and A. Whiten (eds) *Machiavellian Intelligence*. Oxford, Oxford University Press.

Food consumption and brain size in humans

Support for the linkage between brain growth and food types in the hominid lineage comes from the application of life history theory (LHT; see Chapter 3) to the problem by Kaplan and Gangestad (2005). They compared the diet of hunter-gatherers with that of chimpanzees. A condensed version of their findings is shown in Table 6.5.

Table 6.5 Comparison of diet of hunter-gatherers with chimpanzees

Dietary item	Hunter-gatherers (percentage)	Chimps (percentage)
Meat	60 (range 30–80)	2
Extracted foods requiring skill to extract and process	32	3
Easily collected resources (such as fruit and leaves)	8	95

It is obvious from Table 6.5 that the human diet is far more intellectually demanding than that of chimps. We should also note that the foraging and hunting areas covered by hunter-gathering humans (requiring spatial mental mapping) are about 1,000 times that of chimps. Moreover, some of the skills required in human food collection and processing are products of a long learning curve and are not properly mastered until later in life. So in the Ache, for example, while peak fruit gathering is reached by mid- to late teens, the rate of providing extracted resources peaks much later. In relation to hunting, peak rates of food production among the Hivi, Ache and Hazda peoples are not reached until group members reach their thirties, with the productivity of 20-year-olds being only 25–50 per cent of that of the adult maximum (Kaplan et al., 2000; Kaplan and Gangestad, 2005).

Such considerations led Kaplan and Gangestad to suggest that humans have evolved a lifestyle involving a heavy investment in brain tissue early on in life that pays dividends through a long lifespan, which enables humans to learn and acquire skills to pay back their net consumption of calories provided by others in their youth. As Kaplan and Gangestad (2005, p. 81) note:

> Possibly large brains and long lives in humans are coevolved responses to an extreme commitment to learning intensive foraging strategies and a dietary shift towards nutrient-dense but difficult to acquire foods, allowing them to exploit a wide variety of foods and therefore colonize all terrestrial and coastal ecosystems.

Human brain growth continues into adolescence and neuronal connections are still being made and modified for years after they have become fixed in other primates. This has implications for the modular view of the mind: whatever the basis in brain architecture for the existence of these modules, it is clear that they are being modified, calibrated, activated and perhaps learning to communicate with each other for a long time before adulthood. The archaeologist Steve Mithen (1996) has argued that cognitive fluidity exists in concert with the modular hominin brain and is manifested in the communication between preformed modules. The long period of childhood and adolescence enables this communication and hence cognitive fluidity to become established.

6.2.3 Tool use

One of the most obvious differences between the lives of humans and other primates is the reliance of the former on technology. It is true that, in the wild, tool use has been documented in chimps and some populations of orang-utans (chimps, for example, have been observed to use twigs to extract ants from their nests and stones to break nuts), but in both these species existing materials are recruited as tools rather than shaped purposefully for the task.

It is to tool use that we might look then to explain the growth in human brain capacity. If tool use were an important stimulus to brain growth in our evolutionary lineage, then we would expect some association between the sophistication of tool use at any epoch and the brain size of early hominids during the same period. The chain of cause and effect might be that the use of tools early on for such things as catching and processing foods gave survival advantages. From this beginning, more sophisticated tools (sharper points, straighter spears and so on) may have brought about even greater rewards. But more sophisticated tools would require larger brains to conceive and construct them. This would have set up a selective pressure for increases in brain size, a case of 'tools maketh the man'. Such views about brain size and the use of tools were common about 40 years ago and have been revived of late in a modified form (Byrne, 1995). In 1959, Oakley (1959, p. 2) argued that:

> When the immediate forerunners of man acquired the ability to walk upright habitually, their hands became free to make and manipulate tools – activities which in the first place were dependent on adequate powers of mental and bodily co-ordination, but which in turn perhaps increased those powers.

To test the idea that brain growth through evolution was stimulated by tool use, it obviously behoves the investigator to compare the size of brains with the type of technology practised by early hominids at various periods. Brain size can be estimated by the use of endocasts from fossilised skulls. The construction of endocasts involves pouring some substance that sets inside the hollow cavity of a skull and measuring the volume of the cast that results. The volume thus calculated then needs to be scaled for body mass to obtain a measure of the encephalisation quotient (EQ). The EQ of early hominids can then be compared with the state of tool use at any time. Another point of reference here that is useful is the fact that we have accurate information about brain size and tool use for some modern-day great apes such as chimps and orang-utans.

When such comparisons are made, the general thrust of the evidence is to throw doubt on the suggestion that tool use served as a major stimulus to brain growth. The reasons are as follows:

- Early species of the hominid lineage such as *Australopithecus afarensis* that lived about 3.5 million years ago yield higher EQs than modern-day chimps yet no tools have been found at Australopithecine sites. This does not rule out the fact that Australopithecines may have used tools, however, since such tools may have been perishable.
- We begin to find tools in the human record about 2 million years ago associated with the earliest *Homo* species *Homo habilis*. Such tools, however, are very simple and often amount to no more than a stone broken to reveal a sharp edge. Wynn (1988) argues that these tools show no signs of the concept of symmetry, no evidence of a design held in the mind and then imposed on the raw material. Nothing, in short, that reveals any mental sophistication. Wynn estimates that these primitive tools reveal a tool-making competence similar to the abilities of the great apes today. This is a blow to the tool-making hypothesis of cranial enlargement since such hominids had brains of a significantly larger EQ than modern-day chimps.
- A period of rapid brain growth took place during the evolution of *Homo erectus* between 1.5 million and 300,000 years ago. As we might expect if the hypothesis is correct, the tools found associated with these hominids are more regular in shape and size and are more sophisticated, showing, for example, a greater degree of symmetry. One crucial problem, however, is that while brain size grew during this period, tool manufacture remained remarkably conservative. The evolution of tool design failed to keep up with increases in intelligence.
- Between 300,000 years ago and the present, brains did grow slightly in volume but the explosion of technological sophistication has been enormous. From the invention of symbolic culture (cave paintings and so on) about 35,000 years ago, through the Neolithic revolution (invention of agriculture) about 11,000 years ago to the internet of today, technology has been transformed exponentially through relatively small changes in brain volume since 300,000 years ago and virtually no changes in volume since 35,000 years ago.

Reviewing the evidence, Wynn (1988, p. 283) concludes that:

> Given the evidence of brain evolution and the archaeological evidence of technological evolution, I think it

fair to eliminate from consideration the simple scenario in which ability to make better and better tools selected for human intelligence.

6.2.4 Encephalisation and ballistics

Imagine throwing a stone at a moving object. To hit it successfully requires the solution to a series of complex calculations and fine motor tuning. Variables that need to be factored into deciding the velocity, angle and direction of aim include the mass and size of the stone, the direction of movement of the target, and the distance to the target. Humans are quite good at throwing things with some accuracy, and although chimps do throw branches and other objects, their aim seems lamentable compared to humans (see Holloway, 1975). William Calvin (1982) has extended Holloway's insight to propose that the advantages to be gained from accurate throwing, initially perhaps at rival animals gathered around a carcass and then at prey proper, may have stimulated brain growth.

Calvin links his thesis to brain lateralisation, handedness and the origin of language. Whereas the great apes show very little preference for one hand over the other, about 89 per cent of humans write with and throw things with their right hand. This means that circuitry in the left hemisphere of the brain is responsible for this activity. Why this asymmetry started is unclear. One rather speculative idea is that early female hominins cradled their infants on the left-hand side so that their heads were near the comforting heartbeat of the mother, thereby freeing the right hand. The use of the right arm for initially wielding a club to ward off other animals (for example predators) may have compensated early hominins for their lack of speed over open ground as they became increasingly bipedal. Calvin also suggests that the throwing of objects helps to explain some of the problems associated with the function of hand axes. Early pear-shaped Acheulean hand axes (1.5–0.3 million years ago) were made in great numbers by *Homo erectus* and *Homo heidelbergensis*, yet their precise function is still controversial. The simple and obvious answer that they were hand tools faces the problem that their sharp edges would make them difficult to handle when delivering blows. Instead, Calvin (1993) proposes that they were throwing stones whose sharp edges meant that they would not rebound from the hide of an animal. However, Calvin's claim that non-rebounding stones are more damaging since they transfer more momentum to the animal is refuted by simple physics: a rebounding stone actually delivers more force to the target than one that sticks. Whereas this does not invalidate the whole argument, a more compelling case must be made for the axes as missiles. Calvin then advances the novel view that neural circuitry laid down for club wielding (with the right hand) and then developed for ballistic throwing was then ready in place to be coopted for that most human of inventions: language. If correct, this idea would explain why language is primarily associated with the left-hand side of the brain.

6.2.5 The social brain hypothesis

Machiavellian intelligence and the theory of mind

In recent years, several related hypotheses have emerged suggesting that it may be the demands on the social world that have been the main determinant of the growth in primate intelligence. Byrne and Whitten (1988) developed these theories and labelled them together as the 'Machiavellian intelligence hypothesis', named after the Renaissance politician and author Nicolo Machiavelli who epitomised the cunning and intrigue of political life in Italy in the early 16th century. The term **Machiavellian intelligence** was originally inspired by de Waal's comparison of the social strategies employed by chimps and the advice offered to rulers in Machiavelli's book *The Prince*. The essence of the Machiavellian intelligence hypothesis is that primate intelligence allows an individual to serve his or her own interest by interacting with others either cooperatively or manipulatively but without disturbing the overall social cohesion of the g oup. The analogy with human politics is clear: the successful and cynical politician uses his position to further his own ends, while to all appearances serving the people, and without disrupting or bringing into disrepute the elective system. One advantage of linking Machiavellian intelligence to encephalisation is the inbuilt potential for positive feedback. An increase in intelligence will quickly spread as a result of its success over same species competitors, thereby raising the level of intelligence to be reached for the next incremental step needed for further advantages. Positive feedback effects are also observed in predator–prey relationships and were used by Jerison (1973) to explain the increases, as shown in the fossil record, of brain sizes in carnivores and their ungulate prey. The rapid growth of the human brain over a short geological time period points to some sort of positive feedback effect in action.

To investigate whether primates really employ such tactics, we will consider the social groupings of primates and the nature of their intelligence.

The size of a primate group is determined by a number of factors. The minimum group size is determined largely by the need to defend against predators: there is a security in numbers that benefits each individual. The maximum group size is probably determined by both ecological and social factors. The larger the group, the longer the time taken to move en masse from place to place, and the less food there is per individual when it is found. In addition, if the group becomes too large, conflicts over food and status are set up, and the group may split (Dunbar, 1996b). It follows that the benefits of group living must be weighed against its cost.

Social conflicts and costs result from the fact that the neighbour of any individual is a potential competitor for food and for mates. Some sort of order is often maintained by a linear **dominance hierarchy**, those at the top reaping the most rewards. The alpha male in chimpanzees, for example, will often take preference at a new feeding site and will have a good chance of fathering any offspring from a female in oestrus in the group. Beyond these simple predictions, however, social life is complicated, and the success of any individual chimp is the product of a network of coalitions and alliances. This can be seen even in the dominance hierarchy. The alpha male is not simply the chimp with the greatest physical strength: his status may also reflect the status of relatives and allies. His power is thus based on his 'connections'.

We can picture a primate group as a product of centripetal forces resulting from predatory pressures tending to keep the group together, and centrifugal forces emanating from tension and conflict in the group tending to push the group apart. Predation from without and conflict from within both act negatively on the reproductive fitness of an individual, and we can expect evolution to have come up with ways of mitigating these. With regard to conflict, chimps and other primates seem to have hit upon **grooming** as an effective mechanism to reduce intragroup tensions and thus enhance group cohesion. A pair of primates that regularly groom each other are more likely to provide assistance to each other when one is threatened than are non-grooming partners. Grooming seems to serve to maintain friendships, cements alliances and is used to effect reconciliation after a fight.

But Machiavellian intelligence is more than just grooming, and the ambitious primate has other devices to help it to navigate the complex currents of social life. To really succeed in primate politics, deception is needed, and numerous observers have noted that primates will send out signals, such as false warning cries, that can only really be satisfactorily interpreted as being designed to mislead others. One example comes from the observations of Byrne on baboons, *Papio ursinus*. Byrne (1995) noticed that a juvenile male, Paul (A), encountered an adult female, Mel (T), who had just finished the difficult task of digging up a nutritious corm (Figure 6.8). These are desirable food items, and the hard ground probably meant that Paul would be unable to dig his own. Paul looked around and, when assured that no other baboon was watching, let out a scream. Paul's mother (Tool), who was higher ranking than Mel, ran to the rescue and chased Mel away. Paul was left by himself, whereupon he enjoyed eating the abandoned corm.

The significance of this is that the deployment of deceptive tactics demonstrates the ability of some primates to imagine the perspective of others. The interpretation of such observations must be made with great care since, for human observers, it is all too easy to impute intentions that are not there. Nevertheless, a body of evidence is beginning to suggest that some animals are capable of practising deception. This may not sound like a great intellectual feat but, in the animal kingdom, only humans and a few other primate species seem to have this ability. Grooming and deception make great demands on brain power. Social primates must be able to recognise one another, remember who gave favours to whom, who is related to whom and, most demanding of all, consider how a situation would look to another.

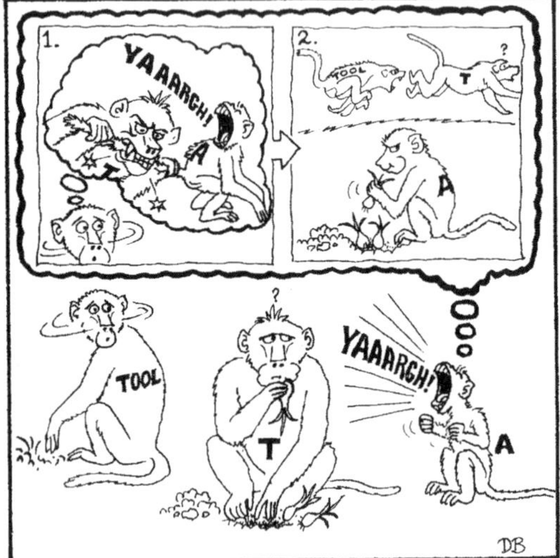

FIGURE 6.8 An interpretation of deceptive behaviour in baboons, *Papio ursinus*

SOURCE: From observations by Byrne, R. W. (1995) *The Thinking Ape*. Oxford, Oxford University Press; drawing by D. Bygott. Used with permission.

Deception involves penetrating the mind of others. **Theory of mind** was a term first used by primatologists when they realised that chimps could solve problems that depended on their appreciating the intentions of another individual, in other words, realising that other objects out there in the world have minds complete with beliefs, intentions and mental states that can be predicted (see also Chapter 15).

Theory of mind

We can conceive of this appreciation of other minds in terms of orders of **intensionality**. A dandelion probably has zero-order intensionality: it is not aware of its own existence; there is no one at home. Self-awareness indicates first-order intensionality. When Descartes began his famous train of sceptical reasoning and pushed doubting to its limit, he arrived at the fact that, if he doubted his own existence, he thereby proved it, since somebody must be doing the doubting, hence 'cogito ergo sum' – 'I think therefore I am'. Second-order intensionality involves self-awareness and the realisation that others are similarly aware. From here on, we can posit an infinite sequence: 'I think' is first order, 'I think, you think' is second, 'I think that you think that I think' is third order and so on. Children acquire second-order intensionality between three and four years of age. Most adults can keep track of about five or six orders of intensionality before they forget who is thinking what. Consider the old Machiavellian trick of sowing a good idea in someone else's mind and then pretending it was their idea so that they will be committed to it (useful when dealing with those on the next rung up in the dominance hierarchy). In such cases, we effectively want someone to believe that we think that they had a good idea ('I think that you think that I think that it was your idea'), which is third-order intensionality.

It is easy to ascribe zero-order intensionality to plants and machines but much harder to decide what has self-awareness, or first-order intensionality. Behaviourism faced this difficulty by treating all animals as machines and thus assuming zero-order intensionality. Some even adopted this approach to humans, but without much success. There are a number of problems in describing awareness in others. There is probably a natural bias in human thinking towards anthropomorphism. Humans have an acute sense of their own existence, our lives are dominated by goals and motives, and it is natural for us, and probably rightly so, to interpret the behaviour of other humans in our own terms. It is all too easy, however, to transpose this framework onto the behaviour of other animals. When a cat offers affection to humans, we are bright enough to spot that this may be just 'cupboard love' and more often than not the cat simply wants feeding. It is easy to interpret this as second-order intensionality: the cat is self-aware, aware that it is hungry and knows that it can convey this to us and so manipulate our minds to set about feeding it. In fact, we are probably far too generous in our attribution of intensionality here. The action of the cat may simply be a learned response. A cat may have first- or even zero-order intensionality, but how can we know?

One ingenious method occurred to Gordon Gallup who, in the 1960s, was a psychologist at the State University of New York. While shaving, Gallup (1970) realised that using a mirror indicates self-awareness. With very little training, humans realise that the image in the mirror is of themselves and can be used to judge and alter their appearance. Most animals, it seems, never appreciate the significance of their own image. Domestic kittens and puppies react as if the image were another individual and then gradually lose interest. Monkeys can use mirrors as a tool to see round corners to solve puzzles presented to them in captivity but never react to their image in a way that indicates self-awareness. One clever test of self-awareness is to place a spot of odourless paint on the hand and forehead of a monkey or ape while it is asleep. When the monkey or ape recovers consciousness, it typically notices the paint on its hand and attempts to remove it. When a monkey is presented with a mirror, it never makes the connection between the spot in the image and the fact that it is on themselves. In contrast, chimpanzees and orang-utans correctly grasp the significance of the image and use the mirror to help to remove the spot of paint. Some gorillas fail the test, but one captive, called Koko, passed easily. More recently, a surprising result has emerged: elephants too can pass the mirror test. A 34-year-old female elephant called 'Happy' at Bronx zoo was able to use a mirror to identify herself and inspect a mark painted on her forehead (Plotnik et al., 2006). The authors of the study suggest that since elephants and primates parted long ago, it is a case of convergent evolution.

Not everyone is convinced, however, that the mirror test reveals some higher order thought process. Just because humans recognise the image as the self does not mean other animals have this same conceptualisation. They may simply be using the mirror as a tool to inspect their own body, a simple extension of more direct inspection for self-grooming purposes – something that many animals do (see Tomasello and Call, 1997; Heyes, 2006).

Moving up the ladder of intensionality, we can now ask whether self-aware animals such as chimps are also aware of other minds, that is, whether they have a theory that other minds exist. As noted earlier, Byrne (1995) has concluded that the observational evidence demonstrates that only chimpanzees, orang-utans and gorillas practise intentional tactical deception, that is, behaviour that can best and parsimoniously be explained by one animal deliberately manipulating the mind of another animal into a false set of beliefs. Evidence on deception in cats and dogs was ruled out as probably being a result of trial and error learning. The evidence from tactical deception suggests, therefore, that only the great apes, some species of baboons and humans are capable of first- and/or second-order intensionality.

Theory of mind was a profound breakthrough for the apes and early hominins. In the case of humans, it has, coupled with language, given us science, literature and religion. The theory of mind may have arisen from the complex social world of early hominins and itself promoted encephalisation, or it may have been a product of relative brain enlargement that developed anyway in relation to ecological factors. Against the promise held out by research into the social world of primates, we must balance the fact that not everyone is convinced by the idea that some primates have second-order intensionality. Tomasello and Call (1997, p. 340), for example, conclude that 'there is no solid evidence that non-human primates understand the intentionality or mental states of others'. These reservations do not, however, invalidate the whole social complexity hypothesis. However chimps represent their social world, it is clear that it makes significant cognitive demands in addition to the demands of foraging and physical survival. These two sets of factors, environment and sociality, may have been inextricably linked in the causation of hominid encephalisation and human intelligence. Some recent work has tested the competing claims of the two theories and does suggest that one set of factors may have been crucially important. The next section examines this issue.

Testing the social brain hypothesis

To test the social brain hypothesis, we obviously need some way of measuring the level of social complexity set by group size and group dynamics, and the level of intelligence possessed by species that forage and live in groups. The social complexity of a group is to some degree indicated by the mean size of the group: the larger the group, the more relationships there are to keep track of, the higher the levels of stress and the greater the all-round level of harassment. Measuring group size is also fairly easy and reliable data exist for a range of primate species.

Measuring the intelligence of an animal is tricky, to say the least. One indirect way is to measure the deviation of brain size from the allometric line of body weight and relative brain size and so calculate the EQ for a given species. But as noted earlier, there are reasons for suspecting that brain size relative to body mass may only be a rough measure of animal intelligence and that we need something more precise.

As a body or organ develops through evolutionary time, it cannot at any stage be rebuilt from scratch. Even if the conditions in which an organism finds itself are very different from those of its ancestry, evolution must make do with what it already has to work on. Not that great feats cannot be achieved by this – a fin can turn into a leg, a leg into a wing and a wing back again into a fin – but evolution is essentially opportunistic, and as a consequence, organisms often bear the scars of their past. The human forearm, the wing of a bat and the hand of a frog, for example, are all based on a five-digit plan reflecting an adaptation from some remote five-digit ancestor. As long ago as the early 1970s, MacLean (1972) argued that the human brain can be divided into three main sections: a primitive core that we have inherited from our reptile-like ancestors; a mid-section that contains areas concerned with sensory perception and integrating bodily functions; and finally an outer layer or **cerebral cortex** that is distinctive to mammals (Figure 6.9).

The word 'cortex' comes from the Latin for 'bark', and it is this crinkly outer layer which lies like a sheet over the cerebrum. It consists largely of nerve cell bodies and unmyelinated fibres (that is, fibres without a white myelin sheath), giving it a grey appearance – hence the phrase 'grey matter' to distinguish it from the white matter beneath. The cortex is only about 3 mm deep in humans. In non-primate mammals, it accounts for about 35 per cent of the total brain volume. In primates, this proportion rises to about 50 per cent for **prosimians** and to about 80 per cent for humans. If we desire some objective measure of animal intelligence, it could be the cortex

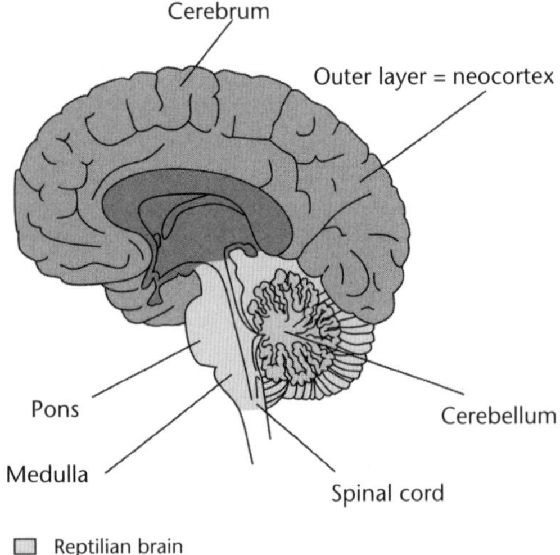

FIGURE 6.9 Triune model of the brain as proposed by MacLean

SOURCE: Adapted from MacLean, P. D. (1972) 'Cerebral evolution and emotional processes: new findings on the striatal complex.' *Annals of the New York Academy of Sciences* **193**: 137–49.

Reptilian brain: Consists of the brain stem, medulla, pons, mesencephalon, basal ganglia, reticular activating system, midbrain and the cerebellum. Its purpose is related to physical survival: digestion, reproduction, breathing, circulation and the execution of the fight or flight response are housed here. Area responsible for basic drives, repetitive and ritualistic forms of behaviour. Involved in 'innate' disposition to establish hierarchies. Also possibly storage of learnt forms of behaviour.

Old mammalian (limbic system): Contains a number of areas concerned with fighting, feeding, self-preservation, sociability, feelings, and affection for offspring. It includes the amygdala (see Chapter 8), the hypothalamus, the mammillary body, the anterior thalamus, the cingulated cortex and the hippocampus. Linked with emotionally charged memories, links emotion and behaviour and so can override the habitual reactions of the reptilian brain. Linked with activities related to food, sex and emotional bonding.

Neomammalian (neocortex): Relatively recent in evolutionary time. Well-developed neocortex found only in higher mammals. Thin sheet of neutrons but highly convoluted. Makes up about two-thirds of brain mass in humans. Receives information from eyes, ears and body wall. Responsible for higher mental functions, well developed in primates, especially humans where it constitutes about five-sixths of the human brain. Processes sensory information and coordinates and plans voluntary movements.

that we need to focus on. The cortex surrounding the cerebellum is often more specifically referred to as the neocortex to distinguish it from other cortical areas of the brain, such as the pyriform cortex and the hippocampal cortex.

Social complexity, neocortical volume and intelligence

If we accept that it may be the neocortex that is the advanced region of the brain concerned with consciousness and thought, it is this region of the brain that should correlate with whatever feature has driven the increase in intelligence in humans and other primates.

To test between environmental and social complexity theories, Robin Dunbar (1993) of the University of Liverpool plotted the ratio of the volume of neocortex to the rest of the brain against various measures of environmental complexity and also against group size. The results were fairly conclusive. He found no relationship between neocortex volume and environmental complexity, but a strong correlation between the size of the neocortex and group size (Figure 6.10).

The correlation observed in Figure 6.10 looks promising for the Machiavellian hypothesis, but neocortex volume is still an indirect measure of intelligence. In an attempt to establish whether the neocortex ratio correlates with Machiavellian intelligence in a more direct way, Byrne (1995) collected data on actual observed instances of Machiavellian intelligence in action.

If one primate deceives another in such a way that it shows some appreciation of the other's mental state, this is taken as an example of Machiavellian intelligence. However, some primates have obviously been studied more than others, and this will tend to increase the number of tactical deceptions observed. Byrne allowed for this effect by calculating a tactical deception index based on the number of studies undertaken. In addition, he was rigorous in excluding episodes that could be reasonably interpreted in other ways (Byrne, 1995). The result is shown in Figure 6.11.

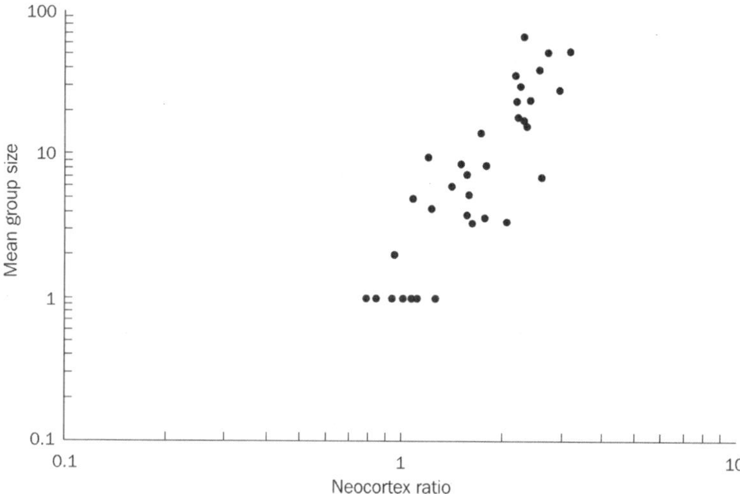

FIGURE 6.10 Plot of group size against neocortex ratio for various species of primate

SOURCE: Adapted from Dunbar, R. I. M. (1993) 'Coeveolution of neocortical size, group size and language in humans.' *Behavioural and Brain Sciences* **16**: 681–735.

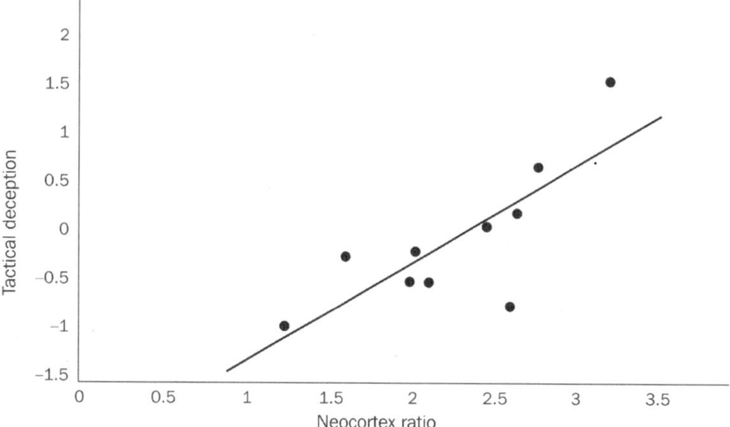

FIGURE 6.11 Relationship between neocortical ratio and index of tactical deception for a variety of primates

SOURCE: Byrne, R. (1995) *The Thinking Ape*. Oxford, Oxford University Press, Figure 14.3, p. 220.

The correlation of 0.77 found in Figure 6.11 is highly significant and offers further support to the idea that neocortical enlargement may have been driven by the advantages to be had from processing and utilising socially useful information.

The social brain hypothesis is not without its critics and faces difficulties in explaining why primates with social systems similar to the great apes, such as capuchins and macaques, are not as intelligent as the great apes. One answer may be that even controlling for group size, the social life of the great apes is more complex and cognitively demanding. But why this should be so is not addressed in the social brain model. There is also the problem of the highly enecephalised but relatively solitary orang-utan – often left out of group size and brain size correlations as an anomaly.

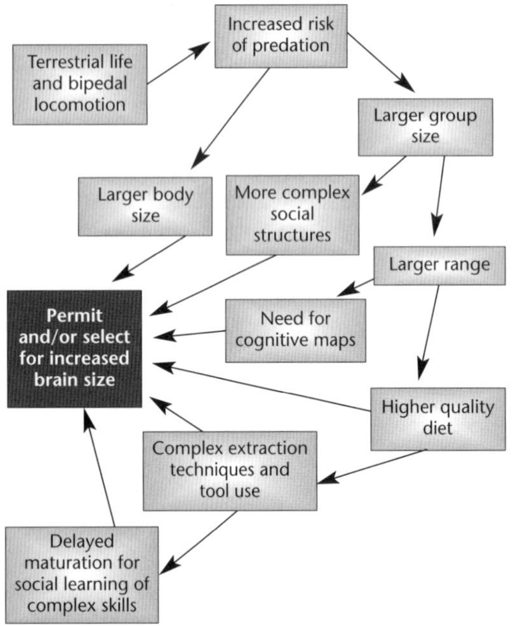

FIGURE 6.12 Conjectured relationship between various factors acting synergistically to exert a selective pressure on brain size in hominins

All single-factor models seem to face anomalies and problems. This suggests a multifactoral approach may be better. In the case of social complexity, for example, although chimpanzees have similar social systems to capuchin monkeys, the successful capuchin monkey only has to outsmart other capuchins, not chimpanzees. The immediate ancestors of chimps were already more encephalised than capuchins, perhaps because of ecological driving forces, and so the chimp line started on the road to increased brain size from a different starting point. We should not therefore expect equalities in encephalisation in relation to social systems between taxa. It is quite possible that facts such as quality of diet, group size, territorial range and tool use acted synergistically to drive up brain size in early hominins (see Figure 6.12).

6.2.6 Hardware–software coevolution: the stimulus of language

From Australopithecinces to modern *Homo sapiens* brain size has roughly doubled every 1.8 million years. This qualifies it for the term 'exponential growth' – growth whereby a quantity doubles every fixed unit of time. Now 1.8 million years might seem like a long time but biologically the evolution of the brain has been remarkably fast. This is why scientists seeking to understand the phenomenon have looked to 'runaway' or 'positive feedback' models to explain it. The idea of sexual selection is a runaway model: females began by preferring clever brains, which meant that men with such brains left more offspring, including daughters with larger brains who preferred even larger male brains and so on.

This can also be understood as positive feedback: where the effect of a growth in something is to make further growth more likely. You can sometimes hear feedback at concerts if a microphone is placed too near a speaker. An initial small sound is picked up by the microphone, amplified and fed to the speaker. The microphone then picks up the sound from the speaker, amplifies it and sends it out with an increased volume. The net result is a runaway effect that leads to an ear-piercing whine.

We can also observe an exponential growth in modern computers. On 11 April 2005 the computer giant Intel placed an offer of a reward of $10,000 on eBay for a single copy of an electronics magazine published in 1965. The copy was much desired since it contained the first announcement of a law of computer growth by the co-founder of Intel, George Moore. The prize was claimed by an Englishman who had kept back copies of the magazine under his floorboards. Moore's law states that the capacity of the latest computers of a given size doubles every 1.5 years. Amazingly, this law has held good over the past 10 years.

The biologist Richard Dawkins has used this law as a metaphor for a view of brain growth, which he calls 'self-feeding coevolution' (Dawkins, 1998). The reasons why computers have grown in power as they have are probably complex, but one of the driving forces must have been the coevolution of software and hardware. Improved software puts a pressure on the hardware to keep up; in turn, new hardware gives rise to greater software possibilities and so on. As an example, the mouse (hardware) stimulated the graphical user interface (software), which led to the popular Windows software. So what could have been the software and hardware for brains? Dawkins suggests that the hardware is of course the brain material – the neurons and their connections; for the software, a number of candidates suggest themselves, perhaps the most obvious being language.

6.2.7 Sexual selection and brain size: the display hypothesis

As we have seen, between about six and three million years ago, our ancestors roamed the African savannah with brains about the size of a modern chimpanzee (450 cc). Then, two million years ago, there began an exponential rise in brain volume that gave rise to modern humans with brains of about 1,300 cc. A tripling of brain size in three million years is rapid by evolutionary standards: in terms of brain power, the hominins left the other primates standing (or rather walking on all fours). One force that can bring about such rapid change is sexual selection (see Chapter 3).

Geoffrey Miller (2000) suggests that such a process has shaped human brains. In this view, humans would have examined potential partners not only to estimate their heath, age, fertility and social status but also to appraise their cognitive skills. Just as the peacock displays its extravagant and gaudy tail to attract peahens, and male bower birds build complex bowers from twigs and colourful feathers to attract females, so humans (especially males) display their intellectual and manual skills through artful displays such as telling complex stories and making artefacts. In today's world this is maintained as performance in art, science and literature.

Miller sees this runaway growth in brain size as beginning with females choosing males that are amusing, inventive and have creative brains. Language accelerates the process since the exchange of information can now be used to judge the suitability of a potential partner. Although brain growth was driven by female choice, both sexes gradually acquired larger brains since brains are needed to decode and appreciate inventive male displays and the genes responsible may not be sex linked anyway. This idea is revisited in Chapter 16.

6.2.8 Genomic imprinting and brain growth

Genomic imprinting is a phenomenon discovered only relatively recently. The traditional view within **Mendelian genetics** was that with the exception of the sex chromosomes (XX in females and XY in males), the chromosomes inherited from each parent were functionally equivalent. Genes could of course be dominant or recessive, but there was no reason to suppose that the operation of normal functional genes could depend upon their parental origin. Indeed, considering the process of inheritance in diploid organisms, it would seem highly unlikely that the fortunes of any gene should depend upon which parent is transmitting it, since as a result of meiotic recombination (see Figure 2.7), a gene transmitted by the mother in one generation could have descended to her from either her father or mother. Similarly, she could pass it to both daughters and sons. We now know, however, that genomic imprinting proves an exception to this picture.

Genomic imprinting entails silencing the operation of a gene (imprinting) when it is inherited from one parent but not the other. At the time of writing, there are about 70 imprinted genes known in humans and estimates suggest that there may be between 100 and 200 mammalian imprinted genes in all (see Davies et al., 2004). A clear example of this is seen in the case of Turner's syndrome, which reveals genes on the X chromosome to be imprinted when they are passed down from the female (see Box 6.1).

Turner's syndrome also provides a fascinating window into the possible fundamental causes of sexual dimorphism in the human brain and the basis of sex-biased mental disorders.

Box 6.1 Turner's syndrome

Turner's syndrome is named after Dr Henry Turner who discovered the condition in the 1930s. It affects about 1 in 2,000 female births but may be as high as 10 per cent in miscarriages. People with this condition have low stature and suffer a variety of problems such as infertility, cardiac abnormalities and impaired cognition. It was pioneering work by David Skuse and his colleagues that led to the elucidation of the role of imprinting in this condition (Skuse et al., 1997). In normal cells, females possess two X chromosomes (XX) compared with the one of males (YX). To compensate for this, one of the X chromosomes is randomly inactivated in each somatic cell – otherwise female cells would get a double dose of whichever proteins the genes on the X chromosome are responsible for. However, during gamete production, occasional faults in meiosis can produce eggs and sperm with no sex chromosomes. If such a gamete were to combine with a normal gamete, we have possible zygotes of the form YpO, XmO and XpO, where O indicates the absence of the expected sex chromosome, Yp implies the Y chromosome came from the father, and Xm and Xp that the X chromosome came from the mother and father respectively. All the YO zygotes and most of the XO ones are not viable but some girls are born XpO or XmO (we should note that the new nomenclature for this is 45X in contrast to the normal karyotype for females as 46XX). Of those girls who are 45X, about 70 per cent are XmO and the other 30 per cent XpO. The advantage of this condition from a research point of view is that in such monosomic (one sex chromosome) cells, the X chromosome is never inactivated. It follows that if any of the genes on the X chromosome are imprinted either from the male or female line, this should give rise to two distinct types of Turner's syndrome. Skuse et al. found that the origin of the X chromosome did indeed seem to affect the type of disorder experienced by patients, as follows:

- 72 per cent of Xm females have more social difficulties than Xp females
- Xm females have more difficulty in suppressing inappropriate social behaviours compared to Xp females
- Four per cent of Xm patients suffer from autism, well above the normal rate of 1 in 10,000 females
- Xm girls score lower than Xp girls on measures of verbal IQ (Scourfield et al., 1997).

One of the more important implications of this work is that it reveals a mechanism to explain differences in social and cognitive skills between normal males and females. There is now considerable evidence that females exceed males in levels of social skills and verbal ability (see Geary, 1998).

It is also clear that males are far more vulnerable than females to disorders involving social adjustment and language (such as autism). This could be explained by the fact that males invariably receive their X chromosome from their mother. But as we have seen, maternal X chromosomes have some of their genes imprinted. If there is no corresponding (homologous) set of genes on the Y chromosome, or these have a lower level of expression, or they are impaired in some way, then such males will have reduced abilities compared to Xp/Xm females that have functioning Xp genes.

The exact mechanism by which genes are imprinted is not yet known. The prevailing hypothesis is that the cytosine nucleotides on the promoter region of the genes become methylated and that this forestalls the transcription on DNA in this region.

Although there are still many uncertainties about the function of imprinting and its interpretation (see Davies et al., 2004 for a review of the difficulties), one quite profound implication of the phenomenon is that it suggests a new mechanism to account for the rapid evolution of human brain size and intelligence. This is that the human brain grew large as a result of escalating genomic conflict propelled by imprinting.

As discussed in Chapter 10, because of asymmetries in nourishment of the embryo, parental care, and degrees of relatedness between genes among siblings dependent on the parental origin of the genes, then genes descended maternally have more to gain than paternal ones from offspring that act cooperatively towards their siblings, and are capable of empathy and restraint. Conversely, paternal genes will gain from a brain system that extracts excessive (from the mother's perspective) resources from the mother and her children, and so is more driven by selfish and impulsive drives. From these considerations, it follows that maternally derived genes would benefit more than paternal ones by finding themselves in an organism with a well-developed neocortex; whereas paternal genes would benefit more than maternal ones from organisms with a less well-developed neocortex but a robust limbic system. The suggestion then is that, as genes for the forebrain evolved, males responded by imprinting their own copies. This hypothesis is consistent with the observed patterns of disorders when damage to imprinted genes occurs. So, for example, mental retardation and speech deficiencies result from damage to maternally active (that is, non-imprinted genes) that guide the construction of the cortex and the striatum (Wagstaff et al., 1992).

Support for these ideas also comes from experiments on mice. *In vitro* fertilisation technology has made it possible to construct androgenetic (ag) and parthenogenetic (pg) 'chimeras'. Androgenetic chimeras contain both their sets of genes from their father; parthenogenetic chimeras have both sets from their mother. By themselves, such chimeras tend to die before about 11 days of development, but adding normal cells at an early stage rescues the development of the embryo. As the mice grow, some interesting patterns of development emerge. First, pg chimeras are retarded in their growth, while ag chimeras grow so quickly that they often have to be delivered by Caesarean section. Second, pg chimera cells are found to be replicating in large numbers on the neocortex and forebrain but few are found in the lower limbic brain areas. In contrast, ag cells are found in large numbers in the limbic brain and the hypothalamus, but few are found in the neocortex (Keverne et al., 1996).

As Badcock (2000) observes, if brain size in the hominid lineage did evolve by escalating genomic conflict between paternally and maternally derived genes, then this is why adaptationist accounts have such a hard time explaining why the hominid brain escalated so rapidly in size above those of other group living primates. On the other hand, we might also ask why such genomic conflict did not take place in other groups to the same extent as that in humans.

6.2.9 Neoteny as a mechanism for brain growth

Although evaluating the different theories purporting to account for why encephalisation occurred is fiendishly difficult, understanding the genetic mechanism by which it occurred may turn out to be a simpler task. Speciation must ultimately rely upon the genetic novelty brought about by mutations. But the problem here is that once a certain level of complexity is reached, such as the primate body plan, major changes in gene structures are likely to be highly damaging. But there is a way forward to exploit the adaptive landscape: small changes to genes involved in the timing of development have the potential to bring about major changes. A prime example of this is known as heterochrony, which is defined as a change to the onset or timing of development such that the appearance of traits in an organism is either accelerated of slowed down compared to that of the premutation ancestor. One specific subset of this process is known as **neoteny**, which is the extension of juvenile features into adulthood.

Neoteny illustrates how ape-like features can be morphed by small and plausible genetic changes into hominin ones. It is well known that the skull shape of a chimpanzee in its juvenile form resembles that of a human, but changes drastically during development. So if the genes that control the maturation of the infant ape skull are suppressed, this will prevent the development of a thick skull and huge teeth, maintaining instead juvenile skull proportions into adulthood.

6.3 Language

The application of evolutionary theory to the origins of human language has always been controversial. The arguments in the 19th century were so heated that, in 1866, one scientific society, the Société de Linguistique de Paris, banned all further communications to it concerning the history of language. Over 140 years later, there are still wide-ranging disagreements over such basic issues as the probable timing of the start of human language and even whether language is a product of natural selection or is merely some emergent property of an increase in brain size. There are those who maintain that language first appeared in the Upper Palaeolithic period about 35,000 years ago, and those who suggest that language arrived with the appearance of *Homo erectus* about two million years ago. There is only room here to glance briefly at some of the main arguments in these debates. We will, however, explore the implications of some recent work on grooming and language.

6.3.1 Natural selection and the evolution of language

The leading exponent of language as a product of natural selection is probably Steven Pinker, a linguist at the Massachusetts Institute of Technology in the USA. Pinker (1994) advances a number of arguments tending to suggest that language has been the outcome of a selective force. In summary, these are:

- Some people are born with a condition in which they make grammatical errors of speech. These disorders are inherited (see also Gopnik et al., 1996)
- Language is associated, although not in a simple way, with certain physical areas of the brain, such as Wernicke's and Broca's areas (see section 6.3.2)
- Complex features of an organism that have been naturally selected, such as the eye of a mammal or the wing of a bird, bear signs of apparent design for specialised functions. Pinker argues that language bears these same types of design feature
- Children acquire language incredibly quickly. Parents provide children with only complete sentences and not rules, but children nevertheless infer rules from these and apply them automatically
- The human vocal tract has been physically tailored to meet the needs of speech. Specifically, humans, unlike chimps, have a larynx low in the throat (Figure 6.13). This allows humans to produce a greatly expanded range of sounds compared with a chimp

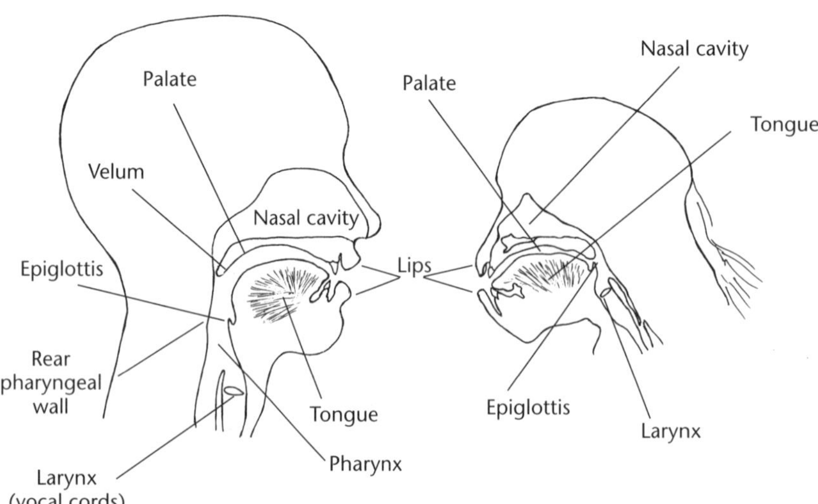

FIGURE 6.13 Comparison of head and neck of a human adult and an adult chimpanzee

SOURCE: Based on Passingham, R. (1988) *The Human Primate*. Oxford, W.H. Freeman.

ate that manner

e would level of ımatical ı correl- hunter-

ts of the ıre such is the all the a wing – and ing at st ask: ıcteris- tionist ıguage ıguage er and ıat the proba- haring guages, the language of children and non-fluent tourists, show that a continuum of language skills and attributes can exist that still have use value. Other animals also show the value of even a limited vocabulary. Vervet monkeys, for example, have alarm calls that distinguish between leopards, eagles and snakes. If language is an adaptation, then it can be expected to be associated with specific areas of the brain, and this indeed is what is found, as discussed below.

6.3.2 Localisation of language function in the human brain

Two milestones in the understanding of the neural basis of language came in the second half of the 19th century with the discovery of areas of the brain linked with language: Broca's area and Wernicke's area (Figure 6.14), named after their discoverers Pierre Paul Broca (1824–80) and Karl Wernicke (1848–1905).

Paul Broca was a neuroanatomist and professor of brain surgery at the University of Paris. In 1861, he performed a postmortem autopsy on a patient nicknamed 'Tan' on account of his inability to utter any words other than 'tan'. Broca showed that Tan had a lesion (caused by syphilis) towards the front of the left hemisphere of his brain. Subsequently, this area became known as **Broca's area** and patients with damage to this region are diagnosed as suffering from Broca's aphasia, characterised by impairments to speech output. If you place your finger on your head just above your left temple, Broca's area lies just beneath the skull.

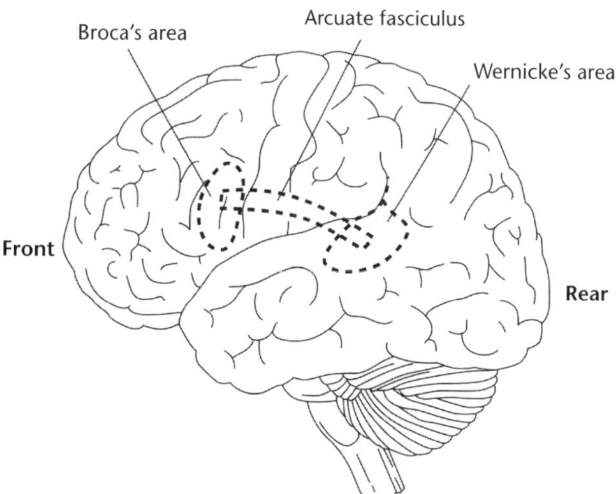

FIGURE 6.14 Broca's and Wernicke's areas
The arculate fasciculus is a tract that connects Wernicke's area with Broca's area. It enables the comprehension centre of Wernicke's area to activate the speech programmes in Broca's area.

A similar disease is Wernicke's aphasia and is linked to damage to an area of the brain situated towards the rear of the left hemisphere in the superior temporal lobe. By placing your finger just above and behind your left ear, you are likely to be pointing to **Wernicke's area**.

Karl Wernicke was a Polish physician who noticed that deficiencies in language comprehension were associated with damage to a different part of the brain to that identified by Broca. Broca's aphasia is characterised by agrammatical sentence utterance: a patient will typically not be able to distinguish between the sentence 'the cow kicked the horse' and 'the horse kicked the cow'. In contrast, people suffering from Wernicke's aphasia can speak fluently and grammatically but the content is meaningless. Moreover, they fail to comprehend why others cannot understand them.

6.3.3 Dating the origin of human language: anatomical evidence

Predictably, if there is debate about whether language is an instinctive evolved trait or a cultural achievement, there will be debate about when it began. If we agree with Pinker that language is a feature of human biology, it must have begun no later than about 200,000 years ago when *Homo sapiens* appeared, and could have been around in earlier hominid species. If we take the cultural achievement view, we must place the origin of language with the origin of symbolic culture, generally around 35,000 years ago. At this point, anatomical evidence may be useful.

A sudden expansion in brain size among our ancestors began about two million years ago with the first member of our genus, *Homo habilis*. By about one million years ago, *Homo erectus* had a brain capacity of about 1,100 cm^3, which is not far from our own of 1,300 cm^3. If language coincided with sudden brain expansion, this points to *Homo erectus* and *Homo habilis* as having had language capabilities. The problem is of course that we do not know whether there is a minimum brain size for language to start. If the minimum were 1,300 cm^3, only early *Homo sapiens* at 200,000 years ago could have had language. We know that language is associated with certain areas of the brain such as Broca's area and Wernicke's area. It has been claimed that an examination of endocasts of fossil skulls shows evidence of the existence of a Broca's area in the cranium of a *Homo habilis* specimen dated at about two million years old (Falk, 1983; Leakey, 1994). There have also been attempts to link asymmetries in fossil crania with the origin of language. Language in most people is associated with the left hemisphere, and, partly as a consequence of packing language circuitry into one half of the brain, the left hemisphere is slightly larger. Holloway (1983) suggests that this too can be detected in a *Homo habilis* specimen. Objections to this have included the fact that even specimens of Australopithecines show brain asymmetries, and the attribution of language to this small-brained ancestor seems improbable.

The most compelling anatomical evidence, however, comes from an examination of the throats of humans, apes and our ancestors (see Figure 6.13 above). The voice box of humans is called the larynx and contains a cartilage, called the Adam's apple, which bulges from the neck. If you feel your Adam's apple, you will notice that it is some distance down from your mouth. Humans have their larynx low in their throats, which allows the space for a large sound chamber, the pharynx, to exist above the vocal cords. In all other primates, the larynx is set high in the throat, which restricts the range of sounds that can be made but does at least allow the animal to breathe and drink at the same time. Humans are in fact born with their larynx high in the throat, which helps the infant to suckle while breathing. After 18 months, the larynx begins to move down the throat, finally reaching the adult position when the child is about 14.

In adult humans, the tongue forms the front wall of the pharangeal cavity, so movement of the tongue enables the size of the pharynx to be altered. In chimps and infants, the higher position of the larynx means that the tongue is unable to control the size of the air chamber immediately above it. The differences in the position of the larynx between humans and primates strongly suggest that we have evolved vocal apparatus to convey a wide range of sounds.

At first glance, it would seem an easy task to determine the position of the larynx in ancestral fossils and thus work out when the enlarged speech-facilitating pharynx was established. The problem is that all this vocal apparatus is composed of soft tissue, which does not fossilise. Clues can, however, be found in the bottom of the skull, the basicranium, which in humans is arched as a result of our specialised vocal equipment and quite unlike the flat shape of other mammals. Following this approach, Laitman (1984) found that a fully fledged basicranium emerged about 350,000 years ago. The earliest *Homo erectus* specimen of about two million years ago has a slightly curved basicranium typical of the larynx position of a modern six-year-old boy.

6.3.4 Grooming coalitions and group size

The encouragement that the Machiavellian hypothesis receives from data on neocortical enlargements prompts a deeper inquiry into why large groups should make greater cognitive demands. As humans, we are acutely aware that large groups quickly resolve into 'cliques' or small subgroups that trade information and help. A similar phenomenon is observed in other primates (but without the mediation of an aural language) when they spend time in fairly stable subgroups grooming one another. If grooming serves as some sort of social cement that binds groups together, we would expect the proportion of time that a primate spends on grooming to be related to group size. Dunbar (1993) tested this prediction using 22 species of primate that live in stable groups. The correlation, shown in Figure 6.15, is at least consistent with the idea that grooming is needed to maintain the cohesion of primate groups. It is important to note that grooming takes place among special subgroups or cliques. Any individual does not distribute grooming time equally among all other members of the total group.

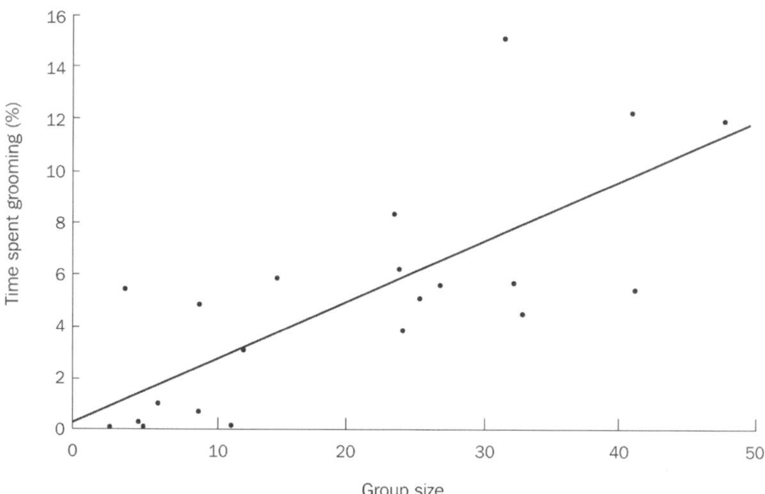

FIGURE 6.15 Mean percentage of time spent grooming versus mean group size for selected species of apes and Old World monkeys

The equation of the regression line is:
$G = -0.772 + 0.287N$.

SOURCE: Dunbar, R. I. M. (1993) 'Coeveolution of neocortical size, group size and language in humans.' *Behavioural and Brain Sciences* **16**: 681–735.

The work of people such as Dunbar, Byrne and Whiten suggests that it is the social complexity of primate life that may have demanded an increase in brain size, and we may reasonably infer that this was also probably a powerful factor in driving up the size of brains in early hominids. One of the benefits of a large brain must surely have been language.

According to Dunbar (1993), the study of grooming and group size offers a way of discovering the origin of language. If we examine once again the plot of group size against neocortical ratio, shown above in Figure 6.10, we can use the equation of the regression line to predict something about the 'natural' size of human groups. The equation is:

$$\log N = 0.093 + 3.389 \log Cr$$

where: N = group size
Cr = neocortical ratio.

The neocortex in humans occupies about 1,006 cm^3 out of a total brain volume of 1,252 cm^3. This gives a Cr of 4.1, which, when inserted into the equation above, predicts a group size for humans of 148, with a 95 per cent confidence interval range of 100–231. We should note that this is the predicted size for a group of hominids living a hunter-gatherer lifestyle. Dunbar and Aiello (1993) assembled evidence, including the size of early Neolithic villages, the number in contemporary hunter-gatherer bands and military companies over the past 300 years, and the number of people with which modern humans can have a genuinely social relationship, to suggest that this figure has some credence. It is difficult to determine what ecological factors drove the size of human groups up from what we must assume were smaller groups of our ancestors, which may have been the size of those of modern chimps, at about 50. It could be that the more open habitat colonised by early hominids was more vulnerable to predation, that early hominids practised a nomadic way of life or that competition existed between different hominid groups.

If we now insert the figure for the size of a human group into the equation represented in Figure 6.15, $G = -0.772 + 0.287N$, we arrive at a predicted grooming time of 42 per cent. Now, grooming is part of the overall time budget of an animal: time spent grooming is time lost for hunting, foraging, looking out for predators and attending to young. Dunbar (1996) suggests that the upper affordable limit for most primates is about 20 per cent, the highest recorded value for an individual species being 19 per cent for baboons, *Papio papio*. Dunbar then proposes that, as the enlargement of the neocortex proceeded hand in hand with increasing group size and hence social complexity, there arose a point at which a more time-efficient means of grooming became essential. Dunbar's thesis is that language evolved as a cheap form of social grooming. Language enabled early *Homo sapiens* to exchange socially valuable information, not of the 'There is a beast down by the lake' sort as is usually supposed, but about each other: who is sleeping with whom, who can be trusted and who cannot. In short, language began as a device to facilitate the exchange of socially useful information – 'gossip' as it is sometimes called.

In support of this thesis, Dunbar and Duncan (1994) suggest that the subject matter of most conversations today is still predominantly of the gossip variety. In one study, conversations that were monitored in a university refectory were for over half the time concerned with social information, and dealing with academic matters for only 20 per cent of the time. The crucial thing about language as an effective grooming device is that it can be used while other activities, such as walking, cooking and eating, are still taking place. Moreover, several groomers can be linked together in a conversation, unlike the pairwise interactions of physical grooming. In fact, if we assume that early humans could only afford the same time as chimps for grooming (about 15 per cent of the time available), then, since the ratio of human group size to chimp group size is about 148:54 or 2.7:1, early humans needed something about 2.7 times more effective than one-to-one grooming. Dunbar and Duncan (1994) suggest that it is no accident that typical human groups forming for conversation or gossip usually number about four. From an individual point of view, interacting with three other people is three times more effective in the use of time than interacting with one, and three is pretty close to 2.7.

Table 6.6 Predicted percentage grooming times for fossil and modern hominins

Taxon	Number in sample	Mean precentage predicted grooming time
Australopithecus	16	18.44
Homo habilis/ rudolfensis	7	22.73
Homo erectus	23	30.97
Archaic *Homo sapiens*	18	37.88
Neanderthals	15	40.46
Modern *Homo sapiens* (female)	120	37.33
Modern *Homo sapiens* (male)	541	40.55

SOURCE: Data from Dunbar, R. I. M. and Aiello, L. C. (1993) 'Neocortex size, group size, and the evolution of language.' *Current Anthropology* **34**(2): 184–93.

Dunbar and Aiello's (1993) analysis can also be used to date, very approximately, the start of human language. If we accept the view that unspecified ecological factors drove up the group size for early hominids, and that this in turn spurred the growth of the neocortex and thus the time spent grooming, there came a point at which the time needed for grooming called for language as a more time-efficient grooming device. Dunbar suggests the 'Rubicon' of grooming time that had to be crossed by language was about 30 per cent. Now, there is an ingenious way to predict grooming time for early hominids. Since grooming time is related to group size and group size is related to neocortical ratio, we can, even though we have no direct idea of the size of the neocortices of early hominids, use the fact that there is an allometric equation linking neocortex size to overall brain size to make some predictions. The results are that a 'Rubicon' of 30 per cent of the time budget translates to a group size of about 107. When this is plotted on a graph of predicted group sizes for various hominid species based on suggestions for the neocortical ratios, it provides a date for the start of language-based grooming for late members of *Homo erectus* and early *Homo sapiens* somewhere between 300,000 to 200,000 years ago. Table 6.6 shows data on predicted grooming times for fossil and modern hominids.

One problem with these data is the high grooming time predicted for Neanderthals. If language provided the answer to the demands for grooming, this tends to suggest that Neanderthals had language, yet evidence from the cultural achievements of Neanderthals would suggest that they were not as advanced, at least in symbolic culture, as was *Homo sapiens*.

Dunbar's prediction is consistent with the idea that language is a distinguishing mark of *Homo sapiens*. It provides a rather later date than some of the anatomical evidence on cranial shapes and topographical features, and Dunbar's ideas are original and provocative. We will summarise a few areas in which the data or methodology has been criticised.

Association between grooming and group size

Figure 6.15 shows the relationship between primate group size and grooming time, which may be an association rather than a causal relationship. It could be that living in larger groups leaves more free time for grooming. Much also depends on the choice of species used to obtain the relationship between grooming time and group size. The inclusion of species that form fission–fusion groups, such as chimpanzees, gives a different regression line. There is even some debate over whether grooming does function to maintain group cohesion.

There are particular problems when Dunbar's method is used to date the origin of language, since there is a three-step process of inferring neocortex size from total brain size, then from neocortex size inferring group size, and from group size deducing grooming time. Any errors in the first or second stages (and there are wide confidence limits) are compounded by the time we get to the last.

Language as grooming

It is not obvious that language used to gossip is a direct equivalent of grooming: one would expect to find some structural similarities in the behaviours. Language appears too complex for grooming, and grammar, if anything, seems as much designed to describe the physical as the social world. It is also not obvious that a conversation group of four really is three times as efficient as speaking to individuals on a one-to-one basis. One-to-one conversations may be more intense and more productive in the sharing of valuable information compared with group gossip. To infer from gossip today an original function for language is a huge leap. We may gossip because our standard of living has given us time. The gossiping of academics and students in university refectories may not be typical of early hunter-gatherers.

Despite these reservations, Dunbar may have hit upon something very important. Language may have a role in supporting social intelligence. To function effectively in groups, we do need to know who are our relatives and friends. This may be obvious with close kin, but we need social information to recognise more distant relatives. In groups in which reciprocal altruism and especially indirect reciprocal altruism are found, we also need to keep track of the reputations of others and ensure that our own reputation is sound. Whether language started as gossip, and information about the physical world was a byproduct, or whether it was the other way around, is still debatable. The social intelligence hypothesis suggests a function for language once it has started. In the next section, we examine some recent evidence that offers insight into how language could ever start to emerge at all.

6.3.5 Mirror neurons and the origin of language

In the mid-1990s, Giacomo Rizzolatti and his research group at the University of Parma in Italy were investigating the behaviour of neurons in the brain of a macaque monkey when they made a remarkable discovery (Gallese et al., 1996; Rizzolatti et al., 2006). The neurons under investigation lay towards the front of the brain in an area known as F5 and known to be associated with hand and mouth movements (Figure 6.16). The procedure involved monitoring the firing (electrical discharge) of individual neurons. Sure enough, as the monkeys performed specific acts, such as grasping a toy or piece of food, so the neurons fired. Then something strange and unexpected was observed: when the monkey observed one of the experimenters grasping something, neurons in the monkey's brain fired in the same way as they would have done if the monkey itself had performed the act. The group called these neurons **mirror neurons** since they mirror (replicate) the behaviour of neurons involved in action (Gallese et al., 1996). Subsequent work has shown that these mirror neurons exhibit the following properties:

- The neurons do not respond simply to object presentation: they require a specific observed action.
- Some neurons are highly selective and specific in the actions they respond to. Response-provoking actions discovered so far include grasping, manipulating, tearing and putting an object on a plate.
- The neurons fire when the action is 'understood' and not simply in response to action-like movement. Hence a hand moving to pick up an object causes the neuron to fire but not a hand moving towards a non-existent object. It is not simply the visual appearance of the target object, however, that is important, since if the object is hidden behind a screen in such a way as the monkey could infer it was there, then the movement of a hand towards it still causes the neuron to fire (Iacoboni et al., 2005).

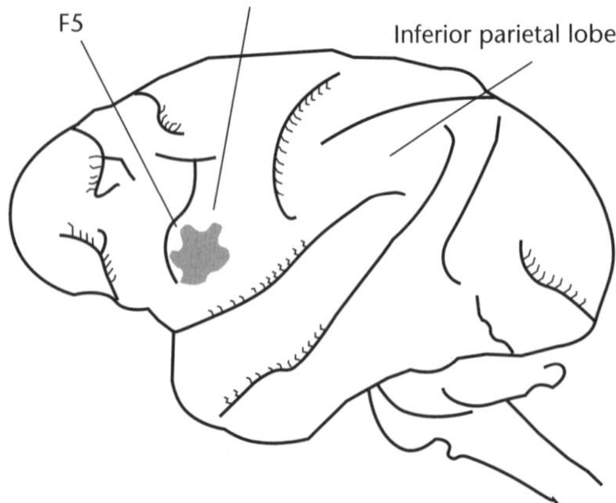

FIGURE 6.16 The brain of a macaque showing the area F5, the location of mirror neurons
This is similar in location to the language areas of the human brain (see Figure 6.14).

Rizzolatti and his colleagues (Rizzolatti et al., 2006) quickly recognised the implications and significance of these findings. They suggest the existence of a mechanism, in monkeys at least, for the direct internalisation and comprehension of the actions and intentions of others. To take a human parallel (assuming they exist in humans as well), imagine yourself observing someone moving towards an object with arm outstretched and hand poised in a gripping position. How do you know what their intentions are? Before the discovery of mirror neurons, most neuroscientists would probably say that you are drawing upon a rapid reasoning process: by comparing sensory evidence with information stored in your memory, you arrive at a realistic prediction about the intentions of the other person. The discovery of mirror neurons suggests an alternative, more rapid process: you understand the actions and intentions of another person because those actions and intentions are also happening inside your own head; your neurons behave as if you yourself were executing the action you are observing.

The obvious question is whether such neurons do exist in humans or not. Ethical considerations obviously forbid the opening of a human skull and the insertion of electrodes (some would question if this is ethical for monkeys as well) and so no human mirror neuron activity has been directly observed. There is plenty of other evidence, however, that suggests that humans do possess a mirror neuron system. In experiments using positron emission tomography (PET), brain activity was imaged while subjects observed an experimenter either grasping an object or, as a control, simply viewing an object. It was found that grasp observation caused activation in the left hemisphere in areas known as the superior temporal sulcus (a fissure in the side lobe of the brain), the inferior parietal lobule and the inferior frontal gyrus (Rizzolatti and Arbib, 1998).

The important point about these areas is that the last two especially correspond to the ventral premotor cortex area in the monkey, or the area F5 where macaque mirror neurons were observed.

It is also highly significant that this activation was noted in the left hemisphere. There are now many neuroscientists who think that area F5 is the monkey homologue of Broca's area in the human brain. In other words, the primate that was our last common ancestor with the macaque (which lived about 25–30 million years ago) had a brain area that evolved to become region F5 in the macaque and Broca's area in humans. So is it a coincidence that mirror neurons are found in both these areas, or is there some deeper evolutionary link between the two? At first inspection, the linkage between a system of action production and recognition and speech is not immediately apparent. But Rizzolatti and Arbib (1998) think there is a profound evolutionary connection between these two areas and that, in short, the human lateral speech circuits were built upon a pre-existing neural substrate of the type observed in the macaque mirror neuron system.

If the mirror neuron system really did form an ancient platform for language development, then a series of evolutionary stages has to be postulated from gestural meaning (inherent in the gesture and communicated by mirror neuron activity) to the abstract meaning in sounds that constitutes language (where the sound of a word has no necessary or obvious connection with its meaning). But since it is suggested that speech evolved from gestural meaning, then a neurophysiological prediction can be made: hand and arm gestures and speech formation may still show some linkage at the level of neural substrate.

There are at least two experiments that support this prediction. Meister et al. (2003) found that the excitability of the hand motor cortex increased when subjects were speaking. Now word articulation recruits the motor cortex on both sides of the brain, but the activation recorded by this technique was linked to the left hemisphere (the side of the language centres). In another study, Gentilucci (2003) asked volunteers to pronounce a syllable ('ba' or 'ga') while observing another individual grasping objects of different size. If gestures (recorded in the mirror neuron system) and language do share an ancient foundation, then one might expect large objects to elicit a higher level of vocalisation. This is exactly what was found: both lip aperture and voice amplitude were greater when the grasp extended to larger objects.

If human speech centres are linked with action recognition, can we observe something similar in macaques? Macaques cannot speak of course, but they do make vocal sounds. Gil-da-Costa et al. (2006) presented macaque monkeys with two types of aural stimuli: recordings of macaque vocalisations, and neutral sounds of non-biological origin but similar in duration, intensity and pitch to the sounds of macaques. PET scans of the brains of these monkeys showed that natural vocalisations evoked significantly higher activity in the tempoparietal area of the superior temporal gyrus (a gyrus is a ridge on the surface of the brain) than non-biological sounds. It is precisely these areas that correspond to the language centres in the human brain.

So there is strong evidence for a linkage between mirror neurons in the brains of macaques and humans, vocalisation and speech. What is now needed is a plausible evolutionary argument to explain how an action recognition system could be adapted and built upon to lead to human language. Some initial considerations look promising. A puzzle for all theories of language evolution is what Arbib (2005) called the 'parity requirement': the meaning for the sender must be the same as that for the receiver. The fact that the mirror neurons do precisely this for gestures (they fire inside the head of an observer, thereby replicating the meaning of the action of the observed) is an encouraging start. From then onwards, the proposed stages of evolution are necessarily speculative. Rizzolatti and Arbib (1998) suggest the following scenario:

- A pre-hominin neuronal system controlling grasping
- A pre-hominin system of mirror neurons in ancestors shared by humans and monkeys
- A pre- or early hominin system of imitation for grasping and action movements. The observer is able to imitate the action observed: this would be a relatively easy step since the very neurons involved in the action are already firing and have to be suppressed in contemporary macaques to avoid physical repetition of the action observed
- A hominin system for recognising the performance of another as a combination of actions already in the understood repertoire
- Proto-sign and proto-speech systems that break out from closed into open repertoires. The beginning of prelinguistic and then linguistic grammatical structures
- A final stage involving cultural evolution to fully developed language.

It is difficult to estimate how reliable this

sequence of events will turn out to be. What is very likely, however, is that the action recognition system described by Rizzolatti and Arbib lies at the heart of many communicative interactions. It is obvious that understanding the intentions of others rests at the core of such capabilities as imitation, empathy, mind-reading (or theory of mind; see above and Chapter 15) and recognising signals of dominance and submission. The distinguished neuroscientist Vilayanur Ramachandran (2000, p.1, accessed 2 March 2007 at http://www.edge.org/3rd_culture/ramachandran/ramachandran_p1.html) is enthusiastic about the potential of mirror neurons:

> I predict that mirror neurons will do for psychology what DNA did for biology: they will provide a unifying framework and help explain a host of mental abilities that have hitherto remained mysterious and inaccessible to experiments.

Ramachandran may be right, but the strong interpretation of mirror neurons supplying instant meaning to the observer faces one enormous problem. If it is suggested that mirror neurons only fire when the movement of an arm is directed towards some meaningful action (the grasping of an object) and replicate this meaning instantly inside the head of an observer, and not when confronted by movement alone, such as a hard moving towards a non-existent object, how does the mirror neuron system 'know' that the former is meaningful? In essence, if meaning is supposedly presented instantly in the brain, how can the system decide to be selective before the action is complete?

The phenomenon suggests some other neural circuitry that processes visual information, establishes meaningful action and then prompts mirror neurons to fire when such actions are observed. Problems like this probably do not undermine the importance of these neurons but they do remind us that complex arguments will be needed to show how mirror neurons lifted primates such as ourselves into the realm of complex linguistic communication.

Summary

- Large brains, their metabolic requirements and the consequent need for prolonged infant care would have favoured increasingly monogamous sexual relationships between our ancestors in the Pleistocene Epoch.
- Possibly the most adequate measure of human brain size is the encephalisation quotient, which measures the ratio of actual brain size to the size of brain expected for an organism of a given size.
- There are numerous theories purporting to explain how and why humans evolved large brains. The real reason may be a synergistic mix of factors and influences.
- The role of social factors in the evolution of primate and human intelligence is increasingly receiving support as part of the Machiavellian intelligence hypothesis.
- Although there is still widespread debate about the origin of language, and even about whether it is a product of natural selection, one intriguing set of ideas suggests that social complexity led to large brains. As group size grew in relation to ecological parameters, language evolved to serve as a grooming device in complex social groups. This line of thought places the origin of language somewhere between 300,000 and 200,000 years ago and identifies a surprisingly significant role for gossip.
- Recent work on mirror neurons suggests a means by which the understanding of gestures may have allowed langauge circuits to develop.

Key Words

- Allometry
- Broca's area
- Cerebral cortex
- Dominance hierarchy
- Encephalisation
- Encephalisation quotient
- Genomic imprinting
- Grooming
- Herbivore
- Intensionality
- Machiavellian intelligence
- Mendelian genetics
- Mirror neurons
- Neoteny
- Prosimian
- Reptilian brain
- Theory of mind
- Wernicke's area

Further reading

Byrne, R. (1995) *The Thinking Ape*. Oxford, Oxford University Press.

A clear exposition of the Machiavellian intelligence hypothesis.

Deacon, T. (1997) *The Symbolic Species*. London, Penguin.

An evolutionary account of the growth of human brains that stresses the importance of the coevolution of language and the brain. Deacon argues that the ability of the mind to construct and hold symbols is key.

Dunbar, R. I. M. (1996) *Grooming, Gossip and the Evolution of Language*, London, Faber and Faber.

A readable and popular account of the evolution of brain size and its possible causes.

Geary, D. C. (1998) *Male, Female: The Evolution of Human Sex Differences*. Washington DC, American Psychological Association.

Geary explains the principles of sexual selection and how these can be used to understand differences between males and females. Provides a good discussion of the evidence for real cognitive differences between males and females.

Harvey, P. H. and Pagel, M. D. (1991) *The Comparative Method in Evolutionary Biology*. Oxford, Oxford University Press.

An excellent book for dealing with the theoretical and methodological problems of constructing phylogenetic trees. Gives a good treatment of allometric equations.

Jones, S., Robert, M. and Pilbeam, D. (1992) *The Cambridge Encyclopedia of Human Evolution*. Cambridge, Cambridge University Press.

Numerous experts have contributed to this book. Thorough and well illustrated with plates and diagrams. Excellent for comparing human and primate evolution.

Pinker, S. (1994) *The Language Instinct*. London, Penguin.

A popular account of the modular and evolutionary approach to language.

Rizzolatti, G., Fogassi, L. and Gallese, V. (2006) 'Mirrors in the mind.' *Scientific American* **295**(5): 30–7.

An accessible article that describes the discovery and significance of mirror neurons. Other articles in this same issue also consider their importance.

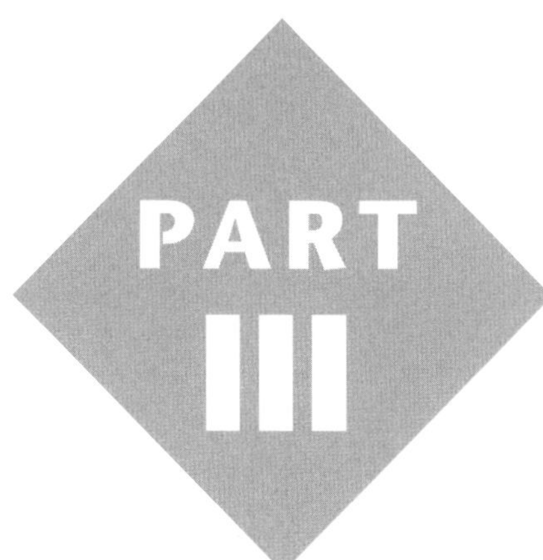

Cognition and Emotion

Modularity, Cognition and Reasoning

Plato ... says in Phaedo that our 'imaginary ideas' arise from the pre-existence of the soul, are not derivable from experience – read monkeys for pre-existence.
(Darwin, 1838, quoted in Gruber, 1974, p. 324)

7.1 The modular mind

7.1.1 Epistemological dilemmas: rationalism or empiricism?

In *The Phaedo*, Plato advances the notion that the mind is born pre-equipped with ideas or ways of structuring experience. Plato was developing the ideas of his mentor Socrates (469–399 BC) who thought that knowledge was essentially a recollection of universal truths that the soul understood in a previous life. In another of his works, *The Meno*, Plato describes how Socrates elicits knowledge from a slave boy. The boy is told of a square of sides one unit long and asked how long the side of a square must be to contain twice the area. Socrates leads the boy to the conclusion that the side must be the length of the diagonal of the original square. If even a slave can deduce the truth of geometry, then the truth must somehow already be there. Both Plato and Socrates believed in the doctrine of innate ideas – we are born knowing certain things. This position is sometimes called 'rationalism' (to distinguish it from **empiricism**) or **nativism**. The doctrine of innate ideas has had a long history (Table 7.1).

In contrast to this rationalist approach stands the empiricist tradition. The English philosopher John Locke, for example, argued that the mind started out as a formless mass and was given structure only by sensory impressions. In a note to himself, Darwin dismissed Locke in the same way as he dispatched Plato's pre-existing soul: 'He who understand baboon would do more toward metaphysics than Locke' (Darwin, 1838, quoted in Gruber, 1974, p. 243).

Table 7.1 Empiricism and rationalism in European thought

Rationalist tradition (certain ideas are innate)	Empiricist tradition (mind forms knowledge only from experience)	Evolutionary epistemology (a rational–empirical hybrid: our mind is born structured by the effects of selection on our ancestors)
Socrates (469–399 BC), Plato (428–348 BC), Descartes (1596–1650), Spinoza (1632–77), Leibniz (1646–1716), Emmauel Kant (1724–1804)	John Locke (1637–1704), George Berkeley (1685–1753), David Hume (1711–76)	Herbert Spencer (1820–1903), Charles Darwin (1809–82), William James (1842–1910)

The debate between rationalism and empiricism is a long-standing one in philosophy (see Kenny, 1986) and the details do not overly concern us here. But Darwin was probably right: the brain at birth is not a formless heap of tissue, nor does it carry a recollection of eternal verities associated with an immortal soul. The brain enters the world already structured by the effects of a few million years of natural selection having acted upon our primate and hominin ancestors. It is therefore born ready shaped, but also eager for experience for calibration and fine-tuning to the world in which it finds itself. Eibl-Eibesfeldt (1989, p. 6) expressed this succinctly:

> The ability to reconstruct a real world from sensory data presupposes a knowledge about this world. This knowledge is based in part on individual experience and, in part, on the achievements of data processing mechanisms, which we inherited as part of phylogenetic adaptations. Knowledge about the world in the latter instance was acquired during the course of evolution. It is, so to speak, a priori – prior to all individual experience – but certainly not prior to all experience.

Evolutionary epistemology

The term for this view of knowledge is evolutionary **epistemology** and was coined by Donald Campbell (1974). Somewhat confusingly, the term can have three distinct meanings corresponding to three levels of analysis: phylogenetic, ontogenetic and cultural. It is the phylogenetic level we are concerned with here – the idea that, during evolution, the mental circuitry of a species becomes attuned to certain inbuilt ways of grasping experience. The ontogenetic level is the idea that the processes of variation and selection go on in the individual mind as it matures: as an organism interacts with the world, concepts and neural connections are selected according to their fitness benefits to the individual mind. One of the most forceful advocates of this position was William James. In the October 1880 issue of *Atlantic Monthly*, he offered an analysis of creative thought in terms of a selective process acting upon a whole variety of conceptions in an individual mind. But these approaches to the construction of individual knowledge lapsed into abeyance until revived by Donald Campbell's (1974) seminal paper. At the cultural level, we have the idea that processes and metaphors from evolutionary biology can be applied to the development of ideas and scientific theories in culture as a whole. Karl Popper (1963), for example, was a selectionist in this sense. His view in *Conjectures and Refutations* was that scientific ideas that are falsified are rejected and, by analogy, become extinct, leaving the fittest to remain in the canon of science. More recently, both Daniel Dennett and Richard Dawkins have explored the issue – the latter coining the phrase 'universal Darwinism' to describe the attempt to apply the idea of natural selection to phenomena outside speciation and phylogeny. This idea is examined in Chapter 16. It is important to realise that the three approaches are analogous but distinct: the truth of any one does not depend on any other – although there have been attempts to link all three. The phylogenetic approach is often linked to the modular view of the mind – the notion that the evolved biological substrate of cognitive activity resides in specialised and dedicated areas of the brain called 'modules'.

Innate biases in learning

Evidence for the existence of domain-specific learning biases in humans came in the 1980s from experiments by Alan Leslie on very young children. Leslie (1982, 1984) showed images of moving objects to young children. When an unrealistic event was simulated, such as a ball (A) moving towards another ball (B), stopping before it reached B and then B moving off, children as young as 6 months showed more interest in this event than realistic ones involving collisions. Similarly, Renée Baillargeon and her colleagues (Baillargeon et al., 1985; Baillargeon, 1987; Baillargeon and DeVos, 1991) have shown that female infants at 18 weeks show surprise when physically impossible displays are presented to them – such as an image sequence showing a lower block removed from a pile of blocks leaving the upper ones poised in mid-air. Results like these are consistent with the idea that the human mind is preformed to some degree and that it arrives with an a priori intuitive physics. It is possible that children have or are programmed to develop a 'theory of body'.

7.1.2 The localisation of brain function

At its simplest interpretation, the notion of the modular brain is simply the unobjectionable view that different components of the brain deal with different specific functions. One of the earliest expressions of the idea came from two phrenologists, Gall and Spruzheim. In 1810, they published *The Anatomy and Physiology of the Nervous System*, in which they described the functions of the cerebral cortex and the corpus callosum. Where phrenology went wrong of course was the belief that the shape of the skull and the position of bumps on its surface could be used to discern intelligence and character.

Since the nineteenth century, studies on people with damage through disease or injury to specific parts of the brain have revealed a whole array of specialised functions. Even rather abstract faculties, such as the ability to make plans and enact behaviour to further them, seem localised in specific brain regions.

On this specific topic, we meet one of the most notorious applications of knowledge of the localisation of brain functions: the surgical procedure called 'lobotomy'. By the 1930s, it was known that damage to prefrontal lobe areas removed the ability of affected individuals to make plans and behave in a manner designed to achieve them. These findings became the basis for the prefrontal lobotomies carried out widely on psychotic patients in the 1950s and 60s. The procedure, critically exposed in the book and film *One Flew Over the Cuckoo's Nest*, produced an apparent calming effect but did not address the root cause of the disorder or cure it in any way. What it did do was reduce the capacity of psychotic individuals to act upon their delusions. One of the key pioneers of this procedure was Antonio Egaz Moniz of the University of Lisbon Medical School. In a decision that remains controversial to this day, Moniz was awarded the Nobel prize for his technique in 1949. When he collected it, he had already been paralysed from a bullet fired in 1939 by an angry lobotomised patient.

In more recent times, the use of the term **modularity** as applied to the mind was initiated by the linguist and philosopher Jerry Fodor. In his now classic monograph *The Modularity of Mind* (1983), Fodor argued that many perceptual and cognitive processes such as vision, hearing and talking were organised and performed with dedicated modules or 'input systems'. Fodor (1983) argued that these modules worked independently of each other and dealt with their own designated inputs from the environment; they were, in other words, 'domain specific'. The important point about Fodor's concept of perceptual modules is that they operate quickly and are relatively immune from interference by memory, experience and reflection. This is illustrated by the intriguing nature of optical illusions (see Box 7.1). In optical illusions, for example, the observer still sees the illusion (that is, the inaccurate conclusion about the image) even when measurements are made and we become conscious and rationally aware of our error. Knowledge and understanding cannot override the processing of the image. For Fodor, such modules had the following properties:

1. *Domain specific:* they are specialised in what inputs they process
2. *Informationally encapsulated:* they are relatively impermeable to other systems

Box 7.1 Some classic optical illusions

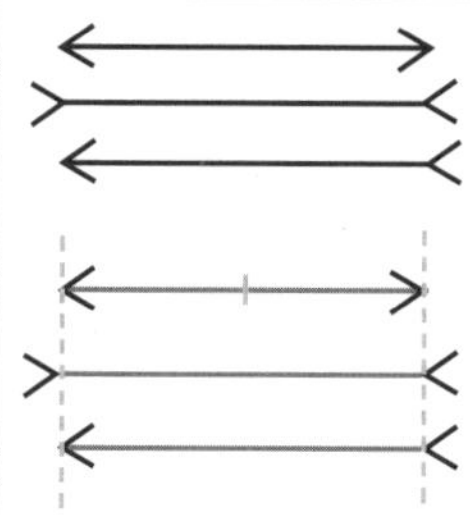

The Muller–Lyer optical illusion
In the arrows opposite, the middle arrow looks longer than the others. The lines below show this is not the case. But the illusion pesists even after this knowledge.

The Adelson chequered square

Probably the most uncanny optical illusion in existence. Square A is exactly the same shade as square B. To convince yourself that A and B are the same, use a scan and sample colours or photocopy this image and cut out the squares A and B and compare in isolation to rest of figure.

Created by Adrian Pickstone based on an original by Edward H. Adelson. Used with permission from Wikipedia.

The white rectangle

The shade of white in the rectangle is exactly the same as the background, yet it probably appears whiter or more distinct as if there were an object there.

3. *Obligate firing*
4. *Fast speed:* they operate quickly – probably as a result of information encapsulation and their mandatory firing, they need not wait to consult other systems
5. *Characteristic ontogeny:* they develop regularly and predictably in normal humans
6. *Correspond to a fixed neural architecture.*

This may look as if the human perceptual system is not well calibrated if it is prone to such errors. However, relatively hard-wired modules like this make good adaptive sense if they are fast. Often humans need a quick response to a threat or danger and the odd mistake is worth making so long as the interpretation errs on the side of caution.

Fodor was insistent, however, that higher level functions such as reasoning were not modular. These central cognitive processes, he argued, are slow and non-mandatory – we can choose, for example, whether to think about an intellectual problem but we cannot choose not to see the illusions shown in Box 7.1.

7.1.3 The manifesto of Tooby and Cosmides

The evolutionary approach to psychology has been much influenced by the work of Tooby and Cosmides and the powerful manifesto on evolutionary psychology that they issued in their chapter in *The Adapted Mind* (Barkow et al., 1992). Many of the issues, in particular the extent to which the brain can be thought of as a set of discrete problem-solving modules, remain controversial, but the approach has met with considerable success as a heuristic model. For Tooby and Cosmides, psychology is to be seen as a branch of biology that studies the structure of brains, how brains process information and how the brain's information-processing mechanisms generate behaviour. The key principles of this paradigm are as follows:

1. The human mind is what the brain does. It is an information-processing device that receives inputs and generates outputs in a manner directly analogous to a computer. In this view, both thought and behaviour are cognitive processes. Consider the example of how to act towards kin. Kin selection theory (see Chapter 9) suggests a

number of factors that must be taken into account in dealing with kin, such as the degree of relatedness of kin to the self, their reproductive value and the costs and benefits of any action undertaken. To behave appropriately requires a cognitive computational programme that factors these parameters into the decision-making. Simple instincts, they argue, will not suffice.

2. The neural circuits that make up the brain were 'designed' by natural selection to solve problems that our ancestors faced in their environment of evolutionary adaptedness (EEA). Such problems were repeatedly encountered and, more importantly, impinged on the survivability of the organism concerned: problems to do with growth, survival, harvesting resources, avoiding predators, finding mates, reproducing and so on.
3. The way in which brains solved the vast array of adaptive problems was not through some general problem-solving device, which would probably be highly inefficient, but through the construction of a set of discrete and functionally specialised problem-solving modules. Each module is capable of responding to and solving a problem only over a restricted domain; hence they are called **domain-specific mental modules**. Tooby and Cosmides (1992) use the analogy of a Swiss army knife with numerous blades and attachments for specific purposes.
4. Cognitive mechanisms that were sculpted during the hundreds of thousands of years that humans spent in a hunter-gatherer lifestyle will not necessarily appear adaptive today: we carry Stone Age minds in modern skulls.
5. Because humanity belongs to one species, all members of which can pool genes with any other member of the opposite sex to create viable offspring, so the variability between mental organs must be limited. These domain-specific mental modules are therefore common to all people, with only superficial intergroup variation. Tooby and Cosmides (1992, p. 68) use an analogy with *Gray's Anatomy*:

   > Just as one can now flip open *Gray's Anatomy* to any page and find an intricately detailed depiction of some part of our evolved species-typical morphology, we anticipate that in 50 or 100 years one will be able to pick up an equivalent reference work for psychology and find in it detailed information-processing descriptions of the multitude of evolved species-typical adaptations of the human mind.

   What follows from this is the fact that the genetic variation that exists between people and peoples will inevitably be minor and have very little effect on the cognitive architecture common to all that constitutes a universal human nature.
6. As with all manifestos, there is an enemy, the enemy in this case being the 'standard social science model'. The only feature of this model that is accepted by Tooby and Cosmides is the idea that genetic variation between racial groups is trivial and insufficient to explain any observed difference in behaviour. From thereon, there is fundamental disagreement. The social science approach, which they admit for this purpose is a conflation of many schools of thought, is taken to suggest that the mental organisation of adults is determined by their culture. As such, the human mind has as its main property merely a capacity for culture. Culture itself rides free from any strong influence from the lives of any specific individuals and certainly free from human nature. The blank slate approach is of course an oversimplification, but Tooby and Cosmides detect in the social science literature an assumption that the human mind is akin to a computer without programmes: it is structured to learn but obtains its programmes from an exterior culture rather than an interior nature.

7.1.4 Some potential candidates for domain-specific modules

Some modules that might be expected to form part of the human mental toolkit are mechanisms for cooperative engagements with kin and non-kin, means by which to detect cheats, parenting, disease avoidance, object permanence and movement, face recognition, learning a language, anticipating the reactions and emotional states of others (theory of mind), self-concept and optimal foraging – to name but a few. A central problem with all of this, of course, is knowing when to stop. At what level of discrimination and finesse have we reached an indivisible module? A useful, albeit conjectured, hierarchical organisation of modules is provided by Geary (1998) (Figure 7.1).

The importance of the social group of modules can be gleaned from the fact that nearly all primates live in complex social groups. Furthermore, social living has almost certainly been a feature of the evolutionary ancestry common to humans and primates over the past 30 million years. The group-level modules enable individuals to function in groups in mutually beneficial ways. The recognition and discrimination of kin provides an important facet of increasing inclusive fitness (see Chapter 2), while a disposition to accept social ideologies would provide a much-needed cement to bind together non-kin for reciprocal exchanges. If early hominins were engaged in intergroup competition, mechanisms to bind groups together, and, moreover, treat in-group and out-group members differently would provide a competitive advantage.

Within the category of individual modules, we can see that, in this schema, language acquisition is expected to be facilitated by specific language centres of the brain, something that is in fact the case. Similarly, all normally functioning humans can be expected to have a theory of mind centre to enable predictions to be made about the behaviour of others. Facial processing is essential for anticipating the mood of others as well as making choices about mating and remembering individuals with whom one may have made social contracts.

Under ecological modules, we note an array of specialised mechanisms for helping the human organism to exploit features of the biological and physical landscape to assist in survival and reproduction. One overall point stressed strongly by Geary is that such modules do not constrain people to behave in fixed patterns along the lines of stimulus–response thinking; predators would soon home in on organisms that behaved in predictable ways. We should instead regard the modules as conferring a flexible response according to the context.

In this view, one can see the purpose of development as enabling the calibration of these modules (setting the start-up conditions) to local social, biological and physical conditions. The modules constrain and bias the type of experiences to which a growing child should attend, but then use the result of the experiences gained to fine-tune the inherent functional mechanisms. Thus, for example, we are born with a disposition to attend to verbal sounds and organise utterances. The language that we finally speak is a result of this cognitive bias coupled with the actual evidence obtained about syntax and vocabulary. This ontogenetic development of modules in relation to the local environment can be thought of as an 'open genetic programme', an open programme being one that takes on board instructions from the environment (Geary, 1998).

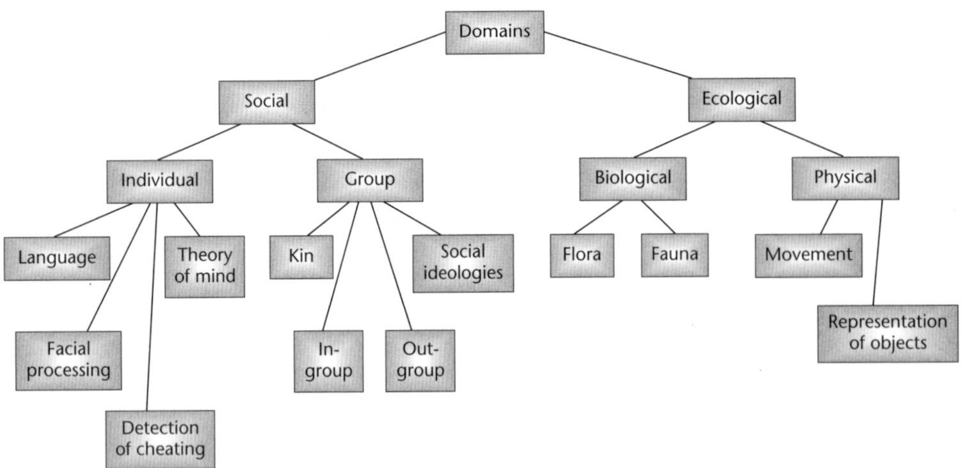

FIGURE 7.1 Some possible evolved domains of the mind

SOURCE: Modified from Geary, D. C. (1998) *Male, Female: The Evolution of Human Sex Differences.* Washington, DC, American Psychological Association, Figure 6.4, p. 180.

It is important to realise that Darwinian psychology does not stand or fall on the existence of discrete modules or their degree of independence and domain specificity. Before Tooby and Cosmides announced their view of the mind, David Marr (1982) made the useful distinction between levels of understanding of how the brain drives behaviour. As Table 7.2 shows, traditional cognitive psychology, with its emphasis on computational systems and problem-solving algorithms, stands midway between the proximate level of neurophysiology (the wiring diagram) and the ultimate level of adaptive responses to selective pressures.

Table 7.2 Levels of explanation in evolutionary neurophysiology

Level	Aims of study at this level	Associated discipline
High level	To elucidate the tasks that the system was 'designed' to achieve	Darwinian psychology
Intermediate level	How the system computes information to achieve these tasks	Cognitive psychology
Lower	How the circuitry is arranged in the brain	Neurophysiology
SOURCE: After Marr, D. (1982) *Vision*. New York, W. H. Freeman.		

In the next section, we will examine how this evolutionary modular approach can throw light on patterns in human reasoning and how thinking at the 'high level' of Table 7.2 can help make sense of the 'intermediate level' below.

7.2 Problems with rational thought

7.2.1 The problem of optimisation

Quite understandably, humans as a species are proud of their powers of rational thought. So much so that it was natural that when mathematicians began to model human behaviour, they tended to assume humans would behave in ways that were optimal from a rational point of view. By the 1960s, the idea of rational optimisation had entered such diverse fields as economics, psychology and animal behaviour. What was common to this endeavour was the idea that the behavioural agents under study (be that human or other animals) acted to maximise returns to themselves in the contexts they were operating in. In economics, for example, humans were expected to maximise net benefits, profits or utility; in psychology, cognition was supposed to be directed towards optimal decision-making; and in behavioural ecology, animals were supposed to behave so as to optimise reproductive fitness or, in the case of optimal foraging theory, calorie intake.

The problem with this approach is that it tended to take for granted almost unlimited cognitive power and access to huge amounts of information. Gigerenzer and Selten (2001) refer to these models as 'unbounded rationality' and argue that many of the underlying assumptions are unrealistic. The realisation in these disciplines that all the relevant information is rarely available led to 'optimisation under constraint' models. Imagine an animal A (human if you like) trying to make a decision about a prospective sexual partner B. If A is trying to decide if B is suitable, she has to factor into her calculations an estimate of the worth of B (looks, resources, health status, social status, age, future prospects) and her estimation of her own worth. If A concludes that her own reproductive value (a measure, for the sake of argument, of all the relevant facts suitably weighted and averaged) is much greater than that of B, she may decide to reject B and search for something better. But the decision to reject also has to take into account the probability of something better than B showing up and also the costs of further searching. The latter consists of time, resources and opportunity costs – instead of searching around for Mr Right, she could have settled down and started to fulfil her Darwinian imperatives with Mr B. In addition, the whole process of making a decision, weighing costs and benefits demands time. Suddenly the whole process is entangled in complexity and leads us to question whether such computations actually take place or whether the organism uses a quicker route.

As optimality models faced these challenges, the work of Amos Tversky and the Nobel laureate Daniel Kahnemann in the 1970s demonstrated that humans are prone to errors of reasoning and that such errors were very revealing about the sort of mental strategies that people actually use to solve problems under conditions of uncertainty.

7.2.2 Errors of thought and reasoning: heuristics and cognitive illusions

Errors in probabilistic reasoning

In an experiment that has become a classic in psychology, L. G. Humphreys (1939) studied the ability of people to predict when one of two light bulbs in front of them lit up. The bulbs were actually programmed to light up randomly but such that one lit up for 80 per cent of the time and the other for 20 per cent of the time. People quickly grasped this frequency and were then asked to score as highly as they could in predicting which bulb would be the next to light up. Subjects typically distributed their predictions in the same ratio as the observed frequencies, so they predicted the lower frequency bulb 20 per cent of the time and the higher one 80 per cent of the time. Although this sounds intuitively correct, it is not the optimal strategy for maximizing the number of correct predictions. This can be seen with a few calculations. Since the process is random, placing 80 per cent of your predictions on a bulb that only lights 80 per cent of the time gives a hit rate of 80 per cent × 80 per cent = 64 per cent. To this we can add the return from placing 20 per cent of your predictions on a bulb that only lights 20 per cent of the time – 20 per cent × 20 per cent = 4 per cent. This gives a total success rate of 68 per cent. A moment's thought shows it to be better to have simply placed all your bets on the 80 per cent bulb, giving a return rate of 80 per cent. Why people do not do this has been the subject of much research. One line of enquiry following the work of Humphreys considered the reward from actually acting rather than being passive.

There are many ideas as to why people exercise suboptimal rationality. Various robust patterns have emerged with corresponding labels such as 'risk aversion' and 'framing effects'. In the case of the bulbs, it may be a tendency of humans to desire to use knowledge and insight once they have gained it rather than remain passive. Barash (2003) suggests that this is why many people invest in the stock market by selecting specific stocks (or paying someone to do so) despite the fact that tracker funds (baskets of shares that simply follow the leading shares on the stock market) outperform the vast majority of managed funds most of the time. Perhaps each believes that they can outperform the average or that some knowledge must surely be of use and preferable to no knowledge.

Another effect is called 'source dependence' and again seems to illustrate the preference for using established knowledge over the uncertain. A good illustration of this is the 'Ellesberg jar'. Test yourself by answering the following question quickly.

Imagine a jar containing 100 balls, 50 red and 50 white; and another jar also containing red and white balls numbering 100 balls in total but in an unknown proportion. You will be given £100 to correctly guess the colour of a ball pulled out at random from one of the two jars. Which jar would you prefer to take your chances on? Most people opt for the first jar, even though the chances are 50/50 for either jar.

In the 1970s, a large body of evidence had accumulated that suggested that ordinary people were rather poor at solving problems that required probabilistic reasoning. Rather poor, that is, compared to professional mathematicians who could solve such problems at their leisure. More worryingly, even professionals such as physicians were not perfect. When medics were given the task of estimating the probability that a patient had a disease based on a scenario where a given test threw up false positives and negatives, their answers were consistently way out (see Box 7.2). Many of these errors that people succumb to have been categorised with names such as the 'conjunction fallacy', the 'base rate fallacy' and 'overconfidence bias'. Thanks to pioneering work by Kahneman and colleagues, the study of such 'cognitive illusions' is now an important part of mainstream psychology (Kahneman et al., 1982). The sections that follow briefly describe the cognitive biases explored by Tversky and Kahneman; later we will examine an evolutionary reinterpretations offered by Gigerenzer and others.

The overconfidence bias

This bias is seen in experiments where subjects provide an answer to a question and then are asked about their level of confidence in their answer. A typical question is: 'Which city has more inhabitants, Hyderabad or Islamabad?'; followed by: 'How confident are you that you answer is correct: 50 per cent, 60 per cent, ... 100 per cent?' Inevitably, people are overconfident about their answers. Of those who state they are 100 per cent confident, only about 80 per cent gave the correct answer. There are few standard or convincing explanations for this effect. One is to evoke another bias, a 'confirmation bias', which says that once people make a choice, they search for further evidence to confirm their decision and overlook counterevidence.

The conjunction bias

The following problem is typical of that used by Tversky and Kahneman (1983).

Linda is 31 years old, single, outspoken and very bright. She majored in philosophy. As a student, she was deeply concerned with issues of discrimination and social justice, and also participated in antinuclear demonstrations.

Which of the two alternatives is more probable:

1. Linda is a bank teller (T).
2. Linda is a bank teller and is active in the feminist movement (T and F).

Of the participants in Tversky and Kahneman's original study, 85 per cent chose 2. This is a fallacy, since the probability of two events (T and F) must be less than one of its constituents. For example, if you throw two dice, then the probability of a six and a four is less than just one being a six. Rationally, it is more likely that Linda is a bank teller than a bank teller and active in the feminist movement. For 2 to be true, Linda has to be a bank teller anyway and it is conceivable that she may not be a feminist. Tversky and Kahneman explained this effect, which has since become widely reported in the literature of a number of disciplines, by positing a 'representativeness heuristic'. Hence, because Linda was described as if she was a feminist and this fitted one part of the description, this biased our better judgement.

Box 7.2 The base rate fallacy and Bayesian reasoning

The **base rate fallacy** has received a lot of attention, possibly, in part at least, because otherwise intelligent and supposedly numerate people, such as medics, who should know better, perform quite badly. One version of this test is phrased as follows:

'If a test to detect a disease whose prevalence is 1/1,000 has a false positive rate of 5 per cent, what is the chance that a person found to have a positive result actually has the disease, assuming that you know nothing about the person's symptoms or signs?'

In one study, 60 students and staff at Harvard Medical School attempted this problem. Nearly half judged the answer to be 0.95 (that is, 95 per cent chance). The average response was 0.56 and only 18 per cent gave 0.02. The correct answer by Bayesian reasoning is 0.02. This is not just of academic interest of course, since many tests for diseases do not have a 100 per cent correct hit rate, and if you test positive, thinking you have a 95 per cent chance of the disease when the real answer may be 2 per cent can have dire consequences.

The origins of a rule for inferring the likelihood of a hypothesis being correct from certain types of data is attributed to the Reverend Thomas Bayes (1702–61). Although Bayes' original formulation is slightly different, a modern expression of the false/positive type of problem is given here.

Let	H	=	a person with the disease
	H′	=	a person without disease
	X	=	a person who tests positive
Then	p(H)	=	base line probability that someone picked randomly will have the disease = 0.001 (that is, 1/1,000)
	p(X/H′)	=	probability of X given H′. In this case, the probability that someone will test positive without the disease = 0.05 (that is, 5 per cent)

p(H′) = probability that someone picked at random does not have the disease = 0.999 (that is, 1 – 0.001)

p(X/H) = probability of X given H. In this case, the probability that someone will test positive if they have the disease = 1

p(H/X) = probability that person will have the disease if testing positive.

According to Bayesian reasoning:

$$p(H/X) = \frac{p(H)p(X/H)}{p(H)p(X/H) + p(H')p(X/H')}$$

Substituting the values above gives:

p (H/X) = 0.001/0.05095 = 0.02 (rounded to two decimal places).

7.3 Cognitive illusions and the adapted mind

So is the human brain, after some two million years in the hominid lineage, a rather poor device at calculating probabilities? We can take some comfort in the fact that we can (thanks to schooling in Bayesian reasoning) eventually work out the right answer. But why is it such an effort, fraught with pitfalls, even for educated people? Also, if some survival advantage were to be attached to this sort of reasoning, then coming up with an answer of 95 per cent (half the respondents in the base rate fallacy experiment) when the actual answer is 2 per cent suggests a catastrophic mismatch between representation and actuality.

Gigerenzer (2000, p. 246) attacks these so-called fallacies by arguing that questions based on single events are inappropriate to test probabilistic reasoning. Another explanation of the effect may be that in both cases the mind is led astray by assumptions that it makes rather than a cognitive error as such. In the bank teller case, for example, it could be that the question is framed so as to suggest Linda is already a bank teller and then we have to estimate if she is also a feminist.

Faced with the phenomena of cognitive illusions, the evolutionary psychologist would probably start to work from two premises:

1. The human mind is well adapted to solving problems, but these problems are the sort encountered in the EEA, not those of abstract mathematics.
2. If problems of the sort that generate cognitive illusions could be recast into more ecologically and socially relevant formats, then performance should improve.

Applying these premises, we can revisit the fallacies described above.

7.3.1 Bounded rationality and adaptive thinking

At the time that notions of unbounded rationality and optimisation under constraints were being propounded, Herbert Simon (1956) advanced the competing idea of **bounded rationality**. Although Simon's and subsequent definitions were not always precise, the concept was destined to capture a more realistic account of how real people make decisions which, although not strictly irrational, are nevertheless not optimising because of the limitations of cognitive power or the incompleteness of information. The concept has been developed and applied in recent years with some success by Gigerenzer (2001) who compares human reason to an adaptive toolbox. The function of this toolbox is to enable the organism to achieve its proximal goals such as finding a mate, avoiding predators, obtaining food, achieving status in a group and so on. The crucial point is that Gigerenzer does not assume that the strategies in the toolbox seek to optimise. The reasons for this assumption are that the number of decisions that an organism has to make is so huge that it would require immense and unrealistic computing power to deal with them all. Furthermore, given the absence of complete information, assumptions have to be made that could generate large errors if optimisation were always sought. In describing this toolbox, Gigerenzer borrows a phrase from Wimsalt in his description of nature: 'a backwoods mechanic and used parts dealer' (quoted in Gigerenzer, 2001).

The analogy is that a backwoods mechanic will have neither a general purpose tool, nor a complete set of tools for all cars nor a complete set of spare parts. Such a mechanic will have to make do with whatever is available for the problem at hand, improvising and adjusting accordingly. If the mind contains, metaphorically speaking, such a toolbox, there are some features that we may expect to be present. They are:

- *Psychological plausibility:* Models will be based on what cognitive capacities humans have rather than assumptions of divine intelligence or pure logic, against which humans will inevitably fall short.
- *Domain specificity:* The rules and devices of the toolbox will be specialised.
- *Ecological (and social) rationality:* The rationality of these domain-specific heuristics is not to be judged against the cannons of pure logic. In this respect, it should come as no surprise if rules such as consistency and optimality are broken. Rather, the success of bounded rationality should be judged by its performance in social and ecological contexts. Humans in particular have to make decisions that are fair and ethical (in the sense of conforming to a group ethos) and that can be defended within a group.

The toolbox will contain **heuristics**, that is, strategies for generating a solution to a problem. Such strategies may consist of simple and economical rules to guide decision-making. Gigerenzer gives two examples: search rules and stopping rules.

- *Search rules:* Searching can be thought of as looking among alternatives for cues to evaluate the choice on offer. Search rules can be illustrated by the so-called 'secretary problem' much investigated by statisticians (see Todd and Miller, 1999). Suppose you have to hire a secretary from 100 applicants. You interview in a random order and either accept or reject. A candidate that is rejected cannot be returned to for hiring. The dilemma of course is this: if you choose too early, you may pick someone of poorer quality than the mean and poorer than many that would follow; if you pick too late, it becomes increasingly likely that the few left may not contain someone suitable. Mathematical demonstration shows that the best solution is to interview a certain proportion, remember the best of them and choose the next who is better than that. The optimum number to interview is 1/e (or 37 per cent) where e is the natural number 2.718... . This is not foolproof but this 37 per cent rule will give the best result about 37 per cent of the time.
- *Stopping rules:* Searches eventually have to stop. A simple combination of a search and stop rule might be to sample a dozen and take the best. Stopping rules need not be cognitive, an emotional system could perform the task. In this respect, love and affection for a sexual partner or children might serve as emotional stopping rules, obliging the organism to stay with its mate and offspring and forestall any further calculations of whether better options lie elsewhere. Such a system could, over a whole life, be highly fitness enhancing compared with the constant abandonment of husbands, wives and children as better options present themselves. As another example of the need for a stopping rule, consider the problem of finding the best deal on a new car. Suppose that there is some price variation for the same model and that the actual price is only to be ascertained by visiting widely dispersed showrooms of dealers and haggling with the salespeople. One solution to the problem of when to stop looking is to make a 'reservation price' and decide when the marginal costs (time effort, travel costs) of further searching equal the marginal expected improvement over the best seen so far. When it starts to cost £300 of travel effort to find a car £300 cheaper than the best seen so far, common sense (ecological rationality) says stop.

Herbert Simon (1990, p. 9) termed the type of examples and solutions given above 'satisficing', which he defined as:

> using experience to construct an expectation of how good a solution we might reasonably achieve, and halting search as soon as a solution is reached that meets the expectation.

These search rules look simple but are they enough for real-life situations that throw up a much greater array of parameters? In mate choice, for example, the organism has to weigh up the quality of mates it encounters, its own quality, search and opportunity costs (the biological clocks of males and females tick away while decisions are delayed), and reactions of potential partners to advances. The crucial point to emphasise again is that the mind does not solve the secretary problem or the cost–benefit analysis of marginal costs mathematically and formally each time. Natural selection solved such problems for us, not by giving us an onboard omnipotent computer but by providing a set of short, fast and simple rules that work reasonably some of the time. A game of chess illustrates the use of this satisficing approach. At any time on the chess board, there are about 30 legal moves available and a game typically involves about 40 moves by each player. Over the game then, from the perspective of one player, there are '30 times itself 40 times' moves or 30^{40}, which is about 1×10^{120}. By comparison, the number of particles in the universe is about 10^{80}. It follows that the human mind cannot possibly evaluate all possible moves. So a good chess player has to make do with some short-cut tricks and some fast but rough forms of reasoning.

The research programme of Gigerenzer's group consists of three approaches. First, to try to identify likely heuristics that could work; second, testing these heuristics in artificial and real-world environments; and third, ascertaining if people actually use them. Gigerenzer's concept of a heuristic must be distinguished from other such uses of this term in the psychological literature. The 'heuristics and biases' research programme of Tversky and Kahneman, for example, has tended to suggest that heuristics are mentally sloppy procedures that usually lead to fallacies (see below). For Gigerenzer et al.'s (1999, p. 28) research programme, however, heuristics are treated more positively as ways 'the human mind can take advantage of the structure of information in the environment to arrive at reasonable conclusions'. In other words, they are rather efficient devices serving the immediate needs of the organism:

> The function of heuristics is not to be coherent. Rather, their function is to make reasonable, adaptive inferences about the real social and physical world given limited time and knowledge. (Gigerenzer et al., 1999, p. 22)

Some examples of potential heuristics that humans may be inherently disposed to employ are:

- *Stop searching:* When a reason for favouring an alternative is found.
- *Sample selection:* Estimate the average and take the next on offer that exceeds the average.
- *Copy prestigious members of a group:* Much research suggests that humans everywhere display a propensity to copy prestigious individuals – a process sometimes known as 'prestige-biased transmission' (see Heinrich et al., 1999). Cues for prestige can easily be gleaned from displays of wealth or attention given to individuals by others. In gatherings, for example, high-status individuals are listened to more attentively, interrupted less and generally treated with deference. The adaptive function of this behaviour is fairly obvious. By copying successful individuals, the copier obtains valuable information about behaviour patterns that have led to success. This form of learning can be very time efficient. Acquiring hunting skills by copying the best hunter is probably more efficient that learning through individual trial and error. Such a mechanism could explain the almost self-perpetuating phenomenon of fame. In contemporary Western cultures, the media spotlight often seems to be trained continually on people for no other reason than that they are themselves often in the media.
- *Conform to the mode:* Running parallel to prestige-biased transmission, conformist transmission is a heuristic that says preferentially copy the most common behaviour in a population. The adaptive value of this strategy is also clear: the most common behaviour must have brought some success and cannot be too fitness reducing, otherwise the group would have died out or abandoned the behaviour. Imagine you are hungry,

walking across a strange landscape and come across a group of people consuming purple fruits but assiduously avoiding the red ones. Which fruits would you eat? Significantly, and as expected, theoretical modelling shows that individuals do rely heavily on this heuristic when cues from the environment are ambiguous or uncertain (Henrich and Boyd, 1998). Numerous studies of social learning in species such as pigeons, rats and guppies suggest that these animals adopt a 'do-what-the-majority-do' strategy (see Laland et al., 1996).

- *Fall in love:* The heuristic need not necessarily be cognitive as this conjecture indicates. The emotion of falling and being in love with someone may serve to bind two individuals together and prevent further searching. Alternatively, it may serve to break apart existing relationships even when a level-headed assessment of costs and benefits would tend to preserve it.
- *The recognition heuristic:* In conditions of uncertainty, the fact that an organism recognises something may yield useful information. For example, rats are known to initially avoid foods they do not recognise; in this respect they exhibit neophobia. They prefer food they have tasted before or can smell in the breath of older rats. The adaptive value of this tactic is obvious. It could also be that making a decision based on something as simple as recognition is also an effective problem-solving strategy for humans. Consider the question: 'Which has the higher value, A or B?', where value could be anything from monetary value, physical size, number and so on. A recognition heuristic might be 'the one you recognise has a higher value'. Goldstein and Gigerenzer (1999) asked students at the University of Chicago and the University of Munich the following question: 'Which US city has more inhabitants; San Diego or San Antonio?' The correct answer (San Diego) was given by 62 per cent of the US students but 100 per cent of the German students. The reason is that all the German students had heard of San Diego but few had heard of San Antonio. All the US students had heard of both. The latter then brought additional information to bear that proved less reliable than the simple recognition heuristic 'if you've heard of it, it is larger'. Goldstein called this heuristic 'ignorance-based decision making'. It could be of value in contexts where recognition is the only cue and further information is unavailable. In this respect, it is hardly surprising that the advertising industry is paid huge sums by companies to promote brand recognition. Advertisements often contain little real information about the product, yet are deemed to be successful if they ensure that a name or logo sticks in the memory of a potential purchaser. When shoppers then have to make a decision between brands A and B and no other information on quality is available, they often choose the one they have heard of before.

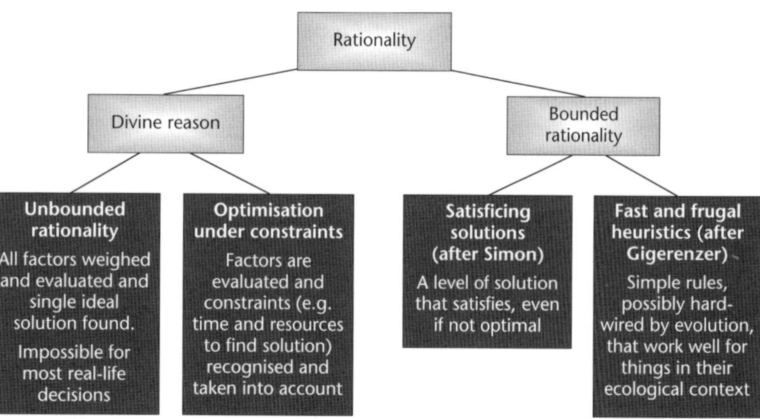

FIGURE 7.2 Some alternative concepts of rationality

7.4 Biases and fallacies revisited

7.4.1 Overconfidence bias revisited

It could be that percentage probability of confidence in a single event is not a primate mind-friendly concept. Gigerenzer et al. (1991) performed the confidence estimation test by asking subjects a number of questions and then to estimate their confidence in their answers. After 50 questions, the subjects were asked to estimate: 'how many of these 50 questions do you think you got right?' When the percentage confidence results for single events were averaged, the usual overconfidence bias was observed – subjects were, on average, about 14 per cent overconfident about each answer. Yet when estimated frequencies of correct answers after batches of 50 questions were analysed, the estimations were remarkably close to the actual frequencies of correct answers. In fact, subjects underestimated by about 3 per cent over all categories. Gigerenzer et al. concluded that the implication of this study is that if the problem is recast in the form that could be expected to correspond to the way the mind actually makes decisions in everyday life, then better performance is observed.

7.4.2 The conjunction fallacy revisited

Hertwig and Gigerenzer (1999) have argued that if we replace the Linda bank teller question with one based on frequencies, then the conjunction fallacy should disappear. In this modification, the question can be rephrased as follows:

There are 100 persons who fit the description above (that is, Linda's; see above):

1. How many of them are bank tellers?
2. How many of them are bank tellers and active in the feminist movement?

Hertwig and Gigerenzer (1999) performed such studies and found that the percentage of subjects committing the conjunction fallacy (that is, T and F) fell from the 80 per cent level observed in Tversky and Kahneman's work to between 10 and 20 per cent.

7.4.3 The base rate fallacy revisited

Part of the problem with the way the disease test question is presented to subjects is that it is unclear what to do with the base rate. To answer correctly, the mind first needs to assume that the person to whom the medical test is administered is randomly selected from the population where the prevalence is 1/1,000. It is obvious that those giving the answer as 0.95 are not doing this.

When the problem is rephrased in a frequency format, the effect on performance is dramatic. Cosmides and Tooby (1996) reformulated the problem as follows:

> One out of 1000 Americans has disease X. A test has developed to detect when a person has disease X. Every time the test is given to a person who has the disease, the test comes out positive. But sometimes the test also comes out positive when it is given to a person who is completely healthy. Specifically, out of every 1000 people who are perfectly healthy, 50 of them test positive for the disease.
>
> Imagine that we have assembled a random sample of 1000 Americans. They were selected by a lottery. Those who conducted the lottery had no information about the health status of any of these people. How many people who test positive for the disease will actually have the disease?
>
> .... out of ....

They also presented the problem in the form of a picture of 100 squares (representing 100 Americans (instead of 1,000 in the original problem) and asked the subjects to answer by colouring the squares. In both these cases, the correct Bayesian answers were far more readily given (Table 7.3).

Table 7.3 Improvements in Bayesian reasoning by frequency presentations of the medical diagnosis problem

Medical diagnosis problem	Number in sample	Bayesian (that is, correct) answers
Original probabilistic version (Casscells et al., 1978)	60	18
Frequency formulation (Cosmides and Tooby, 1996)	50	76
Frequency formulation using pictorial squares (Cosmides and Tooby, 1996)	25	92

SOURCE: Modified from Gigerenzer, G. (2000) *Adaptive Thinking*. New York, Oxford University Press, Table 12.4, p. 253.

The overall conclusion from this approach is that the biases and fallacies identified by cognitive psychologists, and some of the heuristics invoked to explain them, may be an artefact of the experimental procedures. An evolutionary view of the mind predicts that it will be good at solving problems when expressed in frequency terms rather than the fractional probabilities used by statisticians. The idea that 'two people out of ten who ate this meat were sick the next day' is more in keeping with the grain of the mind than 'it is estimated that the probability of sickness following the consumption of this meat is 0.2'.

We have seen that heuristics and biases can sometimes lead to 'cognitive illusions', so called in analogy with visual illusions, where the mind 'sees' something different to that which is the case. The analogy is quite instructive. If we consider again the optical system, it becomes apparent that it was not designed to faithfully present an objective picture of what is 'out there', but rather to present the mind with a useful interpretation. David Marr (1982) added a new impetus to the whole field of vision science when he argued that the visual systems made compensations to perception to enhance the usefulness of the knowledge to be gained. In Box 7.1 above, for example, we see a rectangle when none is there and are convinced the shade of white of the rectangle is different to the background since this is a reasonable assumption. As Workman and Reader (2004) point out, applying this type of illusion to an ecological context, seeing a tiger partially hidden behind long grass as a series of puzzling tiger coloured strips is less useful than seeing the whole tiger.

The next section examines another famous test of reasoning that leads people into errors that could become intelligible when evolutionary reasoning is applied. It is called the Wason selection task.

7.5 Case study: the Wason selection task: cognitive adaptations for social exchange

> Blow, blow, thou winter wind,
> Thou art not so unkind
> As man's ingratitude.
> (Shakespeare, *As You Like It*, 2.7.174)

7.5.1 Logical reasoning and the social contract

This quotation reminds us that our ancestors had to cope with a difficult physical environment and a social world that contained difficult unreciprocating people. We have in-built mechanisms to cope with the cold: we shiver, reduce our exposed surface area and raise hairs from 'goose bumps' to improve insulation. So do we have in-built mechanisms to cope with cheaters? Tooby and Cosmides suggest that we do and that this is reflected in our powers of reasoning. They contend that since hominins have engaged in social interactions over a few hundred thousand years, our brains should have evolved a constellation of cognitive adaptations to social life. If interactions with other humans in our EEA involved exchanges of help and favours (reciprocating altruism), our cognitive algorithms (sequence of thought processes) should be adapted to possess the following abilities:

1. To estimate the costs and benefits of various actions to oneself and to others
2. To store information about the history of past exchanges with other individuals
3. To detect cheaters and be motivated to punish them.

To investigate human cognition, Tooby and Cosmides used a technique called the 'Wason selection task' (Wason, 1966). Wason was interested in Popper's view of science that identified the hallmark of scientific reasoning to be the hypothetico-deductive method. In particular, scientists should test hypotheses by looking for the evidence that would falsify them. Box 7.3 shows the structure of a typical Wason selection task.

Box 7.3 Basic form of Wason's selection task

Context: Part of your new clerical job in the registry of your university is to check that student documents have been processed properly by your previous colleague. The document files of each student have a letter code on the front and a numerical code on the back. One basic and important rule for you to check is:

Rule: If a person has a D code on the front, the numerical code on the back must be a 3.

You suspect that the person you have replaced did not label the files accurately. Examine the four documents below (some showing the front of the file and some the back).

D	F	3	7

Which document(s) would you turn over to test whether any file violates the rule?

The logical structure to the problem in Box 7.3 can be written as:

D	F	3	7
P	not P	Q	not Q

The rule takes the form: 'If P, then Q' (if a D on the front of the file, then 3 on the back). The rule is violated if there is a D on the front but not a 3 on the back, or 'If P and not Q'. Thus we need only examine files D and 7. Tooby and Cosmides then applied the problem to a context that involved social exchange and hence the recognition of benefits and the payment of costs. In such situations, the potential for cheating is exposed. Box 7.4 shows the logical structure of this new setting.

Box 7.4 Wason's selection task in a social exchange context

Context: You are a bouncer in a Boston bar. You will lose your job unless you enforce the following rule:

Rule: If a person is drinking beer, he or she must be over 20 years old.

Information: The cards opposite represent the details of four people in the bar. One side indicates what they are drinking, the other side their age.

Instruction: Indicate only the card(s) you would definitely turn over to see whether any of them are breaking the law.

Beer	Coke	25	16

In this new context, the proportion that chose 'Beer' and '16' (the correct answers) rose to 75 per cent. Tooby and Cosmides explain this improvement in performance by suggesting that the social context evokes a 'search for cheats procedure' in the human mind.

A rival explanation might be that people are simply better at reasoning in a non-abstract context of which they have some experience. Most of us are familiar with the illegality of underage drinking but not with strange rules about student files. To test this, Tooby and Cosmides varied the reasoning tasks as follows:

1. A task that has the same formal structure as the drinking problem but in a totally alien cultural setting.
2. A task in which a concern to detect cheating would in fact lead to logical errors.

With regard to the second task, Tooby and Cosmides found that people were actually led into errors of reasoning by their propensity to look for cheats. With regard to the first challenge, they found that, even if it were couched in unfamiliar cultural terms (Box 7.5), the Wason problem was solved better when the problem entailed costs and benefits. Despite the unfamiliarity of the context, over 70 per cent of subjects were still able to reason correctly and choose 'P' and 'not Q' (Figure 7.3).

Box 7.5 Wason's selection task in unfamiliar context of social exchange

Context and rule: You are part of a tribe where a fundamental rule is that only married men are allowed to eat the aphrodisiac cassava root. Married men are always given a tattoo. All men (married or not) may eat molo nuts – a foodstuff less desirable than cassava root both in terms of taste and effects.

Instruction: Which cards would you turn over to test whether the rule has been violated?

Eats cassava	Eats molo nuts	Has tattoo	No tattoo

The formal expression of this is:

P = Benefit	Not P = Benefit not drawn	Q = Has paid cost	Not Q = Not paid cost

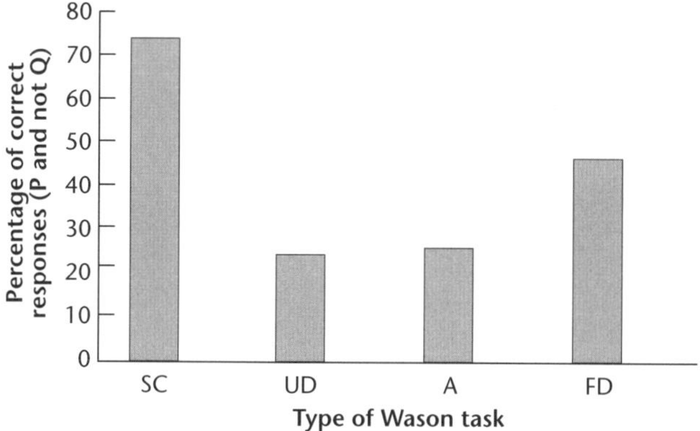

Correct responses

SC = task in the form of a social contract in an unfamiliar setting involving the detection of cheating
UD = unfamiliar context but task descriptive, that is, no cheating connotations
A = task in the form of an abstract logical question (for example, P and not Q)
FD = familiar context but descriptive without cheating implications

FIGURE 7.3 Percentage of correct responses in Wason's selection task according to context

SOURCE: After Tooby, J. and Cosmides, L. (1992) The psychological foundations of culture, in J. H. Barkow, L. Cosmides and J. Tooby (eds) *The Adapted Mind*. Oxford, Oxford University Press.

If the general thrust of Tooby and Cosmides' work is supported, the implications may be profound. Instead of thinking of the human mind as containing a single 'reasoning faculty', it may indeed be better to think of it as a cluster of mechanisms designed to cope with problems that we commonly encountered in the Pleistocene period of our biological evolution. One of these problems was how to reason accurately enough to detect cheating. The critical response to this work has been mixed. For methodological criticisms that still support the social contract hypothesis, see Gigerenzer and Hug (1992). For a more severe criticism of this view, see Davies et al. (1995).

If the detection of cheating were so important in the social environment of our evolution, we might expect that other components of our mental apparatus would be finely tuned to detect cheating. Mealey et al. (1996) investigated whether our memory of faces is enhanced by a knowledge that the face belongs to a cheater. They presented a sample of 124 college students with facial photographs of 36 Caucasian males. Each photograph was supplied with a brief (fictitious) description of the individual, giving details of status and a past history of trustworthy or cheating behaviour. Students were allowed about 10 seconds to inspect each face. Of the 36 pictures seen, 12 were described in the category of trustworthy, 12 as neutral and 12 as threatening or likely to cheat. One week later, the subjects were shown the pictures again (together with new ones) and asked whether they remembered the faces.

The overall finding was that both male and females were more likely to remember a cheating rather than a trustworthy face. The effect was significant for males and females but stronger for males ($p = 0.0261$). This work supports the general notion that our perceptual apparatus is adapted to be efficient at recognising cheaters. Puzzling features, however, remain. In the same study, the authors found that if pictures of high-status males were used, the enhanced recognition of cheaters disappeared for males and was even reversed for female subjects, in that they were now able to recollect trustworthy faces more reliably.

7.5.2 The Wason task and the payoff to the participant

In the tests administered by Cosmides and Tooby described earlier, the selections that correspond to benefits taken (for example drinking beer or eating cassava) and costs not paid (for example under age and having no tattoo) also correspond directly to the 'P' and 'not Q' of formal logic. The whole point of this being that the correct selection of P and not Q cards is greatly facilitated by this social context. Gigerenzer and Hug (1992) extended this idea to test if cheating simply stimulated a formal prepositional logic reasoning device somewhere in the brain or whether reasoning itself was in some way dependent on the perspective of the participant and in particular the cost–benefit payoff to the reasoner. Box 7.6 shows the reformulation of the test.

In this revised scenario, the costs and benefits are unevenly distributed between employee and employer as follows:

- *Employee:* Would be cheated if he or she worked at weekend (P) and did not get a day off (not Q). Hence, this perspective should be especially sensitive to the other side of both these cards. This perspective corresponds to prepositional logic and the formally correct answer of P and not Q.
- *Employer:* Here being cheated means fining an employee who 'did get a day off' (Q) but 'did not work on the weekend' (not P). From this perspective, the other side of both these cards should be most interesting.

Box 7.6 Wason's task: social contract and perspective change

The day off rule: If an employee works on the weekend, that person gets a day off during the week.

The cards below have information about four employees. Each card represents one person. One side of the card tells whether the person worked on the weekend, and the other side tells whether the person got a day off during the week. Indicate only the cards you definitely need to turn over to see if the rule has been violated.

Worked on the weekend	Did get a day off	Did not work on the weekend	Did not get a day off

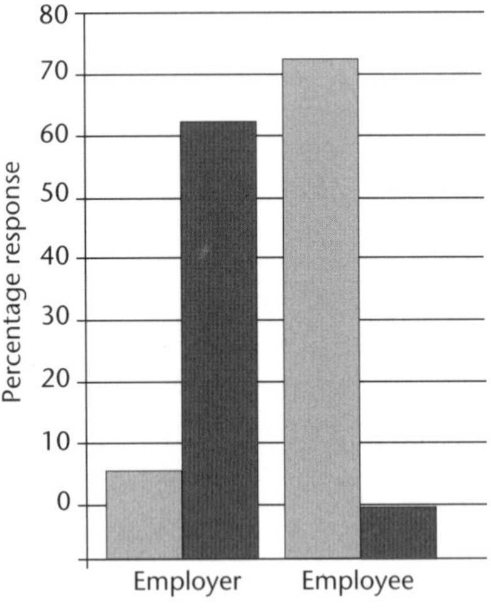

P and not Q means participant turned over 'Worked on the weekend' and 'Did not get a day off' cards

Not P and Q means the participant turned over 'Did not work on the weekend' and 'Did get a day off' cards.

FIGURE 7.4 Percentage responses to Wason's task as revised by Gigerenzer and Hug

SOURCE: Plotted from data in Gigerenzer, G. and Hug, K. (1992) 'Domain-specific reasoning: social contracts, cheating and perspective change.' *Cognition* **43**: 127–71.

In the Wason task, it is noticeable that responses of the sort 'not P' and 'Q' are rarely found. In the experiment carried out by Gigerenzer and Hug, the results are interesting (Figure 7.4).

Commenting on these findings, Gigerenzer (2000, p. 222) notes: 'Our participants were not reasoning with a Kantian moral but with a Machiavellian intelligence.' If this work is robust, it shows the advantages of constructing a theory of cognition that is informed by social interaction rather than trying to fit human reasoning into abstract and logically flawless procedures and then wondering why fallacies occur.

Not all subscribe to Tooby and Cosmides' interpretation of the effect of cheating connotations on performance in the Wason selection task. Oaksford and Chater (1994) propose an 'optimal data selection' theory to account for the findings. They argue that the abstract task (the registry) and the cheating-context task (for example the underage drinking) belong to different domains. The abstract task is, as they say, indicative, meaning that it refers to truth and falsehood – the rule about the filing could be shown to be false. But the underage drinking scenario is deontic (from Greek *deon* = duty) and is concerned with obligations. In this case, finding someone under 20 drinking beer does not prove the rule false but rather that some social infringement has taken place. They advance an interesting explanation for the popularity of the 'Q card' choice in both such tasks. They suggest that in real-life situations, choosing 'Q' (that is, positive confirmation of a generality) may be more informative because of what they call the 'rarity assumption' and the 'expected information gain'. Imagine you are trying to test the validly of your conclusion that black fruits taste sweet. It is more useful to taste yet another black fruit (Q) than to establish the identity of something that tasted bitter (not Q) because there are potentially hundreds of things that could taste bitter that have no bearing on your newly discovered rule. Yet it would be rare (ecologically speaking) to keep finding that black fruits are sweet if there was not some useful basis to the rule. It may be then that this bias to seek positive confirmation is a quite sensible strategy. In which case, choosing 'P and Q' is more 'rational' in terms of expected information gain than the logically correct 'P and not Q'. In the fruits example, we can add the observation that once a hypothesis 'black fruits taste sweet' receives some support, it is adaptive to keep consuming black fruits rather than run a controlled experiment to sample fruits of every possible different colour in the interests of science. Hence a confirmation bias may be adaptive. Oakford and Chater themselves steer clear of applying evolutionary logic to their interpretation but we can see how easily such reasoning fits with ecological rationality. Where the idea of expected information gain does struggle, however, is explaining the biases observed by Gigerenzer and Hug above where cheater detection facilitates reasoning but altruism does not.

The task now for advocates of domain-specific reasoning is to identify the level of abstraction of the mechanisms involved in such reasoning. On the one hand, the results repudiate the idea of a perfectly abstract logical reasoning device. On the other hand, the opposite level of virtually no abstraction with problems solved using memory and familiarity seems to be ruled out by the fact that problems can be solved in unfamiliar contexts. Somewhere along this continuum lies the domain-specific level of abstraction where some content is stripped away but the context (cheating and the perspective of the reasoning) is retained.

7.6 Sex differences in cognition

The evolutionary approach to cognition should have the potential to throw light on sex differences in the way men and women think by identifying perhaps the different adaptive pressures faced by ancestral males and females. Indeed, it has been known for several decades in mainstream psychology that there are consistent sex differences in cognition. Irwin Silverman and Marian Eals (1992) have advanced the idea that such differences may be attributable to differing selective pressure on males and females (as a result of a division of labour) in hominin evolution. Although there were, and are, overlaps in the roles undertaken by hunter-gatherer men and women, the archaeological and anthropological records suggest that men predominantly hunted and women predominantly foraged.

Successful hunting would draw upon cognitive skills associated today with activities such as map reading, maze learning, way finding and the mental transformation of objects (see below). In contrast, successful foraging involves recognising and remembering the location of specific plants in a complex array of vegetation. Today, such cognitive skills are likely to be associated with recalling the spatial configuration of objects. In today's environment then, it is predicted that men should be better at the first set of skills and women better at the second. A number of researchers have investigated this with a series of tests that have become standard (see Box 7.7).

Box 7.7 Three exercises used to test for sex differences in cognition

The mental rotation test
Shepard and Metzler (1971) originally investigated this pheneomenon. Subjects have to decide which object on the right (A, B or C) can be formed by rotating the standard object on the left.

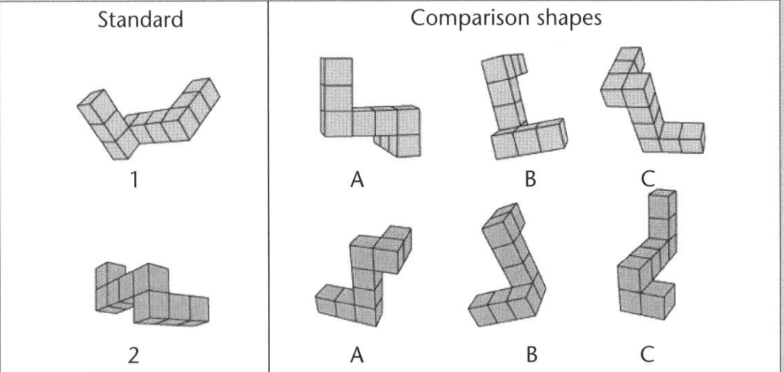

Memory location
Subjects are asked to look at this diagram. Another diagram is presented but with the location of some objects changed. Subjects then have to identify the changes.

SOURCE: Reproduced from Silverman, I. and Eals, M. (1992) Sex differences in spatial abilities: evolutionary theory and data, in J. H. Barkow, L. Cosmides and J. Tooby (eds) *The Adapted Mind*. Oxford, Oxford University Press, Figure 14.3, p. 538.

Water line
Subjects have to draw a line indicating where the water level would be.

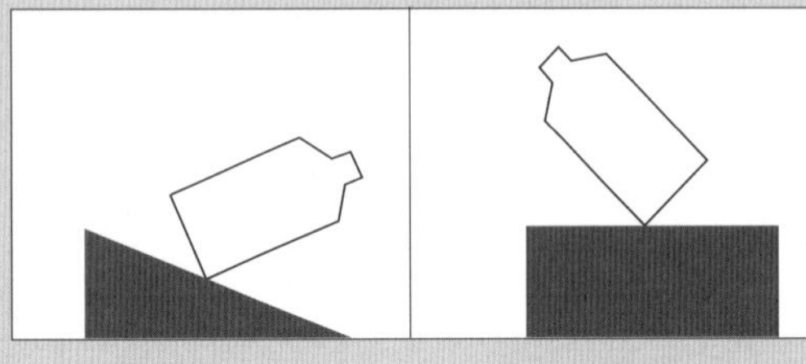

Table 7.4 Summary of results from two separate studies on sex differences in cognitive tests

Silverman and Eals (1992)	Males	Females	Performance	Significance of difference
Object location[1]	83	134	F > M	P < 0.01
Silverman et al. (2000)	**Males**	**Females**	**Performance**	**Significance of difference**
Water line determination[2]	9.91	7.68	M > F	P < 0.001
Mental rotation[3]	25.07	16.50	M > F	P < 0.001

Notes:

1 Points awarded for correct identification of objects added to picture in set time.
2 Numbers show number of correct responses in three minutes to 12 similar water line tests.
3 Numbers show correct responses to 24 tests taken in ten minutes, with one point deducted for each incorrect identification.

SOURCE: Data taken from Silverman, I. and Eals, M. (1992) Sex differences in spatial abilities: evolutionary theory and data, in J. H. Barkow, L. Cosmides and J. Tooby (eds) *The Adapted Mind.* Oxford, Oxford University Press; Silverman, I., Choi, J. et al. (2000) 'Evolved mechanisms underlying wayfinding: further studies on the hunter-gatherer theory of spatial sex differences', *Evolution and Human Behaviour* **21**(3): 201–15.

Silverman and Eals (1992) tested males and females on mental rotation, space relations and object location. In later work, Silverman et al. (2000) examined the performance of males and females on the water line test and also an out-of-doors way-finding task. For the latter, subjects were taken to a dense wooded area and their performance on finding their route back and understanding of their position at various times was measured. Males were found to significantly outperform females on this way-finding task. A digest of their results for the paper-based cognitive tests is shown in Table 7.4.

As the data show, males outperformed females and females outperformed males in skills of proposed relevance to hunting and gathering respectively. This could be evidence for sex differences in domain-specific competences in male and female brains, but as the authors point out, there are other interpretations. Spatial skills, for example, are dependent on the larger domain of allocentric perception competences, that is, the ability of subjects to detach themselves from ingrained egocentric perceptual modes. There could still be sex differences in this but the case for the relationship of differing levels of allocentric perception to hunter-gatherer skills differentiated by sex is weaker. It could also be that sex differences in early experience with toys involving straight lines and angles can help the development of mental rotation skills. Furthermore, young males tend to have a larger home range right from childhood (Gaulin and Hoffman, 1988) and this early experience could foster better way-faring skills or allocentric abilities in general.

7.7 Problems with the modular approach

The rallying call of Tooby and Cosmides has been both effective and controversial, and has sparked off numerous debates. Some of the topics that are still the subject of vigorous questioning are:

- *The nature of the EEA:* Was there a single environment that shaped the human brain, and what were its features? How can we establish what the environment was like?
- *Domain specificity:* Must problem-solving modules be domain specific? Might there not be room for cognitive processes that can be brought to bear on a range of adaptive problems? Might not domain-general processes also have some evolutionary plausibility?
- *Correspondence with neurophysiological structures:* It would be unlikely that modules discovered and investigated by evolutionary psychologists would have a simple mapping structure with neurologi-

cal structures, but what is the relationship between domain-specific modules and the structure of the brain?

- *Behaviour as a result of a cognitive process:* Must all adaptive behaviour be seen as the result of cognitive processes? Because behaviour has to solve a complex problem and provide an adaptive solution, this does not imply that it is driven entirely by mental processes. A physiological system might suffice. The heart, for example, is a highly adaptive organ that has to solve numerous problems to do with pumping rates and the coordination of pressure changes according to the activity of the organism, yet its behaviour is hardly a product of cognition.
- *The superficiality of genetic variation:* Tooby and Cosmides offer a very restricted scope to evolutionary psychology. They posit a world shared by humans with an essentially similar genetic make-up, in which differences between humans are not a significant part of the adaptationist paradigm. No sooner is this thrown down than exceptions accumulate. The first and obvious one is that there must be differences between the architecture of males and females; they must solve, for example, the problem of mate choice in different ways. So now we have two universal human natures. Tooby and Cosmides (1992, p. 80) also concede some, albeit minor, role to variation, noting that there 'may also be some thin film of population-specific or frequency dependent adaptive variation in this intricate universal structure'. David Wilson (1994) offers a constructive criticism of this position and argues the case for incorporating adaptive genetic variation within the scope of evolutionary psychology. Wilson makes the point that, in any population, there may exist a genetic polymorphism such that populations within a single species may have substantial genetic differences that dispose them to behave differently but nevertheless adaptively in their local environment. Many fish species seem to display just this sort of adaptive variation. Males of the bluegill sunfish *(Lepomis macrochirus)*, for example, exist in three forms: 'a parental' form that grows to a large size and defends a nest, a smaller 'sneaker' form and a 'mimic' form that is also small but resembles a female. The sneakers and mimics thrive by dashing into the nest of the larger form at the moment at which he deposits his sperm. Here we have a genetic variation that underlies the shape and behaviour of the phenotype that gives rise to different adaptive solutions (see Wilson, 1994). The very existence of the parental form enables the sneakers and mimics to parasitise their behaviour. The idea of genetic polymorphism may apply to facets of human behaviour. In Chapter 15, for example, we examine the possibility that sociopaths may be genetic polymorphs thriving in an otherwise 'normal' or non-sociopathic world.

Despite these reservations, many now acknowledge that the vision outlined by Tooby and Cosmides has encapsulated a major advance in thinking about the human mind. There remains a lively debate concerning different components of the programme. The domain specificity of mental modules and the assertion that behaviour must be seen as the product of cognitive processes, for example, are rejected by Shapiro and Epstein (1998). Samuels (1998) also rejects the proliferation of mental modules (the 'massive modularity hypothesis') for specific tasks.

The debate will continue but it is clear that the modularity thesis will have to accommodate communication between modules and some degree of phenotypic plasticity. A single genotype might produce a range of phenotypes differing as a result of the environmental cues received during development. It would in fact be a highly efficient and adaptive mechanism that allowed development to be structured by experience in ways to achieve an adaptive fit to local conditions.

Summary

- Evolutionary epistemology views the mind as prestructured by natural selection acting on our ancestors. We are born, therefore, with in-built biases of perception and a priori ways of thinking.
- Studies on perception, memory, cognition and decision-making have traditionally been the province of cognitive psychology. This was a discipline that grew up in the 1960s and tended to concentrate on proximate causes and hence the neural processes and decision-making algorithms that underlie behaviour and cognition. Evolutionary approaches look to ultimate levels of explanation and the sort of problems that the brain had to solve in the environment of evolutionary adaptation. Errors of reasoning and cognitive biases that may now appear suboptimal or even maladaptive may make sense when viewed as adaptations to an ancestral environment or as the result of a trade-off in a mind that has to make compromises to arrive at fast and adaptively efficient (as opposed to logically accurate) solutions. This whole perspective parallels the evolutionary approach to mental disorders explored in Chapters 14 and 15 where it will be shown that one cogent and coherent perspective in evolutionary psychopathology views contemporary disorders in terms of a mismatch between the environment for which the human mind was designed and the world of the 21st century in which it now finds itself.
- The mind is subject to cognitive illusions. A fruitful way of understanding these is to see them as ecologically rational in the environment in which they evolved.
- Two American psychologists, Tooby and Cosmides, have been strong advocates of an approach to evolutionary psychology purporting to show that the human mind is constructed as a series of specialised problem-solving mechanisms or 'domain-specific modules'. Such modules have been shaped by the thousands of years that early humans spent in their EEA and are adaptive in the sense of having been sculpted to solve ancient problems. There are expected to be modules to deal with group membership, the treatment of kin, facial processing and features of the biological and physical environment such as the distinction between flora and fauna and the properties of objects. There is much debate about the rigidity and specificity of these modules. The approach has been successful at least in one area – showing how human reasoning is enhanced and biased by the need to detect cheating in social exchanges.

Key Words

- **Base rate fallacy**
- **Bounded rationality**
- **Domain-specific mental modules**
- **Empiricism**
- **Epistemology**
- **Heuristic**
- **Modularity**
- **Nativism**

Further reading

Barkow, J. H., Cosmides, L. and Tooby, J. (1992) *The Adapted Mind*. Oxford, Oxford University Press.

A highly influential work containing the manifesto of evolutionary psychology by Tooby and Cosmides. Other chapters of interest deal with cognitive adaptations for social exchange, and the psychology of sex and language.

Gigerenzer, G. and Selten, R. (eds) (2001) *Bounded Rationality: The Adaptive Toolbox*. Cambridge, MA, MIT Press.

A series of research papers based on a workshop held in 1999. The concept of the adaptive toolbox is extended to cover reasoning, emotions and social norms. Seeks to promote an interdisciplinary basis for understanding how we reason and feel.

Emotions

Reason is, and ought to be the slave of the passions, and can never pretend to any other office than to serve and obey them.
(Hume, 1739, p. 460)

In the science fiction adventure series *Star Trek*, the perfectly rational and highly intelligent Dr Spock is set up as a contrast with the intelligent but fully human and therefore emotional Captain Kirk. Spock is actually half-Vulcan, half-human and we are invited to witness the struggle inside him between logic and emotion. A moment's thought, however, should convince us that a full and emotionless Vulcan is actually a highly unlikely life form. A being without emotion and motivation (and the two are inseparably conjoined) would show no goal-directed behaviour, it would not feel driven by desires and so, like some computer switched on blinking and waiting for the next instruction, it would do nothing.

Spock embodies the popular idea that emotion and reason must necessarily be in conflict. Outside popular culture, the relationship between emotion and reason has also long troubled philosophers, psychologists and, more recently, neuroscientists. One of the most long-standing metaphors in cultural history has been that of master and slave, with the dangerous and brutish impulses of our emotional life guided by the masterly exercise of reason. Until recently, a common stance was to distrust emotions. The behaviourist Skinner, in his utopian novel *Walden Two* (1976), wrote that 'We all know that emotions are useless and bad for our peace of mind and our blood pressure.' The distinguished psychologist Donald Olding Hebb (1949), although no friend of behaviourism, also argued that emotions tended to produce disorganised behaviour.

Sometimes the pendulum has swung the other way, but this dualistic view of emotions and the intellect still survives even when emotions are accorded higher praise. One of the effects of the Romantic movement in art and literature (beginning roughly about 1790) was to suggest that Enlightenment rationality had for too long repressed the emotions and that they should be allowed to flourish, containing as they do sources of deep natural wisdom, authenticity and creativity. Indeed, within the whole Western tradition of thought, emotion and reason seem to have been cast as polar opposites, compelled to orbit each other like some binary star system, each periodically coming to the fore in cultural approval: the head and the heart, the Apollonian and the Dionysian, Enlightenment rationalism and Romantic emotivism. The quotation from Hume at the start of this chapter is so surprising because it reverses a whole tradition of thinking on the relative positions of reason and emotion. Yet modern research is beginning to suggest that Hume may have been right; moreover, a sharp dividing line between emotion and reason is not that easy to find.

So what are emotions? This is a vexed and tricky question still unresolved by over 100 years of scientific research. For a start, it is not obvious what the observable phenomena are. Does the psychologist concentrate on inner feelings, measurable physiological states, facial expressions or the whole process of stimulus, context and response in which emotions are somehow buried and live?

A common approach (see Eysenk, 2004) is to suggest that emotions are associated with three types of responses:

1. *Behavioural:* such as a prototypic form of expression (usually facial)
2. *Physiological:* a pattern of consistent autonomic changes
3. *Verbal and/or cognitive self-report:* a distinct subjective state.

So, for example, in the case of fear, a subject's eyes may widen, his lips part and eyebrows rise (behavioural); the heart beats faster and sweating may follow (physiological); and finally, the subject describes his or her feelings as terrified or afraid (verbal self-reporting).

The problem is that these three components do not always show high concordance, that is, they are only weakly correlated and are not expressed to the same degree in each individual or between individuals. This is partly to be expected since each subsystem serves different functions and its activation and expression may depend upon the fine details of the context. One function of the behavioural system, for example, is probably to communicate the emotion to others; the physiological system serves to prime the body for a specific course of action such as run, fight or remain motionless; and finally, the internal self-reporting can motivate the individual to change a course of action, continue it or more generally draw upon the emotional feeling to make appropriate choices.

8.1 Some early theories of emotions

8.1.1 *The Expression of the Emotions in Man and Animals* (Darwin, 1872)

Many histories of psychological theories of the emotions start with William James's well-known essay 'What is an emotion', published in 1884. But James's ideas were a development and modification of those of Darwin who explored the same issue in 1872 in his *The Expression of the Emotions in Man and Animals*. Darwin tackled several important questions: why emotions were expressed, why their associated facial expressions took the forms they did (for example why a smile was a raising of the corners of the mouth and not a drooping) and whether emotions were universally expressed and recognised by all races of people. He tackled these questions by reference to three principles (Darwin, 1872):

1. *Serviceable habits:* By this he meant that the expression of emotions originated in movements that were once useful – such as raising the eyebrows when surprised to allow more light (and so more information) to enter the eyes. Or the mouth open threat posture of primates as a possible relic of an attacking position. Darwin suggested that through time these became fixed as a corollary to an emotion even if the action no longer served any function.
2. *Antithesis:* Darwin thought that if a certain frame of mind produced a reaction (such as accounted for by serviceable habits), then the opposite frame of mind might excite an action that is opposite but not necessarily of any use. As an example, Darwin cited astonishment. A normal relaxed and unastonished individual stands with arms relaxed at his side, palms concealed and fingers close together. An astonished reaction, claimed Darwin, is one where both hands are forced forwards, palms outwards and fingers apart. This reaction does not serve any function but is simply the opposite of a state of non-astonishment. Similarly, to avert one's gaze is a sign of social submission, which may be the opposite of a direct gaze, which is a sign of threat.
3. *Direct action of the nervous system:* Reflecting the immature state of physiology in his day, Darwin thought that some reactions were simply the result of nervous forces (such as trembling in great fear or joy or excitement) acting on the body against the will and to no functional purpose.

In this book, Darwin was keen to stress above all two main points: the continuity between humans and other animals, and the universal character of human emotional expression. These two points were instrumental to his broader aim of showing that mankind had a single origin somewhere in Africa (and was not a product of special creation or the separate evolution of different species giving rise to different races), and that man had evolved from more humble life forms. As he wrote towards the end of *The Expression of the Emotions in Man and Animals*:

> I have endeavoured to show in considerable detail that all the chief expressions exhibited by man are the same throughout the world. This fact is interesting as it affords a new argument in favour of the several races being descended from a single branch ... before the period at which the races diverged from each other. (p. 355)

So, for Darwin, facial expressions were universal, innate and inherited from our primate ancestors. In his reasoning, Darwin tended to assume that the universality of emotions implied that they were innate. Logically, of course, this is not a conclusive argument. A universal trait could be the result of humans socially learning the same solution to the same type of problem found everywhere. Surprisingly, Darwin did not actually argue that facial expressions were naturally selected adaptations. He tended to think of them as non-functional vestiges of earlier behaviour that had been passed on and inherited by a Lamarckian mechanism. Fear, for example, may cause our skin to prickle and our hair to stand on end ('like the fretful porcupine'), since this may have helped our mammalian ancestors by making them appear larger, but for modern humans, it was not of much use anymore. Darwin thought that showing emotions to be non-functional helped his more general arguments in favour of the descent of humans. For if emotions were useless relics, this served as a counterargument against creationists who argued that the fine design of our emotional system was a sign of God's wise handiwork.

After some initial interest by Darwin's followers, the long-term influence of *The Expression* was delayed and animal behaviourists and psychologists only really began to take inspiration from Darwin's approach in the 1970s. This is somewhat strange, since in its day the book was very popular, a bestseller in fact, and by then Darwin was a famous scientist.

8.1.2 The James–Lange theory

Following Darwin, the next major theory was advanced independently by William James (1842–1910) in 1884 and by the Danish scientist Carl Lange in 1885. It became known as the **James–Lange theory**, although James was the main proponent. James was particularly interested in emotions that have accompanying bodily reactions. In essence, his theory reverses what was then (and probably still is) the common-sense or popular view that we experience an emotion and our body responds accordingly. So, for example, when we perceive a threat or experience fear, our heartbeat rises because we are afraid. James (1884, p. 189) partially inverted this view and suggested:

> the bodily changes follow directly the PERCEPTION of the existing fact, and that our feelings of the same changes as they occur IS the emotion.

A more simple formulation, as James himself illustrated is: 'it is not "I see a bear, I feel afraid, I run away" but rather "I see a bear, I run away, I feel afraid"'. The James–Lange theory is represented diagrammatically in Figure 8.1.

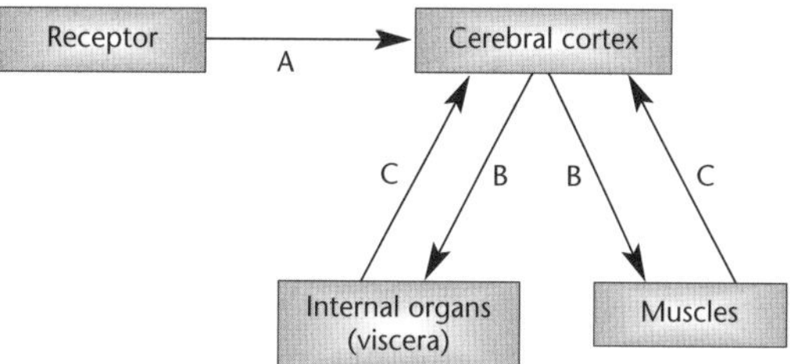

FIGURE 8.1 The James–Lange theory of emotions
The sequence of events according to this theory is A, B, C. In this view, an emotion is an afferent feedback from disturbed organs and muscles.

Subsequent research indicated that reality was not so simple or straightforward as James supposed. Walter Cannon, for example, showed in 1927 that people who have suffered a severance between their viscera and the central nervous system (as in spinal damage) still experience emotions. Other problems were also found, such as the absence of a specific pattern of autonomic response that corresponds to each emotion and the fact that blocking signals from the autonomic system with drugs does not prevent emotions from being felt.

In the light of this, other theories emerged such as 'arousal-interpretation theory' and 'appraisal theory', both of which stressed the importance of the cognitive processing and appraisal of the information gathered by the receptors. After all, we might run from a wild bear in a wood but not from one behind bars in a zoo. Nevertheless, James's theory was invaluable in pointing to the possible role of physiology in the experience of emotions. The eminent neuroscientist Antonio Damasio (2000) thinks that James hit upon a mechanism that is still essential to the understanding of emotions but that he did not develop it fully or adequately consider the role of cognitive evaluation.

8.2 The functionality of emotions

8.2.1 The case for functionality

The difficulty of research into emotions (and the variety of approaches in psychology) is illustrated by the fact that there is as yet no consensus on exactly how many emotions there are, and what emotions should be regarded as primary and what secondary (that is, made up from primary emotions). The behaviourist Watson thought that a child is born with only three types of basic emotion: fear, rage and love. In his early work, Paul Ekman (1973) thought there were six basic emotions that corresponded with six types of facial expression: happiness, surprise, anger, sadness, fear and disgust. Johnson-Laird and Oatley (1992) suggested just five: all those of Ekman minus surprise.

Despite uncertainties about how many emotions we can actually experience, the fact that emotions are likely to serve some adaptive functions should be clear from an examination of some of their characteristics:

- Emotions elicit or are associated with autonomic and endocrine responses.
- Emotions motivate action.
- Emotions are capable of being communicated and so could allow social bonding.
- Pleasant emotions are positively reinforcing.
- Emotions can affect the cognitive evaluation of events and memories.
- Emotions can bias the storage of memories in specific directions.

The functionality of positive emotions is easy to see: it is understandable that natural selection should give us warm feelings for things that are beneficial to our reproductive interests such as good food, a secure environment, sex, supportive friends. Understanding why we behave in ways that communicate (often unintentionally) our emotional state is trickier. Steven Pinker (1997) suggests that there are sound evolutionary reasons for honest displays of an emotional state and that this helps to explain why emotional displays are often linked to parts of human physiology that are not normally under conscious control such as blushing, flushing, blanching, sweating and trembling. So, for example, blushing in public out of shame for some misdemeanour becomes a sure sign to others that the individual recognises that he or she has done something wrong and is experiencing the remorseful prick of conscience. This could be an honest signal that the person will be minded to behave more responsibly in the future. Darwin himself seems to have appreciated the importance of emotions as honest signals, when he wrote:

> The movements of expression give vividness and energy to our spoken word. They reveal the thoughts and intentions of others more truly than do words, which may be falsified. (1872, p. 359)

8.2.2 Evidence of functionality

So far we have speculated that emotions serve functional needs and are adaptations. There are three types of evidence that could potentially lend more credibility to this claim:

1. **Homology**: The idea that the expression of emotions may be shared with our nearest relatives among the primates.
2. *Universality:* If humans in widely different cultures show the same facial displays of emotion and conceive of emotions in similar ways, this lends support for their pan-human biological origins.
3. *Neurophysiological correlates:* If specific parts of the brain can be identified as involved in the processing of emotions, then the presence of this hardware, likely to be shaped by natural selection, is strong evidence for their genetic and hence functional basis.

We will now examine each of these areas of evidence in turn.

8.2.3 Homology

Darwin was one of the first to suggest that there were strong similarities between the sounds and expressions associated with emotions between humans and other primates. As he observed:

> We may confidently believe that laughter, as a sign of pleasure or enjoyment, was practised by our progenitors long before they deserved to be called human; for very many kinds of monkeys, when pleased, utter a reiterated sound, clearly analogous to our laughter, often accompanied by vibratory movements of their jaws or hips. (Darwin, 1872, p. 356)

Darwin went on to speculate that some emotions such as laughter may be shared with our immediate primate relatives, since they were present in our common ancestors, but others such as blushing may have been derived in the human line after it split from other primates and therefore is more recent.

Modern research on homologies between the facial expressions of humans and other primates is still at a fairly early stage. As an illustration, we might consider smiling and laughing. In humans, they are used in similar contexts and it is tempting to regard smiling as a diminutive form of laughing. The nearest correlates to smiling and laughing in chimps are the 'silent bared teeth display' (SBT) and the 'relaxed open mouth display' (ROM) respectively. But van Hoof (1972) found that SBT and ROM were displayed by chimpanzees in different behavioural contexts, with ROM associated with play and SBT connected with affinity or friendship. Van Hoof also speculated that the primate prototype of the human smile was the prosimian or monkey 'grin', which originally functioned as a fear expression. With the evolution of the apes, the expression developed into three forms: SBT1, signalling fear and submission; SBT2, signalling fear, reassurance and appeasement; and SBT3, signalling affection. In humans, he argued, the first function of fear or submission (SBT1) had been lost and the fear function of the second (SBT2) dropped, leaving us with two expressions – smiling and laughing – both signalling affection and play (see Figure 8.2). Preuschoft (1992) built upon this work and by looking at the behaviour before and after the display of these expressions concluded that in Barbary macaques *(Macaca sylvanus)* the SBT display was associated with submission by the sender. Consistent with this, Waller and Dunbar (2005) found that SBT and ROM in chimpanzees were found in dissimilar contexts. In short, they found that rates of affinitive behaviour increased following SBT, suggesting that SBT is a signal of affinity, whereas ROM was observed primarily during play, and play between two individuals was increased in length when ROM was bidirectional. Rates of affinitive behaviour also increased after ROM and this might suggest that both displays assist social bonding and could explain why the two displays have converged in humans.

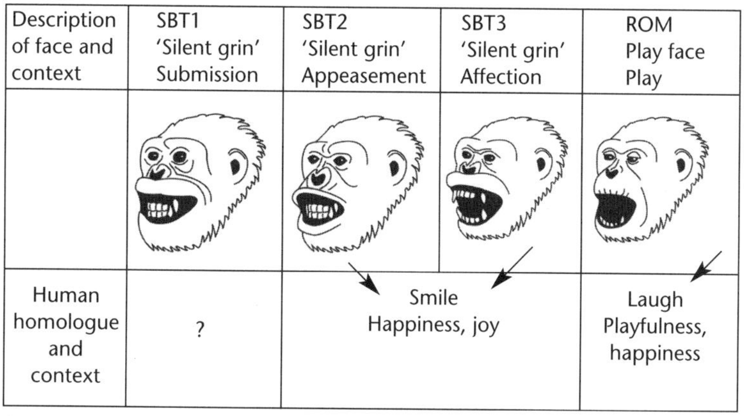

Description of face and context	SBT1 'Silent grin' Submission	SBT2 'Silent grin' Appeasement	SBT3 'Silent grin' Affection	ROM Play face Play
Human homologue and context	?	Smile Happiness, joy		Laugh Playfulness, happiness

SBT = silent bared teeth   ROM = relaxed open mouth

FIGURE 8.2 Chimpanzee–human homologues in facial expressions

SBT2 and SBT3 are likely to have evolved into the human smile. The ROM may be a homologue for the human laugh. It is not clear if SBT1 simply faded out in the human lineage or has transformed into another expression.

SOURCE: Diagrams drawn by Gavin Roberts, based on a series of photographs and description in Van Hooff, J.A.R.A.M. (1971) *Aspects of the Social Behavior and Communication in Human and Higher Primates*. Rotterdam, Bronderoffset.

Another way of testing for similarities between humans and other primates is to make observations on young infants and young chimpanzees in terms of their social development. One of the most widely used assessments is called the 'neonatal behavioural assessment scale' (NBAS). This incorporates observations on how infants orient themselves to objects, their use of resources to regulate their own behaviour, the range of their emotional arousal and signs of stress. Kim Bard (2005) has applied this type of analysis to various groups of human infants and chimpanzees. Some of the chimps were raised in the presence of humans and some raised primarily by their chimp mothers. Bard found that young humans and chimps have much in common. Very young chimpanzees, for example, give smiles and positive vocalisations to familiar faces and even show signs of anger.

8.2.4 Universality

The crucial question in this area is whether our experience of emotions and their expression by gestures and facial movements are culturally determined (the social constructionist view) or universal. Put simply, is the raising of the mouth in the corners (a smile) a sign of happiness or friendliness in all cultures or are there some cultures where a smile might mean 'I am disgusted', 'I am angry' or 'I am afraid'? As we noted, Darwin, for a variety of reasons, was keen to insist upon the universality of facial expressions. In *The Expression of the Emotions*, he made an important contribution by taking photographs of ordinary people and actors staging emotional gestures to show that they are readily recognised by others (Figure 8.3). He then used his extensive network of correspondents overseas to show that similar emotions were displayed by people untouched by Western civilisation.

In recent times, the true heir to this tradition has been Paul Ekman. Ekman realised that testing for the universality and instinctive nature of emotional expression and experience poses numerous experimental problems. One is to disentangle the effects on people of mass culture, with its supply of ready-made images of emotions and their expression, from what might naturally develop. Another is the problem of translating terms for emotions between languages. To overcome this, in the 1960s, Ekman went about interviewing and testing people in a wide variety of cultures. One of his studies concerned the Fore people of the New Guinean highlands. At the time, these people were still living a Stone Age lifestyle without a written language and relatively impervious to the global reach of Western culture. Many of them had never even seen their own faces in a mirror. In an ingenious experimental design, subjects attached an emotional label to a picture by identifying the face of a character in a story such as that of a man unexpectedly attacked by a wild pig. In this example, his subjects identified the character as one that we would say was displaying fear. Ekman then asked these people to act out the emotions of such characters as he filmed them. When the film was shown back in the USA, volunteers easily identified the intended emotion. Later, such studies were extended by Ekman (1973) to include a total of 21 countries. The results were highly significant: despite the numerous differences between such cultures in terms of language, economic development and religious values, there was overwhelming agreement about faces that showed happiness, sadness and disgust. Eventually, Ekman suggested that there were six basic emotions found in all cultures: fear, anger, disgust, sadness, joy and surprise.

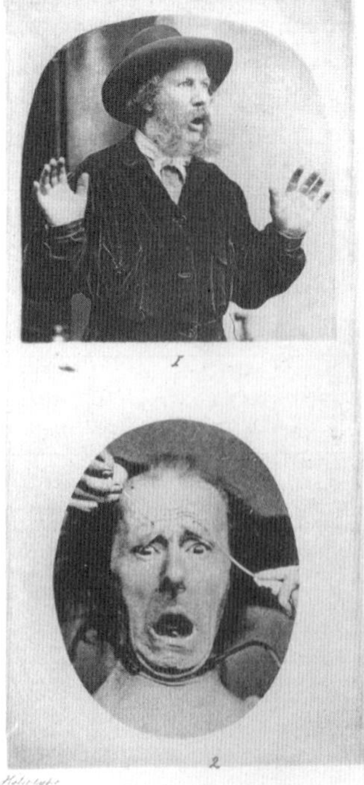

FIGURE 8.3 Facial expressions used by Darwin in *The Expression of the Emotions* (1872) The top photograph shows an actor demonstrating surprise. The lower photograph shows electricity applied to a subject's face to stimulate muscle groups involved in emotional expression – in this case horror

From such studies, it increasingly begins to look as if our emotional life (at least in the way it is expressed facially) is part of a standard development programme. Since this programme develops consistently in all cultures, it would seem reasonable to suppose that it has been shaped by natural selection. However, in arguing for the pan-cultural nature of emotions, we need to distinguish between the expression of the emotion (that is, the response output) and the stimulus that elicited it (the input). Work by Ekman et al. (1987) has indeed shown that there may be a set of universal cross-cultural emotions (such as joy, anger, fear, sadness, surprise and disgust), but the circumstances that are associated with each expression may depend upon the individual's life history and hence the experience of the cultural value attached to different situations and emotions. In other words, the outputs may be pan-cultural but the input system may be flexibly acquired. The work of the behaviourists (although misguided in so many respects) did seem to demonstrate that it is highly unlikely that we have an innate 'fear of a large fanged animal' or 'fear of a snake' reaction, which then prompts the display of fear. It looks more likely that we have innate emotional responses called 'fear' or 'sadness' but the cues for these are socially learned. Babies are not born with a 'fear of a snake' response but they quickly learn it and, more importantly, learn it from watching the reactions of adults (Klinnert et al., 1982).

This system looks like a good compromise between the need to adapt to a changing world (where threats vary from generation to generation) and the need to avoid the time-wasting process of learning afresh through trial and error. Having an innate fear response that alerts the system to a state of readiness, coupled with its linkage to an input that is learned through experience and watching what adults fear, is perhaps the best of both worlds (see also Chapter 14).

This cultural dimension was also captured by Ekman's work. We know from common experience that it is possible to have an emotion and either suppress its facial manifestation or allow its full expression. Ekman (1973) suggests that there are 'display rules' that may govern this decision process and that such rules may be socially conditioned. In one experiment, he filmed American and Japanese students as they were shown both neutral and stress-inducing films. When the experimenter was absent, the facial responses were broadly similar; but when the experimenter entered the room and asked the subjects about their emotions as the film was played again, the Japanese subjects suppressed their facial expressions. Ekman interpreted this as the operation of the Japanese cultural display rule of not showing negative emotions in the presence of an authority figure. However, Ekman did observe 'leakage', in other words, the emotion was never completely suppressed, suggesting that there are two types of circuit. So, by 1973, Ekman had concluded:

> Our neuro-cultural theory postulates a facial affect program, located within the nervous system of all human beings, linking particular facial muscular movements with particular emotions. It offers alternative nonexclusive explanations of the movement of the facial muscles. Our theory holds that the elicitors, the particular events which activate the affect program, are in largest part socially learned and culturally variable ... but that the facial muscular movement which will occur for a particular emotion (if not interfered with by display rules) is dictated by this affect program and is universal. (Ekman, 1973, p. 220)

8.2.5 Neurophysiological correlates

Until recently, research into physiological activity associated with specific emotional states tended to focus on the behaviour of the **autonomic nervous system** (ANS) and the endrocrine system (glands that secrete hormones into the bloodstream). This was partly because such studies were easier to perform than trying to probe the brain directly, and partly because it was obvious that changes in such things as heart rate and hormone levels were involved in some way with emotional responses. The problem that bedevilled this research was that these very systems were also involved in the routine housekeeping functions of the body (such as energy metabolism, tissue repair and homeostasis generally) and so did not offer ideal pointers to the working of the emotional system.

One more direct method has been to examine patients who have suffered damage to specific areas of the brain and try to correlate changes in emotional and cognitive functioning with the areas damaged. One famous, often quoted case of this is the experience of a US railway foreman called Phineas Gage. In 1848, Gage placed a stick of dynamite into a hole drilled into a boulder; as he tamped down the dynamite with an iron bar, it exploded, sending the bar through his left cheek and up through the front of the brain. The bar landed about 100 metres away. Amazingly, Gage survived but was a changed man. Whereas previously he was a thoughtful and industrious person, he now became childish and thoughtless in his behaviour. Later analysis of the damage caused showed that the iron bar had obliterated much of the **orbitofrontal cortex** of the man's brain. This tragic incident was one of the first of many clues that leads neuroanatomists to think that this part of the brain is responsible for integrating signals from the emotional centres below (especially the amygdalae) and sensory information presented to it by the sensory systems and other parts of the cortex.

More recent research has been able to look at the brain under more controlled conditions. There have been two main approaches:

1. *Neuroimaging techniques:* Developments in technology have given neuroscientists a whole array of techniques to study brain activity. In general, these techniques allow researchers to identify which parts of the brain are active under specific circumstances and so, in principle, enable the determination of when and where cognitive processes are occurring. The two primary techniques here are:
   - positron emission tomography (PET), which involves the detection of positrons given off by some atomic isotopes and enables a spatial resolution of brain activity down to 3–4 mm
   - magnetic resonance imaging (MRI and functional MRI), which relies upon the excitement of atoms in the brain by radio waves and the detection of any magnetic changes produced using a large magnet that surrounds the head of the patient under test.
2. *Use of drugs:* It is well known that drugs produce changes in behaviour and feelings. The problem with research in this area is that drugs may affect several areas of the brain and it is not always clear, therefore, which affected area is responsible for or associated with a behavioural or mood change.

Using these techniques has vastly improved our knowledge of how emotional states are related to the activation of different areas of the brain. The next section considers some elementary findings about emotions and brain structure.

8.3 Brain structure

8.3.1 The central nervous system

The nervous system of the human body is divided into central and peripheral divisions. The **central nervous system** (CNS) consists of the brain and spinal cord. The brain (or cerebrum) consists of left and right central hemispheres joined together by a bundle of fibres known as the **corpus callosum** (shown in Figure 8.5). The collection of nerves that connects the CNS with the periphery of the body (skin and visceral organs) is called the peripheral nervous system (PNS). Like all living tissue, the nervous system is made up of cells, in this case neurons and glial cells. **Neurons** are the crucial cells for communication and consist of a central cell body (containing all the main machinery found in most cells such as a nucleus, mitochondria and other organelles), output fibres known as 'axons' and input fibres known as 'dendrites' (Figure 8.4).

Neurons interconnect to form circuits. The typical connection is an axon making contact with the dendrite of another cell at a junction point known as a 'synapse'. The average neuron sends impulses to about 1,000 other neurons. Neurons are divided into motor neurons that conduct impulses towards muscles, sensory neurons that respond to sensory inputs such as light, and interneurons that act as 'go-betweens' or bridges between other neurons.

Slicing through the brain reveals areas of dark (grey) and pale (white) material. The grey matter takes its colour from the tight packing of neuron cell bodies. The white matter takes its hue from the myelin sheaths that insulate the fibres stemming from the grey matter. The brain is divided into three main regions based on their location in the embryo. These are the hindbrain, midbrain and forebrain (Figure 8.5). Box 8.1 shows a further breakdown of the names of cerebral components.

Box 8.1 The central nervous system

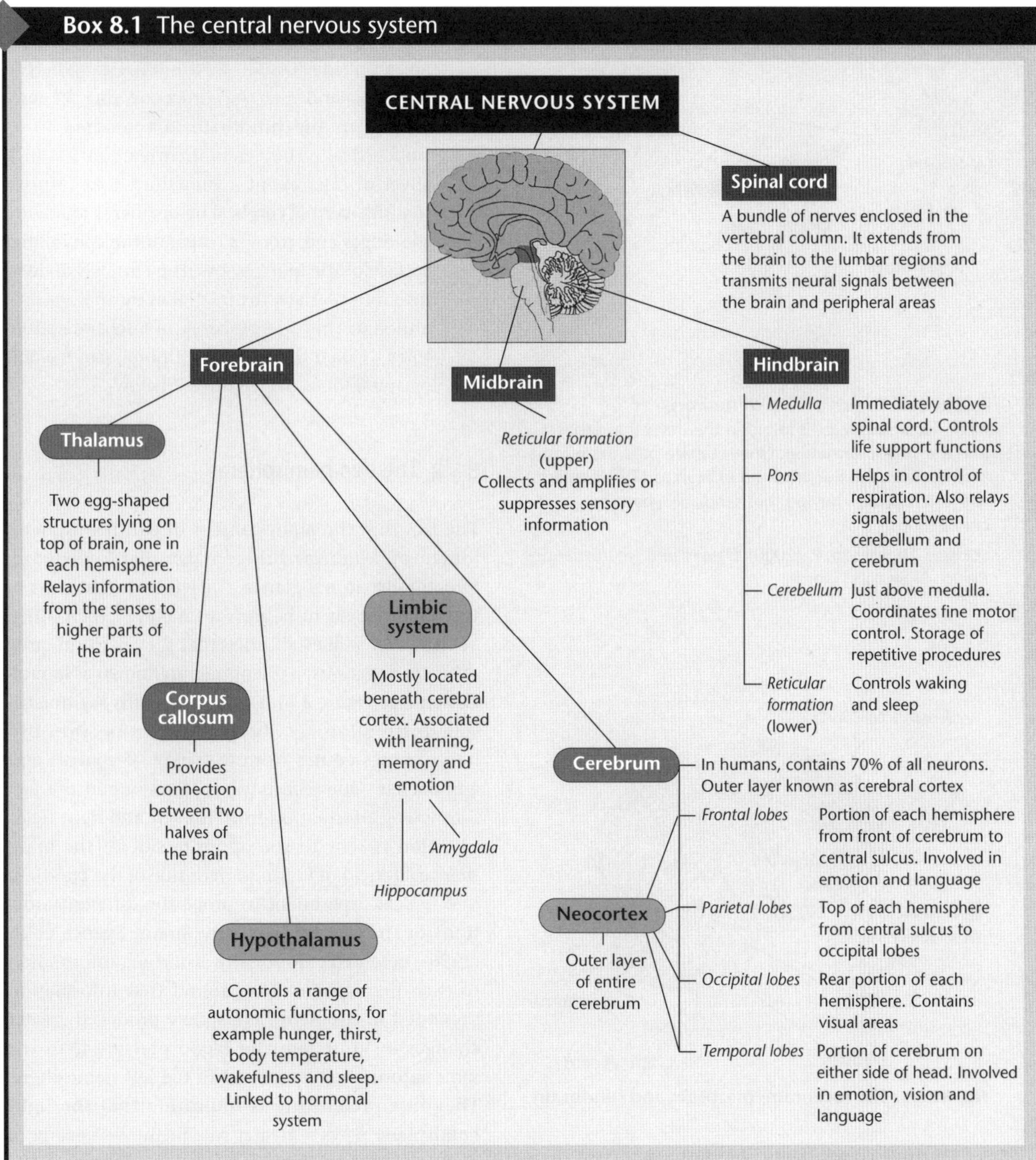

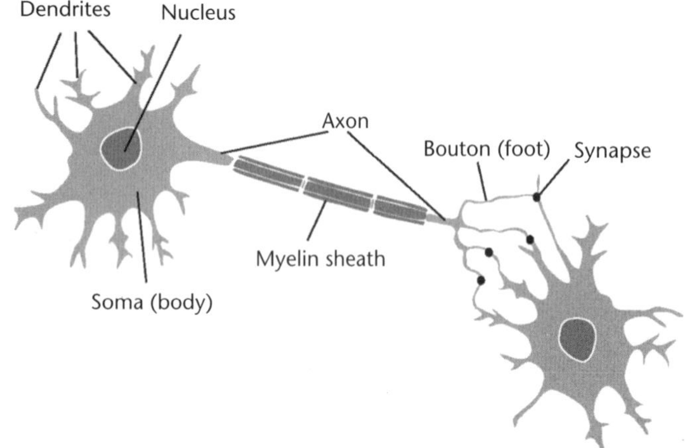

FIGURE 8.4 The structure of neurons
Neurons are nerve cells found in the brain that process and transmit information. One estimate puts the number of neurons in the human brain as high as 10^{12} (a trillion), with each neuron having thousands of synaptic connections.

SOURCE: Thompson, P. (2000) *The Brain: A Neuroscience Primer*. New York, Worth.

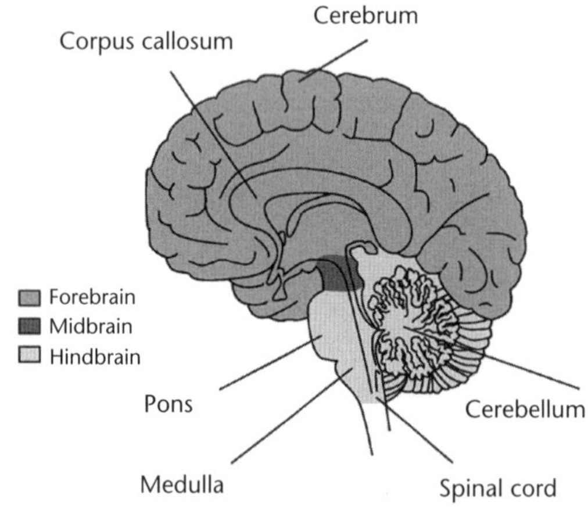

FIGURE 8.5 The forebrain, midbrain and hindbrain

Of these components, the most important division for humans is the forebrain, and the most evolutionary 'advanced' or recent part of this is the cerebral cortex, about 90 per cent of which is called the neocortex. The neocortex (sometimes called isocortex) can be identified by the fact that it is laminated into six layers; the rest of the cortex (or allocortex) has a variable number of layers. The cerebral cortex plays a crucial role in thinking and language. It is made up of grey matter and other grey matter areas outside it, such as the basal ganglia, the amygdala and even the cortex covering the cerebellum in the hindbrain, are referred to as 'subcortical'. The cerebral cortex consists of a sheet of cells about 4 mm thick that forms a mantle over the **cerebral hemispheres** following the ridges and grooves (sulci) of the convoluted surface of the brain below. The cortex is divided into four spatially distinct lobes by anatomists – although they are probably not separate structures – called the frontal, temporal, parietal and occipital lobes (see Figure 8.7 below).

8.3.2 The two hemispheres

The fact that the brain consists of two hemispheres joined by a narrow bridge is familiar knowledge. Structurally, at first glance, the left and right sides of the brain appear to be mirror images of each other. Yet it is now well established that they perform quite different functions; a phenomenon known as **lateralisation** (see Figure 8.6). Evidence for this asymmetrical distribution of functions across the two hemispheres comes from a variety of sources and approaches. One technique is to rely upon the fact that visual information from the left and right fields of vision is sent to the opposite side of the brain (hence left field of vision to right side of the brain). In one typical experiment to probe the different functions of the two halves of the brain, Spence et al. (1996) presented emotionally disturbing information to each hemisphere. They found that information presented to the right hemisphere produced greater changes in heart rate and blood pressure than the same information presented to the left hemisphere. They interpreted this as indicating that the right hemisphere plays a greater role in the processing of emotions than the left.

Studies on patients with a damaged hemisphere are also instructive. Generally, it has been found that damage to the left hemisphere has a major impact on linguistic ability but damage to the right a much smaller impact, suggesting that linguistic ability resides primarily (but not exclusively) in the left hemisphere (see section 6.3.2 on Broca's area and

Wernicke's area). But studies such as these on the abilities of subjects with damaged brains are made difficult by the fact that the two hemispheres pass information back and forth through a connecting bundle of fibres known as the corpus callosum (see Figure 8.5 above). This has meant that researchers have typically had to look for quick responses before each side can inform the other what is going on. However, as a treatment for epilepsy, some patients have undergone surgery to sever the corpus callosum, leaving each side of the brain to operate independently. Studies on these patients have been illuminating. They show, for example, that for these subjects emotional stimuli are only processed accurately when the information is accessed by the right side of the brain.

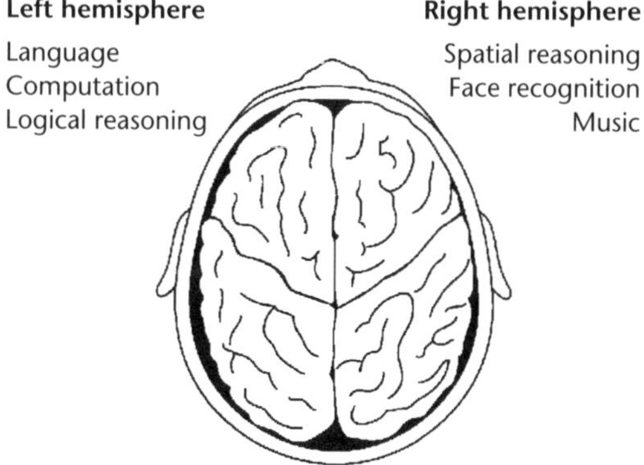

FIGURE 8.6 Lateralisation of some brain functions

More recent research has revealed further layers of complexity. It now seems that the lateral bias in the recognition of emotions might depend upon whether the emotions are positive or negative. It looks like the right side is involved with the recognition and processing of unpleasant emotions but the left side has a role in the experience of more positive and welcome emotions. This is called the 'valence hypothesis' (see Schiff and Lamon, 1994; Springer and Deutch, 1998; Workman et al., 2000).

8.3.3 The amygdala

The fact that each hemisphere is involved in some way in emotional processing is indicated by the presence in both halves of the brain of a small, almond-shaped organ called the **amygdala** (plural amygdalae) (see Figure 8.7). The amygdala is part of the brain's limbic system. Patients with damage to both amygdalae are quite rare but experiments on rats and monkeys have shown the importance of this organ in the mammalian emotional system. As early as 1939, for example, Heinrich Kluver and Paul Bucy found that surgical disruption to the temporal lobes (a part of the brain that controls the amygdalae) of monkeys caused them to show less fear and behave less aggressively. More generally, it seems that damage to the amygdalae of primates greatly compromises their ability to recognise the motivational and reward value of stimuli. Typically, they will attempt to copulate with individuals of other species (non-conspecifics) and attempt to eat foodstuffs normally found unattractive. The emotional reactions of the monkeys were also dulled and flattened – a state termed 'placidity'. So well established is this effect it is sometimes called 'social-emotional agnosia' (that is, lack of knowledge) or the **Kluver–Bucy syndrome** (see also Weiskrantz, 1956). For a short time in the USA, this led to a series of amygdalotomies carried out by 'psychosurgeons' on criminals, with the aim of reducing their fear and anger. More recently, Joseph LeDoux (1997) has shown that the surgical

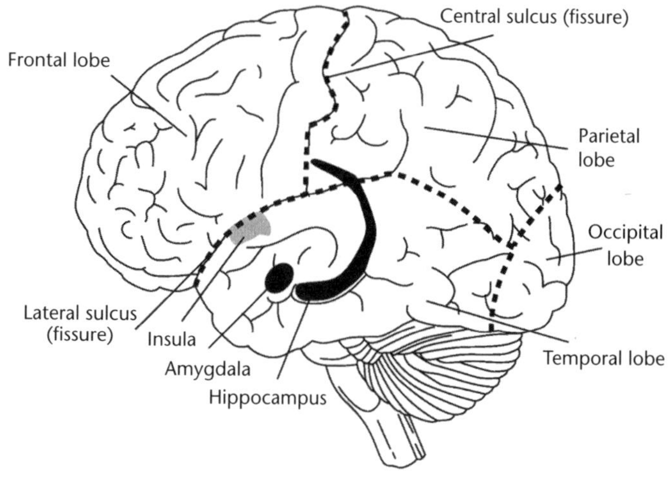

FIGURE 8.7 Location of the amygdala, the hippocampus and the insula

removal of the amygdalae in laboratory rats greatly reduces their rapid response to a fearful situation. Supplementing this there is now considerable evidence that humans who have suffered damage to their amygdalae have impaired recognition of fear and anger. Damasio (2000) describes the remarkable case of a young woman who suffered from Urbach-Wiethe's disease, whereby calcium deposits in her brain gradually destroyed both her amygdalae. In treating this patient, Damasio found that she behaved in a very friendly manner towards all the clinical staff, showed little reserve and, interestingly (as Damsio's co-worker Ralph Adolph found), seemed unable to recognise the emotion of fear in another person's face. In addition, despite having good drawing skills, she was unable to draw a fearful face even though faces displaying other emotions caused no problems.

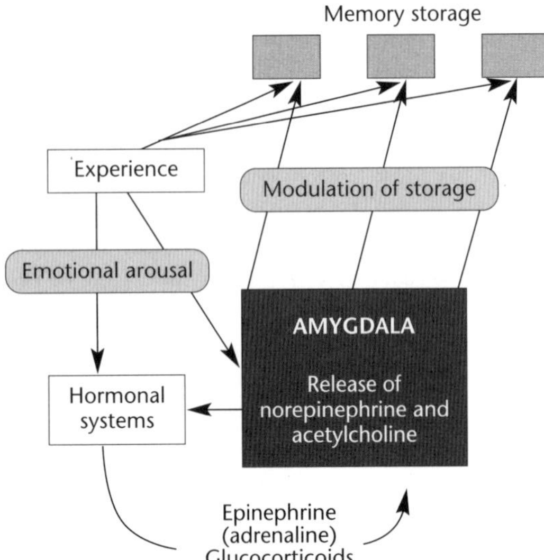

FIGURE 8.8 Schematic diagram of how the amygdala may be involved in memory storage Emotionally arousing experiences can act directly on the amygdala or indirectly through the hormonal systems to cause the amygdala to bias memory storage. It makes adaptive sense that emotionally significant memories are retained.

SOURCE: Based on diagram by McGaugh, J. L., http://darwin.bio.uci.edu/neurobio/Faculty/McGaugh/mcgaugh.htm, accessed December 2006.

The amygdala is also involved in triggering the release of stress hormones such as adrenaline and the glucocorticoids from the adrenal glands (structures that lie just above the kidneys). In the case of adrenaline, this acts upon the muscles to prepare them for action. James McGaugh and colleagues have shown that such hormones also act on memory function (Figure 8.8). This of course makes good adaptive sense: it is important to remember those encounters that caused stress and how we dealt with them (McGaugh et al., 1999).

The low roads and high roads of fear

It is now well established that the amygdalae send signals to virtually every other area of the brain and receive inputs from both the lower thalamic region of the brain and more complex information from the higher cortical centres. These findings have led LeDoux to suggest that there are two types of circuitry involved in fear. One is a quick and dirty or 'low road' response (thalamus to amygdalae) circuit that bypasses the cortex but produces a rapid unthinking and instinctive reaction. The other response proposed by LeDoux is a slower, more considered 'high road' response (thalamus to cortex to amygdalae) that produces a cognitive evaluation of the threat and allows us to respond appropriately. As an illustration, imagine you are walking home alone one dark night and you sense a sudden movement behind you. On hearing the noise behind, you are startled (low road), turn around and evaluate the situation (high road); if it is a domestic cat, you begin to relax, but if it is a menacing stranger, you remain alert.

Evidence in favour of LeDoux's model comes from the study of the orbitofrontal cortex – that part of the cortex that lies just above the eye orbits. The orbitofrontal cortex receives information from the cortex itself and a variety of sensory systems (including taste, odour and visual signals). Significantly, the orbitofrontal cortex is in communication with the limbic system including the amygdalae. In addition, it has been found that some neurons in the primate orbitofrontal cortex respond to the sight of faces (Rolls, 1996). This link, therefore, between the orbitofrontal cortex and the amygdalae provides just one example of how high-level cortical processes (events in the orbitofrontal cortext) may be involved in the emotional response. This is evidence for the 'high road' postulated by Le Doux (see Figure 8.9).

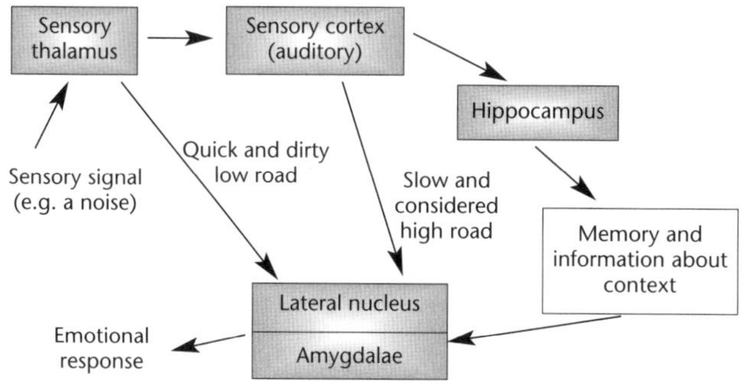

Sensory thalamus: Part of the thalamus. A large mass of grey matter situated deep in the forebrain. It relays to the cerebral cortex information received from a variety of brain regions. A type of final filter for information going to the cortex.

Sensory cortex: Part of the cortex associated with processing sensory information.

Lateral nucleus: One of several divisions in the amygdala. The lateral nucleus receives information from the outside world via the sensory thalamus.

Hippocampus: A brain structure containing storage of explicit memories. Involved with conditioned fear reactions. It stores information about previous disturbing events and their contexts.

FIGURE 8.9 **Pathways of fear**
The amygdalae receive information in two ways. The fast route 'takes no chances': the fear reaction is alerted at the first sign of danger. The slow and considered route is the product of cognitive processing of the fearful stimuli by higher centres of the brain. This signal can then override the fast route if the danger sign turns out to be a false alarm (for example that snake was really a coiled up piece of rope). The hippocampus supplies information about how the local circumstances have related to previous fearful encounters – hence we may be afraid when we revisit the site of a very unpleasant experience even if the experience is not repeated (for example the scene of a mugging).

8.3.4 Mirror neurons and the understanding of emotions

In section 6.3.5, we noted how the recent discovery of mirror neurons suggests a means by which individuals can understand the actions of others. Vittorio Gallese and colleagues have advanced the hypothesis that emotions may also be recognised and understood by a mirror neuron system (Gallese et al., 2004). A crucial area of the brain where some evidence for this has been observed is the insula.

The insula (also known as the lobus insularis or island of Reil) lies within the cerebral cortex (see Figure 8.7 above). It is covered by parts of the frontal, temporal and parietal lobes. The anatomy of the insula reveals two functional subdivisions: towards the front (anterior) lies a 'visceral' sector, and towards the rear a multimodal sector. The anterior sector receives connections from olfactory and gustatory centres; it is also connected to the amygdalae. The posterior portion receives connections from the thalamus that convey interoceptive signals (that is, signals originating within the body) such as emotional and homeostatic information. Damasio (2000) has proposed that the insula, like the amygdalae, is involved in mapping and processing visceral information, giving rise to the conscious experience of an emotion.

Brain imaging studies have shown that the anterior insula is activated by sight of the facial expression of disgust in others. Over 50 years ago, Penfield and Faulk (1955) found that direct electrical stimulation of the anterior insula of patients undergoing neurosurgery brought about reported feelings of nausea and sickness. More recently, Mary Phillips et al. (1997) used fMRI imaging to investigate the response of the amygdalae and insula to fear and disgust respectively. The group found that the anterior insula was activated when an observer witnessed the emotion of disgust on another face, and the more disgust expressed on the observed face, the higher the response. Wicker et al. (2003) also used fMRI to examine the activation of the brains of subjects both exposed to disgusting odours and viewing a film showing others experiencing the emotion of disgust. The experience of pleasing odours was used as a control. Interestingly, the same regions of the anterior insula were activated by both the sight of the emotion of disgust on the faces of others and the personal experience of disgust following the exposure to the unpleasant odour (hence the appropriate title of Wicker et al.'s paper: 'Both of us disgusted in my insula').

All this provides evidence for the existence of structures in the brain that are active during the first- and third-person experience of emotions. Such structures may form the substrate that allows humans to empathise with others, and so could form a platform for a broader range of interpersonal relations and social cognition itself. Defects in this mirror neuron system may underlie conditions where empathy seems deficient such as autism (see section 15.4).

8.4 Emotions and some specific functions

The preceding material has provided evidence that emotions are strong candidates for adaptations serving fitness-enhancing functions: they are universal, are shared with other primates, and are associated with specialised hardware centres in the brain. In this section, we examine the way functionality may be achieved in specific cases.

8.4.1 Emotion, commitment and decision-making

We saw in the last chapter that unbounded rationality is unavailable as a decision-making tool and that instead humans often use fast and frugal heuristics to solve everyday problems. With this in mind, it is worth exploring the notion that emotions aid decision-making. The economist Robert Frank (1988) is a strong advocate of this position. Frank illustrates his approach with the examples of love and guilt. The function of love, he conjectures, is to reinforce the pair bond and so provide a stable unit for cooperative child-rearing. In this view, love serves as a commitment device to enable both sexes to reap the long-term rewards of a stable relationship and resist or ignore the temptations from other members of the opposite sex. If this were the case, then, as Gonzaga and colleagues predicted, feelings of love towards a partner should tend to suppress sexual desire for other attractive potential mates. On the other hand, feelings of sexual desire only towards a partner should be less effective than feelings of love at suppressing desire for others. Gonzaga et al. (2007) placed subjects in these roles by asking them, for example, to write about a partner while trying to suppress thoughts of another. Their predictions were confirmed: feelings of love suppressed thoughts of other attractive potential mates better than feelings of sexual desire.

Findings in support of the motivational function of guilt come from studies on ultimatum games. In these games, volunteers are assigned the role of proposer or respondent. The precise details and rules can be varied, but in typical experiments, the proposers are given a sum of money (for example £10) and told to decide upon a split between themselves and the respondent. The split is the decision of the proposer but if the respondent refuses the allocation, then neither party receives any money. Typically, grossly unfair splits (for example £9 for proposer and £1 for respondent) are refused, even though economic rationality would suggest that the respondent is better off with one pound than none. A study by Ketelaar and Au (2003) found that those proposers who experienced guilt after an unfair divide tended to reverse their behaviour and become more generous a week later. In contrast, few of those who benefited from an unfair divide but experienced no guilt reversed their behaviour. This tends to suggest that guilt can impact upon behaviour, possibly to take more account of longer term benefits from cooperation and mutualism. This is explored further in Chapter 9.

8.4.2 Emotions as superordinate cognitive programmes

Tooby and Cosmides (2005), true to their fondness for computer metaphors, view emotions as types of superordinate programmes that exist to orchestrate other subordinate programmes to achieve the best response to any given situation.

Such programmes are required, they suggest, because the mind has a whole cluster of domain-specific programmes that need activating in a correct sequence otherwise conflicts might occur. This view may explain why emotions have proved so difficult to isolate and study selectively, since as Tooby and Cosmides (2005, p. 53) explain:

> An emotion is not reducible to one category of effects, such as effects on physiology, behavioural inclinations, cognitive appraisals, or feeling states, because it involves evolved instructions for all of them together.

To illustrate their reasoning, Tooby and Cosmides consider fear. The 'front end' of this superordinate programme is to detect a fearful situation. Consider, for example, walking alone at night and suppose elementary perceptual cues indicate noises consistent with being followed or stalked. The emotion programme called fear is activated, which directs a whole cascade of changes such as:

- *Heightened perception and attention to other noises and sights:* threshold shifts, such that less evidence is now needed to interpret other cues as threats
- *Goals and motivation change:* safety now given a higher priority over other goals such as hunger, thirst, finding mates
- *Communication processes initiated:* depending on the context, this could be a cry or facial expression of fear
- *Physiological changes occur:* blood leaves the digestive tract, adrenaline is released, heart rate changes (up if flight reaction, lower if freeze reaction)
- *Behavioural decision rules enacted:* hide, flee, self-defence, remain still.

Table 8.1 Selected research on some postulated functions of emotions

Emotion	Possible function	References
Fear	Primes the body to take action. In panic and agoraphobia, for example, blood supply to the muscles is increased and the mind becomes focused on finding escape routes. Fears that seem unreasonable and disproportionate are called 'phobias'. Phobias may represent ancestral, hard-wired developmental tendencies to fear fitness-relevant objects and organisms (see Chapter 14)	Marks and Nesse (1994) Nesse (2005) Ohman and Mineka (2001) Seligman (1971)
Anger	May provoke extreme acts of destruction. Even spiteful acts may signal to another the costs of provoking anger in this and future situations. Anger may therefore be costly to the actor in the short run, but in the longer run it may be adaptive in modifying future behaviour of others – a 'zero-tolerance' approach. Anger may serve to punish cheaters in prisoners' dilemma-type situations	Fehr and Gaechter (2002)
Sadness and depression	Weakens our motivation to continue with present course of action. May be a means of telling the organism to stop current strategies and conserve resources since they are unrewarding. Could also be a signal that help is required	Hagen (1999) Watson and Andrews (2002) Price et al. (1997)
Jealousy	Forces us to be alert to signs of deception by partners. May activate aggressive behaviours to force the defecting partner back to original relationship or deter partner and new mate from continuing in the new relationship	Buss et al. (1992) Harris (2003) Haselton et al. (2005) Daly et al. (1982)
Love	Love for family members obviously can increase inclusive fitness. Passionate love can further reproductive goals by cementing the pair bond long enough to help raise children	Frank (1988) Ketelaar and Goodie (1998) Nesse (2001)
Disgust	Many believe that disgust is a universal emotion but the targets of disgust are socially determined. In a nutshell, disgust is often the fear of ingesting an undesirable substance. Children acquire a sense of what is disgusting as they grow. In this sense, they are using their parents and adults as guides to what is edible and what is not. Disgust is often elicited even when something unpleasant has merely come into contact with something edible. For example, many people will refuse to drink orange juice from a sterilised bedpan or eat chocolate in the shape of dog faeces. As Pinker observes: 'Disgust is intuitive microbiology.' Sexual disgust (such as the thought of mating with a close relative) may serve to inhibit inbreeding	Pinker (1997) Fessler and Navarrete (2004) Lieberman et al. (2003)
Guilt	A remorseful feeling that follows self-awareness of having been unfair. May serve to drive more cooperative or generous behaviour in future encounters	Ketelaar and Goodie (1998) Ketelaar and Au (2003)

In this whole framework, there are other emotion programmes whose function is not to direct short-term behavioural responses as in the case of fear, but rather to periodically cause revaluation and recalibration of one's own estimation in relation to others. Guilt is probably an example of the latter, which serves to recalibrate the inclination to distribute resources between self and others, such as when favours should be returned. Depression is possibly another example of a condition designed to force a rethink on strategies that are not working and trigger in the depressive a means of curtailing their pursuit (see Chapter 14). Table 8.1 shows how some specific emotions may be linked with evolutionary (fitness-related) functions.

8.4.3 Resolving the paradox of emotions

As Haselton and Ketelaar (2006) note, emotions do seem to pose a paradox. Decades of research confirm that they are probably universal features of human nature, they are found in our primate relatives, and they are correlated with distinct areas of brain activity. If, as seems likely, they are adaptations, it would be odd if they evolved to disrupt decision-making and cloud our judgements on important matters. Yet this is often the experience of all of us when we struggle to remain objective, dispassionate and wrestle with emotions we perceive to be irrational, disturbing and unruly. This is the paradox of emotions explored in the hybrid human-Vulcan *Star Trek* character Dr Spock: clear-sighted logic versus unruly passion. The way round this paradox is perhaps twofold. First, we should realise that our emotional system was designed for conditions very different to modern environments. Consequently, the system may appear suboptimal or irrational in the context of modern lifestyles. This is a constant refrain within evolutionary psychology and we will return to it in Chapters 14 and 15 when discussing fears and phobias. Second, evolution operates to maximise our reproductive fitness and not our experience of well-being (although the latter can sometimes be coopted to serve the former). It may be, then, that saddling organisms with unpleasant emotions such as an easily activated rapid fear response, a constant level of anxiety, or periodic bouts of jealousy, makes adaptive sense (see Chapter 14 on the smoke detector principle). Moreover, as conscious complex organisms, we have our subjective personal goals (contentment, security, peace of mind and so on) that are different to the dictates of our emotions. Possibly then, the heart of this paradox lies in the tripartite tension between the pull of ancestral emotions, personal goals and modern environments.

Such reasoning may also help to explain why happiness is so hard to obtain, for it would be unlikely that natural selection would have designed a state called 'happiness' that could be experienced for any great length of time. Temporary human happiness (contentment, satisfaction) is often linked with the achievement of goals that further reproductive fitness. For our hunter-gatherer ancestors, this would be reliable sources of food, shelter from the elements, freedom from the risk of predators, a supportive social group, high status in the group, mating opportunities, healthy and loving children and so on. But consider the fate of someone who had all these but was still driven by the urge for more, who experienced some nagging discontentment that only more wives, a better husband or a safer environment would satisfy. In some situations, this might pay off and someone driven like this might leave more offspring than someone content with their lot. This could explain why happiness is a fleeting condition, why it correlates with wealth but only slightly, why, like the rainbow, it recedes as fast as we chase it. When Thomas Jefferson wrote in the American Declaration of Independence that people have the right to 'Life, liberty and the pursuit of happiness' he was perhaps right to stress 'pursuit'.

Summary

- Darwin was one of the first to explore the idea that humans share emotional reactions with other animals. He suggested that human emotions were experienced universally.

- William James and Carl Lange argued that emotions may also be the experiences of physiological reactions that take place on our body that occur prior to any conscious evaluation of the emotion-inducing event.
- There is strong evidence that emotions are natural adaptations that served and serve specific fitness-related functions. Evidence comes from their homology with other animals, their universal expression in humans, and their localisation in specific bits of brain hardware.
- The brain is lateralised: different functions happen in different halves. A highly important part of the brain involved in emotional reactions is the amygdala. Humans have two amygdalae: one in each half of the brain. The amygdala receives information from both ancient and recent parts of the brain; it is also (like the hippocampus) responsible for modulating memory storage.
- There is evidence that mirror neurons reside in the insula. Such neurons fire when a subject experiences disgust, or witnesses the emotion of disgust in the facial expression of a third party. Such a direct communication of an emotional state may underlie the ability of humans to empathise with each other.
- Emotions can influence reasoning and decision-making. Some of the simple heuristic rules explored in Chapter 7 may be affected by the emotional system.
- At first glance emotions seem to pose a paradox: why do we experience feelings that were presumably laid down by natural selection for our own benefit as disturbing and disruptive? The answer may lie in the fact that natural selection provided fitness-enhancing structures for a lifestyle different from today, and that enhancing fitness does not mean humans will remain in a state of happiness.

Key Words

- **Amygdala**
- **Autonomic nervous system**
- **Central nervous system**
- **Cerebral hemisphere**
- **Corpus callosum**
- **Homology**
- **Hypothalamus**
- **James–Lange theory**
- **Kluver–Bucy syndrome**
- **Lateralisation**
- **Limbic system**
- **Neocortex**
- **Neuron**
- **Orbitofrontal cortex**

Further reading

Damasio, A. (2000) *The Feeling of What Happens*. London, Vintage.

A popular account of some recent ideas about emotions, with an emphasis on Damasio's own discoveries and ideas.

Oatley, K. (2004) *Emotions. A Brief History*. Oxford, Blackwell.

A useful and readable work that explains how the study of emotion has changed over time.

Cooperation and Conflict

Kin Selection and Altruism

The only thing I'm high on is love ... love for my son and daughters. Yes, a little LSD is all I need.
Marge Simpson (1995) in "Home Sweet Home-Diddly-Dum-Doddily"

In Chapter 2, we saw how William Hamilton and others expanded Darwin's original ideas on fitness into the concept of inclusive fitness – the idea that reproductive success can be measured by the propagation of genes into direct and indirect offspring. We also saw that this concept partly solved the riddle of altruism: 'selfishness' at the level of genes can give rise to altruistic acts of their vehicles if those acts benefit the genes themselves. Put very simply, **Hamilton's rule** suggests that the more closely we are related to another individual (the higher the r value), the more likely we are to be kind towards them. The first part of this chapter looks at the application of this idea to make sense of kin-directed human behaviour in traditional and modern cultures. It is obvious, however, that humans behave altruistically even to non-kin. Such behaviour calls for a separate set of concepts and other modes of explanation. The understanding of human non-kin directed altruism has been greatly clarified through the use of concepts such as mutualism, reciprocity and game theory. Consequently, the second half of this chapter is devoted to exploring these notions.

9.1 Kin and parental certainty

Hamilton showed that kin-directed altruism is favoured between individuals who are closely related (that is, high r values). This raises the question of how humans can tell if another is related to them in a relationship such as brother, sister or offspring. The solution to sibling identification was probably twofold. One involved a simple rule of the form 'if you have grown up with someone from childhood in the same family, they are probably a brother or a sister' (as we shall see in Chapter 13, one consequence of the application of this rule is probably to instigate a negative desire towards mating, that is, co-socialised children grow up not to find each other sexually attractive). Another solution may have used the sense of smell: there is increasing evidence that humans are able to react towards others using odour as a cue for genetic similarity and difference (see Chapter 13). In terms of offspring, however, at the heart of family relationships there lies an asymmetry in parental certainty, sometimes captured by the phrase 'mommy's babies, daddy's maybes'. Mothers can be sure of their parentage, while fathers have to make do with the assumption that they are truly the father.

Given the asymmetries in parental certainty, it might be expected that men's investment in their offspring might be influenced by any available cues of mate fidelity and the resemblance of the child to the father. A number of studies support these expectations (see Platek et al., 2003). Recently, Apicella and Marlowe (2004) conducted a questionnaire-based study of males with children. They asked the men questions about their investment in their children, the resemblance of children to themselves, and the perceived faithfulness of their wife or partner. Men reported greater investment in their children if they perceived them having a greater resemblance to themselves. Men also claimed to place greater investment in their children if they perceived the children's mother as being more faithful. The concept of parental certainty would also suggest that males will attend more closely to cues of resemblance than females (who are assured of their parental status). To test if males and females are affected differently by a physical resemblance to a child, Platek et al. (2004) took photographs of volunteer male and female subjects and blended their images with those of children's faces, creating 'self-morphs'. By this means, the researchers were able to produce children's faces that resembled the subject. They then presented these faces, among other non-resembling faces, to the subjects. Subjects were asked a series of questions eliciting positive and negative reactions such as: 'Which child would you adopt?' 'Which child would you most resent paying child support to?' As predicted, they found that males were more likely than females to select self-morphs in relation to positive questions. There was, however, no difference between males and females in response to negative questions.

In another study, Rebecca Burch and Gordon Gallup (2000) studied the behaviour of 55 men convicted of spouse abuse. They found that the severity of injuries suffered by the female partner was inversely proportional to the perceived resemblance between the father and his biological children. Significantly, in families where stepchildren were present, female injuries were worse but there was no relationship between resemblance of the father to his biologically unrelated children and spouse injury. Burch and Gallup (2000) suggest that paternal resemblance may act as a cue to assess fidelity and paternity.

9.2 Sibling affection and r values

It is an easily observed fact that although sibling rivalry exists, siblings often form close bonds with each other that last throughout life. This could be taken as an obvious example of kin selection: humans taking an altruistic interest in their kin. The proximate mechanism may be co-socialisation (and the triggering of affection towards predicted siblings) or the tendency of parents to encourage cooperation among their offspring.

But in addition to any postulated developmental mechanism, in all cultures there are social conventions that bind families together both legally and morally. Evolutionary imperatives and social conventions are of course not mutually exclusive (in particular, the latter may be the means of achieving the former), nevertheless, it would be instructive to

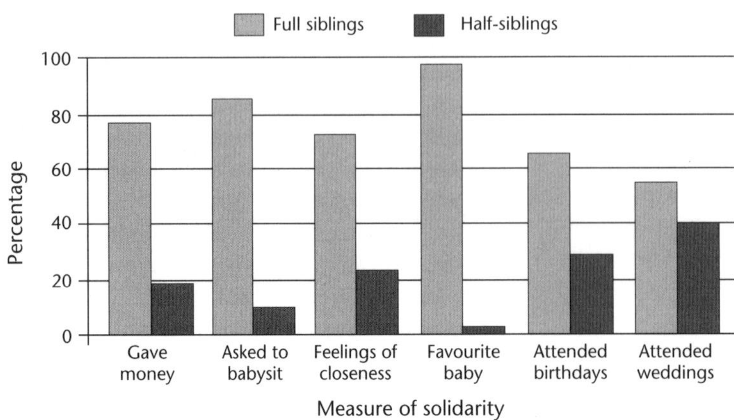

FIGURE 9.1 Measures of affection and solidarity between full and half-siblings in a religious community

SOURCE: Data taken from Jankowiak, W. and Diderich, M. (2000) 'Sibling solidarity in a polygamous community in the USA: unpacking inclusive fitness.' *Evolution and Human Behaviour* **21**: 125–39, p. 133.

disentangle the relative effects of biological factors and cultural norms. In relation to this, a study by two sociologists William Jankowiak and Monique Diderich (2000) on a polygamous Mormon community based at Angel Park in the southwest of the USA has provided some interesting findings. In these communities, the head of the family is officially the father who takes several wives. The official dogma on family life (reinforced and promulgated through schooling and church services) is that social cooperation and harmony within the family are highly important and that the children of all the father's wives (which will be full siblings and half-siblings) have equal status. These researchers interviewed 70 adults who had grown up in such families and recorded various measures of how close they felt to their full and half-siblings.

As Figure 9.1 shows, more affection was felt for full than half-siblings, despite the family philosophy that all should be equal. The proximate mechanism for this effect is not clear. It is likely, however, that mothers (possibly unconsciously) encourage more affection and cooperation between their own biological children than between her children and those of her husband's other wives.

9.3 Discriminating grandparental solicitude

> *Os filhos das minhas filhas, meus netos são.*
> *Os filhos dos meus filhos, podem ser ou não.*
> My daughter's children, my grandchildren are.
> My son's children, may be or not.
> (Old Portuguese proverb)

The old Portuguese proverb (it rhymes in Portuguese) carries more than a grain of truth. Within evolutionary theory, the very existence of grandparents has been explained in terms of the advantages of investing in second-generation kin. Given the asymmetries in parental certainty between men and women, we might expect differences in the investment of grandparents in their grandchildren according to levels of paternal and hence grandpaternal certainty. If we trace our ancestors backwards, then every link through a male introduces some level of uncertainty. As an illustration, a grandmother is certain that her grandchildren through her daughter are hers but less certain of her grandchildren through her son. A moment's thought should reveal that the order of certainty becomes MGM > (MGF or PGM) > PGF (see Figure 9.2).

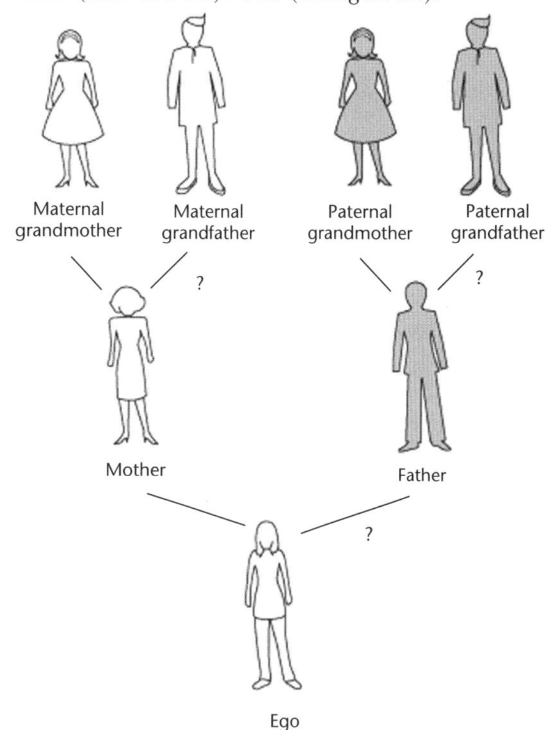

FIGURE 9.2 Differing degrees of grandparental certainty
Question marks show potential links of uncertainty.

The term 'investment' can be captured by a multitude of measures such as emotional closeness, frequency of contact, value of gifts, value of bequests and so on. In a study performed on 120 undergraduates, Todd DeKay (1995) found the following order in grandparental investment as evidenced by the measures of closeness, time spent together and gifts received: maternal grandmother > maternal grandfather > paternal grandmother > paternal grandfather. What is interesting is that maternal grandfathers (MGF) were consistently rated higher than paternal grandmothers (PGM) on all measures, even though the degree of grandparental certainty ostensibly looks to be the same (with one male link). DeKay explained this by suggesting that infidelity rates were higher in the parent's generation compared to the grandparent's generation. Hence, there will be more uncertainty associated with the most recent male link.

A similar study was carried out by Harald Euler and Barbara Weitzel (1996) on German subjects. The sample size here was larger and 603 volunteers were identified who had all four grandparents alive until they reached the age of seven. The subjects then estimated the solicitude they received from each grandparent. The results are shown in Figure 9.3.

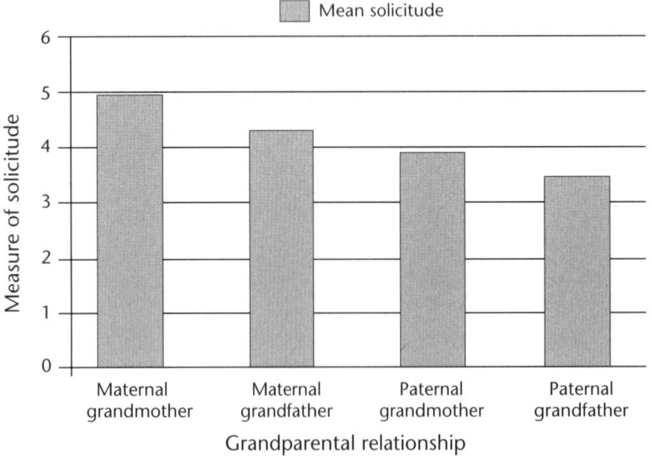

FIGURE 9.3 Solicitude of grandparents as experienced by grandchildren

SOURCE: Data from Euler, H. A. and Weitzel, B. (1996) 'Discriminating grandparental solicitude as reproductive strategy.' *Human Nature* **7**: 39–59.

Table 9.1 Comparative grandparental care from a study on two Greek communities

Care received from grandparents for urban Greeks	Significance
Maternal GM > paternal GM	P = 0.008
Maternal GF > paternal GF	P = 0.0619 (NS)
Paternal GM > paternal GF	P = 0.0197
Maternal GM > maternal GF	P = 0.0003
Care received from grandparents for rural Greeks	**Significance**
Paternal GM > maternal GM	P = 0.0492
Paternal GF > maternal GF	P = 0.0013
Paternal GM > paternal GF	P = 0.000
Maternal GM > maternal GF	P = 0.0001

SOURCE: Data from Pashos, A. (2000) 'Does paternal uncertainty explain discriminative grandparental solicitude.' *Evolution and Human Behaviour* **21**: 97–111.

Euler and Weitzel were able to rule out the possible effect of residential location on these estimations by showing that there was hardly any difference between residential distance measures for the four types of grandparent.

However, there are other factors that influence **grandparental solicitude**. Salmon (1999), for example, found that birth order affected the quality of the relationship between grandparents and their grandchildren: specifically, children with a middle born father or mother had less contact with their grandparents than those of first or last born parents. The type of culture is also important. Alexander Pashos (2000) found differences between rural and urban Greek communities in the relationship between grandparents and their grandchildren. Table 9.1 shows a sample of his results.

Within the rural Greeks, we notice a pronounced patrilateral bias not found among urban Greeks or in the other studies previously cited. It is important to note that the bias applied only to males in this sample. Such results call for a different type of explanation to the parental certainty hypothesis. Pashos observed that overall rural Greek males rated grandparental solicitude higher than females did. He attributed this to the preference in this society for male offspring and hence grandsons. In rural Greece, there is also a high frequency of **patrilocal** residence and so paternal grandparents generally reside closer to their grandchildren than maternal grandparents. This whole study illustrates the need to take into account societal factors other than kinship and **paternity certainty** in predicting family relationships. Of course, this does not rule out the possibility that there will be sound evolutionary explanations for the preference for male offspring and patrilocality themselves.

9.4 The distribution of wealth: inheritance and kin investment

Wealth is, in a sense, the embodiment of indirect **reciprocity**. A coin or a note is a physical symbol that you are owed something for your previous efforts. You may have exchanged your labour for money with an employer, the money that was handed to you can now purchase the labour of someone else and so it goes on. You could of course hand over your money to someone else, such as a friend or a relative, without any obvious future personal reward. Such is the situation when people make bequests in their wills. But we don't hand over our hard-earned wealth beyond the grave at random, and selectionist thinking can be applied in this context too. First, we will examine patterns in how people divide up their estate, and then show how inheritance rules and practices are related to marriage systems. This in itself provides an interesting example of how a cultural practice (rules of inheritance) can be related to genetic self-interest.

9.4.1 Inheritance of wealth: practice in a contemporary Western culture

In the legal system of most Western cultures, an individual has considerable freedom of choice in deciding how to distribute their wealth in bequests to relatives and friends following death. It is reasonable to suppose, for the same reasons outlined above, that in our evolutionary past, the allocation of resources following death would have significantly affected an individual's inclusive fitness. So are there vestiges of any adaptive preferences in behaviour today? In attempting to answer this, Smith et al. (1987) analysed the bequests of a random sample of 1,000 individuals recorded in the Probate Department of the Supreme Court of British Columbia. Their data were used to test four predictions:

1. Individuals will leave more of their estate to kin and spouses than to unrelated individuals. This is hardly a surprising prediction from general knowledge but it follows from kin selection theory, in the sense that resources left with kin can improve inclusive fitness; resources left with genetically unrelated spouses may still find their way to kin eventually.
2. Individuals will leave more of their estate to close kin (high r) than to distant kin (low r).
3. Individuals will leave more of their estate to offspring than to siblings. Although a son or daughter have the same r value of 0.5 as a brother or sister of the deceased, it is likely that a bequest to, say, a son will enhance his reproductive value more than that to a brother. If a bequest were to be made to a brother or sister, they are both likely to be past reproductive age, or in a position where wealth has a limited effect on reproduction.
4. There will be a male bias in bequests to children of wealthy individuals and a female bias to children of poorer individuals. The logic of this prediction for male offspring is similar to Hartung's hypothesis discussed below: wealthy sons were once better able to secure more wives and hence more offspring compared to poor sons, whereas wealth probably had a smaller effects on the reproductive output of daughters. The female bias is predicted from the fact that in mildly polygynous mating (which may have represented our ancestral state), poor sons may not secure any wives, and so produce no grandchildren, but poor daughters may mate with a man who already has a wife.

Predictions 1–3 were confirmed by looking at how the percentage of bequests was distributed according to closeness of kin (Table 9.2).

Table 9.2 Bequests made to relatives as a percentage of total estate for 1,000 people in British Columbia

Coefficient of relatedness (r)	Beneficiary	Percentage of estate (mean)
0.0	Spouse	36.9
0.5	Sons	19.2
0.5	Daughters	19.4
0.5	Brothers	3.2
0.5	Sisters	4.8
0.25	Nephews and nieces	5.1
0.25	Grandchildren	3.2
0.125	Cousins	0.6
0	Non-kin	7.7

SOURCE: Adapted from Smith, M., Kish, B. J. and Crawford, C. B. (1987) 'Inheritance of wealth as human kin investment.' *Ethology and Sociobiology* **8**: 171–82.

To test prediction 4, the estates of the 1,000 deceased individuals were split into four wealth bands (Table 9.3). Again, the data support kin selection theory. Wealthy parents tended to bequeath a significantly larger portion of their estate to sons compared to daughters, and vice versa for poorer parents.

Table 9.3 Distribution of estate to offspring according to sex

Value of estate (E) in Canadian $	Percentage bequeathed to sons	Percentage bequeathed to daughters	Greatest percentage to	Significance
E < 20,350	9.9	19.4	Daughters	$P < 0.01$
E 20,350–52,900	14.6	18.6	Daughters	NS
E 52,900–110,850	21.9	24.6	Daughters	NS
E > 110,850	30.2	15.1	Sons	$P < 0.01$

SOURCE: Adapted from Smith, M., Kish, B. J. and Crawford, C. B. (1987) 'Inheritance of wealth as human kin investment.' *Ethology and Sociobiology* **8**: 171–82.

9.4.2 Inheritance rules and marriage systems

Kin selection theory tells us that parents will 'invest' biological resources in their kin because kin represent the passport that will carry their genes into future generations.

Hartung (1976) postulated that where wealth correlates with reproductive success, then parents should pass on their wealth to male rather than female offspring. This follows from the fact that wealth can be used by a male, where mating is polygynous, to increase his number of wives and hence number of children. In monogamous mating contexts, the bias in favour of males should be less pronounced: it is no use leaving resources preferentially to sons if the culture does not permit more than one wife (although it may help them secure remarriage should they lose a first wife). Hartung (1982) tested these ideas in a survey of 411 cultures. He found a strong male bias in inheritance where polygyny was common (Table 9.4).

Table 9.4 Male bias in inheritance system of a culture according to mating behaviour

Mating system	Percentage cultures with strong male bias
Monogamy	58
Limited polygyny	80
General polygyny	97

SOURCE: Data summarised from Hartung, J. (1982) 'Polygyny and the inheritance of wealth.' *Current Anthropology* **23**: 1–12.

Hartung did take pains to address the problem of independence of cultures and hence data points (so-called Galton's problem), but Mace and Cowlishaw (1996) have suggested that a phylogenetic approach provides a better series of controls. They traced back 261 cultures in 11 language families to their inferred ancestral state of mating behaviour. They then looked for cases where the marriage system, inheritance rules, or both changed as the cultures evolved. Where change did occur, the end states of inheritance bias and marriage system after the transition were recorded. The data are shown in Table 9.5.

Table 9.5 End states of marriage system and inheritance rules following a cultural transition

End state of inheritance rule	End state of marriage system		
	Monogamy (%)	*Limited polygyny (%)*	*General polygyny (%)*
Strong male bias	25	44	87
Weak or no bias	75	56	13
No. of cultures	20	18	16

SOURCE: Data summarised from Mace, R. and Cowlishaw, G. (1996) 'Cross-cultural patterns of marriage and inheritance: a phylogenetic approach.' *Ethology and Sociobiology* **17**: 87–97.

The work of Mace and Cowlishaw supports the idea that marriage patterns and inheritance rules evolve together in ways that are adaptive. The evolution of monogamy is strongly associated with an absence of sex-biased inheritance rules. In contrast, where general polygyny evolves, it is usually accompanied by a strong male bias in inheritance. It is worth noting that whereas Hartung found that monogamy was roughly equally associated with male bias and absence of male bias, this revised phylogenetic approach showed that an evolved association between monogamy and strong male bias is less common (25 per cent). In summary, if you live in a culture where polygyny is common and you are rich, it would serve your biological interests to leave your wealth to sons: with wealth they have a chance to obtain more wives and so produce more grandchildren. If marriage and sexual relations are largely monogamous, it is not so crucial.

The studies of Mace and Cowlishaw (1996) and Smith et al. (1987) are worth comparing in that they illustrate how evolutionary psychology may illuminate different facets of human cultural behaviour. In the Mace and Cowlishaw study, it is the cultural norms and rules that reflect fitness-maximising behaviour. This is evidence that is germane to the much larger question of the relationship between genes and culture pursued in the next section.

It may seem as if male inheritance only benefits the inclusive fitness of fathers, but mothers benefit too from wealth passing to a son who thereby becomes reproductively successful. The Smith et al. study shows how in societies where cultural conventions are more flexible, individuals still behave through 'free choice' so as to maximise their inclusive fitness. It is interesting that the bias to male offspring from wealthy parents probably had adaptive value 100,000 years ago, and, in polygynous cultures, may still have, but probably has little long-term effect in Western cultures where monogamy is legally enforced.

9.4.3 Paternity certainty, patrilineality and matrilineality

The passing on of wealth has traditionally been divided by anthropologists into two systems: patrilineal and matrilineal. In patrilineal systems, wealth is passed down the paternal line, usually from father to son; in matrilineal systems, the wealth is passed down from mothers to sons or daughters or more often from fathers to their sister's sons.

In about 10 per cent of well-documented societies, men have obligations of care towards their sister's children. In these same societies, the men have few obligations towards their wife's children. If we compare the existence of matrilineal societies with patrilineal, some significant patterns emerge:

- There are no societies recorded where a wife is obliged to look after her brother's children.
- There are no societies where a man is obliged to take care of his brother's children.
- There are no societies where women are obliged to take primary care of their sister's children.

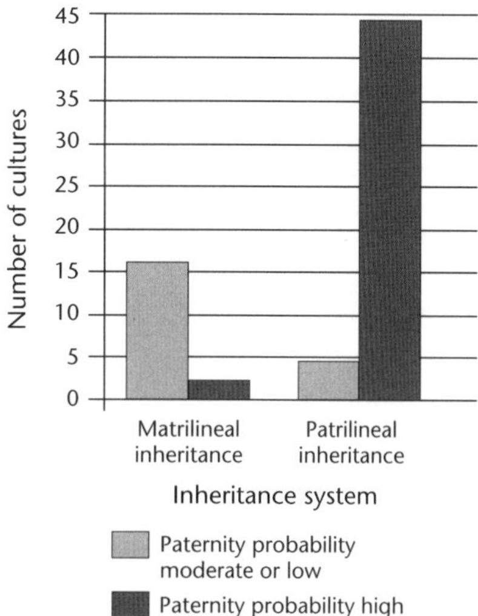

FIGURE 9.4 Distribution of matrilineal or patrilineal inheritance according to paternity certainty

SOURCE: Data plotted from Hartung, J. (1985b) 'Matrilineal inheritance: new theory and analysis.' *Behavioral and Brain Science* **8**: 661–8.

To understand this pattern and the conditions tending to promote matrilineality, we can consider the degree of relatedness in families where there is paternal uncertainty. The degree of relatedness of a man to his son is 0.5p, where p = the probability of paternity (recall that the coefficient of genetic relatedness between parent and offspring is 0.5; see section 2.5.2). If p = 1 (the assumed father is really the biological father of the child), then r = 0.5; but if p = 0, then the r value between father and son is zero. A man's relatedness to his sister (by the same mother) is 0.25 + 0.25p. The 0.25 is a reflection of the fact that the man is reasonably assured that he and his sister share the same mother; the fact that p is multiplied by 0.25 is a reflection of the fact they may have different fathers. In this case, if p = 1, then r between brother and sister = 0.5; but if p = 0, then the r value is still finite at 0.25. Following the same logic, we can see that a man's relatedness to his sister's children is 0.5(0.25 + 0.25p). It follows that a man is more related to his sister's children if:

$0.5(0.25 + 0.25p) > 0.5p$ (that is, when $p < 0.33$).

In a study of 1985, John Hartung looked at the distribution of matrilineal and patrilineal inheritance in 70 well-documented cultures. His results are shown in Figure 9.4.

Hartung's results are consistent with the idea that men will invest in their sister's children in cultures where paternity certainty is low, since in these cultures, men can still be assured of an r value of at least 0.25 with the beneficiaries of their wealth.

9.5 Reciprocal altruism

9.5.1 Mutualism, parasitism, altruism and spite

So many words in science have a precise meaning different from the looser meaning of their everyday usage. In the main, this does not matter too much, but in the case of the concept of altruism, it has led to many unfortunate misunderstandings. The problem arises because any act that seems to imply that animals are not behaving selfishly has been called altruistic. In most cases, however, the acts are not altruistic at all because either the payoffs are not so obvious or are delayed. Similarly, the phase 'selfish gene' coined by Dawkins has led to the criticism that biologists are simply being anthropomorphic. To rescue these useful terms, some clarification is sorely needed. Figure 9.5 shows a matrix illustrating the essential difference between altruism, mutualism, selfishness and spite. These terms are now discussed in more detail.

Initiator \ *Recipient*	Gains	Loses
Gains	Mutualism or reciprocity	Selfishness, for example parasitism
Loses	Altruism	Spite

FIGURE 9.5 Matrix of relations defining mutualism, altruism, selfishness and spite

SOURCE: Adapted from Barash, D. (1982) *Sociobiology and Behaviour.* New York, Elsevier.

Parasitism

Parasites are organisms that benefit from a relationship with their host at the expense of the host organism. Often we find the genes of one organism manipulating the behaviour of the other. The cold virus, for example, not only invades your system and subverts its functions to produce copies of itself, but it also manipulates you into helping it spread further by persuading your lungs to expel an aerosol of droplets containing the virus at a great velocity. Dawkins referred to such effects as 'the extended phenotype'. In nest parasitism, the cuckoo manipulates the builder of the nest into giving aid to a completely different species. In these cases, the 'altruism' of the donor is extracted by manipulation, and the donor gains nothing. We could view this as cuckoo genes reaching out beyond the cuckoo vehicle into the behaviour of the nest owner. Dawkins summed this up in his 'central theorem of the extended phenotype':

> An animal's behaviour tends to maximise the survival of the genes 'for' that behaviour whether or not those genes happen to be in the body of the particular animal performing it. (Dawkins, 1982, p. 233)

Mutualism

Mutualism is becoming the preferred term to **symbiosis**. We can distinguish two types of mutualistic behaviour: interspecific (between two or more species) and intraspecific (within a single species). Some species form interspecific mutualistic partnerships because individuals of each have specialised skills that can be used by the other. Aphids have highly specialised mouths for sucking sap from plants. In some species, this is so effective that droplets of nutrient-rich liquid pass out of the rear end of the aphids undigested. Some species of ant take advantage of this by 'milking' the aphids in the same way as a farmer keeps a herd of cattle. The aphids are protected from their natural enemies by the ants, which look after their eggs, feed the young aphids and then carry them to the grazing area. The ants 'milk' the aphids by stroking their rears to stimulate the flow of sugar-rich fluid. Both sides gain: the ants could not extract sap so quickly without the aphids, and the aphids are cosseted and protected from their natural predators by the ants.

In the case of intraspecific mutualism, two or more individuals of the same species cooperate and each gains a net benefit. If two lionesses cooperate, their chance of capturing a prey is probably more than twice that of each individual. Sharing half the meat then becomes a net benefit to each. In fact, lionesses in a pride are related as are the male lions that cooperate in taking over a pride, so cooperation is favoured by kin selection and mutualism.

Altruism

Kin-directed altruism was examined in Chapter 2. Further examples of altruism and reciprocal altruism are the main subjects of this chapter.

Spite

Spite has proved an elusive phenomenon to observe in the natural world. This is hardly surprising since it is difficult to see how genes for this behaviour could spread if they suffer a net cost, a cost moreover not outweighed by any inclusive fitness gains since, by definition, the recipient also suffers a cost. Indeed, if we consider Hamilton's equation ($rb > c$), then this can be simply phrased as 'an activity will flourish if the net benefits (rb, where here r can be regarded as the modifier of the value of the benefit) exceeds the cost (c)'. If we think of benefit as gene frequency increase over a given period of time and cost as gene frequency decrease, then it is obvious that only when benefits exceed cost can genes for this behaviour flourish (see section 2.4). Now, in the case of spite, we have a negative benefit and a positive cost – the initiator presumably expends some energy to inflict damage on another – and so it would seem impossible for spite to ever emerge. This explains why spite had been almost impossible to authenticate in animal societies. Indeed, what often seems to be spite often turns out to be good old selfishness. Male bower birds, for example, will sometimes go out of their way to wreck the carefully constructed bowers of other males. This may look like spite, but, as Hamilton argued, destroying a rival's bower reduces the reproductive success of competitors and leaves the field more open to the destructive male (Hamilton, 1970).

But there are two possibilities proposed by Wilson and Hamilton. Wilson (1975) suggested that spite could emerge if a third party related to the imitator benefited from the costs suffered by the other two parties, athough strictly speaking, this could be seen as a form of indirect altruism. In contrast, Hamilton (1970) proposed the intriguing idea that spite could spread in the gene pool if r was negative. In the equation $rb > c$, if both b and r are negative, then the condition for the spread of genes can once again be met (a negative times a negative gives a positive). The value of r can be negative if the recipient is less likely to share any gene of the initiator than the average for the general population. A spiteful gene can then flourish since it reduces the frequency of competing alleles in the gene pool. This system, however, requires some unlikely conditions: negative relatedness and very accurate kin discrimination. Such conditions are unlikely to be found in mammalian societies, but there is some evidence that they may be met in a few insect societies where Hamiltonian spite remains a distinct, albeit much debated, possibility (see Foster et al., 2001).

The presence of spite in humans probably requires a higher level explanation. One possible explanation is that spiteful behaviour brings maladaptive consequences for humans because it represents a miscalculation of the effects of certain courses of action. The threat of spite could bring rewards: 'if you don't do what I want, we will both suffer' may sometimes work, but if the person threatened calls the bluff of the aggressor, then it becomes maladaptive.

9.5.2 Reciprocal altruism or time-delayed discrete mutualism

Acts can often appear altruistic in the sense of Figure 9.5 above, when in fact we simply have not taken a sufficiently long-term view of the situation. Trivers (1971) was one of the first to argue that altruism could occur between unrelated individuals through a process he termed 'reciprocal altruism', which is really a more refined version of the maxim: 'you scratch my back now and I'll scratch yours later'. We are really looking for genes that by cooperating with each other enhance their own survival and reproductive success through their own vehicles. In kin selection, an individual may help another in the belief that the helping gene is present in the recipient. In the case of reciprocal altruism, aid is given to another in the hope that it will be returned.

We can now see that there is a rather fine line between our definition of mutualism and that of reciprocal altruism. The most useful distinction is that mutualism involves a series of constant reliances. An extreme form of this would be lichen, which are composed of an alga and fungus in an inextricable symbiosis. Similarly, the bacteria in your gut that help you digest food are mutualistic in that the exchange of food products is virtually constant. We can think of reciprocal altruism as a sort of time-delayed mutualism. An exchange takes place in which it seems that the beneficiary gains and the initiator loses according to Figure 9.5 above. What we expect, however, is that in cases of reciprocal altruism, the favour is returned at a later date. Game theory (see below) has also been used to distinguish between reciprocal altruism and mutualism, with the suggestion that rewards and punishments (for cheating) have different values.

9.5.3 Conditions for the existence of reciprocal altruism

We would expect the following conditions to obtain if reciprocal altruism is to be found:

1. An animal performing an altruistic act must have a reasonable chance of meeting the recipient again to receive reciprocation. This would imply that the animals should be reasonably long-lived and live in stable groups to meet each other repeatedly.
2. Reciprocal altruists must be able to recognise each other and detect cheats who receive the benefit of altruism but give nothing back in turn. If defectors cannot be detected, a group of reciprocal altruists would be extremely vulnerable to take-over by cheaters. Codes of membership for many human groups, such as the right accent and the right clothes, and the initiation rituals and signals of secret societies, could serve this function.
3. The ratio 'cost to donor/benefit to receiver' must be low. The higher this ratio, the greater must be the certainty of reciprocation. This is sometimes called 'gains in trade' by economists – when something of low value to the donor is exchanged for something from the recipient that the donor values highly; in return, the recipient receives something that they value highly and gives away something of low value for them.

Although humans spring to mind as obvious candidates, species practising reciprocal altruism need not be highly intelligent.

Examples of reciprocal altruism

One of the best documented examples concerns vampire bats *(Desmodus rotundus)*. These were studied by Wilkinson (1984, 1990), who found that vampire bats, on returning to their roost, often regurgitate blood into the mouths of roost mates. Such bats live in stable groups of related and unrelated individuals. A blood meal is not always easy to find. On a typical night, about 7 per cent of adults and 33 per cent of juveniles under two years old fail to find a meal. After about two or three days, the bats reach starvation point. It might be thought that regurgitation is an example of kin selection and undoubtedly some of this is occurring. But the exponential decay of loss of body weight prior to starvation suggests that the conditions for reciprocal altruism could be present.

Figure 9.6 shows weight loss against time. In essence, the time lost by the donor is less than the time gained by the benefactor. The bats' mode of life also means they constantly encounter each other. Wilkinson conducted experiments whereby a group a bats was formed from two natural clusters. Nearly all the bats were unrelated. They were fed nightly from plastic bottles. Each night one bat was removed at random and deprived of food. Wilkinson noticed that on returning to the cage, it was fed by other bats from its original natural group. Reciprocal partnerships between pairs of bats were also noticed.

Reciprocal altruism has also been documented in Gelada baboons. Dunbar (1980) found a positive correlation between support given by one female Gelada baboon to another and the likelihood that it will be returned. Evidence that reciprocal altruism may be at work in chimpanzee groups has been forthcoming from the work of de Waal (1997) on captive chimpanzees. De Waal found that if chimp A had groomed chimp B up to two hours before feeding, then B was far more likely to share food with A than if it had not been groomed. Interestingly, B was equally likely to get food from A whether A had groomed it or not. De Waal's results suggest that grooming serves as a sort of service that is repaid later.

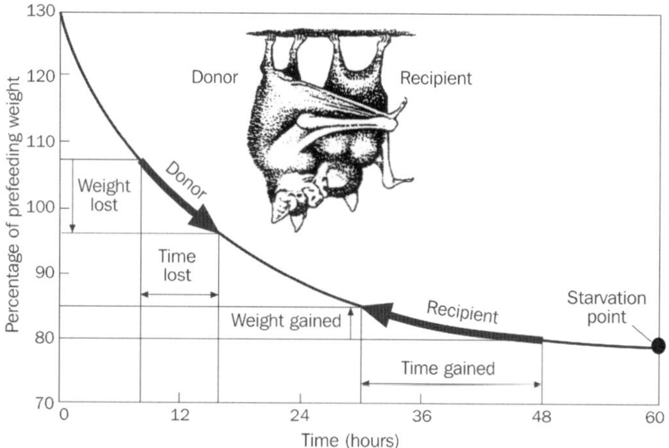

FIGURE 9.6 Food sharing in vampire bats shown in terms of a cost–benefit analysis
In this diagram, the donor loses about 12 per cent of its prefeeding body mass and about 6 hours of time before starvation. The recipient gains, however, 18 hours of time and 5 per cent of prefeeding weight. The fact that time gained is so much greater than time lost favours reciprocal altruism.

SOURCE: Adapted with permission from a drawing by Patricia Wynne in Wilkinson, G. S. (1990) 'Food sharing in vampire bats.' *Scientific American* **262**: 76–82.

Despite the examples cited above, it is fair to say that reciprocal altruism within a species, for all its attractions as a means of securing mutual benefits, is not widely observed among non-human animals. The most enthusiastic reciprocal altruists are humans. The very act of working for a company or a public body is an act of reciprocal altruism: we give our labour for a month in the expectation that we will be paid at the end. Without reciprocal altruism, civilised life would be impossible. Reciprocal altruism is the basis for trade and barter in human societies. It is known that early humans often traded stone tools over great distances. Presumably they were not exchanged for other stone tools because there would be no point. Possibly, one group had access to raw materials, and the skills to knap stone-made tools, and exchanged them for skins or food with another group that had access to different resources and skills. Generally, reciprocal altruism flourishes when there are differences in needs and different capabilities to meet them. This is why it is often observed between species (where these asymmetries are greater). So, as Dawkins points out, flowers need pollinating but since they can't fly, they 'pay' bees with nectar to do it. Birds called honeyguides can find bees' nests but lack the strength to break into them. Honey badgers (ratels) have the strength to break into nests but lack wings to find them. The solution? The honeyguides have a special flight that leads the ratels to the nest and the two species share the spoils (Dawkins, 2006).

But, returning to humans, reciprocal altruists face one gigantic problem: how to ensure that a favour given to someone is likely to be returned at a later date. Such problems have been modelled in game theory and to this we now turn.

9.6 Game theory and the prisoners' dilemma

It is a safe enough bet that the early replicators were entirely selfish. From this primal state, the step towards kin-directed altruism is a short one and it is easy to envisage how this could have occurred: care given to direct offspring could move to indirect offspring and all the benefits of inclusive fitness would follow. It is also easy to see how reciprocal altruism benefits both parties once it is established. A favour can be given to another individual at small cost to oneself on the understanding that the favour will be returned. There are conditions that need to be set for this, such as the ability to recognise who donated the favour, a good chance of meeting that individual again and so on, but these are not unlikely conditions. The big problem, however, is to account for the origin of cooperative behaviour, given that in the very first interaction, it would pay to act selfishly. The problem of the evolutionary emergence of cooperation is highly pertinent to humans. Humans spend a great deal of time cooperating and delaying immediate gratification for future rewards. In this section, we will explore some models that show that cooperation may be a fitness-maximising strategy. If this is so, then the simplistic linkage of nature–bad and culture–good is exploded. We may be caring and morally sensitive creatures by virtue of our biology.

9.6.1 The prisoners' dilemma

In many real-life situations, our best course of behaviour (in terms of bringing rewards) is often dependent upon the behaviour of others; but the problem is that we do not necessarily know how others will behave. Situations like this have been modelled using **game theory** (Axelrod and Hamilton, 1981). A good starting point to investigate the moral basis of behaviour is a game known as the **prisoners' dilemma**.

The word 'prisoner' relates to one context where the logic of this game could apply. If two suspects are apprehended at the scene of one of a series of crimes, one tactic the police could adopt is to separate the individuals and question them independently. If the overall evidence the police gathered is flimsy and a successful prosecution will rely upon a confession, each suspect may be offered the promise of a lighter sentence if they turn 'king's evidence' and inform (defect) on the other. If they both defect, then they implicate each other and both receive a full jail sentence. If both cooperate and refuse to be tempted to defect, then, given the lack of evidence, each receives a smaller sentence for a minor part of the crime. Figure 9.7 shows a payoff matrix for this type of scenario.

The game is entirely hypothetical but it illustrates that 'rational behaviour' (in the sense of maximising returns to oneself) can result in the least favourable outcome; if both parties defect or inform, they are each worse off. They are in effect punished for failing to cooperate with each other. The dilemma then is whether to cooperate or defect. Defection will bring the best rewards if the other cooperates, but each prisoner does not know what the other will do. It is difficult to see how cooperation could evolve in this system. One might imagine that suspects would confer before the crime and agree to both cooperate but this still begs the question of why cooperate and not defect. The game of prisoners' dilemma was first formalised in 1950 by Flood, Dresher and Tucker (see Ridley, 1993). It is a situation that applies to many interactions in life. As Ridley points out, the gigantic trees in tropical rainforests are products of prisoners' dilemmas: if only they would cooperate and agree to, say, not grow over eight metres, all would be able to put more energy into reproduction and less into growing gigantic trunks to tower over their neighbours. But they can't. Human folly is also often the result of prisoners' dilemma situations. The arms race between the superpowers in the postwar period left both America and the former Soviet Union worse off. In all these cases, the protagonists were locked into one form or another of the prisoners' dilemma. Another example from non-human animals concerns cleaner fish. These are small fish that swim inside the mouths of a larger fish and clean the teeth of the host and remove food debris, ectoparasites and so on. It is tempting for the cleaner to take a nip out of the larger fish and swim off – thereby gaining some useful calories; the temptation for the larger fish is to forego cleaning and swallow the cleaner.

	Player B **Cooperate** 'It was neither of us'	**Defect** 'It was him'
Cooperate *Player A*	R = Reward = 1 year	S = Sucker's payoff = 5 years
Defect	T = Temptation to defect = 0 year	P = Punishment = 4 years

FIGURE 9.7 Payoff matrix for two prisoners caught in a dilemma
Note that the values are the payoffs to Player A.

The prisoners' dilemma problem has relevance to human behaviour in that many human interactions in the past must have taken this form. It might be expected then that the long evolution of humans and other animals should have given natural selection time to solve the problem. In essence, we are looking for an **evolutionary stable strategy** (ESS) – a strategy which, if pursued in a population, is resistant to displacement by an alternative strategy. Headway was made in relation to this problem when it was realised that social life is akin not to one chance encounter between individuals of the prisoners' dilemma form but to many repeated interactions. The problem then became to explain what strategy each should pursue if the game is played over and over again. It then transpires that defection is not always the best policy.

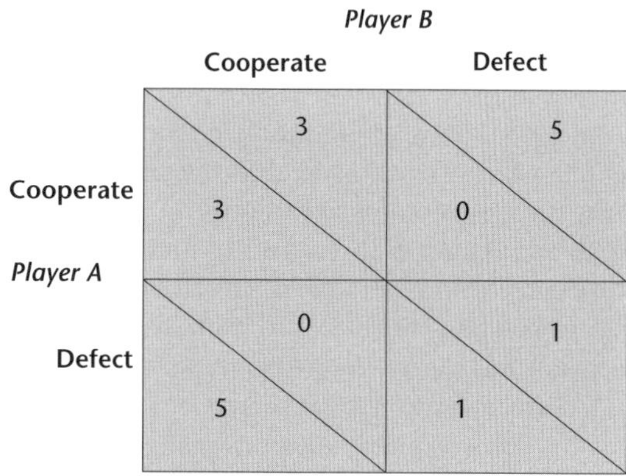

FIGURE 9.8 Prisoners' dilemma values for cooperation and defection
Values are now shown for each player. The convention for such games became R = reward, T = temptation (to defect), P = punishment, S = sucker's payoff. In the figure, R = 3, T = 5, P = 1, S = 0.

9.6.2 Tit for tat

To investigate this, it is best to elevate the problem to a more abstract level. Figure 9.8 shows a representation of the dilemma using values for reward and punishment used by a political economist called Robert Axelrod. When we are faced with such situations in life, we have a wide range of options. We could play meek and mild and always cooperate. In the face of defection, we would then 'turn the other cheek' and continue to cooperate to our ultimate detriment. Another option is to always play rough ('hawkish') and constantly defect. To compare the strategies over time, imagine a population of individuals that interact only once, then the strategy 'always cooperate' is easily displaced by a mutant 'always defect'. A single defector introduced into a population of cooperators will thrive, the population of cooperators will slump and become extinct because a cooperator will always lose on meeting a defector. Eventually, the population will be composed of all defectors. As a group, they will not do as well as the cooperators, but selection does not act for the good of the group. It is even possible that this defecting strategy will lead the species to extinction. When all the population is composed of defectors, this is an ESS: it is resistant to invasion by a mutant cooperator.

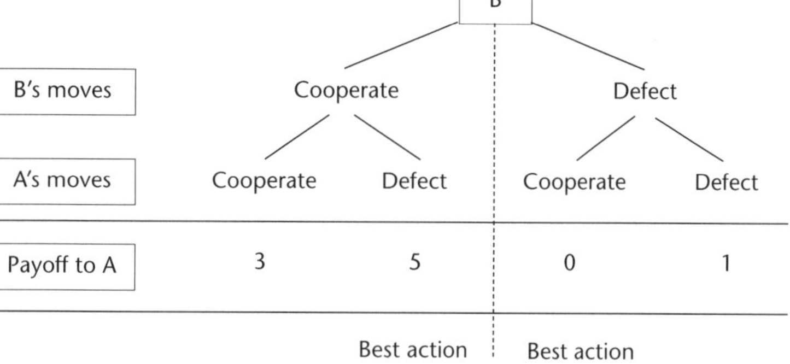

FIGURE 9.9 A decision tree for a prisoners' dilemma scenario
The best action for A is to defect if A does not know in advance how B will play.

Using these figures, we can construct a decision tree for A in the light of what B may do (see Figure 9.9). An inspection of this shows that the best course of action for A is to defect; since A and B are interchangeable, it follows that this is also the best course of action for B. Here we get to the heart of the dilemma: individually rational action leads to joint calamity when both parties pursue it (Figure 9.8).

In the early 1970s, John Maynard Smith began to explore the potential of game theory – of which the prisoners' dilemma is part – to explain animal conflict (Maynard Smith, 1974). His ideas about 'hawk and dove strategies' could have been applied to the prisoners' dilemma but since they belonged to biology and the prisoners' dilemma belonged (at the time) to economics, they were ignored. In fact, one of Maynard Smith's strategies, 'the retaliator', is similar to the strategy of 'tit for tat' that was to do so well in later prisoners' dilemma tournaments. A similar convergence in thinking was taking place as Robert Trivers started working upon his ideas on reciprocal altruism in 1971.

In 1979, Robert Axelrod realised that the prisoners' dilemma only yielded defect as the rational move if the game was only played once. He argued that cooperation could evolve if pairs of individuals in a prisoners' dilemma met repeatedly. It could follow that, from repeated interactions, both would learn to cooperate and so reap the greater reward. Axelrod set up a tournament to test the success of strategies for playing the prisoners' dilemma many times. He invited academics from all over the world to submit a strategy that would compete with the others in many repeated rounds. Sixty-two programs were submitted and let loose on Axelrod's computer. The important features of this first tournament are as follows:

1. Each pair that meets would be very likely to meet again. In other words, future encounters are important.
2. Each strategy would be matched against each other and itself.
3. The scoring points were set as follows : R = 3, T = 5, S = 0, P = 1. It was decided that the convention $R > (T + S)/2$ should be adopted, which means that mutual cooperation yields greater rewards than alternately sharing the payoffs from 'treachery' and 'sucker' moves.

The strategy that won was submitted by Anatol Rapaport – a Canadian who had studied game theory and its application to the arms race. The strategy was a simple one called **tit for tat**. This strategy says:

1. Cooperate on the first move.
2. Never be the first to defect.
3. In the face of defection, retaliate in the next move but then cooperate if the other player returns to cooperation.

Although tit for tat won the overall contest, it never won a single match. The reason is easy to see. The best a tit for tat strategist can do is draw with its opponent. In the face of a defector, it loses the first encounter but then scores as many points afterwards. By definition, it can never strike out to take the lead because it is never the first to defect. Tit for tat wins overall because even if it loses it is never far behind. The problem with 'nastier' strategies is that they have to play each other. If players constantly defect on each other, their total reward is low. For a time, it looked like tit for tat could serve as a model for moral behaviour: it does pay to cooperate after all. Tit for tat might win a general contest but could it ever invade an aboriginal population of selfish defectors? The answer is yes if a few tit for tat strategists appeared and met each other. For a while, it looked like tit for tat could demonstrate the possibility of the evolution of human morality: niceness could prosper.

Problems with tit for tat

This strategy seemed to hold high hopes for modelling the behaviour of humans and other animals. After the publication of these results in Axelrod's book, *The Evolution of Cooperation* (1984), criticism began to mount. It became clear that tit for tat was not an ESS and its success was sensitive to the precise details of the rules Axelrod had set. In other words, it was possible to devise a strategy that would beat tit for tat. In fact, in the original tournament, the more forgiving strategy of tit for two tats ('Slow to be angered and quick to forgive', as the singer Joan Armatrading sang) would have done better. Strangely, this was the very strategy that Axelrod offered as an example when notifying others of the tournament but no one opted to submit it – perhaps thinking they could do better. Also if the rules were such that the competition was based on elimination, tit for tat would quickly be eliminated.

Another problem is that tit for tat is sensitive to errors. Whether or not an opponent has defected or cooperated on the previous move requires the transfer of a message. In the real world, as opposed to the cyberworld of computer tournaments, messages become corrupted and occasional mistakes are made. At an error rate in signalling of 1 per cent, tit for tat still came out on top. At 10 per cent, tit for tat is no longer the champion. Under these circumstances, tit for two tats or generous tit for tat (GTT) can prevail. Here the strategy is to forgive up to two defections. This prevents the effect of noise or mistakes leading to mutual and constant recrimination. But this strategy has its problems too, since GTT allows cooperators to flourish. Cooperators will do as well as GTT and it is conceivable that their numbers might drift to a position where they form the majority. It is now that defectors could strike. They can gain from cooperators who then dwindle in number. If there are not enough GTTs left, then the population will become all defecting. So playing tit for tat may drive away defectors if communication is clear, but if errors are made, GTT begins to flourish. GTT allows cooperators to grow in number, defectors can then thrive upon cooperators and we could be left with a population of all defectors.

The central problem with all these games is that it is difficult to design a strategy that would always win whatever the others in the pool. If only the others are known, we could design a winner but this is not the case. The theoretical promise of using prisoners' dilemma-type games to model the evolution of altruism out of an initial pool of selfish behaviour has been set back by such problems.

9.6.3 Applications of game theory

The theory of games has found widespread application in modelling animal and human behaviour, the latter in such diverse fields as sexual strategies, economic decision-making and political thinking (see Barash, 2003). Two examples will be given below.

Mutualism

Game theory now allows us, theoretically at least, to distinguish between mutualism and reciprocal altruism. It can be imagined that in mutualism, the rewards of cooperation exceed those of treachery (Figure 9.10).

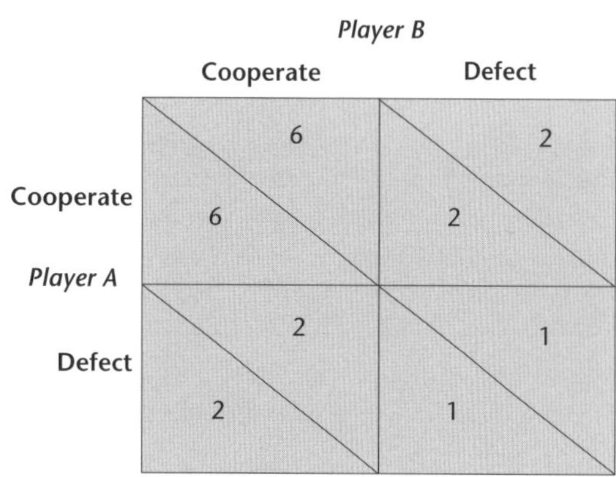

FIGURE 9.10 **Hypothetical payoff matrix for mutualism**
R = 6, S = 2, T = 2, P = 1.

In this situation, it does not pay to defect. For example, if one lioness does not help another in the hunt, there may be little food for either of them. A variety of studies on cooperative hunting behaviour, such as that practised by lions and African wild dogs, endorse the conclusion that such behaviour is mutualistic (Creel and Creel, 1995). It is a fairly safe assumption that early hominins engaged in a variety of mutualistic interactions. The cooperation required to bring back large game to feed dependent relatives (especially children) would have encouraged mutualistic behaviour. It is possible in fact that the greatest threat to our ancestors was the existence of other hominin groups tightly bound in a system of mutualistic cooperation and willing to instigate violent conflict. This threat may have driven up group size to the levels discussed in Chapter 6 and left its mark on the human mind in the form of in-group and out-group reasoning.

Box 9.1 Application of reciprocal altruism to trench warfare in the First World War

Axelrod (1984) drew upon the work of the sociologist Ashworth (1980) to apply the concept of prisoners' dilemma contests to trench warfare on the Western front in the First World War. Axelrod and Ashworth argue that the diaries of the infantrymen on both British and German sides show that an unofficial policy of 'live and let live' emerged between troops that faced each other for prolonged periods.

As is well known, the war was one essentially of attrition. If one side restrained, that is, chose to cooperate, the other would gain an advantage by killing more troops ($T > S$). On the other hand, if both sides defected and attacked, although both suffered losses, troops on both sides would regard this as preferable to their side alone losing troops. In other words, $T > P$ and $P > S$. If both sides cooperated and restrained fire, a stalemate resulted and so

$R < T$. But from the viewpoint of troops on both sides, mutual restraint or cooperation is preferable to both sides defecting (attacking) since, if both sides lose roughly equal numbers of troops, there is no relative gain – something less desirable than no relative gain without the risk of loss of life, that is, $P < R$. It also follows that $R > (T + S)/2$ since both sides would prefer to remain alive and in stalemate than to take equally a half share of $(T + S)$. The conditions for an iterated prisoners' dilemma were met because $T > R > P > S$ and $R > (T + S)/2$.

Axelrod claims that eyewitness reports from the infantry show clear evidence of tit for tat behaviour. Riflemen, for instance, would aim to miss each other but hit some other non-human target accurately to show their aim was good. If casualties were caused, then an equal number were exacted on the other side. The artillery often refrained from shelling the supply roads behind the enemy lines, since it would be straightforward for the other side to retaliate and both would be deprived of fresh rations ($R > P$). High command suppressed such truces whenever it could; several soldiers were court-martialled and whole battalions were punished. The action that finally ended the reciprocal cooperation took the form of random raids ordered by the officers. The raiders were ordered to kill or capture the enemy and, unlike the aim of a rifle or shell, such raids were difficult to fake.

FIGURE 9.11 A ration party of the Royal Irish Rifles in a communication trench during the Battle of the Somme
The date is believed to be 1 July, 1916. According to the British sociologist Tony Ashworth, troops facing each other along the Western Front in the First World War would sometimes adopt a policy of live and let live. Robert Axelrod sees this as an example of a tit for tat cooperative game strategy.
SOURCE: Wikipedia Commons.

Environmental problems and the prisoners' dilemma

It would be foolish to imagine that all human interactions resolve into prisoners' dilemmas but the diversity of situations to which the model can be applied is quite surprising. The abuse of environmental assets for individual gain is one such situation.

In 1968, the biologist Garrett Hardin coined the phrase **tragedy of the commons** as a metaphor for the nature of many environmental problems (Hardin, 1968). Hardin used the idea of the free grazing of common land such as found in medieval Europe to convey the result of individuals each maximising their own self-interest. Suppose three herdsmen each graze three cattle on a piece of land over which they have grazing rights. Equilibrium will be established between the rate of growth of grass and soil formation and the loss of grass and soil by grazing. At a certain level of grazing, the equilibrium is stable and cattle can be fattened without irreversible damage to the environment. But then it occurs to one of the herdsmen that he could place an extra beast out to graze. All the herdsmen would suffer to some degree, in the sense that their animals would not reach their former size, but the loss to the individual herdsman would be more than compensated by the possession of an extra animal. The tragedy arises because all the herdsmen think like this. The result is overgrazing and a collapse of that ecosystem.

Forty years after Hardin's original conception, we can recognise that this is another form of the prisoners' dilemma. By maximising his own utility, each herdsman is in effect defecting. When they all defect, they are all worse off. Hardin's idea is actually more useful as a metaphor than as a realistic description of how the commons were managed in the Middle Ages, but, looking around, we can observe the tragedy of the commons in action in relation to many environmental problems of the 20th century. For years, the seas have been treated as belonging to nobody, and therefore have been regarded as a suitable dumping ground for pollutants. In some places, overfishing has devastated fish stocks. Similarly, when we burn fossil fuels, we extract considerable benefit from the energy released and pass the cost on to the global commons to be shared by others. To use the terminology of economists, the costs are externalised and the polluter only pays a fraction of the true damage cost. To overcome the tragedy of the commons, and make the polluter pay the full costs of the damage caused, is the problem of problems in green economics. According to Pearce et al. (1989), we need to internalise the externalities.

Interpreted in terms of game theory, we need to adjust the rewards and penalties to encourage people and institutions to cooperate rather than defect. If, for example, carbon dioxide emissions or car usage were heavily taxed, the temptation to defect would not be so great.

Differences between prisoners' dilemma and social dilemma situations

Situations where the cost of defecting (not paying taxes, overgrazing, overfishing, releasing CO_2) are borne by the wider community are sometimes called 'social dilemmas', and although they may appear similar, there are some important differences between prisoners' dilemma and social dilemma situations. In the former, the cost is borne by one player, while in the latter, the cost is shared among many, including the player. Moreover, in prisoners' dilemma situations, defection or cooperation is not always anonymous. Your actions may be recorded by the other player and filed away as part of your reputation. Prima facie, then, it looks like social dilemmas will be less conducive to cooperation than prisoners' dilemmas and social dilemmas will need to be accompanied by group exhortations to act in the group interest. We are all probably good reciprocal altruists and cooperators among our circle of acquaintances, but people need more coercive measures to avoid defecting and dropping the costs onto a wider anonymous group – hence fines for not paying taxes, and fishing quotas. Politicians periodically have to remind people about the common good. As J. F. Kennedy said in his inaugural address of 1961: 'my fellow Americans: ask not what your country can do for you – ask what you can do for your country.'

Ridley applies game theory to environmental problems and draws out the political lesson (by his own admission 'suddenly and rashly') that one solution to the prisoners' dilemma is to establish ownership rights over the commons and confer them on individuals or groups. According to Ridley (1996), through ownership, and effective communication within the owning group, comes the incentive to cooperate out of self-interest. The conclusion from Ridley's extrapolation from biological game theory to the free-market political economy will not be to everyone's taste, but it represents one serious contribution to the problem of how to channel self-interest to the greater good.

More generally, it is interesting how commentators on the tragedy of the commons infer different solutions depending on their political orientation. Free-market conservatives such as Ridley use it as an illustration of the need for private ownership: if only the metaphorical shepherds owned their own bit of the resource, then they would ensure, out of self-interest, that it was managed properly. Meanwhile, liberals argue the need for a central regulating agency to curb excessive self-interest that might damage the common good.

9.6.4 Indirect reciprocity and reputation

Ridley's analysis highlights the importance of effective communication to encourage cooperation.

Recent work on reciprocal altruism confirms this point in a novel way. We have seen that reciprocal altruism can be expected to operate when animals have a good chance of meeting again to return favours. Such situations arise where animals gather in small colonies. Early human groups were almost certainly like this. But the commonest criticism of this model of reciprocal altruism is that real-life encounters are often between individuals who have little chance of meeting again. So can the idea of reciprocal altruism be extended to cover non-repeated exchanges between individuals?

Nowak and Sigmund (1998) think that it can. They found, using computer simulations, that altruism could spread through a population of players who had little chance of meeting again so long as they were able to observe instances of altruism in others. The simulation involved creating donors who would help the recipient if the recipient was observed to have helped others in previous exchanges. The logic of this manoeuvre is that a donor will be motivated to donate help to someone with a track record of altruism since help is likely to be fed back indirectly in the future. But more importantly, each act of altruism increases the 'image score' of the donor in the minds of others and so increases the probability that she or he will be the recipient of help in the future from other observers. The problem with large groups is that any player will know the image score of only a fraction of the population. By making some simplifying assumptions, Nowak and Sigmund derived a remarkable relationship. They set initial conditions such that a player who was observed to have helped in the last encounter was awarded a score of +1 and that a player who had defected was awarded 0. In the next encounter, the new player defected on those with a score of 0 but cooperated with those on a score of +1. If we call the fraction of the population that any player knows the score of q, the cost to the donor c and the benefit to the recipient b, then they found that cooperative behaviour can become an ESS when $q > c/b$. That is:

$$\text{The probability of one player knowing the image score of the other} > \frac{\text{Cost to donor}}{\text{Benefit to recipient}}$$

The interesting feature of this relationship is its similarity to Hamilton's equation, whereby altruism spreads if $r > c/b$ (see Chapter 2). The similarity between the equations may be nothing more than a coincidence, and certainly in humans our proclivity to cooperate or not is based on something subtler than the last observed encounter. Nevertheless, this work is important in that it shows reciprocal altruism can operate without repeated exchanges between the same pairs so long as sufficient information flow takes place.

9.7 Game theory and the moral passions

Given the conditions under which we would expect reciprocal altruism to flourish – repeated encounters and sufficient cognitive ability to recognise helpers and cheaters – we could expect to find reciprocal altruism and indirect reciprocity practised frequently in human populations. It is also likely that interactions that favoured altruists, and hence strategies such as tit for tat, were common in the life of early hominins. It is possible, then, that such interactions have left their mark on the mental life of humans perhaps in the form of distinct modules dealing with emotions. If so, we should find that we are keen to cooperate, but equally determined to punish defectors; we should remember those who gave us favours and not forget those who cheated. Evidence for the importance of detecting cheaters in our psychological make-up was provided in Chapter 7 when the modular mind was considered. This section explores further the possibility that there is some evidence that humans are indeed 'wired' with developmental genetic algorithms that are adapted to a life of reciprocal social exchanges.

Trivers (1985) has argued that the practice of reciprocal altruism formed such an important feature of human evolution that it has left its mark on our emotional system. Humans constantly exchange goods and help, over long periods of time with numerous individuals. Calculating the costs and benefits and deciding how to act require complex cognitive and psychological mechanisms. Trivers suggests that a number of features of our emotional response to social life can be related to the calculus of reciprocal exchanges, and our emotional response serves to cement the system together for the benefit of each individual. Some typical emotional responses that are amenable to this sort of analysis are guilt, moralistic aggression, gratitude and sympathy. In addition, we have a highly codified system of justice to ensure fair play.

One of the hallmarks of a civilised society is taken to be an independent and objective judicial system. A point made by moral philosophers (for example John Rawls), that can be tested by introspection among other ways, is that we judge a situation fair when individuals can endorse it without knowing what position they occupy in the outcome. In other words, the strong sense of justice that humans possess is the desire that benefits and penalties have been distributed optimally.

We experience moral outrage and are motivated to seek retribution when an altruistic act is not reciprocated. Selection may thus have favoured a show of aggression when cheating is discovered in order to coerce the cheater back into line. If we are the cheater in a reciprocal exchange, by, say, not returning a favour or not discharging an obligation, we may, if detected, be cut off from future exchanges that would have been to our benefit. So perhaps guilt is a reaction designed to motivate the cheater to compensate for misdeeds and behave reciprocally afterwards. In this way, it serves as a counterbalance against the tempting option to defect in a prisoners' dilemma. To encourage reciprocation in the first place we have a sense of sympathy. A sense of sympathy motivates an individual to an altruistic act and gratitude for a favour received provides a sense that a debt is owed and favours must be returned. Figure 9.12 shows a conjectured matrix for the emotional reactions that are likely to be associated with various moves in prisoners' dilemma-type scenarios. This topic is explored more deeply in Chapter 17.

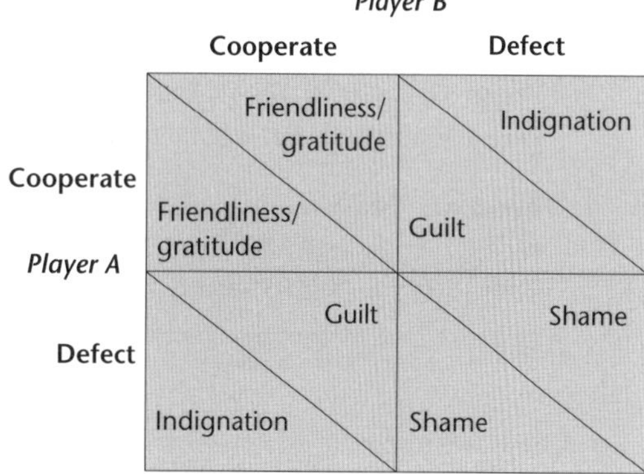

FIGURE 9.12 Some possible emotional reactions to moves in the prisoners' dilemma
It is suggested that guilt, for example, may accompany defection as a means of persuading the defectors to return to cooperation in future encounters.

Summary

- Altruism can be understood biologically in terms of kin selection and reciprocal altruism. In the former, genes direct individuals to help other individuals that are related and so share genes in common. In the latter, individuals help unrelated individuals in the expectation of favours returned. Kin selection and reciprocal altruism offer insights into the biological basis of altruism in humans.
- There is evidence that our closeness to relatives and our altruistic inclinations towards them are moderated according to their coefficient of genetic relatedness.
- Game theory shows that strategies involving cooperation such as tit for tat could offer a model for the evolution of human morality.
- The undoubted importance of altruism and cooperation in the environment of evolutionary adaptation made it imperative that cheaters could be detected. The need to detect cheaters could have shaped our emotional life such as our sense of justice and our experience of gratitude and sympathy.

Key Words

- Evolutionary stable strategy (ESS)
- Game theory
- Grandparental solicitude
- Hamilton's rule
- Mutualism
- Paternity certainty
- Patrilocal
- Prisoners' dilemma
- Reciprocity
- Symbiosis
- Tit for tat
- Tragedy of the commons

Further reading

Barash, D. P. (2003) *The Survival Game*. New York, Henry Holt.

Readable introduction to game theory and its manifold applications.

Ridley, M. (1996) *The Origins of Virtue*. London, Viking.

Discusses game theory and human cooperation. Interesting and controversial application of game theory to politics and environmental issues.

Conflict Within Families and Other Groups

I know indeed what evil I intend to do
But stronger than all my afterthoughts is my fury,
Fury that brings upon mortals the greatest evils.
(Euripides, *The Medea*, *c.*431 BC)

The Greeks knew a thing or two about dysfunctional families, and it is not surprising that, on several occasions, Freud turned to their myths and legends to find labels for what he supposed were problems of the human psyche. He posited, for example, an Oedipus complex, whereby males subconsciously desire the death of their father in order to sleep with their mother, and an Electra complex involving the secret desires of daughters for their fathers. Most evolutionary psychologists have not been very impressed by these ideas, seeing them as both unlikely and misleading. Much human behaviour, especially that involving conflict, is, however, maladaptive. In *The Medea*, Jason abandons his wife Medea for a more desirable bride. Medea, in her anger, kills the bride's father, the bride, and even her own children by Jason. Medea can be taken to signify both a wronged woman's fury and the dark and inexplicable irrational forces in human nature. *The Medea* is of course a work of fiction, but people do such things. We need not, however, invoke a Medea complex; Darwinism (as you might expect) also has something to say about murder and infanticide.

This chapter, then, is concerned with the application of evolutionary theory to understanding interactions involving conflict between offspring and their parents, between siblings and between spouses. To tackle these problems, theoretical perspectives established in earlier chapters will be deployed. In Chapter 2, it was noted that the existence of altruism posed a special problem for Darwinian theory and that the breakthrough came with Hamilton's notion of inclusive fitness. The concept of inclusive fitness also helps us to understand conflict between related individuals. It seems obvious that, since offspring contain the genes of their parents, parents will be bound to look after their genetic investment. But parents have loyalties that are divided between care for their current offspring and the need to maintain their own health in order to produce future offspring. It follows that offspring may demand more care than parents are willing to give, and in this situation, we should expect to observe a mixture of altruism and conflict. It is also easy to see that partners in a sexual relationship may have different interests and strategies, and it is these conflicting interests that help to explain violence and strife within marriages.

One extreme manifestation of human conflict is homicide. From an analytical point of view, homicide has the advantage that considerable statistical information is available. Two American psychologists, Margo Wilson and Martin Daly, have pioneered the use of homicide statistics to test evolutionary hypotheses, and this chapter also considers their work.

10.1 Parent–offspring interactions: some basic theory

10.1.1 Parental altruism

Everyone is aware that the animal world abounds

with examples of parents (often mothers) making great sacrifices in their efforts to protect and nurture their young. Perhaps the most extreme form of motherly care can be seen in some spider species (of which there are about 10) where the young eat their mother at the end of the brood care period. In the case of *Stegodyphus mimosarum*, the spiderlings do not eat each other or spiders of other species, yet they devour their mother with relish; it seems that this 'gerontophagy' serves as the mother's final act of parental care. From a Darwinian perspective, parents will cherish their progeny because their inclusive fitness is thereby increased. In turn, parents will be loved by their offspring because they can provide help to increase the fitness of the offspring. Parents will also be loved, although with less fervour, because they can increase the inclusive fitness of current offspring by producing more siblings. Altruism between parents and offspring is thus covered by Hamilton's theory of inclusive fitness. Parents will donate help b at a cost c to themselves as long as:

$$\frac{b}{c} > \frac{1}{r}$$

where r = coefficient of relatedness between parents and offspring. For diploid outbred offspring (that is, parents who are not related), $r = 0.5$.

10.1.2 Parent–offspring conflict and sibling rivalry

The widespread occurrence of parental care, and the clear biological function it served, probably diverted biologists for many years from the fact that conflict between parents and offspring may also have a biological basis. It awaited the work of Trivers to point out that **parent–offspring conflict** is also predicted from evolutionary theory (Trivers, 1974). The theory of such conflicts is worth examining for the light it throws on human behaviour, especially **maternal–fetal conflict**.

To follow Trivers' original argument, suppose one parent (the mother) gives birth and care to one infant each breeding season and that the infant needs and benefits from the care. The problem faced by the parent is when to cease providing care. A theoretical solution requires that we examine the costs and benefits of parental care to the parent (P) and the offspring (A). Figure 10.1 shows how r values are distributed between parent and offspring.

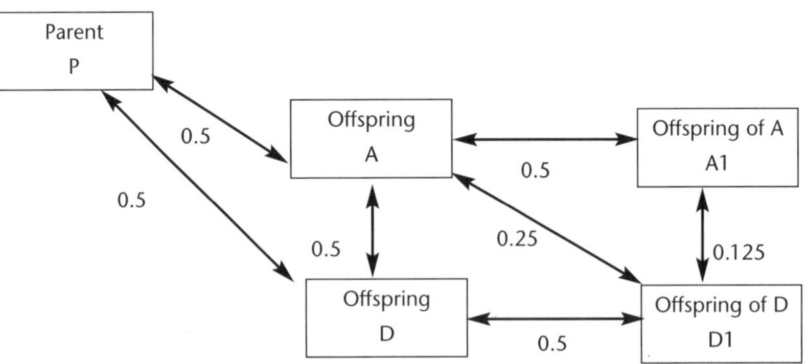

FIGURE 10.1 Coefficients of genetic relatedness (r values) between parents and offspring, between siblings and between nieces or nephews

Suppose we define the benefit to the parent in terms of units of probability that A will survive and reproduce, and the cost to the parent as units of reduced probability that it can produce another offspring D that will survive and reproduce. As A matures, the law of diminishing returns dictates that further care to A will bring fewer benefits than investing in a new offspring D. Eventually, the interests of parent P (or equivalently the helper genes carried by P) are best served by withdrawing from A and investing in D. Although the parent ultimately cares equally for A and D (they are both related to the parent with $r = 0.5$), A is concerned about D only half as much as it is concerned about itself, since the r value between A and A is 1 but that between A and D is 0.5. Similarly, A is twice as concerned about its own offspring than about the offspring of D. The upshot of all this is that A and its parent P will have a conflict about how much care is diverted to A. Offspring A will demand more care from P than it is in the interests of P to give. P will desire to give less help to A in order to focus resources on D. The analysis predicts that offspring will reach an age at which they will demand more investment than their parents are willing to give.

It has proved difficult to test Trivers' theory quantitatively because of the difficulties of quantifying costs and benefits. There is qualitative evidence, however, that is in broad agreement with theoretical expectations. One prediction from conflict theory is that offspring will prefer parents to direct resources at themselves (especially when they are young) rather than expending effort in producing more offspring. Young human parents often jokingly complain that 'kids are the best contraceptives'. Among chimpanzees, this may literally be true. Tutin (1979) noted that immature chimps often attempt to interrupt the copulations of their parents.

As the age of a parent increases, and it nears the end of its reproductive life, so its chances of producing any more offspring reduce and the costs to itself of giving care to existing offspring decrease. It should follow that older parents are more willing to invest in their young than younger ones. This seems to be supported by the fact that the abortion rate in mothers declines with the mother's age, although other factors may also be at work here (see Tullberg and Lummaa, 2001).

The logic of genetic relatedness dictates that individuals will care for themselves and their offspring more than they will for their siblings and the offspring of their siblings. At its extreme, this asymmetry could result in **siblicide**, which is in fact widespread in nature. In some species of eagle, the mother normally lays two eggs, even though, in nearly all cases, only one offspring survives. The mother perhaps lays the second egg as an insurance policy against infertility in a single egg. On hatching, the elder chick kills the younger (see Mock and Parker, 1997).

The ideas introduced by Trivers can help to explain parent–offspring conflict and sibling rivalry. In the latter case, siblings will seek to divert the flow of parental resources from a sibling to themselves. Some evidence in support of this comes from the fact that tantrums are used by infants in a number of species, including baboons, chimpanzees and zebras, as an apparent attempt to increase maternal investment in themselves (Barrett and Dunbar, 1994).

Frank Sulloway (1996) has developed these ideas into a remarkable theory of birth order personality differences in human families. He argues that competition between siblings for parental investment leads children to occupy different personality niches to secure their needs. First-borns, he suggests, try to maintain their initial advantage by aligning their interests with their parents and take advantage of their greater age than later children by assuming authority over them. They tend to adopt their parent's values and are more conservative, conventional and responsible than later born siblings. In contrast, later born children develop traits that enable access to parental resources despite the presence of older siblings. To achieve this, they are typically more flexible and imaginative than their elder siblings, less conservative and conventional, and more likely to be the rebels in the family.

This interesting line of research has generated a large volume of literature, some of it supportive (Davis, 1997; Saroglou and Fiasse, 2003) and some of it less so (Jefferson et al., 1998; Beer and Horn, 2000). One problem in these studies is that the comparison of children's personalities between families introduces lots of confounding effects such as differences in socioeconomic status, class and race. Recently, Healey and Ellis (2007) used a within-family methodology and found personality differences between first- and second-born children in agreement with Sulloway's model.

The work of Trivers has stimulated a fresh look at behaviour that was once thought of as being unproblematic. In pregnancy, for example, it is easy to assume that the interests of the mother and fetus are virtually identical and that the considerable investment by the mother in the fetus is a clear case of kin-directed altruism. Work by Haig (1993) and others, examined in the next section, shows that the situation is more complicated and provides a useful testing ground for Trivers' theory of conflict.

10.2 Maternal–fetal conflict

One of the most intimate relationships in the natural world must surely be that between a mother and her embryonic developing child. The mother is the life support system for the fetus: she provides it with oxygen from every breath she takes and food from every meal she eats. It is tempting to think that, in this precarious state, the interests of the mother and fetus must be identical, and that conflicts of the sort examined above can only arise after the birth of the child, when the mother will soon be in a position to produce more offspring. Even here, however, gene-level thinking brings some surprises. The Harvard biologist David Haig (1993) has applied the thinking of Trivers to this situation and has produced his own theory of genetic conflicts in human pregnancy.

The crucial point to appreciate in Haig's analysis is that the fetus and the mother do not carry identical genes: genes that are in the fetus may not be in the mother if they were paternally derived. Even if they are maternally derived (and therefore present in the mother), they have only a 50 per cent chance of appearing in future offspring of the same mother. It is quite feasible, then, argues Haig, for fetal genes to be selected to draw more resources from the mother than is optimal for the mother's health or optimal from the mother's point of view of distributing resources among her current and future offspring. We will consider four examples in which Haig's theory has met with empirical support.

10.2.1 Conflicts over glucose supplied to the fetus

When a non-pregnant woman eats a meal high in carbohydrates, her blood sugar level rises rapidly, but then falls as the insulin secreted in response to the raised glucose level causes the liver to store the excess glucose as glycogen for later use. In contrast to this, a similar meal taken during late pregnancy will cause the maternal blood glucose level to rise to a higher peak than before and, moreover, stay high for a greater length of time, despite the fact that the insulin level is also higher. In effect, the mother is less sensitive to insulin and compensates, but not entirely, by raising her insulin level. This is puzzling. Why should a woman develop a reduced sensitivity to insulin and then bear the cost of having to produce more?

Haig's theory of genetic conflict suggests an answer to this problem. It is in the interests of the fetus to extract more blood sugar for itself than it is optimal for the mother to give. The mother is concerned for her own survival after giving birth and, in addition, is more concerned for existing and future offspring than is the fetus. The fetus can send signals to the mother via the **placenta** just as the mother can send signals to the fetus. Haig suggests that one result of this is that, in late pregnancy, the placenta produces allocrine hormones that decrease the sensitivity of the mother to her own insulin, thereby allowing the glucose level to rise to benefit the growth of the fetus. The mother responds in this tug-of-war by increasing her insulin level. Further evidence in favour of this theory is that the placenta possesses insulin receptors and that, in response to a high insulin level, it produces enzymes that act to degrade insulin and thus disable the mother's counterattack. We can picture the escalation of measure and counter-measure as resulting in a pair of forces acting upon the level of some parameter such as glucose, each attempting to move it in an opposite direction towards two opposed optimum levels (Figure 10.2).

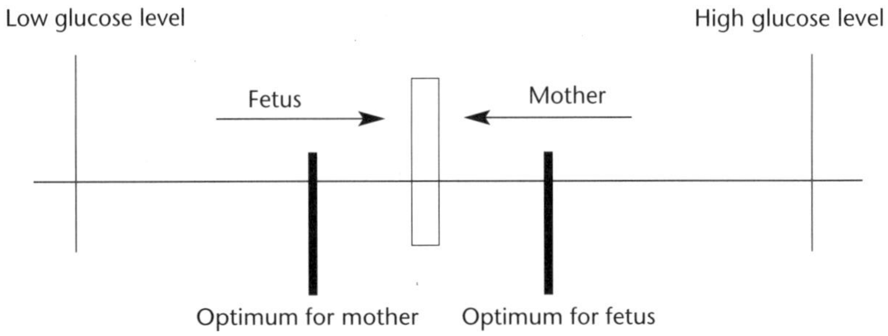

FIGURE 10.2 Schematic representation of the effort by the mother and fetus to drive the blood glucose level to different optima
The central cursor can be thought of as moving on a sliding scale subject to pressure from the mother and fetus.

10.2.2 Conflicts over decision to miscarry

Roberts and Lowe (1975) estimated that about 78 per cent of all human conceptions never make it to full term. Most conceptions miscarry before the 12th week and many before the first missed menstruation. When spontaneous abortions are karyotyped (that is, the structure of their chromosomes examined under an optical microscope), they are seen to have genetic abnormalities. Early miscarriages seem to be the result of some sort of quality control mechanism employed by the mother to terminate a pregnancy before she has committed significant resources. If this is the case, there must be a threshold of quality below which a mother will attempt to terminate the pregnancy and save resources for future offspring and above which she will accept the pregnancy to full term. Fetal genes will also have an interest in the decision of whether or not to terminate. If the quality is extremely low, it may be better for genes to 'abandon hope' of survival in the current fetus and 'hope' for survival (with a 50 per cent chance of being represented) in a future offspring. In this case, it would be preferable from the point of view of the fetus not to jeopardise the chances of future viable offspring by dragging from the mother resources that will be wasted.

But here's the rub: the cut-off point for quality will be different in the mother and the fetus. In essence, the fetus is predicted to respond to a lower quality cut-off point than the mother does. It is significant, then, that the maintenance of pregnancy in the first few weeks of gestation is dependent on maternal progesterone. Later on, the fetus gains control of its fate by releasing human chorionic gonadotrophin into the maternal bloodstream. This hormone effectively stimulates the release of progesterone and serves to maintain the pregnancy.

One would expect that the older the mother, the lower the quality she will still accept as she nears the end of her reproductive life. It may be no accident, therefore, that births with genetic abnormalities increase in frequency with the age of the mother. This increase is sometimes referred to as 'relaxed filtering stringency' (RFS). The idea that RFS represents an adaptive maternal strategy was suggested over a decade ago (for example see Kloss and Nesse, 1992) but without decisive evidence.

Recently, Markus Neuhauser and Sven Krackow (2006) of the Institute of Medical Informatics, Biometry and Epidemiology at University Hospital Essen in Germany have advanced several testable hypotheses about the way RFS increases with age. They examined births of Down syndrome (DS) babies over the period 1953–72 (before abortion commonly provided a route to termination). They looked at the mother's age but also whether she had children before the Down syndrome child and the age gap between pregnancies. From a reproductive fitness point of view, a mechanism that screens for defective embryos and discards them is adaptive so long as the compound costs of losing a potentially fertile (albeit damaged) child, delaying reproduction and the risk of not producing further children are outweighed by the higher fitness of a future normal embryo. Neuhauser and Krackow argue that, from this, it is to be expected that the stringency of the filtering mechanism should relax as the number of pre-existing children becomes larger. This follows from the fact that the incremental loss in overall fitness of aborting a defective embryo is smaller according to how many children the mother already has. Related to this reasoning is the suggestion that DS incidence should increase more rapidly with age for mothers giving birth for the first time (that is, primiparae) compared to mothers who already have children. The reasoning behind this hypothesis is similar: if an elderly mother is carrying an embryo for the first time, this may be her only chance of reproducing at all and so such a child represents a significant proportion of her overall reproductive output (although DS children suffer from a variety of abnormalities, they are fertile). Figure 10.3 shows how DS incidence increases with age for primiparae and all births, showing that this hypothesis is supported.

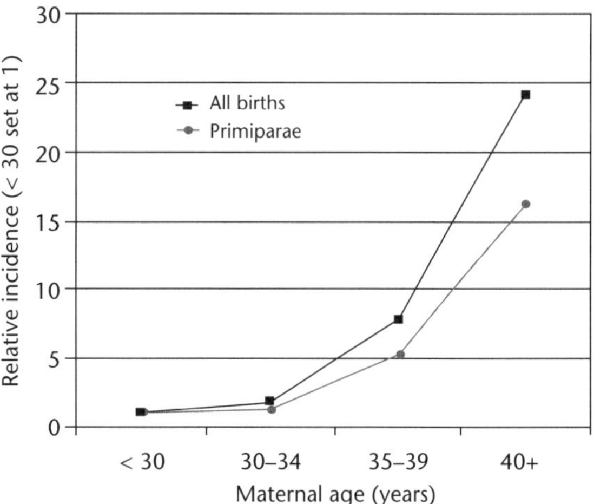

FIGURE 10.3 Relative incidence of DS according to age of mother and whether birth is primiparae or not

SOURCE: Redrawn from data in Neuhauser, M. and Krackow, S. (2006) 'Adaptive-filtering of trisomy 21: risk of Down syndrome depends on family size and age of previous child.' *Naturwissenschaften* DOI 10.1007/s00114-006-0165-3.

In accordance with predictions, they also found that rates of DS fell significantly in older mothers (35 years and older) as the number of previous children increased (see Figure 10.4). The change was not significant, however, for mothers below 35 years of age.

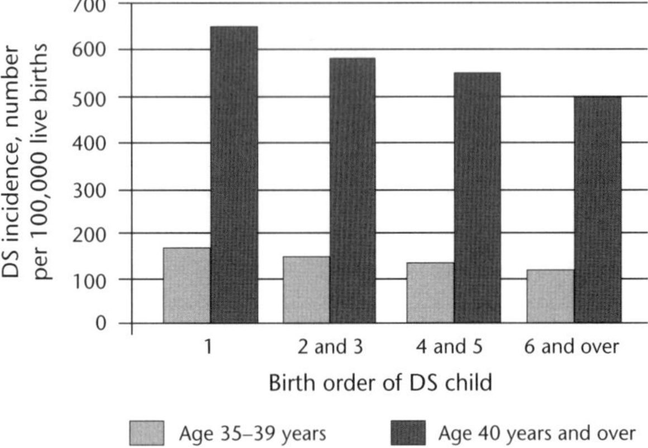

FIGURE 10.4 The incidence of DS in relation to two maternal age categories and birth order of the DS child

The researchers also predicted that the filtering stringency should increase the younger the youngest child already in the family. This follows from the idea that a new child will compete for resources with an existing child of close age, and a defective child will impact more strongly on the welfare of an existing young child than a normal child. In support of this, Neuhauser and Krackow (2006) did find that DS births were significantly spaced further apart from the previous child than normal births.

10.2.3 Conflicts over blood supply

The placenta obtains nutrients from the bloodstream of the mother, and, as we have seen, conflicts arise over the amount of glucose in the mother's bloodstream. Maternal blood, however, supplies a whole range of other nutrients, and, roughly speaking, the rate at which they can be extracted by the embryo will be proportional to the flow of blood to the placenta. Conflict theory predicts that the fetus will act to stimulate more blood to the placenta than is optimal for the mother. The flow of blood along an artery is given by a simple equation:

$$\text{Flow} = \frac{\text{Pressure difference}}{\text{Vascular resistance}}$$

It follows that the fetus could increase its supply of nutrients by either dilating the arteries of the mother and thus reducing the vascular resistance, or increasing the blood pressure in the maternal supply. Maternal genes are predicted to act to counter both of these manoeuvres. Haig (1993) reviews a range of evidence and concludes that hypertension (raised blood pressure) in pregnancy is one result of maternal–fetal conflict. Significantly, for example, high birth weight is positively correlated with maternal blood pressure, whereas low birth weight is associated with hypotension (low blood pressure).

10.2.4 Conflict after parturition

As does the fetus, so a young infant may also behave to extract more resources from a mother than is optimal for the mother to provide. The presence of benzodiazepines (substances that inhibit neurotransmission and thus serve as sedatives) in human breast milk may also be a response of the mother to excessive infant demands. At the level of the genes, we could expect maternally derived genes to be less demanding than those that are paternally derived, since paternal genes are probably not present in the mother and will be selected to exploit the mother for their own advantage. Evidence in favour of this prediction comes from conditions in which genetic abnormalities are observed. In Prader-Willi syndrome, infants have a maternal copy of a region of a chromosome known as 15q11–13 (to be read as 'chromosome 15, long arm, bands 11 to 13') but lack a paternal copy. These unfortunate infants have a weak cry and a poor sucking response, as well as being generally inactive and sleepy. In contrast to this, children with Angelman syndrome have paternal copies of 15q11–13 but no maternal ones. In these cases, sucking is prolonged but poorly coordinated. Children are highly active when awake and suffer sleeping difficulties. It looks here as if we have a system in which paternally derived genes demand excessive resources from the mother and are restrained by maternally derived genes.

These two conditions illustrate the general

phenomenon of genomic imprinting, a situation in which a gene behaves differently depending on whether it was contributed by the father or the mother. A good example concerns the Igf-2 gene, which produces an insulin-like growth factor that promotes embryonic development. In mice and humans, copies of the Igf-2 gene from the mother are switched off and copies from the father left on. This is consistent with evolutionary conflict theory. It is in the interests of the paternally derived Igf-2 gene to demand more resources than the mother is prepared to give. The reason for this stems from the theory considered earlier: the mother wishes to reserve her resources for future offspring, but the current paternal Igf-2 gene has no guarantee, unless the mother is 100 per cent monogamous, which is unlikely, that it will appear in future offspring. In this light, as Pagel (1998, p. 19) neatly summarises, 'genomic conflict is one cost of infidelity'. Given the excessive demands of the paternal Igf-2 gene, the best countertactic of the mother is to switch off her Igf-2 copy and cease its contribution to the production of the growth factor. Although genomic imprinting provides evidence for selection at the level of the gene, all such cases are unlikely to be explained by conflict theory. Iwasa, for example, has recently advanced a 'dose compensation' explanation (see Pagel, 1998).

Haig (1993) himself points out that the presence of conflict does not in itself imply system instability or failure. A useful analogy is that of evenly matched tug-of-war teams: as long as both sides keep pulling, the system is stable. However, the system collapses when one side lets go. In this light, both Prader-Willi syndrome and Angelman syndrome can be seen as resulting from one side of the conflict not operating, with disastrous effects. This whole scenario reminds us that evolution is all about compromise.

10.3 Human violence and homicide

The application of adaptationist thinking to human conflict within families was pioneered by Martin Daly and Margo Wilson in the 1980s. Daly and Wilson completed their PhDs in animal behaviour in the early 1970s and were inspired by Wilson's (1975) *Sociobiology* to apply evolutionary theorising to step-families. Since then, they have written numerous articles and books dealing with human conflict in the form of homicide.

Before we examine their work, it is important to understand the way in which adaptive reasoning can be applied in this area. In modern nations, homicide is, for the most part, damaging to the fitness of an individual. The perpetrators are likely to be found and either incarcerated or executed. Many homicides also have a negative effect on inclusive fitness since it is often one relative who kills another. To cap it all, many homicides are followed by suicide, which is hardly fitness-maximising. At first sight, it would seem bleak territory on which to erect adaptationist arguments, and, unsurprisingly, homicide has usually been regarded as a result of inherent human wickedness, a failure of social upbringing or the result of some sort of pathological condition. The originality of the work of Daly and Wilson has been to realise that, amid all these causes, there may also be the effect of psychological mechanisms that can be understood in selectionist terms.

Here, evolutionary psychology differs from the traditional approaches of behavioural ecology and sociobiology. Sociobiologists, when studying the behaviour of animals other than humans, looked for the adaptive significance of contemporary behaviour. With some qualifications, it was assumed that behaviour should be optimal with respect to fitness. The evolutionary psychologist, however, looks for motivation mechanisms and 'domain-specific' mind modules that were shaped during the environment of evolutionary adaptedness (the Pleistocene Epoch when the genus *Homo* appeared; see Chapter 5). The output of these modules is adaptive only on average and largely in the environment in which they were shaped. Like all mechanisms, they can suffer problems of calibration. A jealous rage that once served the interests of an early male in driving away would-be suitors from his wife may translate in the 20th century to a man murdering his wife's lover and receiving a life sentence.

Homicide is a gruesome business, but the serious nature of the crime makes statistics on killings more reliable than those on probably any other act of violence. It is against patterns of homicidal statistics that predictions from evolutionary psychology can be tested.

10.3.1 Infanticide

Infanticide is not uncommon in the animal kingdom. The males of such animals as lions and lemurs will often kill the offspring of unrelated males to bring a female back into oestrus. Among the vertebrates, it is mostly males who kill infants, but where males are the limiting factor, the roles are reversed. Among the marsh birds called jacanas, polyandry is often found, and sex roles are often inverted. A large territorial female may have several nests containing her eggs in her territory, each nest presided over by a dutiful male. If one female displaces another and takes over her harem of males, she sets about methodically breaking the eggs of her predecessor. In these examples, infanticide can be understood as a means of increasing the fitness of the killer.

We may find such instances distasteful, but we rightly regard examples of human infanticide as being even more horrific. In most countries, infanticide is illegal, and debates continue about the legitimacy of feticide (the destruction of a fetus in the uterus). It could, however, be that in the environment in which early humans evolved, infanticide in some circumstances represented an adaptive strategy. If we can judge the reproductive experiences of early humans by those of modern-day hunter-gatherers, we begin to appreciate the tremendous strain that raising children probably entailed. Infant mortality would be high, fertility would be low, partly as a result of the prolonged feeding of infants, and the best that most women could hope for would be two or three children after a lifetime of hard work. Under these conditions, raising a child that was defective and had little chance of reaching sexual maturity, or a child that was for some reason denied the support of a father or close family, would be an enormous burden and contrary to a woman's reproductive interests. At a purely pragmatic level, infanticide may have sometimes been the best option to maximise the lifetime reproductive value of a woman. One might term the withdrawal of support by the father the 'Medea effect'.

If infanticide did once (as it still does in some cultures) represent a strategy for preserving future reproductive value, we might expect the frequency of infanticide to decrease as the mother's age increases. This follows from the fact that, as a mother ages, the number of future offspring she can raise falls and so each one becomes more valuable. Daly and Wilson (1988a) present evidence that this effect is at work both among the Ayoreo Indians and modern-day Canadians. Figure 10.5 shows infanticide in relation to the age of the biological mother.

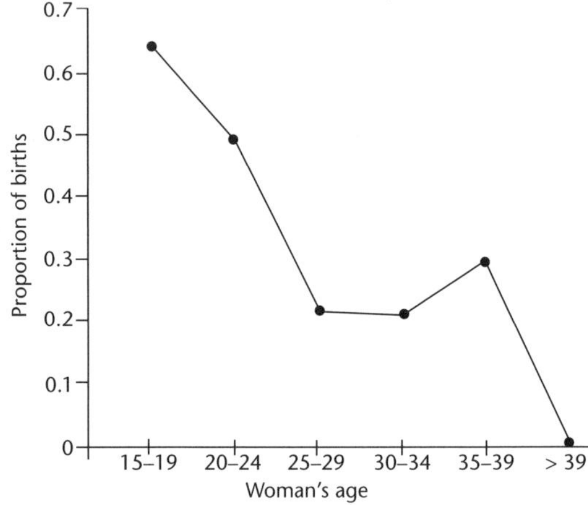

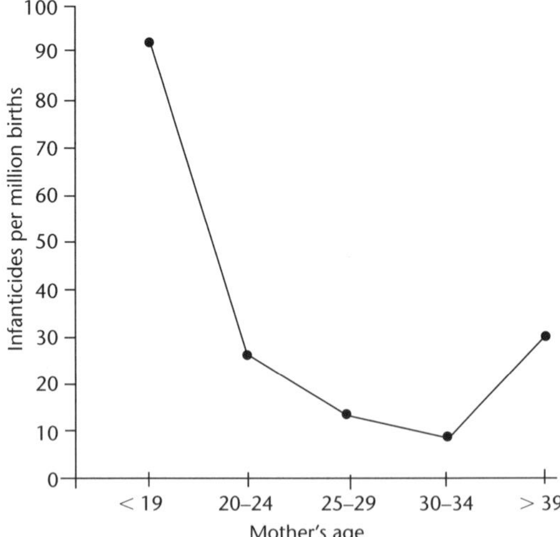

FIGURE 10.5 Rate of infanticide versus age of biological mother

SOURCE: Adapted from Daly, M. and Wilson, M. (1988a) *Homicide.* New York, Aldine De Gruyter.

The results from Figure 10.5 are in agreement with predictions, but it is difficult to rule out other effects. Women may become better mothers as they age, learning from experience; younger mothers may suffer more social stress. The effect could thus be one of socially learned skills and culturally specific stress factors rather than an adaptive response.

Reproductive value of offspring and infanticide

In the analysis above, Daly and Wilson derived testable predictions from the way in which the reproductive value of the parent varied as a function of age. The child too has reproductive value as the carrier of the parent's genes and as a potential source of grandchildren. From these considerations, predictions can also be derived.

From a gene-centred perspective, children are of value to their parents because they have the potential to breed and continue to project copies of genes into future generations. This dispassionate approach suggests that the value of a child would increase up to puberty and decrease as the child approached the end of its reproductive life. This follows from the fact that the chances of, for example, a 10-year-old girl reaching sexual maturity at 16 (a typical age for the first menstruation in a child in a hunter-gatherer society) are greater than those of a 2-year-old. The 10-year-old has benefited from 10 years of parental investment and has only 6 more years to survive the hazards of life to reach 16, whereas the 2-year-old has 14 years to go. The variation in valuation will be more marked and steep in cultures with a high infant mortality rate, but even in industrialised countries, where infant mortality has dropped dramatically over the past 100 years, the effect will be present.

From the way in which the reproductive value of offspring varies with age, Daly and Wilson (1988a)

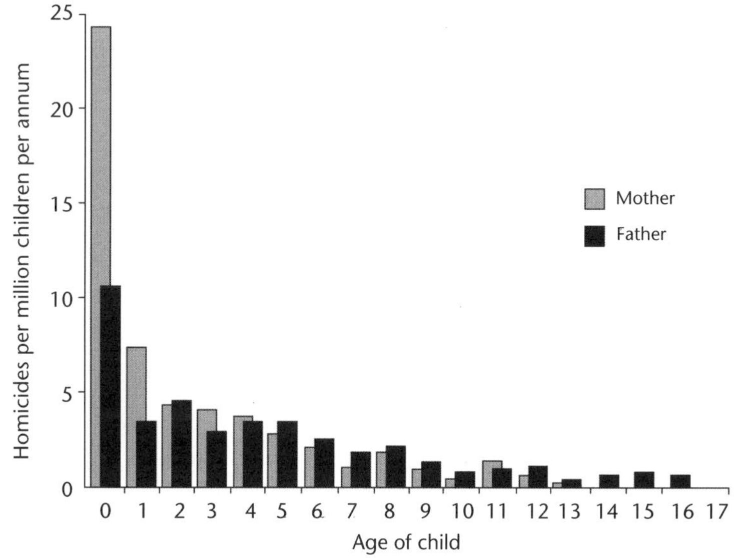

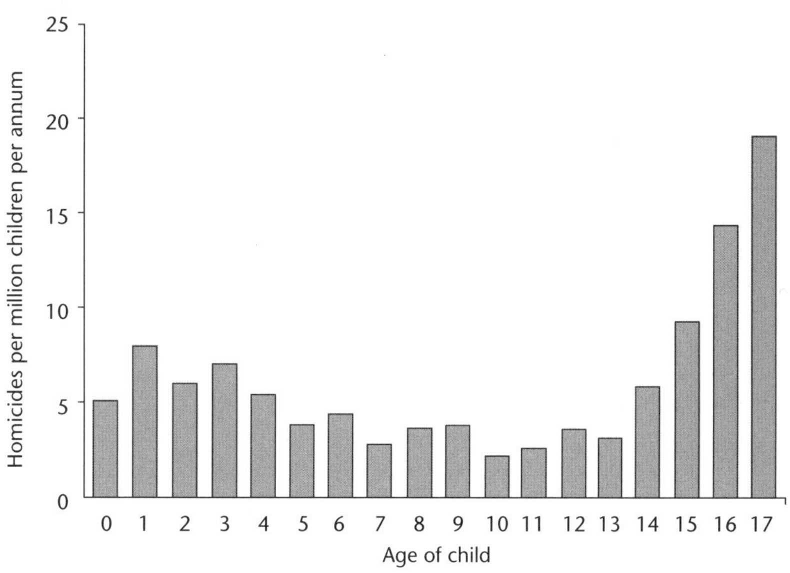

FIGURE 10.6 **Child's risk of homicide**

SOURCE: From Daly, M. and Wilson, M. (1988a) *Homicide*. New York, Aldine De Gruyter.

made the following predictions about parental psychology:

- In conflicts, parents will be careful in aggressive encounters with their offspring according to their reproductive value. The filicide rates should fall as the age of the child increases from zero to adolescence.
- In the environment of evolutionary adaptation, the greatest increase in reproductive value would have occurred in the first year. The filicide rate is consequently predicted to fall rapidly after the first year of life.
- Child homicide by non-relatives is not expected to vary with age in the same way as filicide since the children of other parents are neutral in respect of their reproductive value to an unrelated offender.

Figure 10.6 shows data for childhood homicide in Canada in the period 1974–83 as used by Daly and Wilson to test these predictions. The results are consistent but do not rule out other interpretations. One could posit, for example, an 'irritation factor' that varies with the child's age. The lower risk for fathers may then represent a lower time of exposure, and the fall in risk with age may be a product of coming to terms with the difficulties of child-rearing. This theory would predict a higher level of homicide by parents against adolescents, considering the conflicts between teenagers and parents – but then again teenagers are harder to kill. The decline in child homicide with the age of the child also applies to homicides initiated by step-parents, who, presumably, have no interest anyway in the reproductive value of their stepchildren (see Figure 10.6 above). This too indicates that other factors may be at work.

Infanticide and step-parents: the Cinderella syndrome

Popular traditional culture abounds with tales of wicked step-parents who fail to provide proper care for their children. The story familiar to many in the West is that of Cinderella. Cinderella's biological mother has died and her father has remarried. Cinderella is raised in the house of her stepmother, who has two children of her own by a former marriage. Despite the obvious virtues of Cinderella compared with her 'ugly sisters', she is treated harshly (Figure 10.7). Fortunately, a way out of her plight is provided by the deus ex machina of a charming prince. It would be wrong to try to construct a scientific argument on the basis of a fairy story, but it is worth noting that cultures all around the globe have their own variant of the Cinderella myth. The essence of such stories is that stepfathers and stepmothers are wicked and not to be trusted. Again, this proves little by itself since all the stories could have descended from a common archetype. Such is the bad image of step-parents in traditional folklore that, faced with a high divorce and remarriage rate in Western countries, children's books now go out of their way to show step-parents in a more positive light.

FIGURE 10. 7 Cinderella

The stories may, however, reflect a fundamental component of the human experience: the raising of children by non-genetic parents. Throughout human history, step-parents must have played a role in the raising of young when, for a variety of reasons, one parent was killed or left the family unit. A prediction that follows from selectionist thinking about parenting is that parental solicitude should be discriminating with respect to the offspring's contribution to the reproductive interests of the parent. In stepfamilies, it would follow that parents should be more protective of their biological children than their stepchildren. We are not reliant on folk tales to test these predictions.

Daly and Wilson (1988a, 1988b) supply a range of cross-cultural statistical evidence showing that children in stepfamilies are injured or killed in disproportionate numbers compared with those in their biological families. The effect seems to be extremely robust. As Daly and Wilson (1988a, p. 7) remark:

> Having a step-parent has turned out to be the most powerful epidemiological risk factor for severe child maltreatment yet discovered.

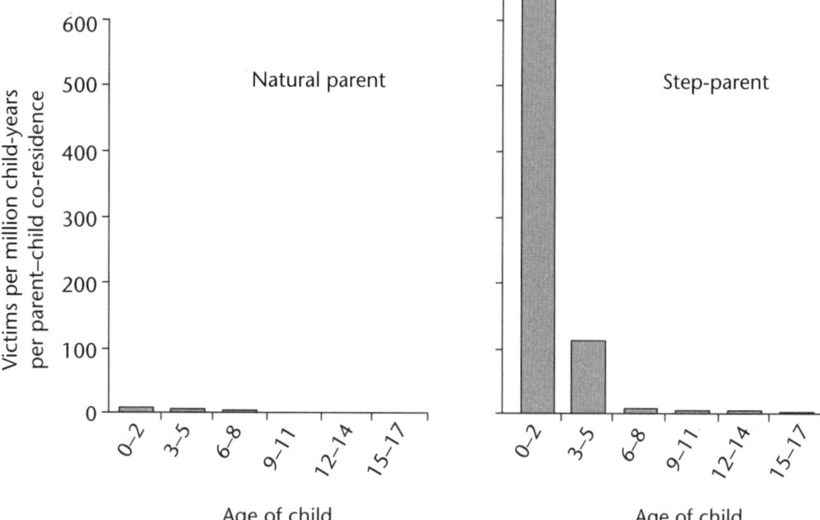

FIGURE 10.8 Risk of a child being killed by a step-parent compared with a natural parent in relation to the child's age

SOURCE: From Daly, M. and Wilson, M. (1988a) *Homicide*. New York, Aldine De Gruyter.

A child living with one or more step-parents in the USA in 1976, for example, was 100 times more likely to be fatally abused than a child of the same age living with his or her biological parents. This effect is also illustrated by the Canadian data for 1973–84 (Figure 10.8).

These findings have, understandably, been met with shock and incredulity, and there have been numerous attempts to show they are not sound. One obvious objection is that families containing step-parents may be unrepresentative of family life since, by definition, step-parents are the products of some breakdown of an existing family unit. The previous family unit may have failed because of violent behaviour by one of the parents, and hence the sample is not representative. The problem with this objection is that, within stepfamilies containing biological and stepchildren, it is the non-biological offspring of one of the partners who tend to be hurt. The effect has also been noted in other cultures. Flinn (1988) documented the interactions between men and their children in a village in Trinidad. He found significantly more interactions involving conflict between fathers and their stepchildren than between fathers and their genetic offspring. There are still other interpretations of course: stepchildren may resent the intrusion of an outsider and thus promote conflict. Again, however, we must face the prospect that Daly and Wilson may have touched upon an evolutionary derived component of the human psyche.

More recent research by Grant Harris and colleagues has tended to confirm the general thrust of the findings of Daly and Wilson. They used data from a Canadian police service database to compare abuse and infanticide by genetic and non-genetic parents (Harris et al., 2007).

One interesting focus of their work was cases of filicide by stepmothers. They reasoned that solicitude for a genetically unrelated child carries greater relative cost for a stepmother than a stepfather. This follows from two considerations: first, women experience less variability in reproductive success and have a potentially lower reproductive rate than males and so care for an unrelated child reduces her future fitness more than it would for investment by a stepfather; and second, fathers have greater parental uncertainty anyway and so the contrast between putative genetic offspring and stepchildren is less for a father than a mother. From this, the group predicted that stepmothers would show evidence of reduced levels of parental solicitude for non-genetic offspring than stepfathers. To measure this, they devised a measure of abuse and neglect by factoring in evidence on malnourishment, lack of care, failure to provide proper clothing, and physical abuse, prior to the death of the child. An extract from their findings is shown in Table 10.1. It can be seen that abuse from stepmothers is indeed higher than that from stepfathers and genetic mothers.

It hardly needs saying that this does not cast a shadow on all stepfamilies or that violence in these circumstances is somehow excusable. Fortunately, homicide is still an extreme rarity; it is probably more of a puzzle to explain, in selectionist terms, why so many stepfamilies do work and so many children receive love and affection from non-biological parents. Crime statistics can be depressing, so it is good to realise that counter-Darwinian (or at least not obviously selected) behaviour is also part of being human.

Table 10.1 Abuse and neglect of children by genetic and non-genetic parents

	Genetic mother	Genetic father	Stepmother	Stepfather
Abuse/neglect score of victim	0.15 +/– 0.13	0.03 +/– 0.04	1.0 +/– 0.9	0.16 +/– 0.12

NOTE: Level of abuse is a measure devised by the authors based on a variety of parameters. Figures show mean +/–95 per cent confidence intervals. Any mean lying beyond the confidence intervals of another is significantly different.

SOURCE: Data from Harris et al. (2007). 'Children killed by genetic parents versus stepparents.' *Evolution and Human Behaviour* **28**(2): 85–95.

10.4 Human sexual conflicts

> The cuckoo then, on every tree,
> Mocks married men; for this sings he
> Cuckoo, Cuckoo, cuckoo; O word of fear,
> Unpleasing to a married ear.
> (Shakespeare, *Love's Labours Lost*, 5.2.885)

Shakespeare was a good psychologist. The risk of **cuckoldry** – the 'word of fear' – has damaging repercussions for the male, far more so than for the female. In this final section, we examine how this asymmetry in sexual interests has conditioned and moulded conflict between the sexes.

10.4.1 Marriage as a reproductive contract: control of female sexuality

Marriage is a cross-cultural phenomenon and, although ceremonies differ in their details and marriage law varies across cultures, marriage has a set of predictable features across virtually all societies. Marriage entails or confers:

- Mutual obligations between husband and wife
- Rights of sexual access that are usually but not always exclusive of others
- The legitimisation of children
- An expectation that the marriage will last.

Viewed through Darwinian eyes, marriage begins to look like a reproductive contract. This is seen at its clearest when fears arise that the contract has been breached, such as when one male is cuckolded by another. In reproductive terms, the consequences of cuckoldry are more serious for the male than the female: the male risks donating parental investment to offspring who are not his own. This male predicament is reflected in laws dealing with the response of the male to incidents where he finds his partner has been unfaithful. In the USA, a man who kills another caught in the act of adultery with his wife is often given the lesser sentence of manslaughter rather than murder. This is a pattern found in numerous countries: sentences tend to be more lenient for acts of violence committed by a man who finds his wife in flagrante delicto. The law carries the assumption that, in these circumstances, a reasonable man cannot be held totally responsible for his actions.

Violence after the event may serve as a threat to deter would-be philanderers or to rein in the affections of a wife whose gaze may be wandering. Before the event, however, males of many species guard the sexuality of their partners. Research on animals has shown that, in numerous species where parental investment is common, males have evolved anti-cuckoldry techniques. Male swallows, for example, follow their mates closely while they are fertile, but when incubation begins, mate guarding ceases and males pursue neighbouring fertile females. When the same males perceive that the threat of cuckoldry is high, such as when they are experimentally temporarily removed, they seem to compensate by increasing the frequency of copulation with their partner. Indeed, male swallows do have something to fear: Moller (1987) estimated that about 25 per cent of the nestlings of communal swallows may be the result of extrapair copulation.

If anti-cuckoldry tactics have evolved in avian brains where members live in colonies, both birds are ostensibly monogamous and both sexes contribute to parental care, might we not expect similar concerns to have evolved in humans? The answer, according to Wilson and Daly (1992), is a resounding yes. Numerous cultural practices can be interpreted as reflecting anxieties about paternity. A few of the more obvious cases are described below.

Veiling, chaperoning and purdah

Obscuring a woman's body and facial features as well as ensuring that she is always accompanied when she travels are common practices in patriarchal societies and can be seen as ways of restricting the sexual access of women to other men (Figure 10.9). It is usually only practised on women of reproductive age, children and postmenopausal women being excluded.

Foot binding

This was once used in China, partly as a display by a man that he was wealthy enough to be able to dispense with the labour of his wife. It also serves as a way to restrict her mobility and hence her sexual freedom.

Genital mutilation

Unlike male circumcision, female genital mutilation is specifically designed to reduce the sexual activity of the victim. Practices range from partial or complete clitoridectomy to infibulation. Girls aged 13–18 of the Sabiny tribe in Uganda are taken to a village clearing where their clitorises are sliced off. It is estimated that about 6,000 girls are mutilated in this way each day in Africa and that more than 65 million women and girls now alive in Africa have been 'circumcised' (Hosken, 1979). The practice is condemned by some governments, such as that of Uganda, but the Sabiny women resent attempts to ban the practice as cultural interference. In some African countries, infibulation involves the suturing of the opening of the labia majora, opening them up again on marriage. Infibulation makes sexual intercourse impossible, and so is a guarantee that a bride is intact. Women are cut open on marriage, and more cuts are needed for the delivery of a child. In all, female genital mutilation is to be found in 23 countries worldwide. If there is any practice that sinks the pretensions of cultural relativism, it must be this.

Women as men's legal property

FIGURE 10.9 A Bengali Muslim woman covered in a hijab
Although an injunction to modesty in female dress appears in the Koran, the wearing of a hijab (Arabic, curtain) in public is not an essential requirement of Islam, and its cultural origins are obscure. The hijab, worn by girls after puberty, functions to discourage amorous male advances.

Studies on the history of European adultery laws reveal the double standards that have until recently operated. It seems that, before recent reforms, the reproductive capacity of a female was treated as a commodity that men could own and exchange. Some of the salient features of these laws are as follows:

1. Laws tended to define adultery in terms of the marital status of the woman, the marital status of the man usually being ignored.
2. Adultery often took the form of a property violation, the victim being the husband, who might be entitled to damages or some other recompense.
3. If a wife was adulterous, this represented clear grounds for divorce; if a husband was adulterous, divorce was a rarer consequence.

4. As late as 1973, Englishmen could legally restrain wives who were intent on leaving them.

10.4.2 Jealousy and violence

> Not poppy, nor mandragora,
> Nor all the drowsy syrups of the world
> Shall ever medicine thee to that sweet sleep
> Which thou owedst yesterday.
> (Shakespeare, *Othello*, 3.3.333)

Iago is one of the most evil of Shakespeare's characters, and he understood well the power of jealousy acting on the human mind. Tormented by his suspicions, Othello first kills his wife Desdemona and then himself. In the light of selectionist thinking, the emotion of jealousy in males is an adaptive response to the risk that past and future parental investment may be 'wasted' on offspring that are not their biological progeny. In females, given the certainty of maternity, jealousy should be related to the fact that a male partner may be expending resources elsewhere when they could be devoted to herself and her offspring. Men also lose the maternal investment that they would otherwise gain for their offspring, since this is directed at a child who is not the true offspring of the male. It is to be expected that jealousy and its consequences would be asymmetrically distributed between the sexes. We should also expect it to be a particularly strong emotion in men, since humans show a higher level of paternal investment than any other of the 200 species of primates. In a near-monogamous mating system, men have more to lose than women.

To test for sexual differences in the experience of jealousy, Buss et al. (1992) issued questionnaires to undergraduates at the University of Michigan asking them to rank the level of distress caused by either the sexual or the emotional infidelity of a partner. The result suggested that men tend to be more concerned about sexual infidelity and women more about emotional infidelity.

The same effect was observed when subjects were 'wired up' and tested for physiological responses to the suggestion that they imagine their partner behaving unfaithfully either sexually or emotionally. The difference was less marked for women, but men consistently and significantly showed heightened distress to thoughts of sexual, compared with emotional, infidelity.

Such effects are what would be predicted from an evolutionary model of the emotions. Men would tend to be more concerned about the sexual activity of their partners, since it is through extrapair sex that a male's investment is threatened. Women, on the other hand, should be less concerned about the physical act of sex per se than about any emotional ties that might lead her partner and his investment away from herself.

Following the suggestions of Buss et al (1992), the evidence supporting sex differences in human jealousy has typically come from three types of study:

1. Forced choice studies where men and women are obliged to state which situation (sexual infidelity or emotional infidelity) they would find most distressing.
2. Physiological response studies where physiological parameters (such as heart rate, blood pressure and palm sweating) are monitored as subjects are asked to imagine various types of jealousy-inducing scenarios (Harris, 2003).
3. Continuous rating scale questionnaire studies where men and women grade their response to situations that invoke jealousy.

Reviewing such studies, Harris (2003) cast doubt on whether the evidence really did support predictions from evolutionary psychology as advanced by Buss and others. In essence, she concluded that in some studies the differences were small or even reversed. More recently, Robert Pietrzak et al. (2002) have suggested that such ambiguities in the literature may be a result of the fact that different tests have been applied at different times to different subject groups. Accordingly, this group set about applying all three types of test to the same group of subjects. This time, the three tests gave consistent findings. On the forced choice experiment, males reported more distress at the thought of sexual infidelity and females more distress at the thought of emotional infidelity. The physiological measures used by the group were heart rate, skin conductance, skin temperature and frowning activity. Relative to measures of these variables when neutral imagery was used, men demonstrated a greater increase than women in heart rate, skin conductance and frowning during sexual infidelity imagery. In contrast, and as predicted, women were more responsive than men on all these measures in response to images of emotional infidelity. Figure 10.10 shows the results for four measures on the continuous rating scale measurements in relation to sexual and emotional infidelity.

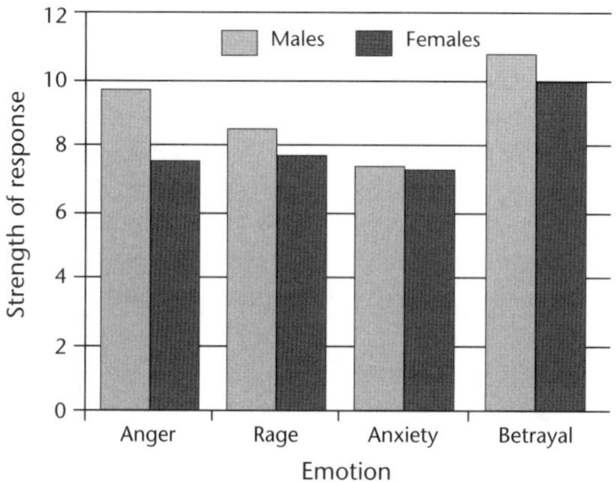

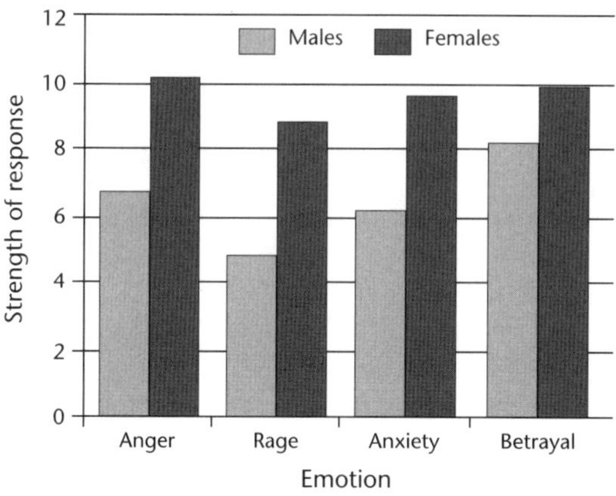

Notes
(a) For sexual infidelity, differences (M > F) were significant for anger ($p < 0.001$), rage ($p < 0.001$) and betrayal ($p < 0.01$).
(b) For emotional infidelity, differences (F > M) were significant for anger ($p < 0.01$) and anxiety ($p < 0.02$).

FIGURE 10.10 Differences between males and females in their emotional response to sexual and emotional infidelity scenarios

SOURCE: Plotted from data in Pietrzak et al. (2002) 'Sex differences in human jealousy: a coordinated study of forced-choice, continuous rating-scale, and physiological responses on the same subjects.' *Evolution and Human Behaviour* **23**(2): 83–95.

Although the results were internally consistent and as predicted from evolutionary expectations, the drawback remains that the investigation was carried out on a narrow subject group: US college undergraduates from a largely middle-class background. It remains to be seen if similar findings are obtained across other age and social class groups and other cultures.

10.4.3 Divorce and remarriage

A failed marriage, thankfully, rarely ends up in the homicide statistics, but patterns of divorce and remarriage also provide a convenient testing ground for evolutionary hypotheses. Divorce statistics are in close agreement with findings on published advertisements in which men seek looks and youthfulness while women are more concerned with status and resources (see Chapter 12). Buckle et al. (1996) examined statistics for divorce and remarriage for populations in Canada, England and Wales. They found that, in stating grounds for divorce, men were more concerned about adultery while women were more concerned about cruelty. This is interpreted in the language of evolutionary psychology as paternity certainty and care for offspring respectively. This seems to be a robust finding and Figure 10.11 shows the grounds for divorce in England and Wales in 1995.

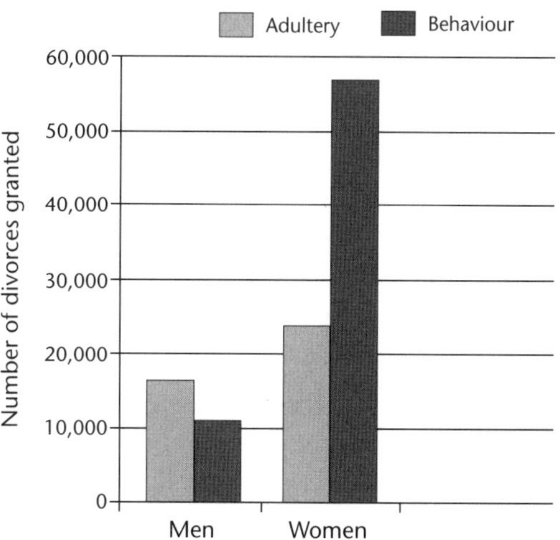

FIGURE 10.11 Percentage of divorces as a function of the grounds given by male and female petitioners for England and Wales, 1995

Women were also much more likely to terminate a marriage at an early stage than men; in fact, over 90 per cent of all divorces in England and Wales (mostly from 1974 to 1989) were initiated by females under 25. This is just what would be expected in the light of the shorter reproductive span of females compared with that of males. If a woman suspects that she has made a bad decision and she wants children, it is imperative to dissolve the marriage in the early stages. Waiting reduces her reproductive potential and her future marriageability. The opposite effect is observed for males: as they get older, they are more likely to seek a divorce (Figure 10.12). Men remain fertile for longer and could raise a second family with a younger wife. Predictably, on remarrying, men seek wives about six years younger than themselves, whereas for first marriage the figure is only two years.

The consistency of such data with evolutionary expectations reinforces the idea of marriage as a 'reproductive contract' in which men and women pursue different strategies to optimise their reproductive fitness. Humans are not the only animals to dissolve the pair bond in the case of childlessness. Ringdoves, for example, are generally monogamous during the breeding season but experience a 'divorce rate' of about 25 per cent. The major reason seems to be infertility: if the couple fail to produce chicks, they separate and look for other mates (Erickson and Zenone, 1976).

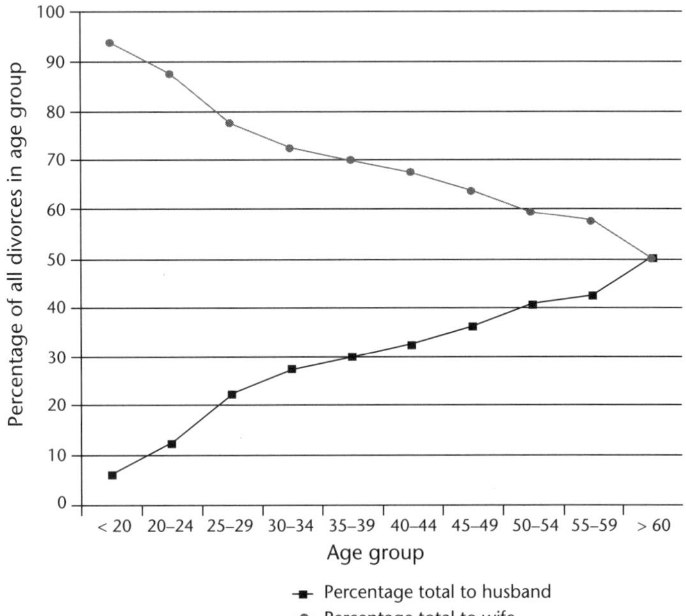

FIGURE 10.12 Percentage of divorces granted in England and Wales, 1995

SOURCE: Office of Population Census and Surveys.

Summary

- A knowledge of genetic relatedness (*r*) is useful in helping to understand both altruistic and antagonistic behaviour. Conflict will arise when the reproductive interests of two individuals differ. This is predicted between sexual partners and even between parents and offspring, since offspring may demand resources at the expense of current and future siblings.
- Maternal–fetal conflict provides a good example of the dynamic balance that is reached when the optima for resource allocation differ between two genomes.
- Human violence resulting in death provides a set of data that allows predictions from evolutionary psychology to be tested. There is evidence, for example, that the risk of a child being killed by its mother declines as the ages of the child and the mother increase. These facts are in keeping with the concept of reproductive value, but there may be other factors at work.

- Whereas kinship is usually predicted to reduce levels of violent behaviour, there were probably circumstances in the environment of evolutionary adaptation of early hominids in which, for example, infanticide served as an adaptive strategy to maximise the lifetime reproductive value of a parent. One of the most robust findings of the work of Daly and Wilson is that stepchildren experience a much higher risk of infanticide that genetic children. Psychological mechanisms that were once adaptive in the life history of early hominids may, when activated, now give rise to maladaptive consequences in cultures in which homicide is punished. Daly and Wilson have published pioneering work on homicide, but the nature of the statistics used makes it difficult to control for all the social and cultural factors at work.
- Much of the coercion and violence exercised by men over women can be understood in terms of paternity assurance. Marriage, from a Darwinian perspective, can be regarded as a reproductive contract between men and women to serve similar but, in important ways, different interests. Men have probably been selected to monitor and jealously guard the sexuality of their partners, whereas it is probable that women are more concerned about resources. Jealousy is an emotion experienced by both sexes but with more violent repercussions in cases where men suspect sexual infidelity on the part of their partner. Divorce statistics are consistent with the contract hypothesis and reflect age-related fertility differences between men and women.

Key Words

- **Cuckoldry**
- **Infanticide**
- **Maternal–fetal conflict**
- **Parent–offspring conflict**
- **Placenta**
- **Siblicide**

Further reading

Daly, M. and Wilson, M. (1988a) *Homicide*. New York, Aldine de Gruyter.
Daly, M. and Wilson, M. (1998) *The Truth About Cinderella*. London, Orion.

Both these books contain the ground-breaking interpretations of crime statistics made famous by Daly and Wilson.

Mating and Mate Choice

Human Sexual Behaviour: Mating Systems and Mating Strategies

We must acknowledge, as it seems to me, that man with all his noble qualities, with sympathy which feels for the most debased, with benevolence which extends not only to other men but to the humblest living creature, with his god-like intellect which has penetrated into the movements and constitution of the solar system – with all these exalted powers – Man still bears in his bodily frame the indelible stamp of his lowly origin.
(Darwin, 1883, p. 619)

Man consists of two parts, his mind and his body, only the body has more fun.
(Woody Allen, 1975, *Love and Death*)

The first part of this chapter examines the physical and historical evidence on human sexuality by employing perspectives often found in biological or physical anthropology, perspectives that routinely focus on Darwin's 'indelible stamp'. Its goal is to establish the species-typical mating strategies employed by human males and females. It has already been established in Chapter 3 that males and females, by virtue of basic differences in biology, are likely to have different interests, so it should come as no surprise to find that these strategies are different. We should also expect these strategies to vary with local conditions, since optimally adaptive behaviour is likely to be context sensitive. Consequently, the second half of this chapter looks at how male and female mating inclinations vary with social circumstances.

11.1 Contemporary traditional or preindustrial societies

11.1.1 Cultural distribution of mating systems

Humans living in today's industrial or more developed countries are living in conditions far removed from those prevailing in environments where the basic plan of the human genotype was forged. In addition, many such cultures are strongly influenced by relatively recent ideologies and belief systems, such as Judaism, Christianity and Islam, with their injunctions in favour of or against specific mating systems (Judaism, Christianity and Islam prohibit polyandry, for example, Christianity and (by and large) Judaism prohibit polygyny, but polygyny is permitted under Islam). If we want to ascertain the sexual behaviour of humans before our mode of life was transformed by industrialisation, or before our heads were filled with strict religious and political ideologies, it makes sense to look at traditional hunter-gatherer cultures that still exist, or cultures that have been relatively immune to Western influence and maintain their traditional patterns of life. This is not to suggest that such cultures are primitive or have no ideas of their own, but we could be assured that such cultures have not been subject to the mass persuasion systems of the Church and state found in the West.

A broad sweep of different human societies reveals that, in many, the sexual behaviour observed departs from the monogamy advocated (at least in a legal sense) in most Western cultures (Figure 11.1).

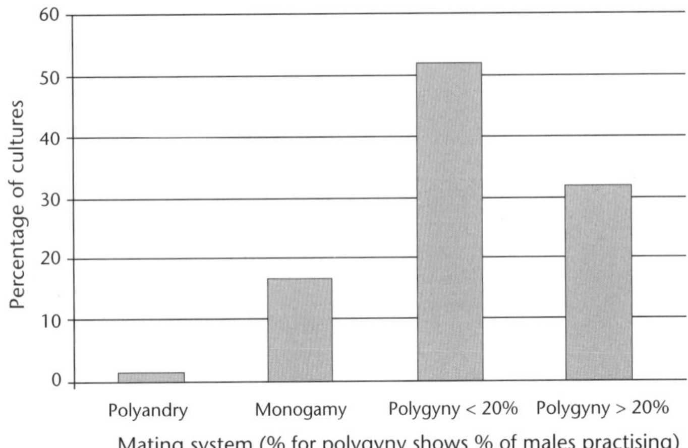

FIGURE 11.1 Human mating systems in 186 traditional cultures prior to Western influence

SOURCE: Data from Murdock, G. P. and White, D. R. (1969) 'Standard cross-cultural sample.' *Ethnology* **9**: 329–69.

Another fruitful line of inquiry is to look at contemporary hunter-gatherer societies. Here we at least know that people face ecological problems similar to those of our ancestors 100,000 years ago. The often harsh existence of hunter-gatherers also places restrictions on the sort of social development that can take place. Patterns of mating behaviour cannot drift too far from that which is optimal for survival.

11.1.2 Hunter-gatherer societies

It is a sobering thought that at least 50,000 generations of differential survival and reproduction took place in a foraging way of life. No hominin subsisted in any other way until the invention of agriculture about 10,000 years ago. In studying hunter-gatherers today, however, we meet a set of problems different from those prevailing in the Pleistocene Epoch. Although humans have been hunter-gatherers for 99 per cent of their time on the planet, contemporary hunter-gatherers are mostly confined to marginal environments, such as deserts or sub-Arctic tundra, rather than the more hospitable environments that have, for obvious reasons, been the subject of development and colonisation. Contemporary hunter-gatherers may therefore be 'survivors' from the inroads of agriculture, and we must be wary that they may not be entirely representative of what hunter-gathering was once like.

Bearing such problems in mind, we note that studies, such as those by Howell (1979) on the !Kung San people, tend to reveal a pattern of mild polygyny, males having a slightly higher variance in reproductive success than females because of serial polygyny resulting from remarriage or simultaneous polygyny.

Foley (1992) argues that the distribution of modern hunter-gatherers in tropical Africa is confined to areas of low and high rainfall where large mammals are not particularly abundant. It is probable therefore that contemporary hunter-gatherers live in environments depleted in large herbivores and carnivores, relative to those of our ancestors, and we must bear in mind that hunting was once probably more important than it is today.

It turns out that food supply and the role of hunting in obtaining food are important factors in understanding the mating strategies of hunter-gatherers. The Ache people of Paraguay were hunter-gatherers until 1971, when they were enticed to live on a government reservation. Studies by Hill and Kaplan (1988) showed that men would often donate meat to women in exchange for sex, high-ranking men gaining most from this practice. In this case, it seems that some Ache men achieved polygyny through meat-induced adulterous affairs. As a general rule, however, it is highly likely that a foraging way of life, especially where hunting produced an important part of the diet, never really sustained a high degree of polygyny.

The reasons for this are basically twofold and fairly simple. First, hunting large animals is risky and needs a combination of cooperation and luck. Given the prolonged period of gestation and nurturing for human infants, hunting is carried out by males, and the cooperation needed means that male rivalry must be kept within strict limits. Following a kill, the meat must be shared between all those who helped and also with other unsuccessful groups, along the lines of reciprocal altruism. If a high degree of polygyny prevailed in such groups, the sexual rivalry would militate against such altruism. In fact, the equitable sharing of hunted food is characteristic of hunter-gatherers and totally unlike that of other social hunting species where after a kill there is a free-for-all. Second, even if there were a surplus after sharing, meat is difficult to store. It is hard to see how, in a foraging culture, sufficient wealth or resources could ever be accumulated by one man to support a sizeable harem. Predictably, polygyny has been found to be pronounced or common in very few known hunter-gatherer societies. In most hunter-gatherer groups, men will have one or at most two wives (Figure 11.2).

The evidence above on the difficulty of accumulating resources and the need for cooperative hunting does not rule out the possibility that, if resources were more abundant in the late Pliocene Epoch and thus easier to accumulate by a few men, a higher degree of polygyny might have been found. Foley (1996) takes this view, arguing that the social structures we observe in today's hunter-gatherers are not vestigial but represent novel adaptations to a resource-depleted, post-Pliocene environment. Foley argues that, as the Neolithic revolution left the hunter-gatherers outmoded, it was the agriculturists who would have maintained the ancestral social system of the Pliocene Epoch, consisting of 'polygynous family groups linked by alliances of male kin organised patrilineality' (Foley, 1996, p. 108). If Foley is right, there may be some merit in examining social systems of early agricultural communities (see Box 11.1).

FIGURE 11.2 Yanomami Amerindians washing vegetables
The Yanomami, living in the rainforests of southern Venezuela and northern Brazil, are tribal people practising gardening and hunting. They number about 15,000 individuals and have no written language or formally coded laws. Yanomami males are often polygynous and this results in violent conflicts between males as they compete for wives.

Box 11.1 Power, wealth and sex in early civilisation

In the USA, Mildred Dickemann, John Hartung and Laura Betzig have all pioneered the Darwinian approach to human history. Betzig (1982, 1986) examined six civilisations of early history: Babylon, Egypt, India, China, the Incas and the Aztecs. She found that, in all of them, an accumulation of power and wealth by a ruling elite coincided with the prodigious sexual activity of the rulers. All these societies at some stage of their history were ruled by male despots or emperors who kept harems (Figure 11.3). The harems of these male rulers were vigorously defended and guarded by eunuchs, extreme penalties being meted out to any subject who had sexual relations with the ruler's concubines. This degree of polygyny, consisting of hundreds or even thousands of wives, would be unimaginable in hunter-gatherer societies. Economically, such harems were made possible by the Neolithic revolution that enabled the taxation of the majority to support the retinue of a minority. In short, harems were favoured by social inequality. As Betzig (1992, p. 310) concludes: 'across space and time, polygyny has overlapped with despotism, monogamy with egalitarianism'.

In seeking to explain this phenomenon, it could be argued that such harems were displays of wealth – items of conspicuous consumption that gave pleasure to the ruler as well as broadcasting his power. Each of these reasons may offer a partial explanation, but several features of the regulation of the harems do not fit easily into such conventional accounts. Betzig shows how the structure of the harem seems to be designed to ensure the maximum fertility of the women concerned as well as ensuring absolute confidence of paternity for the despot. In some cases, wet nurses were employed, thereby allowing the harem women to resume ovulation soon after the birth of a child. In the civilisations of Peru, India and China, under the emperors Atahallpa, Udayania and Fei-ti respectively, great care was taken to procure only virgins for the harem. In the Tang Dynasty of China, careful records of the dates of conception and menstruation of the women were kept as a means of ascertaining their fertility. In this light, harems appear as breeding factories designed to maximise the propagation of an emperor's genes.

Betzig (1992) also studied the aristocracy of the early Roman Empire. Roman marriage was monogamous by law, but men still found ways to secure extrapair copulations. Historical sources such as Tacitus and Seutonius consistently speak of the voracious sexual appetite of the early emperors and how they were provided with virgins and concubines. Wealthy Romans also kept male and female slaves, even though few of the female slaves had real jobs around the household. Betzig rejects the suggestion that female slaves were employed to breed more slaves, since male slaves were forced to remain celibate and pregnant female slaves did not command a higher price. Betzig argues that female slaves were used by noblemen to breed their children. The fact that slaves born in a Roman household were often freed with an endowment of wealth (unlike the slaves used in mines) suggests that noblemen were freeing their own offspring.

Exactly why extreme polygyny faded is not certain. The extreme polygyny of harems seems to represent an interlude between the end of hunter-gathering and the spread of democracy. Alexander (1979) posits that intense polygyny was destabilised by the inability of the ruling polygynists to command the loyalty of deprived and frustrated foot soldiers. Ridley (1993) suggests that the rise of democracy allowed the ordinary man to express his resentment at the sexual excess of others. Even today in some countries, the sexual excesses of men elected to powerful positions can bring about their downfall. Whatever the historical reasons, no doubt complex, it is fairly clear that the extreme polygyny practised by ancient despots is not typical of the human condition for most of its history. Such evidence does show, however, how males can act opportunistically to achieve extreme polygyny in conditions favourable to their reproductive interests.

FIGURE 11.3 ***Favourites of the Harem* (Constantinople, *c.*1900)**
Throughout history, powerful men have employed harems and concubines. As well as serving as a status symbol for the male and providing him with sexual pleasure, they seem to be expressly designed to ensure the propagation of his genes.
SOURCE: Library of Congress Prints and Photographic Division, LC-USZ61-939.

11.2 Physical comparisons between humans and other primates

> The next time a new species of primate is discovered we should be able to deduce its social behaviour by examining its testes and dimorphism in body and canine size. (Reynolds and Harvey, 1994, p. 66)

The above claim is indeed a remarkable one, since it suggests that complex social characteristics can be inferred from simple measurable parameters. If the claim is reliable, the same technique should also, since humans have a shared ancestry with the primates, throw considerable light on ancestral human sexual behaviour. We will now examine the potential of this claim. We will show how it applies to primate sexual behaviour and then to human.

11.2.1 Body size dimorphism

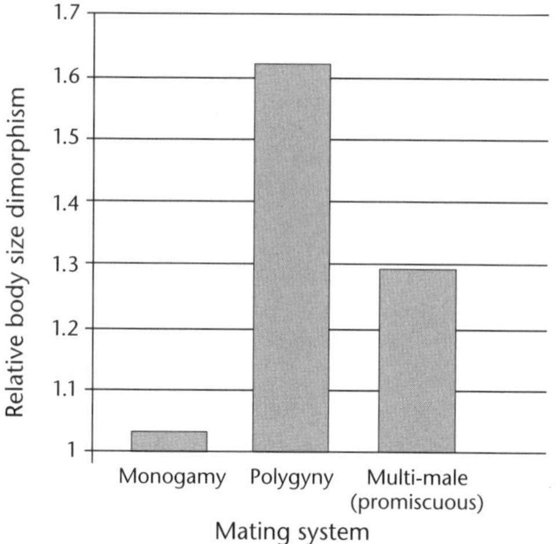

FIGURE 11.4 Body size dimorphism versus mating system for various species of primate

SOURCE: Adapted from Harvey, P. H. and Bradbury, J. W. (1991) Sexual selection, in J. R. Krebs and N. B. Davies (eds) *Behavioural Ecology*. Oxford, Blackwell Scientific.

Figure 11.4 shows how body size dimorphism (adult body weight of male divided by adult body weight of female) varies in relation to breeding system for primates. This variation is consistent with the idea that the large differences between measures of dimorphism in monogamous and polygynous (single-male) contexts can be explained by the more intense competition between males for females in the latter. Even in multi-male groups, some competition is observed to secure a place in dominance hierarchies.

11.2.2 Testis size

The significance of **testis** size is that it indicates the degree of sperm competition (see Chapter 3) in the species. In the 1970s, the biologist R. V. Short suggested that the difference in testis size for primates could be understood in terms of the intensity of sperm competition. To obtain reliable indicators, testis size has to be controlled for body weight since larger mammals will generally have larger testes in order to produce enough testosterone for the larger volume of blood in the animal, and a larger volume of ejaculate to counteract the dilution effect of the larger reproductive tract of the female.

When these effects are controlled for, and relative testis size is measured, the results support the suggestion of Short that relatively larger testes are selected for in multi-male groups where sperm competition will take place in the reproductive tract of the female (Figure 11.5). A single male in a harem does not need to produce as much sperm as a male in a multi-male group since, for him, the battle has already been won through some combination of body size and canine size, and rival sperm are unlikely to be a threat. In contrast, in promiscuous multi-male chimpanzee groups, females will mate with several males each day when in **oestrus**. It is possible that the sexual swellings that advertise oestrus in many female primates, which males find irresistible, actually promote sperm competition, since the best way for a female to produce a son who is a good sperm competitor is to encourage competition among his potential fathers.

A rival hypothesis to sperm competition is that of sperm depletion. The argument here is that large testes are needed by males who engage in frequent sexual activity to replenish depleted supplies of sperm. This hypothesis fails when tested on bird species. From Figure 11.6, we can see that relative testis size in polyandrous birds is higher than in leking species. Yet in polyandry, the number of females per male is low, but in leking species males will mate with many females.

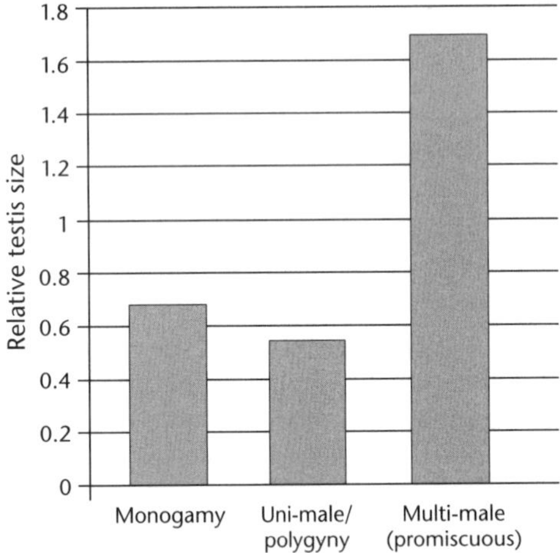

FIGURE 11.5 Relative testis size versus mating system for various species of primates
The y-axis refers to deviations from the allometric line of testis size against body weight. Thus, a value of 1 would indicate a testis size expected for an average primate of a given body weight. Values above 1 indicate that testes are larger than would be expected. The values are for genera rather than individual species. The difference between monogamous and single-male groups is not significant, but both differ significantly from multi-male groups (see Harcourt et al., 1981).

SOURCE: Adapted from Harvey, P. H. and Bradbury, J. W. (1991) Sexual selection, in J. R. Krebs and N. B. Davies (eds) *Behavioural Ecology*. Oxford, Blackwell Scientific.

A better explanation for these differences is that, in polyandry, high levels of sperm competition are found since a female is impregnated by more than one male. The word **lek** comes from the Swedish word meaning play. In behavioural ecology, it refers to species where a male will display before females, and females will then choose the most successful male. That male will then copulate with many females and the females will ignore other males. Lekking species include sage grouse, black grouse and peacocks (who display their train). Leks entail low levels of sperm competition since females mate with only the one successful male. We should also note from Figure 11.6 that male birds in monogamous colonies have larger testes than those in solitary monogamous breeding situations. Birkhead and Moller (1992) suggest that colonies provide ample opportunities for 'extrapair copulation'. Males and females may be ostensibly monogamous in terms of caring for offspring, but both are willing to undertake adulterous affairs, hence the need for larger testicles in the male.

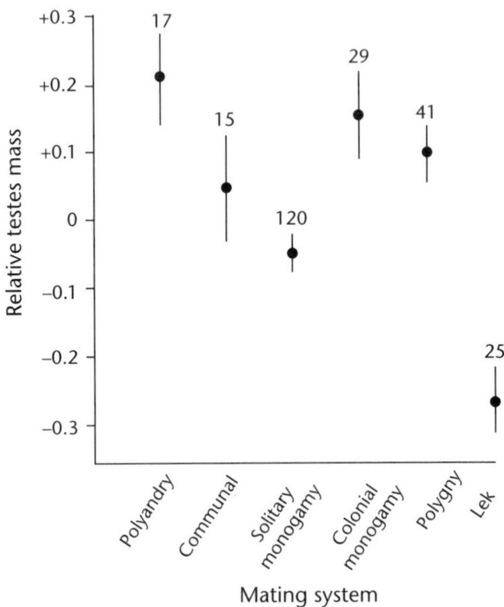

FIGURE 11.6 Relative testes mass in relation to mating systems for various species of birds
The y-axis represents the deviation from the value expected for birds of a given mass. A high plus value indicates that testes are heavier than would be expected for birds of this body weight. A minus value indicates testes are smaller than would be expected for birds of a given average body weight.

SOURCE: After Birkhead, T. R. and Moller, A. P. (1992) *Sperm Competition in Birds: Evolutionary Causes and Consequences*. London, Academic Press.

11.2.3 Testis size and bodily dimorphism applied to humans

Diamond (1991, p. 62) has called the theory of testis size and sperm competition 'one of the triumphs of modern physical anthropology'. As we have seen, the theory has great explanatory power, and we will now apply it to humans.

Table 11.1 shows some key data on testis size and bodily dimorphism for humans and the great apes. The fact that men are slightly heavier than women could reflect a number of features of our evolutionary ancestry. It could indicate the protective role of men in open savannah environments, it could be the result of food gathering specialisation whereby men hunted and women gathered, or it could reflect male competition for females in uni-male or multi-male groups. The dimorphism for humans is mild, however, compared with that for gorillas; this would indicate that *Homo sapiens* did not evolve in a system of uni-male harem mating.

Also, if early humans routinely competed to control groups of females, we would not only expect a higher level of body size dimorphism but also smaller testes. The testes of gorillas relative to their body size are less than half the average for humans. On the other hand, if early humans had behaved like chimpanzees in multi-male groups, we would expect larger testes. In fact, if human males had the same relative size of testis as chimps, their testes would be roughly as large as tennis balls. Figure 11.7 shows a comparative and schematic representation of human, gorilla, orang-utan and chimp males. The size of the large circle relative to the female shows the degree of sexual dimorphism in the species. The length of the arrows and the pair of dark shapes show the relative size of the penises and testes of the males. One unexplained feature of male morphology is the large size of the human penis.

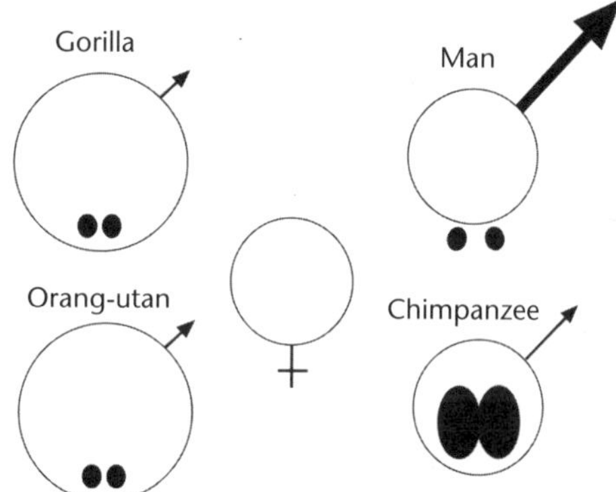

FIGURE 11.7 Body size dimorphism: female's view of males, illustrating the relative size of the body (circle), penis (arrow) and testes (dark ovals)

SOURCE: Adapted and redrawn from Short, R. V. and Balban, E. (eds) (1994) *The Differences Between the Sexes*. Cambridge, Cambridge University Press.

Table 11.1 Physical characteristics of humans and the great apes in relation to mating and reproduction

Species	Male body weight (kg)	Female body weight (kg)	Dimorphism M/F	Mating system	Weight of testes (g)	Weight of testes as % of body weight	Approx. number of sperm per ejaculate (x 10^7)	Estimated global population
Humans *(Homo sapiens)*	70	63	1.1	Monogamy and polygyny?	25–50	0.04–0.08	25	More than 6,000,000,000
Common chimps *(Pan troglodytes)*	40	30	1.3	Multi-male in promiscuous groups	120	0.3	60	Fewer than 110,000
Orang-utan *(Pongo pygmaeus)*	84	38	2.2	Uni-male Temporary liaisons	35	0.05	7	Fewer than 25,000
Gorilla *(Gorilla gorilla)*	160	89	1.8	Uni-male Polygyny	30	0.02	5	Fewer than 120,000

NOTE: The last column shows the precarious state of the natural populations of our nearest relatives.

SOURCE: Data from Harcourt, A. H., Harvey, P. H. et al. (1981) 'Testis weight, body weight and breeding system in primates.' *Nature* **293**: 55–7; Foley, R. A. (1989) The evolution of hominid social behaviour, in V. Standen and R. A. Foley (eds) *Comparative Socioecology*. Oxford, Blackwell Scientific; Warner, H., Martin, D. E. et al. (1974) 'Electroejacualtion of the great apes.' *Annals of Biomedical Engineering* **2**: 419–32.

From a comparison of human testis size with other primates, Short (1994, p. 13) concludes that we are not 'inherently monogamous ... neither are we adapted to a multimale promiscuous mating system'. His view is that 'we are basically a polygynous primate in which the polygyny usually takes the form of serial monogamy'. This is to examine the problem from a male perspective of course. The fact that humans do have relatively large testicles (larger than gorillas) does imply that sperm competition has been a feature of our ancestry. Now 'it takes two to tango', as the saying goes; so if sperm have competed in our past, then women must have mated (polyandrously) with more than one man. But as we saw in Chapters 3 and 5, infant humans need biparental care and this is facilitated by a strong pair bond. The way to reconcile these differing indications might be to conclude that ancestral human mating was ostensibly monogamous but plagued by 'adultery' or covert extrapair copulations. Not so different, if we take a clinical view, from the situation in many societies today.

The evidence from immunology

Recent immunological research has also thrown light on the likely patterns of mating among early humans. Nunn et al. (2000) examined the number of white blood cells in 41 species of non-human primate and found that the greater the number of mating partners that females had, the higher the white blood cell count. They reasoned that the basis for this was that 'promiscuous' species were susceptible to sexually transmitted diseases and so required a more complex immune system. The authors observed that the immune system of humans was more like the polygamous gorilla *(Gorilla gorilla)* and the monogamous gibbon *(Hylobates lar)* than the chimpanzee; suggesting that humans are marginally polygamous (uni-male, multi-female) and marginally monogamous (uni-male, uni-female) rather than multi-male–multi-female in their mating behaviour.

The size of male testes and bodily dimorphism may yield useful information about patterns of human mating, but it could also be expected that the female body might hold some clues about the mating propensities of early *Homo sapiens*. One notable aspect of human female sexuality is that women are continually sexually receptive throughout the **ovarian cycle**. Another is that the precise moment of **ovulation** is concealed from both male and female. The implications of this are explored in Box 11.2.

Box 11.2 Concealed ovulation

Couples who are desperate to have a baby know only too well that it is difficult to pinpoint the precise time at which a woman ovulates. Ovulation is concealed from both the woman and the man. Clear testimony to this is the lucrative business of manufacturing contraceptives on the one hand and test kits for ovulation on the other. Yet humans are part of a minority among mammals in this respect. In most mammalian species, oestrus is announced with an explosive fanfare of signals. In the case of female baboons, for example, the skin around the vagina swells and turns bright red, she emits distinctive odours and, just to reinforce the point, she presents her rear to any male she happens to fancy.

The concealment of ovulation is a feature that humans share with about 32 other species of primate, but for at least 18 other species, ovulation is advertised boldly and conspicuously. We can confidently expect concealment and advertisement to have some adaptive significance, and a number of intriguing theories have been proposed to account for both. Concealed ovulation (**sexual crypsis**) in humans has led to a great variety of hypotheses. The problem is not a trivial one, since the concealment of ovulation only serves to make sexual activity even more inefficient. As sexual activity is costly to an organism (in terms of energy and time expended, exposure to risks of transmission of disease and vulnerability to predators), there must be some sound evolutionary reasons for masking the time when sex could be more productive. It is likely therefore that concealed ovulation is part of a female's mating strategy.

A variety of theories has been proposed:

- One is the 'sex for food' hypothesis. Given the ubiquity of prostitutes in history, it has been suggested by Hill (1982) that sexual crypsis allowed early female hominids to barter sex for food. If ovulation were conspicuous, a male would know when a gift in exchange for sex would bring him some reproductive advantage, and he could withhold such gifts if the woman were not fertile. Concealing ovulation allows women to be in a position almost constantly to exchange sex (with the prospect of paternity) for resources. Hill argues that there is ample evidence in the ethnographic literature that human males trade resources such as meat for sex. Women gain not only valuable nourishment, but also the opportunity to copulate with some of the best males in the group, since males good at provisioning are also presumably able in other respects. There is some evidence that, in hunter-gatherer societies, this still happens, in that women will exchange sex with the best hunter in return for meat (Hill and Kaplan, 1988).
- An alternative to this is the 'anticontraceptive hypothesis'. Burley (1979) has suggested that sexual crypsis evolved after women had developed enough cognitive ability to associate conception with copulation. If women were aware of ovulation and could thus refrain from sex to reduce the risks of childbirth to themselves, they would leave fewer descendants than those who were unaware and hence could not exercise this choice. In this view, concealed ovulation is the product of genes outwitting consciousness, the triumph of matter over mind.
- Benshoof and Thornhill (1979) suggest that, by concealing ovulation from her ostensibly monogamous partner, a female can mate with another male (who may appear to have superior traits) without alerting her first mate. They suggest that a woman may unconsciously 'know' when she is ovulating, and this could help her to assess when an extramarital liaison would be rewarding. In this scenario, sexual crypsis becomes a strategy employed by the female to enable her to extract paternal care from a male who thinks that she will bear his children while enjoying the ability to choose what she regards as the best genetic father. All males are potentially genetic fathers, but not all males are able to provide paternal care; some, for example, may already be paired off.
- Richard Alexander and Katherine Noonan at the University of Michigan have proposed a view of sexual crypsis as a tactic developed by women to divert men from a strategy of low-investing, competitive polygyny towards a more caring and high-investing monogamy. For a male to be assured of paternity, he needs to remain in close proximity to his partner for prolonged periods of time. It is no use a man wandering off to have sex with another woman, since she may not be ovulating and his wife back at home may become the subject of the attentions of like-minded philandering males. So the male stays at home, finds his partner constantly desirable and has the reward of a high degree of confidence in his paternity. This is the so-called 'daddy at home' hypothesis (Alexander and Noonan, 1979).
- It is common in primates to find males killing offspring that are not their own in order to bring the bereaved mother into oestrus again. Support for the need for females to evolve a countermeasure against this comes from studies on the 'postconception oestrus' of grey langurs *(Presbytis entellus)* and red colobus monkeys *(Colobus badius)*. In both of these species, an oestrus signal is given even when the females are pregnant. This could be interpreted as a measure to confuse males; if so, it is one of the few examples of 'dishonest signals' given by females of their reproductive status. With this in mind, Hrdy (1979) has suggested that sexual crypsis served to confuse the issue of paternity for early hominids and thus prevent infanticide by males who, if they were confident that they were not the genetic fathers of offspring, could engage in this practice. This is sometimes known as the 'nice daddy' theory.

There is clearly no shortage of ideas to account for concealed ovulation, and the literature is vast and difficult. Even a superficial glance at primate sexual behaviour shows that concealed ovulation can be found in a variety of mating systems, such as those of monogamous night monkeys, polygynous langurs and multi-male vervets. It is entirely feasible, therefore, that the various hypotheses are not entirely exclusive. One, such as the 'nice daddy' theory, could explain the origin of concealment, and another, for example the 'daddy at home', its maintenance. In disentangling these arguments, much progress has been made by two Swedish biologists, Birgitta Sillen-Tullberg and Anders Moller (1993). By looking at the probable phylogeny of the anthropoid primates, they claim to be able to evaluate the various hypotheses.

The crucial test is of course when concealed ovulation first appeared. Using procedures to disentangle primitive, derived and convergent characteristics, Sillen-Tullberg and Moller constructed a phylogenetic tree of changes in the visual signs of ovulation. They concluded that the primitive state for all primates was probably slight signs of ovulation, and that concealment had evolved independently 8–11 times (the range indicating the effect of slightly different assumptions in the modelling).

Sillen-Tullberg and Moller draw three sets of conclusion from their work:

1. Ovulatory signals disappear more often in a non-monogamous context. This tends to support Hrdy's (1979) infanticide hypothesis.
2. The fact that, once monogamy has been established, ovulatory signals do not usually disappear throws doubt on the 'daddy at home' hypothesis of Alexander and Noonan (1979) as the origin of concealed ovulation.
3. Monogamy evolves more often in lineages that lack ovulatory signals. The lack of ovulatory signals could be an important condition for the emergence of monogamy.

In short, if we accept the conclusions of Sillen-Tullberg and Moller, it seems that concealed ovulation may have changed and even reversed its function during primate evolution. It began as a female strategy to allow ancestral hominid women to mate with many males yet remain secure in the knowledge that the confusion over paternity would protect their infants from infanticide. Once concealed ovulation was established, a woman could then choose a resourceful and caring male and use concealed ovulation to entice him to stay with her to provide care. Women became continually sexually active, and men correspondingly found women continually desirable – even without signs of ovulation. It is an interesting thought that the continual sexual receptivity of human females and the high frequency of non-reproductive sex in which humans engage began as a female strategy to outwit men. An ancestral male, anxious to secure his paternity, could no longer afford the time budget to guard a group of females and prevent sexual access by other males. Instead, he was forced into a monogamous relationship, providing care for his partner and what he thought were his own offspring, in return expecting sexual fidelity from his wife. This in turn led to a male psychology that was particularly sensitive to any evidence of unfaithfulness, with repercussions for the emotional life of men. Such is the origin of the strength of sexual jealousy that today it still accounts for a good deal of human conflict (see Chapter 10).

11.3 Pluralistic sexual strategies

Chapters 3 and 5 showed that humans seem to have adaptations and life history features associated with monogamy, such as altricial offspring, strong pair bonds and highly selective mate choice criteria. On the other hand, some evidence, such as ethnographic data, physical sexual dimorphism and male testis size, points towards simultaneous or serial polygyny or even moderate polygynandry. Some evolutionary thinkers such as David Buss and David Schmitt try to reconcile these findings by suggesting that humans are pluralistic in their mating strategies, with separate design features for long- and short-term mating. This approach is sometimes called the **sexual strategies theory** (SST).

In this framework, long-term mating is characterised by a heavy investment by both partners, loving feelings and a prolonged courtship period. Short-term mating is conceived as a brief encounter or one-night stand or extrapair mating.

11.3.1 Sex differences in long- and short-term mating strategies

It is to be suspected that there will be sex differences in the structure of short- and long-term mating strategies consistent with what we have already discussed about differences in male and female reproductive biology and the differential costs and benefits of alternate strategies to both sexes. Table 11.2 shows some expectations.

Table 11.2 Differences in male and female long- and short-term mating strategies

	Males will seek	Females will seek
Long term	Signs of youth and fertility Desirable physical appearance, especially waist to hip ratio Signs of postmarital sexual fidelity	Status Resources Ambition Maturity Kindness Willingness to invest resources Dependability
Short term	Large number of partners Short time delay between meeting and intercourse Men will accept lower acceptable standards than women for a mating opportunity Extramarital affairs sought	Dominance Secondary sexual characteristics related to testosterone, indicating high genetic quality

The full reasoning behind this table is further explored in Chapter 12, but here we may note that the features listed derive from expectations about how each sex can further their reproductive sex. A woman cannot physically raise more than a few children in her lifetime, so she will look for signs of genetic quality in her mates and the ability of her mates to provide resources to care for her offspring. Men will be alert to signs of fertility (since the fertility of women falls more rapidly with age than that of men) and fidelity (since males face the problem of paternity certainty; see Chapter 10). Men will also be interested in a larger number of partners than women since male reproductive output is not constrained by their biology but by the number of women he can persuade to have sex with him.

Evidence for one point in Table 11.2 – the fact that men actively seek more partners than women – comes from a remarkably thorough survey, sampling the responses of over 16,000 people across 52 nations and 6 continents, which was coordinated by Schmitt. Figure 11.8 shows differences in the number of desired sexual partners between men and women over increasing periods of time.

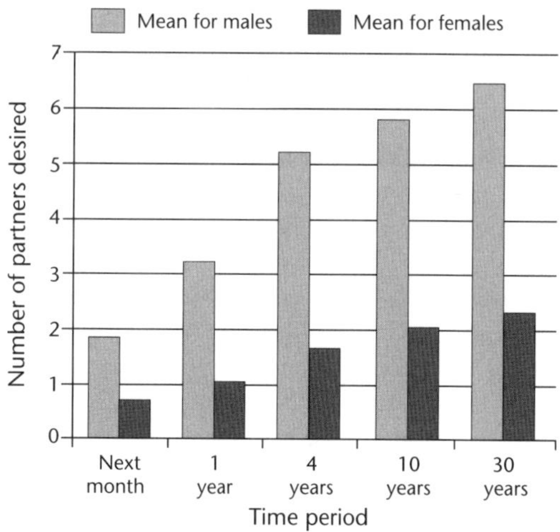

FIGURE 11.8 A comparison of the number of sexual partners desired over different time intervals for males and females using a large cross-cultural sample
The differences were significant ($p < 0.001$) at each interval.

SOURCE: Data from Schmitt, D. (2003) 'Universal sex differences in the desire for sexual variety: tests from 52 nations, 6 continents and 13 islands.' *Journal of Personality and Social Psychology* **85**(1): 85–104.

Figure 11.9 also shows that there was some cultural variation in this measure but, overwhelmingly, men desired more partners than women.

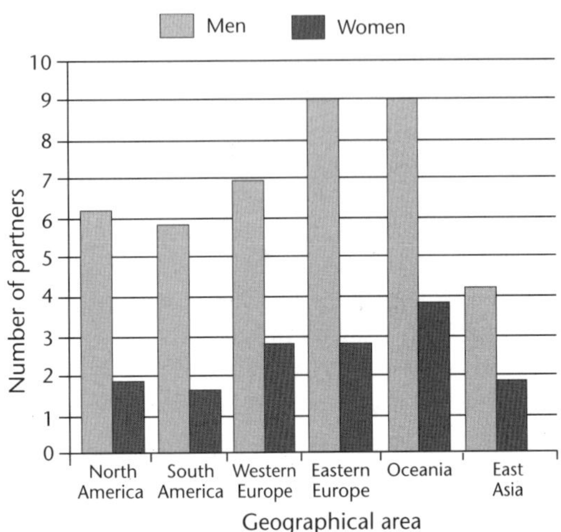

FIGURE 11.9 Mean number of sexual partners desired by men and women over 30 years

SOURCE: Schmitt, D. (2003) 'Universal sex differences in the desire for sexual variety: tests from 52 nations, 6 continents and 13 islands.' *Journal of Personality and Social Psychology* **85**(1): 85–104, Figure 1, p. 92.

11.3.2 Sexual strategies in relation to context

Schmitt used something called the 'sociosexual orientation inventory' (SOI), as devised by Simpson and Gangestad (1991), to investigate how differences in mating strategies between men and women relate to a range of cultural parameters. The inventory poses a series of questions and the responses are quantified to give an overall index of sociosexuality. A low score is characteristic of subjects who favour monogamous relationships, prolonged courtship, and make a heavy investment in long-term relationships. A high score is associated with individuals who are more promiscuous, rapidly proceed to sexual intercourse after meeting someone, and are less interested in romantic closeness.

As part of a massive international project, Schmitt and his collaborators collected data from volunteers in 48 nations covering a wide range of ethnic, geographical and linguistic groups to test a range of hypotheses concerning sexual behaviour (Schmitt, 2005). We discuss three of the key hypotheses:

1. *Men should have a higher sociosexuality than women across all cultures.*

This hypothesis follows from basic biological differences between man and women in terms of potential maximum number of offspring, period of fertility, and levels of parental investment. Basically, as discussed previously, ancestral men would have benefited more than women from pursuing strategies we now associate with high sociosexuality.

2. *Cultures with lower operational sex ratios (where sex ratio is defined as number of marriageable men/number of marriageable women) should be associated with higher female sociosexuality. Conversely, higher operational sex ratios should be associated with lower female sociosexuality scores.*

The logic of this prediction follows from Pedersen's (1991) insight that members of the opposite sex compete to display characteristics desired by the opposite sex. If men are relatively scarce, women must increase their levels of sociosexuality to satisfy the requirements of men (who, being in short supply, can 'call the shots' and so be choosy) in order to compete effectively. If, on the other hand, the operational sex ratio rises above one, females become the scarcer resource and so we might expect sexual behaviour to be 'female shifted', that is, a move towards longer term strategies with careful choice of fewer partners. One can think of this phenomenon as analogous to the laws of supply and demand in economics. If the supply is short, the price shifts towards the interests of the seller (that is, rises); if the supply is readily available, with lots of suppliers offering a high volume of goods, the price shifts towards the interests of the consumer (that is, it falls) (see Figure 11.10).

3. *Early childhood experiences have some impact on adult mating strategies.*

We have phrased this vaguely, since there are two theories in this area that make opposite predictions. The basic idea behind both is that early childhood experience provides information about social and physical environments that enable the adult to shift his or her mating strategy in directions that are most adaptive to these conditions. In the lifespan model developed by Belsky et al. (1991), high-stress juvenile environments (typified by insensitive parenting, economic hardship and a harsh physical environment) are predicted to lead to insecure attachments (in the sense of Bowlby; see Chapter 14), higher levels of sociosexuality and early sexual maturity. In contrast, lower levels of stress and more secure environments are predicted to lead to lower levels of sociosexuality. A similar set of ideas was proposed by Chisholm (1996), who suggested that cultures with high mortality rates and unpredictable resources should push optimal mating strategies towards early reproduction, insecure attachments and high levels of sociosexuality. In lower stress cultures with more abundant resources, Chisholm's model predicts later reproduction and lower sexuality. The logic of both these sets of ideas is that if environments are harsh, unpredictable, stressful, and resources scare, then it pays not to adopt a high investment strategy, since such investments may be lost or impossible to sustain. Better instead to produce more offspring earlier and with lower levels of investment in the assumption that some will survive.

There has been a great deal of research supporting the Belsky/Chisholm model. It has been found, for example, that girls from father-absent homes reach puberty before girls living in families with their biological fathers (Surbey, 1990; Wierson et al., 1993). Ellis and Graber (2000) investigated the effect of maternal psychopathology and the arrival of a stepfather (or mother's boyfriend) on pubertal timing for 87 American girls. Consistent with the Belsky/Chisholm model, they found that girls who were exposed to father absence and familial stress reached puberty earlier than girls from stable homes. A significant part of this study was also the fact that the mother's mental health was also implicated in early puberty. This suggests the possibility that maternal psychopathology is a significant cause of family stress and hence father absence.

Science is often at its most exciting when two rival theories make contrasting predictions about similar phenomena. At such times, we can then construct a crucial experiment that can help to decide which theory is to be rejected and which supported. It transpires that sociosexuality and its cultural determinant provides an instance of this. This is because, in contrast to the development attachment theories of Belsky and Chisholm, we have Gangestad and Simpson's (2000) theory of strategic pluralism. According to this perspective, demanding local environments make the need for biparental care even more acute. From this it would follow that the importance of fidelity and investment in offspring should increase, leading to lower levels of sociosexuality as relationships shift to more monogamous patterns.

Figure 11.10 and Tables 11.3 and 11.4 show how the data collected by Schmitt and colleagues bear on the hypotheses above. In relation to hypothesis 1, it was found that men's sociosexuality exceeded that of women in every culture sampled. Table 11.3 shows a sample of just a few of the nations studied.

Figure 11.10 shows how women's sociosexuality relates to the operational sex ratio. As the graph shows, as the sex ratio (men/women) falls below 100 (100 is defined as a ratio of one or equal numbers of males and females), so women's sociosexuality tends to rise. We can conclude that hypothesis 2 is supported.

Table 11.3 Sex differences in sociosexuality among six nations

Nation	Male sociosexuality	Female sociosexuality	t value	Significance of difference
Argentina	55.52	30.10	7.73	$P < 0.001$
Australia	46.52	30.73	7.17	$P < 0.01$
Netherlands	50.51	31.56	5.78	$P < 0.01$
UK	57.38	29.60	9.53	$P < 0.95$
USA	48.03	29.24	19.07	$P < 0.01$
Zimbabwe	34.80	13.98	5.92	$P < 0.001$

SOURCE: Adapted from Schmitt, D. (2005) 'Sociosexuality from Argentina to Zimbabwe: a 48-nation study of sex, culture, and strategies of human mating.' *Behavioural and Brain Sciences* **28**: 247–311, Table 6, p. 299.

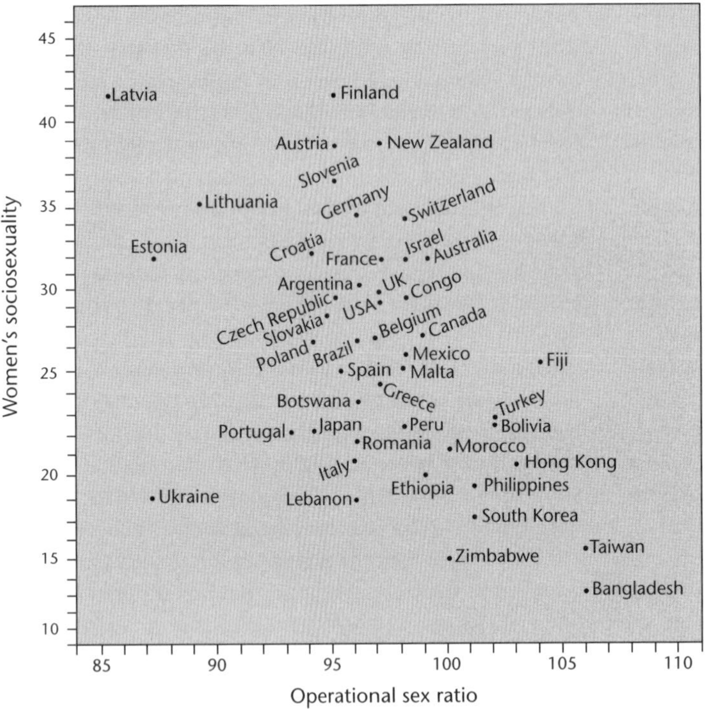

FIGURE 11.10 Levels of women's sociosexuality in relation to operational sex ratios across 48 nations ($r = -0.45$, $p < 0.001$)

SOURCE: Schmitt, D. (2005) 'Sociosexuality from Argentina to Zimbabwe: a 48-nation study of sex, culture, and strategies of human mating.' *Behavioural and Brain Sciences* **28**: 247–311, Figure 9.2, p. 278.

Table 11.4 shows how the data bears on hypothesis 3.

As Table 11.4 shows, the data provides support for the theory of strategic pluralism and is inconsistent with the developmental attachment theories.

The study of Schmitt provides a fascinating insight into how mating strategy might respond to environmental conditions. But as Schmitt himself observes, there are various notes of caution to be sounded. The SOI score for each subject is based on self-reporting, albeit anonymously, and the subjects (as is often the case in such surveys) were primarily college students. There are also alternative explanations for the data obtained. In Figure 11.10, for example, the decrease in women's sociosexuality as the operational sex ratio rises may be a result of men constraining female sexuality through various fidelity and chastity-enforcement measures rather than a shift in male and female psychology as women become the scarce resource. In regard to the apparent refutation of the Belsky/Chisholm theory, we should note the important point that measures of stress were aggregate cultural measures (often from UN data on

Table 11.4 Correlations of SOI with selected cultural variables

Sociocultural variable	Predictions from developmental attachment theory (Belsky/Chisholm)	Predictions from theory of strategic pluralism (Gangestad and Simpson)	Observed correlation of sociocultural variable with SOI	Bearing of data
Familial stress, for example infant mortality rate	Positive correlation	Negative correlation	–0.51 (p < 0.01)	Support for strategic pluralism
Economic resources, for example human development index	Negative correlation	Positive correlation	0.39 (p < 0.01)	Support for strategic pluralism
Mortality, for example life expectancy	Negative correlation	Positive correlation	0.38 (p < 0.01)	Support for strategic pluralism
Early reproduction, for example teenage pregnancy rate	Positive correlation	Negative correlation	–0.36 (p < 0.01)	Support for strategic pluralism

individual nations) rather than real stress in actual families. The fact that the subjects were college students also indicates their own families may not have been typical.

Despite the fact that the strategies of men and women are different and respond flexibly and differently to social context, it is obvious that both sexes have a highly refined sense of sexual attractiveness. We do not mate at random: we choose our partners carefully. Coupled with this, there is strong agreement on what features are attractive in males and females. As we evolved away from the ancestral uni-male groups of the Australopithecines, we also invented standards of male and female beauty. The evolutionary logic of such standards is the subject of the next chapter.

Summary

- An evolutionary approach to human sexuality helps us to understand the mating strategies pursued by ancestral and contemporary males and females.
- Humans are like many mammals, in the sense that females are the sex that limits the reproductive success of males. The large harems created for the exclusive use of ancient despots and emperors show that, in some conditions, males can behave opportunistically to achieve an extreme degree of polygyny.
- Such harems, however, probably represent an aberration of the 'normal' mating behaviour characteristic of the human species. The extended period of postnatal care needed for human infants, and the need for males to be confident of paternity, probably ensured that early *Homo sapiens* were monogamous or only mildly polygynous.
- Such a conclusion is reinforced by interpretations of data on sexual dimorphism and male testis size. The fact that males are slightly larger than females tends to suggest some degree of intrasexual competition among males for mates. Human testes are too large, however, to point to a uni-male mating system, such as found among gorillas, and too small to be consistent with multi-male and multi-female 'promiscuous' mating groups.
- There are some enigmatic features of female sexuality, such as the concealment of ovulation, that have received much theoretical attention. The balance of current theories suggests that sexual crypsis (concealed ovulation) began among early hominid females in a uni-male setting, enabling them to choose desirable males for mating without the risk of infanticide from suspicious males. In effect, sexual crypsis served as a strategy whereby women resisted signalling to men the period of maximum fertility and thereby confused paternity estimations. Once established, it is suggested that women could then use sexual crypsis to extract more care from a male.
- Basic biological differences between men and women can be used to predict the existence of differences in mating strategies and behaviour. One typical result is the prediction, frequently confirmed, that men will desire more sexual partners in any given period than women.
- The sociosexuality of women can be expected to vary according to the sex ratio. There is evidence that as the ratio of men to women falls, women's sociosexuality shifts in a male direction.

Key Words

- Lek
- Oestrus
- Ovarian cycle
- Ovulation
- Sexual crypsis
- Sexual strategies theory (SST)
- Testis

Further reading

Baker, R. R. and Bellis, M. A. (1995) *Human Sperm Competition*. London, Chapman & Hall.

Makes some controversial claims about human sexuality based on unusual and original research.

Betzig, L. (ed.) (1997) *Human Nature: A Critical Reader*. Oxford, Oxford University Press.

A useful book containing numerous original articles on human sexuality, together with a retrospective critique by the original authors.

Ridley, M. (1993) *The Red Queen*. London, Viking.

A delightful book that explores the nature of sexual selection and its application to humans.

Short, R. and Potts, M. (1999) *Ever Since Adam and Eve: The Evolution of Human Sexuality*. Cambridge, Cambridge University Press.

A superbly illustrated and authoritative work. A sensitive account of the evolution and significance of human sexuality.

Human Mate Choice: the Evolutionary Logic of Sexual Desire

(W)hen on thee I gaze never so little,
Bereft am I of all power of utterance,
My tongue is useless.

There rushes at once through my flesh tingling fire,
My eyes are deprived of all power of vision,
My ears hear nothing by sounds of winds roaring,
And all is blackness ...

A dread trembling o'erwhelms me ... and in my madness
Dead I seem almost.
(Sappho of Lesbos c.600BC, translated by E. M. Cox)

The onset of sexual desire exerts a powerful force over human lives. One of its most obvious features is that it is highly discriminating – we have strong preferences about who will and who will not suffice as a potential partner in both the short and longer term. Furthermore, however idealised our view of romantic love, there is abundant evidence that humans employ a variety of hard-headed criteria in assessing the desirability of a mate, including income, occupation, intelligence, age and, perhaps above all, physical appearance. The appearance of the human body is an important reservoir of information and has exerted a potent influence over the whole of our culture. Western art since the time of the Greeks and certainly since the Renaissance has celebrated the beauty, grace and symbolic significance of the human body. We are enthralled and fascinated by the appearance of members of our own species. Whether there are universal and cross-cultural standards in the aesthetics of the body is still a debatable point, but certainly in Western culture, there exists a multimillion dollar industry to help us better to shape our bodies towards our ideals or disguise the effects of age. Corporate advertising worked out long ago that one of the best ways to market a product is to associate it with a handsome specimen of one or both sexes. Where nature falls short, the surgeon's knife can be recruited to the cause and the statistics on cosmetic surgery are a testament to the West's near-obsession with physical beauty. Table 12.1 shows the top five cosmetic surgical procedures for men and women in the USA carried out in 2005.

So where does this aesthetic sense come from? Somewhat surprisingly, Darwin himself, reviewing standards in a variety of cultures, concluded that there was no universal standard for beauty in the human body. A common modern response to the question is to repeat the mantra 'beauty is in the eye of the beholder' and while this does convey a fragment of the truth, it is also misleading. If standards did vary enormously between individuals, then anyone could become a glamour or fashion model; yet common sense tells us that attractive people are viewed to be attractive by a large consensus. Furthermore, some recent meta-analyses of the literature on attractiveness have shown that there is massive agreement in attractiveness ratings within and between cultures (Langlois et al., 2000). Beauty does not seem to reside in the eye of the beholder, nor is it simply a response to culturally induced norms.

To a Darwinian, the way to proceed on this question is clear enough: our perceptual apparatus will be

Table 12.1 The top five cosmetic surgery procedures in the USA for 2005

Top five for women		Top five for men	
Procedure	*Number of procedures carried out in 2005*	*Procedure*	*Number of procedures carried out in 2005*
Breast augmentation	291,350	Nose reshaping	99,680
Liposuction	287,932	Hair transplant	39,244
Nose reshaping	198,732	Liposuction	35,673
Eyelid surgery	197,709	Eyelid surgery	32,988
Tummy tuck	128,874	Breast reduction	16,275

SOURCE: American Society of Plastic Surgeons (2007) *Breast Surgery Statistics 1996*. Retrieved August 2007 from http://www.plasticsurgery.org/media/statistics/1996statistics.cfm.

designed to respond positively to features that are honest indicators of fitness – for aesthetics read reproductive potential. We are descended from ancestors who made wise choices in selecting their partners and, to the extent that we have inherited their desires and inclinations, we can expect any criteria we use in choosing a mate to be fitness enhancing. This chapter is focused around both these issues: the aesthetic judgement exercised, and the decision-making criteria used, when males and females choose a sexual partner.

12.1 Evolution and sexual desire: some expectations and approaches

Darwinians view attractiveness in terms of reproductive fitness rather than as the relationship between an object, an observer and some abstract Platonic form. Features that are positive indicators of reproductive fitness in a potential mate should be viewed as attractive by males and females. In this sense, beauty is more than skin-deep – it is to be found in the 'eye' of the genes. Despite the mild degree of polygyny indicated by the evidence presented in Chapter 11 and the few cases of opportunistic extreme polygyny, it is clear that, in most relationships, men and women make an appreciable investment of time and energy. Consequently, both sexes should be choosy about future partners, but in different ways.

Of all the features used in appraising a potential mate, two in particular have produced robust empirical findings that reveal inherent differences between male and female taste. These are physical attractiveness and the status of males. In the case of male status, the application of the principles established in Chapters 3 and 11 predicts that, since females make a heavy investment in raising young and biparental care is needed following birth, females will be attracted to males who show signs of being able to bring resources to the relationship. This ability could be expected to be indicated by the dominance and status of a male within the group. Dominance could be selected for by intrasexual selection if males compete with each other, and by intersexual selection if females exert a preference for dominant males. A crude indication of dominance would be size, and we have already noted that humans are mildly dimorphic. In the complex social groups of early humans, however, there was bound to be a whole set of parameters, such as strength, intelligence, alliances and resource-holding and provisioning capabilities that indicated the social status of the male. Some of these would be subtle and context dependent, and a female would be best served by a perceptual apparatus that enabled her to assess rank and status using context-specific cues and signals.

If females respond to indicators of potential provisioning and status, males should be attracted to females that appear fecund and physically capable of caring for children. Since the period of female fertility (roughly 13–45 years of age) occupies a narrower age band than that of the male (13–65 years), we would also expect the age of prospective partners to be evaluated differently by each sex. Men should be fussier about age than women and hence rate physical features that correlate with youth and fertility higher on a scale of importance than should women. Indeed, studies have shown that attractiveness ratings decline with age for both men and women, but that these effects are more pronounced for women than men

(McLellan and McKelvie, 1993). This seems to be a double standard but one that may have deep evolutionary foundations: the preference for youthfulness in women is to be expected not only because youthful women are more fertile, but also because younger women have a longer period of fertility ahead of them than older women and could therefore produce more offspring by the admiring male.

To test these expectations, we can examine human preferences using data from a number of sources and types of study:

- What people say about their desires in response to questionnaires
- What people look for when they advertise for a partner
- Medical evidence on whether features deemed to be attractive really do correlate with health and fecundity
- The use of stimulus pictures (line drawings and photographs) to elicit body shape and facial preferences.

In the sections that follow, each of these approaches will be employed.

12.2 Questionnaire approaches

12.2.1 Cross-cultural comparisons

The use of a questionnaire on sexual desire in one culture lays itself open to the objection that responses reflect cultural practices and the norms of socialisation rather than universal constants of human nature. In an effort to circumvent this problem, David Buss (1989) conducted a questionnaire survey of men and women in 37 different cultures across Africa, Europe, North America, Oceania and South America, and hence across a wide diversity of religious, ethnic, racial and economic groups. As might be expected, numerous problems were encountered with collecting such data, but Buss's work remains one of the most comprehensive attempts so far to examine the sensitivity of expressed mating preferences to cultural variation. From the general considerations noted above, Buss tested several hypotheses (Table 12.2).

Table 12.2 Predictions on cross-cultural mate choice preferences

Prediction	Adaptive significance in relation to reproductive success
Women should value earning potential in a mate more highly than men do	The likelihood of a woman's offspring surviving and their subsequent health can be increased by allocation of resources to the woman and her children
Men should value physical attractiveness more highly than women do	The fitness and reproductive potential of a female is more heavily influenced by age than for a man. Attractiveness is a strong indicator of age and fertility
Men, on the whole, are likely to prefer women younger than themselves	Men reach sexual maturity later than women do. Also as above
Men will value chastity in a partner more so than women	'Mommy's babies, daddy's maybes'. For a male to have raised a child not his own would have been, and still is, highly damaging to his reproductive fitness
Women should rate ambition and drive in a prospective partner more highly than men do in their partners	Ambition and drive are linked to the ability to secure resources and offer protection, both of which would be fitness enhancing to a woman

SOURCE: Buss, D. M. (1989) 'Sex differences in human mate preferences: evolutionary hypotheses tested in 37 cultures.' *Behavioural and Brain Sciences* **12**: 1–49.

The results in terms of the number of cultures in which there was a significant difference ($P < 0.05$) between the qualities addressed in each hypothesis above are shown in Table 12.3.

The results show moderate to strong support for all the hypotheses. Data on age difference also allow a calculation of mean age preferences for mating. On average, men prefer to marry women who are 24.83 years old when they are 27.49 years old, that is, 2.66 years younger than themselves. Women, on the other hand, prefer to marry men who are 3.42 years older. Interestingly, there are data available for 27 of the 33 countries sampled for actual ages of marrying; the actual difference is that men marry women 2.99 years younger than themselves. The fact that these figures agree so closely suggests that, at least in terms of age, preferences and practice are reassuringly similar. The reasons for these consistent and near-universal differences in mating age have been the subject of much speculation, with no consensus as yet emerging. One possibility is that age in a male is an indicator of status since status usually rises with age.

Table 12.3 Number of cultures supporting or otherwise hypotheses on gender differences in mate preference

Hypothesis	Number of cultures supporting hypothesis *(percentage of total in brackets)*	Number of cultures contrary (con) to hypothesis or result not significant (NS) *(percentage of total in brackets)*
Women value earning potential in a partner more than men do	36 (97%)	1 NS (3%)
Men value physical attributes more than women do	34 (92%)	3 NS (8%)
Women value ambition and industriousness more than men do	29 (78%)	3 con (8%) 5 NS (13%)
Men value chastity more than women do	23 (62%)	14 NS (38%)
Men prefer women younger than themselves	37 (100%)	0 (0%)

SOURCE: Data from Buss, D. M. (1989) 'Sex differences in human mate preferences: evolutionary hypotheses tested in 37 cultures.' *Behavioural and Brain Sciences* **12**: 1–49.

12.2.2 Urgency in copulation

If we accept the reliability of questionnaire studies such as those carried out by Buss and others, it seems that the human mind is sexually dimorphic in the psychology of sexual attraction. With this in mind, Symons and Ellis (1989, p. 133) hypothesised that males and females would respond differently to the following question:

> If you had the opportunity to copulate with an anonymous member of the opposite sex who was as physically attractive as your spouse but no more so and as competent a lover as your spouse but no more so, and there was no risk of discovery, disease or pregnancy and no chance of forming a durable liaison, and the copulation was a substitute for an act of marital intercourse, not an addition, would you do it?

The prediction was that since sexual novelty benefits a male's reproductive interests more than that of a female, males would be more inclined to answer yes to this question. The answers were in categories of 'Certainly would', 'Probably would', 'Probably not' and 'Certainly not'. The results are shown in Figure 12.1 and offer support for this contention.

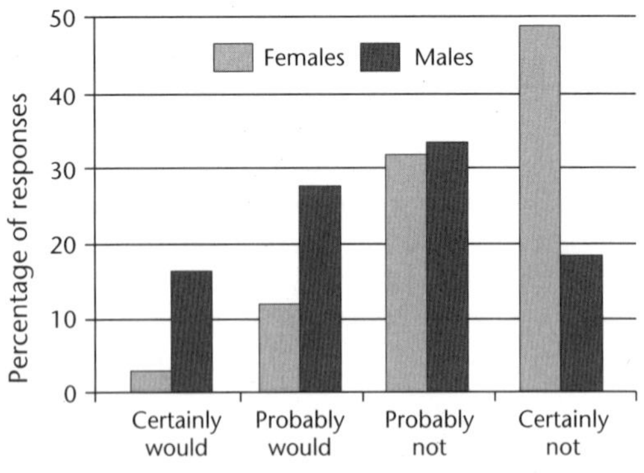

FIGURE 12.1 Responses to questions on casual sex expressed as a percentage of the respondents answering
Number in sample: female = 122, male = 87.

SOURCE: Data from Symons, D. and Ellis, B. (1989) Human male–female differences in sexual desire, in A. E. Rasa, C. Vogel and E. Voland (eds) *The Sociobiology of Sexual and Reproductive Strategies*. London, Chapman & Hall.

12.2.3 A qualified parental investment model: the effect of levels of involvement

Such questionnaire studies have consistently shown clear differences in the psychology of sexual attraction, but they have also revealed some degree of convergence. Of the top ten criteria for a good long-term partner listed by subjects in a study by Buss and Barnes (1986), seven out of the ten were the same for men and women, although they were of course given different priorities. These were kindness and understanding, intelligence, personality, health, adaptability, creativity and graduate status. We could explain this by arguing that the dimorphism in taste is only slight because of the effect of near-monogamous mating and the large investment made by both males and females in living together and raising children. A subtle way of probing what gender differences might exist beneath the consensus was applied in a study by Kenrick et al. (1990). They argued that humans are different from most other mammals in that there is a large range in the possible amount of investment made by couples in any relationship. It could range, in the case of a male, from a few drops of sperm in a one-night stand to a lifelong commitment to a spouse and the raising of children. With this in mind, Kenrick et al. investigated whether the criteria for choosing a mate varied with the level of involvement in a relationship. The investment implications of different levels of involvement can be expected to vary according to gender (Table 12.4).

Table 12.4 Degree of involvement and expected implications for investment according to gender

Level of involvement	Date	Sexual relations	Steady dating	Marriage
Investment by female	Low	High	High	High
Investment by male	Low	Low	High	High

It is obvious that progressing from dating to sexual activity is associated with a sharper rise in potential investment implications for women than for men. To study the effect of this, Kenrick and colleagues asked 93 undergraduate students to consider the importance of 24 criteria (physical attractiveness, status and so on) in accepting a mate at the four different levels of involvement. Figure 12.2 shows the findings for earning capacity.

In terms of all the criteria assessed, a number of patterns emerged. As expected, significant differences began to appear at the sexual level of involvement for characteristics defining family values and health, females finding these more important than males once the level of sexual commitment is reached. Women rated status, however, more highly than did men at all levels of involvement.

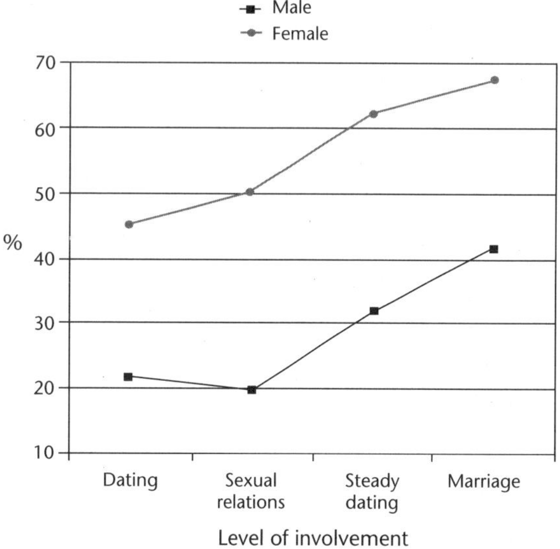

FIGURE 12.2 Minimum level of acceptable earning capacity at different levels of involvement
A level of, for example, 67 per cent for females at marriage indicates that, to be acceptable, a male must earn more than at least 67 per cent of other males but could earn less than 37 per cent of all males.

SOURCE: Data from Kenrick, D. T., Sadalla, E. et al. (1990) 'Evolution, traits, and the stages of human courtship: qualifying the parental investment model.' *Journal of Personality* **58**: 97–116.

It should be noted that this study was of a rather narrow profile of the population in one culture, and it is curious that, for some characteristics, the significance of the gender difference declined from steady dating to marriage. The robustness of these findings in the face of different social conventions and across different population profiles remains to be established.

There are clearly problems with many studies

based on questionnaires, particularly when unselective samples are used. One sometimes gets the impression that American undergraduates are constantly being plagued by interviewees asking about their sex lives. Nevertheless, the findings tend to be in agreement with evolutionary expectations. If, as social science critics would say, responses are conditioned by social norms, we still have the problem of explaining why so many social norms correspond with evolutionary predictions.

12.3 The use of published advertisements

An intriguing way to gather information on mating preferences is to inspect the content of 'lonely hearts' advertisements in the personal columns of newspapers and magazines. A typical advertisement reads:

> Single prof. male, 38, graduate, non-smoker, seeks younger slim woman for friendship and romance.

Notice that the advertisement offers information about the advertiser as well as his preferences for a mate. Such information carries some advantages over questionnaire response surveys, in that it is less intrusive and less subject to the well-known phenomenon that interviewees will tend to comply with what they take to be the expectations of the questioner. Moreover, the data are 'serious', in that they represent the attempts of real people to secure real partners. Against this must be placed the fact that the data are selective and probably do not represent a survey across the entire population profile. Greenless and McGrew (1994) examined 1,599 such advertisements in the columns of *Private Eye* magazine. The results for physical appearance and financial security are shown in Figure 12.3.

The results are consistent with the questionnaire surveys of Buss and others, and support the following hypotheses:

1. Women more than men seek cues to financial security
2. Men more than women offer financial security
3. Women more than men advertise traits of physical appearance
4. Men more than women seek indications of physical appearance.

The work of Greenless and McGrew has been repeated many times, with similar findings. Figure 12.4 gives details of the findings of a survey by Dunbar (1995a) showing the interesting result that women become less demanding of resources as they age.

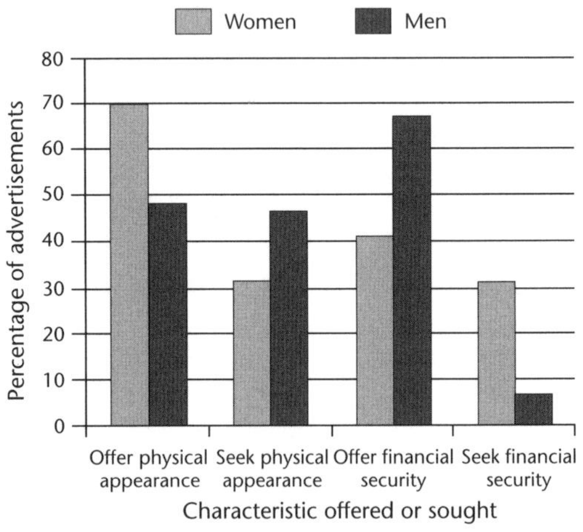

FIGURE 12.3 Percentage of advertisers seeking and offering physical appearance and financial security

SOURCE: Data from Greenless, I. A. and McGrew, W. C. (1994) 'Sex and age differences in preferences and tactics of mate attraction: analysis of published advertisements.' *Ethology and Sociobiology* **15**: 59–72.

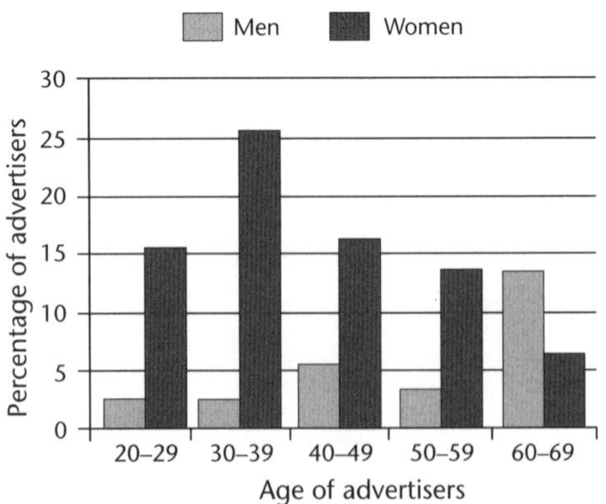

FIGURE 12.4 Features of 'lonely hearts' advertisements: those seeking resources according to age group

SOURCE: Adapted from Dunbar, R. I. M. (1995a) 'Are you lonesome tonight.' *New Scientist* **145**(1964): 12–16.

It also appears that men of high status seek, and are able to attract, women of higher reproductive value. At the anecdotal level, most people know of ageing male rock stars or celebrities who marry women many years their junior. In a study of a computer dating service in Germany, Grammer (1992) found a positive correlation between the income of men using the service and the number of years separating their age from the (younger) women they sought. The reversal of the trend shown in Figure 12.4 for the age group 60–69 may indicate the fact that more women are alive than men in this category. Reproduction is not really an issue now, but men may be able to choose women with resources. Also women in this age category have less fertility potential to trade for resources.

Studies on the content of advertisements give no indication of how successful they may be in attracting a mate. Men may advertise resources but there is no indication from the advertisement alone that this is a successful strategy in terms of number of responses to the advertisement. To probe this issue further, Lucila de Sousa Campos and colleagues counted how actual responses to such advertisements in a Brazilian newspaper (the *Folha de Sao Paulo*) varied with the qualities offered. They also examined the effect of age of advertisers on response rate and the content of the advertisements (de Sousa Campos, 2002). Their results were in keeping with evolutionary expectations: they found that the number of responses to female advertisements decreased with the age of the female, while the responses for male advertisers increased with their age (Figure 12.5).

This pattern may be a response to the fact that female fertility and so sexual desirability declines with age, while that of males remains steadier. Males also tend to acquire more resources as they age and become fewer in number (biasing the sex ratio towards females).

In terms of the content of the advertisements, de Sousa Campos et al. found that females became less demanding in terms of the age and occupational status of a potential partner as they themselves aged. In contrast, males became more demanding as they aged and more interested in childless partners. Furthermore, as men grew older, they sought females younger than themselves by an increasing number of years. Females preferred slightly older men but this difference only decreased slightly with the woman's age.

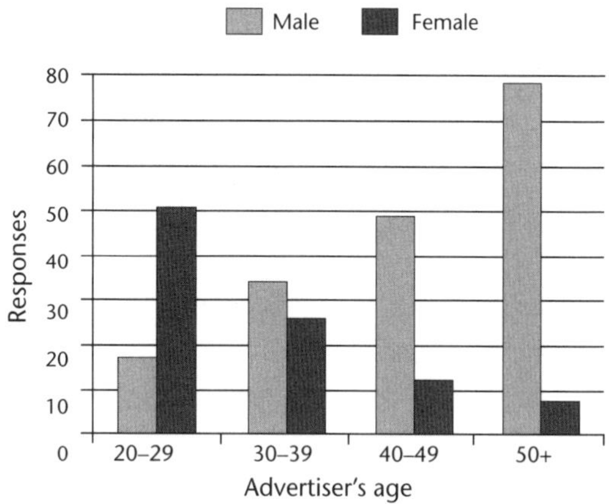

FIGURE 12.5 Responses to personal advertisements as a function of the advertiser's age

SOURCE: Data from de Sousa Campos et al. (2002) 'Sex differences in mate selection strategies: Content analyses and responses to personal advertisements in Brazil.' *Evolution and Human Behaviour* **23**(5): 395–406.

12.3.1 Origin of mate choice preferences: evolutionary psychology or structural powerlessness

Studies on expressed preferences reveal nothing about the ontogeny of those preferences. Some social scientists have proposed an alternative to adaptive explanations, which Buss and Barnes (1986) call the 'structural powerlessness and sex role socialisation' hypothesis. This suggests that since, in patriarchal societies, women have less access to power and wealth than men, the chief way in which a woman can attain status and acquire resources is to marry up the social ladder (hypergamy) and trade looks for status. The hypothesis fails on the first count, however, of explaining why preferences are remarkably similar across a wide range of cultures. If it is argued that all these cultures share the same features of patriarchy, we then need an even grander theory to explain this. A more serious problem is that the structural powerlessness model makes a prediction at variance with the facts. If women seek high-status males to advance their own standing, it follows that women should be less selective with regard to status and wealth as their own premarital wealth and power increases. The evidence, however, suggests otherwise: high-status women still value high-status men. Buss (1994) found that women with a high income tend to value the financial status of men even more than do women on a lower income.

12.4 The use of stimulus pictures to investigate body shape preferences

12.4.1 Waist to hip ratios (WHRs): male assessment of females

It is clear that good looks are important to men seeking a partner, but what type of looks should be preferred? The belief that there is a vast variability in notions of beauty between cultures in time and place has tended to thwart the scientific search for universal and adaptive norms for beauty. Reference is often made to the rather fleshy nude women appearing in paintings by Titian and Rubens, which are compared with modern-day models to emphasise the changing ideals of attractiveness (although whether or not the artists were attempting to depict an ideal is questionable). In 1993, however, Devendra Singh, a psychologist working at the University of Texas, published some important work suggesting that there may be some universals in what the sexes find attractive. Singh (1993) argued that two conditions must be met by any universal ideal. First, there must be some plausible linkage of features designated as attractive to physiological mechanisms regulating some component of reproductive fitness, and hence a positive correlation between variation in attractiveness and variation in reproductive potential; in other words, attraction must equate with fitness. Second, males should possess mechanisms to judge such features, and these should be assigned a high degree of importance in the estimation of attractiveness.

Singh (1993) argues that the distribution of body fat on the waist and hips meets the conditions above. More specifically, he suggests that the **waist to hip ratio** (WHR) is an important indicator of fitness and attractiveness. The WHR (that is, waist measurement divided by hip circumference) for healthy premenopausal women usually lies between 0.67 and 0.80, whereas for men, it usually lies between 0.85 and 0.95. It is well known that obesity is associated with a higher than average health risk, but what is more surprising is that the distribution of fat in obese women, and hence the WHR, is also a crucial factor in predicting their health status. Singh collected a body

Table 12.5 Research pertaining to reproductive and health implications of high female WHRs and varying BMI values

Female WHR and related health and reproductive effects	Reference
Healthy premenopausal women have WHRs in range 0.67–0.80	Lanska et al., 1985
WHR approaches male range after menopause	Arechiga et al., 2001
Elevated WHR is a risk factor for cardiovascular disorders, adult onset diabetes, hypertension, cancer, ovarian cancer, breast cancer and gall bladder disease	Huang et al., 1999, Misra and Vikram, 2003
Women with higher WHRs have more irregular menstrual cycles	van Hoof et al., 2000
Women participating in donor insemination programmes have lower probability of conception if their WHR is above 0.8 (after age and BMI are controlled for)	Zaadstra et al., 1993
Women suffering from polycystic ovary syndrome who have impaired oestrogen production have higher WHRs than average. When such women are given hormone treatment with an oestrogen-progestagen compound, their WHRs decrease over time	Pasquali et al., 1999
Female BMI and related health and reproductive effects	**Reference**
High BMI (that is, obesity) associated with complications in pregnancy, menstrual irregularities and infertility	National Heart, Lung and Blood Institute, 1998
Low BMI related to menstrual difficulties and non-ovulation	DeSouza and Metzger, 1991
Very low BMI (anorexia nervosa) associated with higher miscarriage rates, higher premature birth rate and lower birth weight	Bulik et al., 1994

of evidence to show that women with a WHR below 0.85 tended to be in a lower risk category compared with women with a WHR above 0.85 for a range of disorders such as heart disease, diabetes, gallbladder disease and selected carcinomas. Since Singh's original papers, medical evidence does suggest that WHR is associated with the health status of females. Table 12.5 shows a sample of this research and also includes **body mass index** (BMI) – another variable thought to be responsible for attractiveness ratings (see below and Figure 12.6).

If the WHR does have an adaptive significance, it should not be subject to the vagaries of fashion. To test this, Singh examined the statistics of Miss America winners and *Playboy* centrefold models.

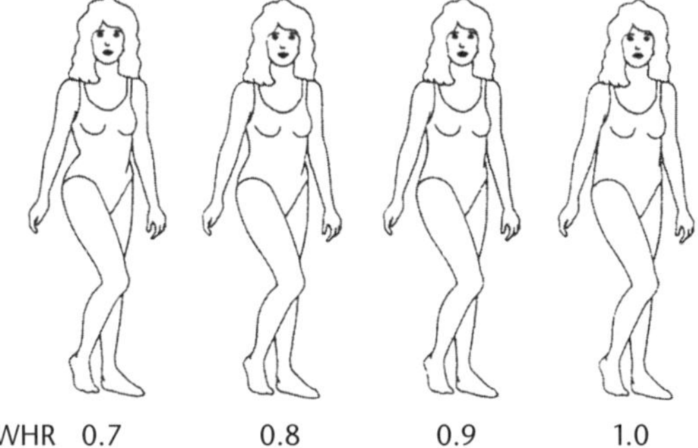

FIGURE 12.6 Stimulus figures given to subjects in Singh's study

A series of four such figures was presented in three categories: underweight, normal and overweight. Only the normal weight category is shown here.

SOURCE: After Singh, D. (1993) 'Adaptive significance of female attractiveness.' *Journal of Personality and Social Psychology* **65**: 293–307.

The results confirm the widespread suspicion that the weight of fashionable models has fallen over the past 60 years. What is equally significant is the fact that the WHR has remained relatively constant. This suggests that while weight as an indicator of attraction is, to some degree, subject to fashionable change, the WHR is far more resilient. It follows that the WHR is a possible candidate for a universal norm of female beauty. It is stable among women deemed to be attractive, and it correlates with physical health and fertility. The next step is to ascertain whether it is a factor in the assessment of beauty.

To test perceptions of attractiveness in relation to weight and WHR, Singh (1993) presented subjects with a series of line drawings (Figure 12.6), with the instruction that they rank the figures in order of attractiveness. The results were unambiguous: for each category of weight, the lowest WHR was found to be most attractive. In addition, an overweight woman with a low WHR was found to be more attractive than a thin woman with a high WHR. This again suggests that attractiveness is more strongly correlated with the distribution of body fat, as it affects the WHR, than with overall weight per se.

12.4.2 Waist to hip ratios (WHRs): female assessment of males

The prevalence of magazines depicting women in various states of undress would seem to imply that men are more easily aroused by visual stimuli than women. Magazines devoted to pictures of men exist but cater largely for the homosexual market rather than for serious viewing by women. This does not, however, mean that women are insensitive to male physique. Singh (1995) also applied the concept of WHR to the female assessment of male attractiveness. It is well established that fat distribution in humans is sexually dimorphic and that, among the hominoids, this is a feature unique to humans. After puberty, women deposit fat preferentially around the buttocks and thighs, while men deposit fat in the upper body regions, such as the abdomen, shoulders and nape of the neck. Such gynoid (female) and android (male) body shapes vary surprisingly little with climate and race. Singh reviewed evidence to show that the WHR is correlated with other aspects of human physiology, such as health and hormone levels (Figure 12.7).

Singh presented women with line drawings (Figure 12.8) of men in various weight categories and with varying WHRs, asking them to rate the men's attractiveness. In all body weight categories, figures with a female-like, low WHR were rated as being least attractive. The most attractive figure was found to be in the normal weight category with a WHR of 0.9 (Figure 12.8). Interestingly, when females were asked to assign attributes such as health, ambition and intelligence, they tended to assign high values to the high WHR categories. Men with WHRs of 0.9 not only look better, but are also assumed to be brighter and healthier.

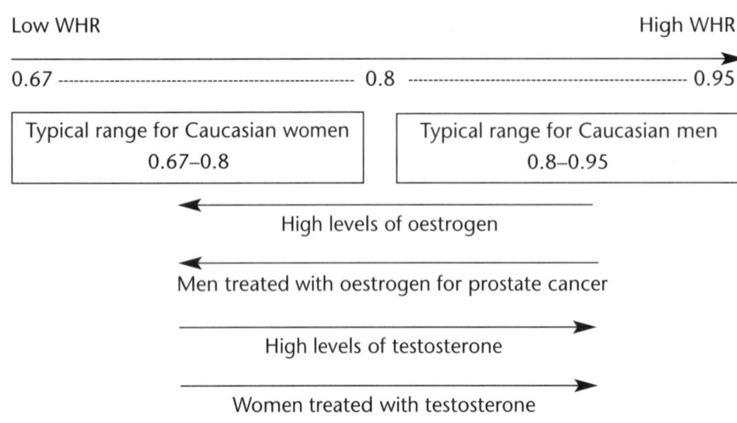

FIGURE 12.7 Distribution of WHR according to gender and hormone levels
Low WHRs are indicative of high oestrogen levels, high WHRs indicate high levels of testosterone.

SOURCE: Data from Singh, D. (1995) 'Female judgement of male attractiveness and desirability for relationships: role of waist-to-hip ratios and financial status.' *Journal of Personality and Social Psychology* **69**(6): 1089–101.

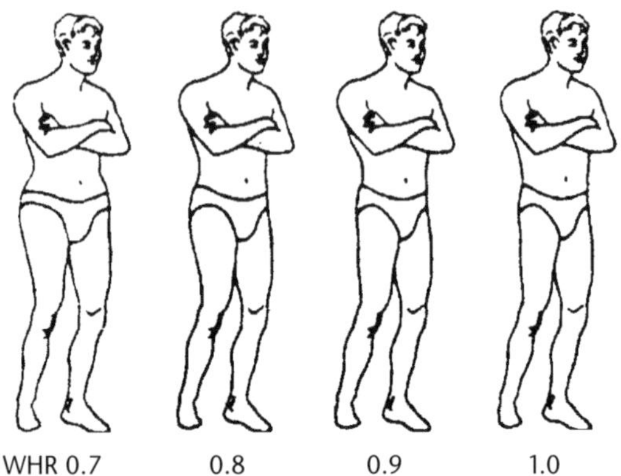

FIGURE 12.8 Stimulus pictures of men with various WHRs in the normal weight category

SOURCE: Singh, D. (1995a) 'Female judgement of male attractiveness and desirability for relationships: role of waist-to-hip ratios and financial status.' *Journal of Personality and Social Psychology* **69**(6): 1089–101.

Singh's initial work on male assessment of females was carried out on Caucasian males. To test whether different cultures shared the same WHR preference, Singh and Luis (1995) applied the same procedures to a group of Indonesian men recently arrived in the USA (94 per cent of whom were of Chinese descent) and a group of African-American men. The findings were virtually identical to those of the Caucasian study. The normal weight group was found to be the more attractive overall and again, within all groups, a female WHR of 0.7 was found to be most attractive. Women reached conclusions similar to those of men when evaluating the line drawings. It appeared to Singh that neither ethnicity nor gender significantly affected the WHR dimension of attractiveness.

12.4.3 The cultural variability of attractiveness judgements

Learning from the media

Singh's work has aroused a great deal of interest and some controversy. One possible objection is that men are exposed to a barrage of images of women from the media, so preferences may be culturally determined. It is probable that many men learn their cues for what counts as attractive from the behaviours of other men and respond positively to the features of media-generated stars and icons.

Dugatin (1996) has reported the action of social cues on mate choice even in guppies. Female guppies are known to prefer males with orange colouration, but they have been observed to choose a male with less colouration if other females have already chosen him. The exact function of this is unclear, but such copying may save a female guppy time and thus reduce the energy costs and risks involved in finding a mate. The logic, although not exactly foolproof, may be that if several females have already chosen a male, he must have desirable qualities.

To test the possible influence of cultural cues, Yu and Shepard (1998) presented the same images used in Singh's study on male preferences to one of the few remaining cultures that exists in isolation from Western influences, the Matsigenka indigenous people in southeast Peru. The results were strikingly different from those of Singh's study of American subjects. The most attractive figure emerged as the one labelled by Westerners as overweight, with a WHR of 0.9. The figure found to be most attractive by

Singh's subjects (of normal weight, with a WHR of 0.7) was labelled by a typical Matsigenka male as having had fever and having lost weight around the waist. In the same study, Yu and Shepard also examined the preferences of South American indigenous people who had been exposed to Western influence and found male preferences to be more in keeping with US standards.

Yu and Shepard (1998, p. 32) conclude that many so-called cross-cultural tests in evolutionary psychology may 'have only reflected the pervasiveness of western media'. They also point out that an adaptationist explanation is still possible. In traditional societies, physical appearance may be less important in mate choice, since individuals are well known to each other, and couples have direct information, such as age and health status, about each other. In Westernised countries, a daily exposure to strangers may have sharpened our need to make judgements using visual cues.

Even if WHR ideals do seem to remain steady in a given culture, other aspects of body shape seem variable in their estimation. As Swami and Furnham (2007) observe, ideals for female body shape seem to change through time even in the same culture. After the First World War, an ideal of female body shape emerged that was androgynous in form (de-emphasising the waist and breast). This is sometimes explained as women asserting their equality with men by emphasising the masculinity of their body shape (women over 30 were given the vote for the first time in the UK in 1918; in the USA the presidential election of 1920 was the first time all women in America could exercise the vote). This trend is illustrated by the fact that the mean of the vital statistics of Miss America winners in the 1920s was about 32–25–35. Interestingly, this gives a WHR of 0.71, but the breast measurements are much lower than those found in the 1950s. The icon of female beauty in the 1950s was Marilyn Monroe who had a curvier hourglass figure than the ideals of the 1920s. In the 1960s, the height of Miss America winners increased while their weight decreased. Such was this trend that, by the 1980s, the majority of Miss America winners had a BMI of less than 18.5, a figure considered clinically underweight.

WHR preferences in a hunter-gatherer culture

To test if male preferences for a female WHR of around 0.7 is universal and culturally invariant, Marlowe and Wetsman (2001) presented drawings of women to men of the Hadza people. The Hadza are hunter-gatherers living in Tanzania. In this culture, labour is sexually divided: women dig for tubers and gather wild berries and fruit, while men search for honey and hunt for game. Marlowe and Wetsman were mindful of the problems that beset many of these stimulus image experiments, namely, separating WHR and BMI effects. The problem is that higher WHRs also suggest a higher BMI and so a higher overall level of fatness. In their study, they found that Hadza men preferred a higher WHR as most attractive compared to that chosen by US subjects (see Figure 12.9). They argue that several considerations could explain these findings. First, among hunter-gatherer tribes, where women's work is energetically expensive, thinness could indicate poor health due to parasites and disease, compromising their ability to gather food and tend to children. Second, in all cultures, a high WHR is associated with pregnancy and is therefore expected to be undesirable in a prospective mate. But the effect of any indication of pregnancy will be mediated by the total fertility rate (TFR) of women in each culture. In the USA, for example the TFR is 2 (meaning that the average women raises 2 children), but among the Hadza, the TFR is 6.2. The lower the TFR, the more important it is to attend to signs of pregnancy: in the USA, a man could typically raise only 1 more child with an already pregnant woman, but a Hadza male has the potential to sire another 5.2 children with an already pregnant woman. Marlowe and Wetsman (2001) reason that this effect may help explain why a high WHR would be more damaging to the reproductive prospects of an American male and hence would be regarded as less attractive compared to a Hadza male.

In a more recent study, Marlowe and colleagues took into account the different body shapes of native Hadza women and typical US women. Compared to Caucasian women, Hadza women have buttocks that protrude more prominently. This may be significant, since estimates of WHR from a frontal image depend on assumptions about body shape, so that, for example, for the same frontal WHR, more protruding buttocks will lower the overall WHR if measured circumferentially. Marlowe et al. (2005) presented US and Hadza men with side images as well as frontal images. A sample of the results is shown in Table 12.6.

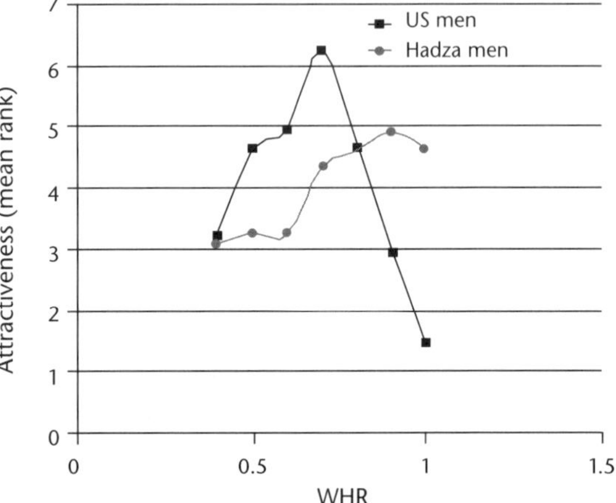

FIGURE 12.9 Attractiveness of female body shape in relation to WHR judged from line drawings by US and Hadza males

SOURCE: Plotted from data in Marlowe, F. W. and Wetsman, A. (2001) 'Preferred waist-to-hip ratio and ecology.' *Personality and Individual Differences* **30**: 481–9, p. 483.

Table 12.6 Comparison of preferred frontal and side-view WHRs for US and Hadza men

	US males	Hadza males
Frontal WHR	0.70	0.90
Side-view WHR	0.65	0.63
Weighted mean and hence overall WHR	0.68	0.79

SOURCE: Data taken from Marlowe, F. W., Apicella, C. L. et al. (2005) 'Men's preferences for women's profile waist-to-hip ratio in two societies.' *Evolution and Human Behaviour* **26**(6): 458–69.

It is notable that in this study, differences in optimally attractive WHR remain between US and Hadza men (0.68 v. 0.79) but this is a smaller difference than found in previous studies (Marlowe and Wetsman, 2001). Another significant finding from this study was the fact that the mean WHR of Hadza women is higher than the mean WHR for US women (see Table 12.7).

Table 12.7 A comparison of mean WHR for samples of Hadza and American women

Sample population	Age (range)	Mean WHR +/– 1 SD
Hadza women	17–82	0.83 +/– 0.06
Hadza women	17–24	0.79 +/– 0.04
Young American students	18–23	0.73 +/– 0.04
American nurses	23–50	0.74 +/– 0.08

SOURCE: Data taken from Marlowe, F. W., Apicella, C. L. et al. (2005) 'Men's preferences for women's profile waist-to-hip ratio in two societies.' *Evolution and Human Behaviour* **26**(6): 458–69.

There may be genetic reasons for the differences in US and Hadza female WHRs, themselves reflecting adaptive pressure resulting from differing levels of optimality in different environments. As Marlowe et al. (2005) conjecture, Hadza women may have higher WHRs than US women as a result of the need for a larger gut to digest the bulky fibrous tubers that figure prominently in the Hadza diet. Or a more male WHR shape (signifying a relatively smaller pelvis) may reflect the increased importance of mobility over longer distances for Hadza women compared to Caucasian women. Whatever the underlying causes of these differences, it is interesting that the WHR preferences of US and Hadza men correspond quite closely with the means of young women in their respective populations (compare Table 12.6 with Table 12.7). Marlowe et al.'s interpretation of these findings is such to suggest that a fixed WHR preference is not a human universal, but rather that men adjust their evaluation of ideal WHRs according to local ecologies. This could consist of a simple mean tracking programme such that men alter their perceptions of ideal according to local averages. Such a mechanism would be consistent with other work that shows a preference of averages in faces and other objects (see Halberstadt and Rhodes, 2000).

WHR and BMI: confounding effects and cultural variability

One criticism of Singh's work on WHRs is that it ignores the crucial role played by the BMI in judgements of attractiveness. The BMI is the height of a person in metres divided by the weight in kilograms squared (Figure 12.10).

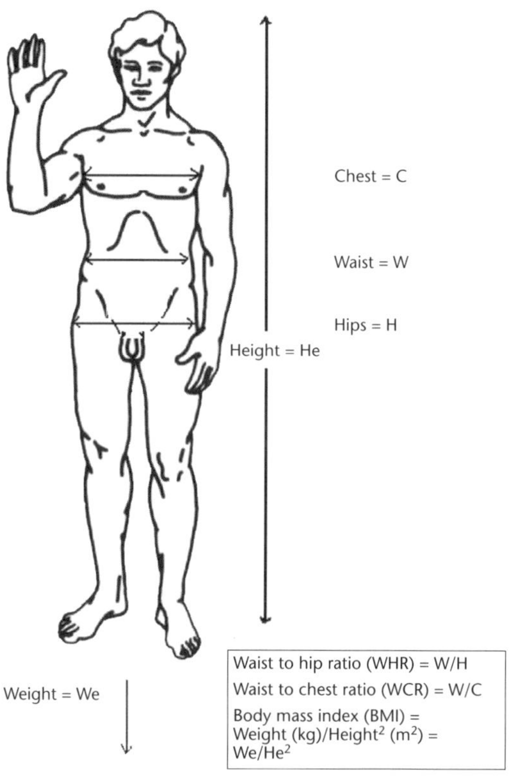

FIGURE 12.10 WHR, BMI and WCR

BMI also has health implications (see Table 12.5 above) and so may be expected to influence attractiveness judgements. For Caucasian women in Western Europe and the USA, the optimal BMI (from a health point of view) is estimated at 19–20 kg/m^2 (Tovee and Cornelissen, 2001). The 'normal' range for BMI is 18.5–30 and it serves as a rough measure of whether someone is overweight or underweight (see Figure 12.11).

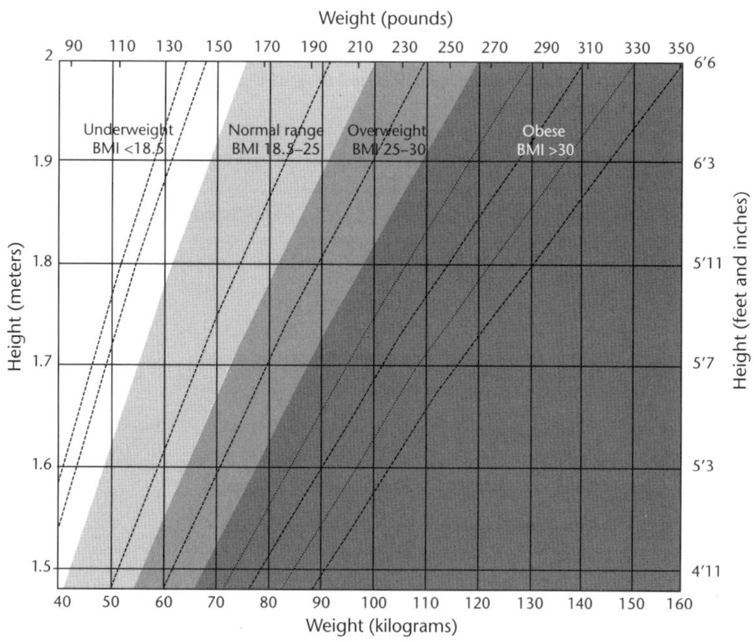

FIGURE 12.11 BMI values for humans of differing heights and weights

SOURCE: Adapted with permission from Wikipedia.

There is now much controversy in the literature about the relative effects of WHR and BMI on attractiveness ratings. It has been suggested that the problem with Singh's original line drawings is that as WHR is changed, then so too is the overall body mass of the figure. In Singh's overweight group of line drawings, for example, altering the width of the waist also alters the apparent BMI, making it difficult to establish if the determination of attractiveness comes from the variation in WHR or BMI or both (Tovee and Cornelissen, 1999).

In a study on the Shiwiar people of the Ecuadorian Upper Amazon, Laurence Sugiyama (2004) employed an experimental procedure based on Singh's original drawings designed to isolate effects of BMI and WHR. He concluded that Shiwiar males preferred females with a BMI higher than local averages. Thus, if WHR and BMI were not independently assessed, this would translate to high WHR figures, since such figures give the impression of high BMI values. When differences in body weight are controlled, Shiwiar men exhibited a preference for lower than locally average female WHR figures.

In another study, Streeter and McBurney (2003) designed an experimental procedure in which they claimed the effects of weight and WHR could be isolated. They manipulated a photograph of a woman to obtain a whole range of hip, waist and chest measurements. Subjects were also asked to estimate the weight of the woman. The authors found that although weight accounted for 66 per cent of the total variance in attractiveness estimations, the effect of WHR was still clearly apparent. For both male and female judges, the attractiveness rating of the photograph peaked at a WHR of 0.7.

Other work that demonstrates the variability of BMI preferences comes from Martin Tovee and colleagues (Tovee et al., 2006). They looked at the WHR and BMI preferences of four groups of males: 100 British Caucasians, 35 ethnic Zulus, 52 Zulus who had immigrated to Britain and 60 British men of African descent. Some interesting results were found for the relationship between BMI and perceived attractiveness according to the ethnic group (Figure 12.12).

As can be seen from Figure 12.12a, for UK Caucasians, the attractiveness of BMI-related images rose and fell, peaking at around 20 kg/m2 – a figure near the optimum for health and fertility; but for native Zulu people, the attractiveness remained high, above a BMI of 20 kg/m2. Figure 12.12b shows that Britons of African descent have a similar response pattern to Caucasian Britons, while Zulus who have recently immigrated to Britain had a response somewhere in between native Zulus and native Britons. Tovee et al. explain the preference for high BMI figures among native South African Zulu men by pointing to the different health associations with BMI in the UK and rural South Africa (see Table 12. 8). In essence, a high BMI carries positive connotations and implications in rural South Africa but potentially negative ones in the UK.

From such findings and those on WHR in the same study, Tovee et al. conclude that this provides evidence for flexible attractiveness preferences that shift according to local conditions and corresponding health associations. It is interesting to note that the recent Zulu immigrant to the UK has a response midway between native Zulus and resident Britons. This shows that the change in preferences is not a genetic drift of any sort but a phenotypic change in the cognitive appraisals of body shape. The human brain may therefore be receptive to cues about what constitutes optimality or normality in the local population. One way

(a) UK Caucasians and South African Zulus compared

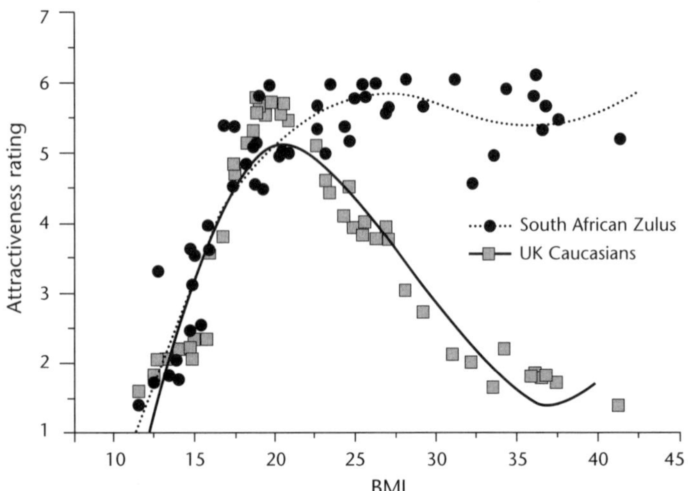

(b) Zulus recently migrated to the UK and Britons of African origin compared

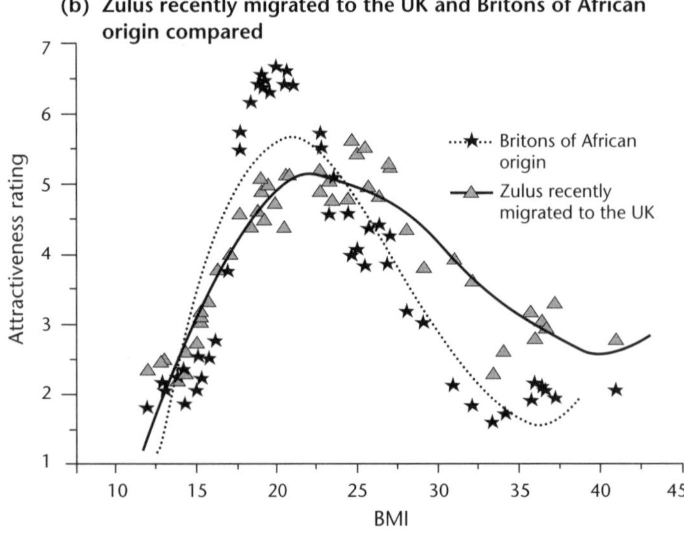

FIGURE 12.12 Comparison of the attractiveness ratings by three groups of observers according to BMI

Each point represents the average rating for the group for a specific BMI body image.

SOURCE: Taken with permission from Tovee, M. J., Swami, V. et al. (2006) 'Changing perceptions of attractiveness as observers are exposed to different culture.' *Evolution and Human Behaviour* **27**(6): 443–57, Figure 1, p. 448.

of achieving this is through imitation: recent arrivals to a new cultural context imitate the norms prevailing there.

Table 12.8 Suggested associations between BMI and SES in two different cultures

	UK	Rural South Africa
High BMI	Associated with low SES, poverty and poor diet	Associated with high SES
Low BMI	Associated with lower levels of cancer, higher SES and longer term female health	Thought to signal weight loss due to disease and parasite infections

Further evidence that the aesthetic judgement of male and female body shapes is related to SES comes from the work of Swami and Tovee (2005a, 2005b) on British and Malaysian subjects. They found that BMI accounted for most of the variance in attractiveness of female body images and that lower SES groups preferred higher BMI figures (Table 12.9).

In later work, Swami and Tovee (2006) even found that the degree of hunger in the same subject group could influence attractiveness judgements. In a study using photographic images as stimuli, they found that hungry men preferred women of higher BMI than those preferred by satiated men. As these authors note, such a study provides evidence for a proximate mechanism for the integration and expression of cultural norms at the individual level. One intriguing hypothesis that could account for such hunger-related changes and other aspects of cultural variation in aesthetic judgements is the 'environmental security hypothesis' advanced by Pettijohn and Tesser (1999). This theory adopts a behavioural ecological approach to explain the context-dependent nature of attraction. It proposes that when environmental conditions are uncertain, variable or harsh, then individuals will be attracted to others with more mature characteristics. The reasoning behind this is that maturity is likely to be associated with the ability to cope with threatening situations and difficult conditions (you don't reach maturity if you can't). Now, since BMI increases with age (as does WHR), this could explain why hungry men prefer women of higher BMI compared to the preferences of satiated men. It could also account for the way body weight and size preferences vary with the SES of various cultures. In effect, the hungry affective state is used as a cue for what the local environment is like.

The interpretation of why the preferred BMI rises with lower SES groups is still open to question. It may be that it falls as societies develop and become exposed to Western standards, although this then still leaves open the question as to why a lower BMI

Table 12.9 The effect of decreasing SES on male preferences for female body shape ('female figures') and female preference for male body shape ('male figures') in British and Malaysian groups

SES	Study site and group	Variance in attractiveness accounted for by BMI		Variance accounted for by WHR		Peak BMI for attractiveness	
		Female figures	*Male figures*	*Female figures*	*Male figures*	*Females*	*Males*
High	British	84.1	37.8	7.4	12.3	20.85	20.86
Low	Rural Sabah	76.9	73.7	1.6	4.8	22.78	24.09

NOTE: The peak BMI increases as the SES level decreases for both female judgements of men and male judgements of women. Note also that these authors found BMI to be more important than WHR as a determinant of attractiveness for all groups. Some of the variance in attractiveness for men was also accounted for by waist to chest ratio (not shown here).

SOURCE: Adapted from Swami, V. and Tovee, M. J. (2005a) 'Male physical attractiveness in Britain and Malaysia: a cross-cultural study.' *Body Image* **2**: 383–93, Swami, V. and Tovee, M. J. (2005b) 'Female physical attractiveness in Britain and Malaysia: a cross-cultural study.' *Body Image* **2**: 115–28.

is associated with more affluent cultures; or it may be an adaptive response to cues of health and fertility in resource-poor cultures.

This cultural variability in ideals of male and female body shape does pose a challenge to the strict nativist view of EP that our preferences were adapted to conditions in past environments. We could posit mental mechanisms that respond to what is adaptive in local contexts but this assumes that there was sufficient variability in conditions and food availability in the Old Stone Age for this context-sensitive mechanism to evolve.

One way of resolving the issue of the relative contribution of WHR and BMI to attractiveness may be to see them as part of a nested hierarchy of cues or a series of filters of decreasing bandwidths. WHR could serve as a broad pass filter to initially distinguish male from female, pregnant from non-pregnant females and then young from old females. The BMI effect could then kick in to give some context-specific information about the health and nutritional status of the person.

12.4.4 The female breast

In Western cultures, and probably many more, there is a fascination with the enlarged mammary glands of the human female. There are strong cultural mores about when they can be revealed, or should be concealed or only half-revealed. The femininity of a woman is strongly associated with her breasts. Women sometimes pay large sums of money and experience much discomfort to have them reduced or enlarged. In 1996, for example, in the USA, there were 87,700 breast augmentations and 64 per cent of these were women in the age range 18–34. In the same year, there were 57,679 breast reductions, with 45 per cent of these in the 18–34 age range (American Society of Plastic Surgeons, 2007).

There is agreement between the sexes that they are objects of some importance, but what are they for? Most people would take this to be a pointless question, since it is obvious that they are there to provide infants with milk. A consideration of the facts below forces a rethink on this issue:

- Breasts are strongly sexually dimorphic and appear at puberty
- Permanently enlarged breasts are not found among any other primates: most primates have enlarged breasts only during pregnancy and lactation
- Large breasts, although attractive to males, interfere with locomotion, and women athletes engaged in running sports tend to have small breasts
- There is some cross-cultural variation in breast morphology but with no obvious ecological correlates
- The size of a woman's breasts bears very little relationship to her ability to lactate. Women could supply the necessary nutrition to a baby with much smaller breasts.

It looks, then, as if breasts have not been shaped by natural selection, women would move better without them and their permanently enlarged state is not essential in order to supply milk. Acting as a storage device for fat is a possibility, but storage around the waist would be mechanically more efficient. Breasts are thus prime candidates for good genes or runaway sexual selection.

Some studies have shown that breast symmetry correlates with fertility, which suggests a role in the honest advertisement of good genes. The fact that breast size is not negatively correlated with **asymmetry** runs counter to this, however, since a sexual trait that increases in size should become more asymmetrical as the demands of size growth take their toll on symmetry (Thornhill and Gangestad, 1994). Another idea is that breasts (like the male beard) serve as a sign of reproductive value. They develop rapidly at puberty and so signal the onset of reproductive age. It is advantageous for men to mate with women of this age and so they developed a preference for enlarged breasts (Barber, 1995; Marlowe, 1998). Until more conclusive evidence is forthcoming, the consensus seems to be that they have been sexually selected, but the precise mechanism is not certain.

12.5 Facial attractiveness: honest signals, symmetry and averageness

12.5.1 Honest signals

In Chapter 3, we noted how one of the theories for the origin of sex emphasised the need of organisms to keep one step ahead of invading parasites. It is easy to forget that our bodies are host to millions of microorganisms. Our skin, hair and guts are teeming with them: the number of microbes you carry around with you is larger than the number of cells in your own body. Most of these are symbionts (that is, they act mutualistically with your own cells; see section 9.5.1), of course, but parasites also abound. In the coevolutionary struggle with parasites, there is no optimum solution to be found: each adaptation is only temporary, and each solution sets up a selective pressure for an anti-solution (Thornhill and Gangestad, 1993). Parasites generally use the resources of the host – nutrients, enzymes, proteins and so on – for their own reproductive purposes. It is probable that, in any population, the majority of **pathogens** will be best adapted to the most common biochemical pathways of their hosts. Sex is one defence against this: by outbreeding, parents produce offspring that are different from themselves; any parasites that were successful with parents may be less so with their offspring, since the offspring will have a whole new array of proteins and a different defence system.

It follows that the greater the genetic polymorphism at the population level – that is, the more alleles at any given locus on the genome of the species – the more likely it is that at least some individuals will stay ahead of parasites. If genetic variation confers fitness on a population, the same thing may be true at the level of an individual: the more heterozygous an individual, the more variability to be found in its genome, the more varied the proteins that are produced and thus the harder it will be for a parasite to exploit its host efficiently. But organisms face a problem: genetic variation that is good from the point of view of excluding parasites could be disruptive if the variation departed from what was adaptive for any given set of conditions. Just as too much inbreeding may cause a depression of fitness by generating homozygosity for recessive alleles, and homozygosity may depress fitness by allowing parasites to exploit the lack of variation in proteins, so excessive outbreeding may introduce less well-adapted variations. There is clearly a balance to be struck.

On one side of the equation, sexual selection leads us to believe that mates will choose each other taking into account the resistance to parasites that a potential partner possesses and hence could potentially confer on its offspring. Following this line of reasoning, Thornhill and Gangestad (1993) suggest that four sorts of preference might be expected to evolve through sexual selection in an environment where parasites are prevalent:

1. A preference for heterozygosity
2. A preference for parasite-resistant alleles
3. A preference for indicators of development stability are hence signs of a well-functioning genome
4. Preferences for handicaps that can only be afforded by parasite-resistant individuals, that is, honest signals of genetic value. Such handicaps may include hormones that suppress the immune system, and hence the bearers of such handicaps must have sound immune systems to be able to survive.

The question arises of how a choosy partner will be able to estimate the degree of heterozygosity, the state of the immune system in a potential mate, and signs of developmental stability. We noted in Chapter 3 that there is evidence of both mice and humans being attracted to smells that indicate differences in the major histocompatibility complex (MHC) between mates (see also section 13.3.3), but there may be more obvious ways. If males send out honest signals to females in the form of secondary sexual characteristics, well-adapted males can advertise their worth. Alternatively, they can reveal a handicap that they are able to bear as a result of an efficient immune system. It transpires that testosterone could serve as just such a handicap: it suppresses the immune system such that only well-adapted males can tolerate a high level of it, additionally revealing its presence by affecting the growth of secondary sexual display characteristics. Figure 12.13 shows how the system is conjectured to work.

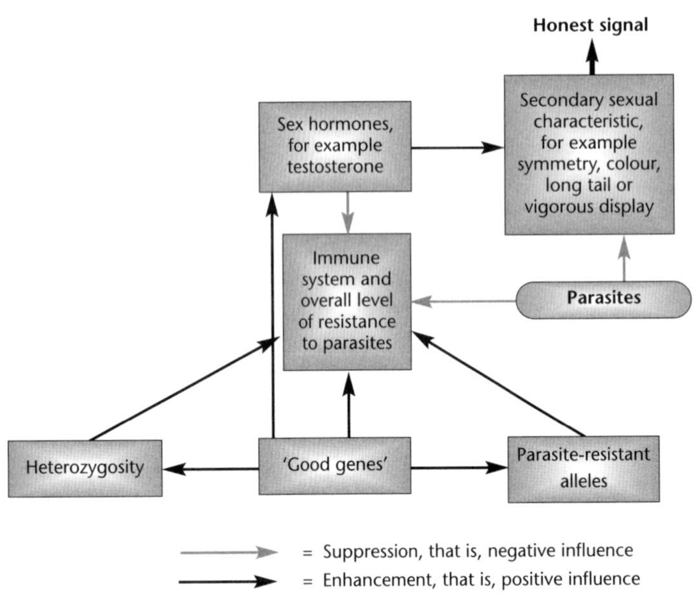

FIGURE 12.13 Conjectured relationship between resistance to parasites, secondary display characteristics and honest signals

The system shown in Figure 12.13 results in the outcome that only fit males can afford to suppress their immune system with a high level of testosterone. Consequently, they display characteristics that indicate this fact. Evidence in support of this scheme for non-human animals comes from the work of Saino et al. (1997) on barn swallows. The length of tail of a barn swallow seems to be an honest advertisement of testosterone. Males with short tails were not only less attractive to females, but also, when injected with testosterone, suffered a higher mortality than their long-tailed counterparts. As Figure 12.13 shows, symmetry may be used as a cue for good genes and this is addressed in the next section.

12.5.2 Symmetry and fluctuating asymmetry

Over the past 20 years, there has been a massive increase in our understanding of the factors that influence the attractiveness of faces. One of the factors investigated has been symmetry. If you stand in front of a mirror and shield each half of your face in turn, you will notice that, unless you are very lucky, your face is not perfectly symmetrical. But does symmetry enhance attractiveness? Early research on this issue was clouded by the fact that when faces were made perfectly symmetrical (by simply presenting two mirror images of each half of the face), artefacts of this procedure, such as shading effects, adversely affected the viewers' judgements. Better techniques have reliably shown that symmetry does indeed enhance attractiveness (Grammer and Thornhill, 1994; Mealey et al., 1999; Perrett et al., 1999; Scheib et al., 1999). Furthermore, the importance of symmetry extends to the rest of the body. Gangestad and Thornhill (1994) found a positive correlation between the symmetry of men, measured on such features as foot length, ear length, hand breadth and so on, and estimations by women of their attractiveness. Work by Tovee et al. (2000) has demonstrated the importance of female body symmetry in influencing attractiveness ratings. In a forced choice/two alternatives experiment involving the simultaneous presentation of an image of a female body together with its copy morphed to increase symmetry, there was a significant preference for the symmetrical image over the normal image. If we follow the evidence on symmetry, the obvious question that arises is why symmetry should be attractive. The most probable explanation is that it requires a sound metabolism and a good deal of physiological precision to grow perfectly symmetrical features. The development of bilateral characters such as feet, wings, fins and so on is open to a variety of stressful influences, such as parasite infection and poor nutrition, which result in an asymmetrical finished product. On this basis, the extent of symmetry observed in such characters may serve as an honest signal of phenotypic and genotypic quality.

Research on the importance of symmetry often uses the concept of **fluctuating asymmetry** (FA). FA refers to bilateral characters (that is, one feature on each side of the body) for which the population mean of asymmetry (right measure minus left measure) is zero, variability about the mean is nearly normal, and the degree of asymmetry in an organism is not under direct genetic control but may fluctuate from one generation to the next. So FA might be usefully measured from, say, ear measurements but not bicep measurements, since it is perfectly normal that right-handed people (the majority) will have slightly larger biceps on the right arm, but there is no reason to expect any systematic difference in ear sizes across the population. There is now abundant evidence that FA is increased by mutations, parasitic infections and environmental stress, so FA consequently becomes a negative indicator of phenotypic quality (Manning et al., 1996). The measurement of absolute and relative fluctuating asymmetry is shown in Figure 12.14.

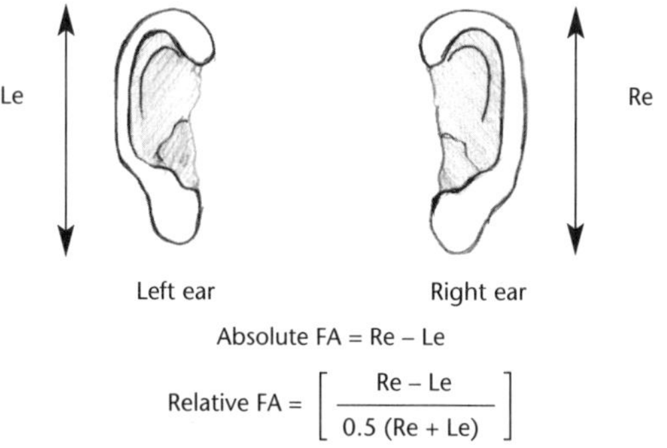

FIGURE 12.14 Measurement of relative and absolute fluctuating asymmetry
Relative FA is usually taken as the positive value of the relative difference between right- and left-sided measurements.

Manning has suggested that if body size dimorphism in humans is a result of sexual selection, with females preferring taller males, it could be expected that body size is also an indication of phenotypic quality. Now, phenotypic quality, from the evidence presented above, can also be expected to be negatively correlated with the degree of asymmetry in an organism: the larger the degree of asymmetry (that is, the higher the FA value), the poorer the quality of the organism. On this basis, we should find a weaker relationship for female body size since body size in females has not been the subject of a sexual selection pressure. The preliminary findings from a study on just 70 adult males offer some support for this (Manning, 1995). When the average relative FA of four traits was compared with body size, a positive correlation was observed for females and a negative one for males.

Manning et al. (1997) have also suggested that since the maintenance of symmetry requires metabolic energy, those males who have 'energy-thrifty genotypes' are better placed to maintain a low FA. Those males with high resting metabolic rates should show greater signs of asymmetry, since less energy is available to divert to symmetry maintenance. A preliminary study of 30 males supported this prediction. There now seems to be much evidence that high levels of FA negatively influence attractiveness (see Little et al., 2000).

12.5.3 Averageness

In 1883, Galton wondered if certain groups had distinctive facial characteristics. To investigate this, he took photographs of two groups: criminals and (curiously) vegetarians. Galton overlaid the photographic plates of each subject, creating a composite face for each group. His efforts to extract distinguishing features of criminals and vegetarians proved fruitless, but he did notice that the composite or 'averaged' faces were quite attractive. Over the past 20 years, computer-assisted technology has enabled this line of research to proceed apace. Using a computerised technique of merging faces, Langlois and Roggman (1990) found that not only were composite faces more attractive than individual ones, but also the more faces that went into making a composite, the more attractive, up to a point, the face became. A face composed of 32 faces, for example, was more attractive than one composed of two.

But why should composite faces be attractive at all? Symons (1979) suggested that the average was rated as attractive because the average of any trait would tend to be optimally adapted for any trait, since the mean of a distribution presumably represents the best solution to the adaptive problem. Thornhill and Gangestad (1993) add to this the fact that protein heterozygosity tends to be highest in individuals who exhibit the average expression of heritable traits that are continuously distributed. It could be, then, that **facial averageness** correlates with attractiveness because averageness is an indication of resistance to parasites.

The experimental problem to overcome in such studies is the fact that there are at least two variables at work: averageness and symmetry. Composite faces also tend to be more symmetrical since asymmetries are ironed out. Symmetry, as already noted, is a reliable cue to physiological fitness that takes the form of resistance to disease and a lack of harmful mutations that compromise development stability. Yet there must also be other factors at work, for whereas symmetrical faces are regarded as being more attractive than faces with gross asymmetry, computer-generated, perfectly symmetrical faces are not perceived as being as attractive as natural faces with a slight asymmetry (see Perrett et al., 1994). There is now some consensus that averageness and symmetry both work independently to enhance attractiveness (for a review, see Rhodes, 2006).

12.5.4 The enigma of the beard

In a culture in which shaving is routinely practised, it is easy to forget that one feature that is strongly sexually dimorphic is facial hair. The beard emerges at puberty and thus looks like a prime candidate for sexual selection. The function of the male beard is a subject of much dispute and speculation and one that shows how easy it is to weave just so stories that may not hold together. As yet, however, there is no consensus on whether runaway (Fisherian) or good genes sexual selection has been at work. The fact that adult males sport beards is not likely to be the result of natural selection, since children and women make do without them. Darwin (1871) speculated that the nearly hairless state of a women's body, compared to that of men, was probably a product of atypical (since selection came from the male) sexual selection in the distant past. Barber (1995) suggested that beards were signals to women of maturity and social dominance. In the EEA, a man's beard would have been difficult to trim (also, why bother?) and so its size would have increased with age. Since a man's social status also tends to increase with age, so the beard became a signal of maturity and social standing. Women who found such men attractive did better than those who did not. The problem here is that such a theory might suggest that women should still find beards attractive, but the evidence on this topic is ambiguous. In some studies, the rating of men's physical attractiveness increased with the quantity of facial hair (Pellegrini, 1973; Hatfield and Sprecher, 1986). But other studies have revealed negative attractiveness ratings of beards (Kenny and Fletcher, 1973; Feinman and Gill, 1977).

In some cultures, such as the native Andean Indians of Bolivia and Peru, men lack facial hair. It may be that beards are absent from cultures where there is a higher risk of parasite infection. Since parasites alter the appearance of facial skin, women might have demanded hairless faces to ensure that men honestly signalled their health status. The presence or lack of hair is not, however, obviously correlated with any ecological factors. A beard could serve as a dishonest signal by seeming to enlarge the size of the chin, a large chin generally being thought to indicate a high testosterone level.

It may be that beards once served as a weak signal of a good prospective mate, but a signal easily manipulated by cultural trends. Indeed, there is plenty of evidence linking the popularity of beards to changes in fashion and the behaviour of role models. The ancient Egyptians found beards on ordinary people unattractive. Roman emperors tended to be clean-shaven until the emperor Hadrian set a trend in the second century AD by growing a beard. This was partly to express his interest in Greek philosophy (he studied at Athens and Greek philosophers wore beards) and possibly, in part, as a way of concealing some facial blemishes. During and after World War I, mass propaganda successful sought to persuade people of the hygienic benefits of removing beards, and the clean-shaven state has been common among Western men ever since. Beards re-emerged in the counterculture of the 1960s as a deliberate gesture of rebelling against convention.

12.5.5 Other aspects of attractive faces

Eibl-Eibesfeldt (1989) proposed that the qualities that males find attractive in human females, such as a small, upturned nose, large eyes and a small chin, correspond to 'infantile' features. Such features are found in the young of other creatures that both sexes find attractive or 'cute'. It may be that women have developed these features to evoke the caring response that males feel towards their young. In effect, women's faces have targeted in on the perceptual bias in male brains. Other features in women's faces that men find attractive are also associated with signs of youth. As women age, increasing levels of testosterone stimulate hair growth. Significantly, men tend to find hairless faces in women more attractive – as evidenced by the efforts of women to remove facial hair. Hair colour also tends to darken with age, and the attractiveness of blonde hair could be that it is a reliable sign of youth. Skin darkens with age and with pregnancy. There is some evidence that men find paler skin more attractive, although the effect is often masked by the fashion for sunbathing (see Barber, 1995, for a review of this area).

Some of these conjectures, and some of the information presented so far, are summarised in Table 12.10.

Table 12.10 Facial features and possible function in natural or sexual selection

Facial feature in male or female	Possible role in selection
Lack of facial hair in women and attempts to remove facial hair	Indicates youth
Neotenous (childlike) features regarded as attractive in women	Could elicit nurturant response in men. Also small chin and nose indicates low levels of testosterone
Symmetry in male and female faces regarded as attractive	Symmetry is an indicator of physiological precision, protein heterozygosity and hence resistance to or freedom from pathogens
Large chin and prominent cheekbones in men	Such features indicate high levels of testosterone. Testosterone may serve as a handicap in that it suppresses the immune system
Hair on face of men (attractive?)	Could be runaway sexual selection. Indicates maturity. Enlarges appearance of chin
Average faces attractive	Averageness correlates with symmetry. Average could indicate heterozygosity and resistance to pathogens. It could also indicate an optimum level for a continuously distributed trait. Or it may appeal to inbuilt cognitive preferences for the familiar

Attractiveness and health

The supposed adaptive functions of attractiveness in Table 12.10 can be used to generate many hypotheses, such as a connection between health and attractiveness, fertility and attractiveness, testosterone and masculinity and so on. Some of these have already been tested. Shackelford and Larsen (1999), for example, report a small-scale study suggesting that facial attractiveness may correlate with good health. But other studies have found it difficult to show that average or highly symmetrical faces are also indications of a healthier or more fertile person. This is partly to be expected, given the advent of medical technologies that can compensate for any slight fitness depressions.

Until recently, it has been assumed that highly masculine faces (broad chin and high cheekbones) will be products of high testosterone levels. Recently, more direct evidence has been forthcoming. Ian Penton-Voak and Jennie Chen (2004) took facial photographs of 50 Caucasian men whose testosterone levels (from samples of saliva) were also measured. The group was then divided into 25 high and 25 low testosterone males. Composite facial images were then constructed from the faces in each group. When presented with these images, volunteer observers tended to significantly judge the high testosterone faces as more masculine than those in the low testosterone group.

Since there is sexual dimorphism in face shape, it is possible to enhance the femininity or masculinity of any face by manipulating features such as chin size and cheekbone prominence using computer imaging. Perrett et al. (1998) did just this, looking for the effects of masculinisation and feminisation on perceptions of attractiveness. They found that feminising a female face increased its attractiveness, a preference they saw as contrary to the averageness hypothesis: female faces became more rather than less attractive if they departed from the average in a feminine direction. One surprising result of initial work on this phenomenon was that when an average male shape was feminised (Figure 12.15), it too became more attractive. Later work by this same group and others has pointed towards a hormone-mediated female mate choice strategy, a subject considered in the next section.

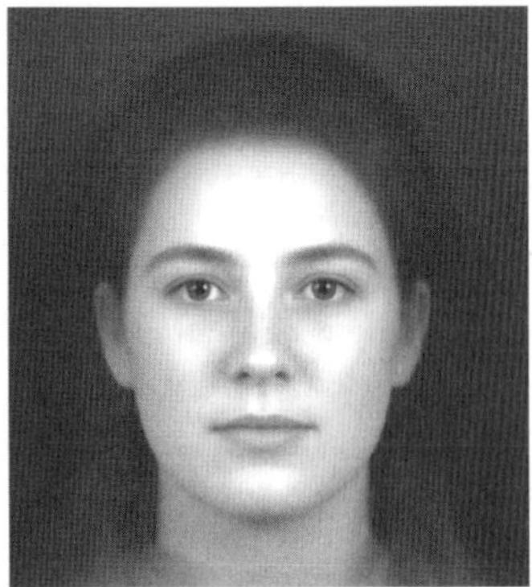

Average Caucasian female

Average Caucasian male

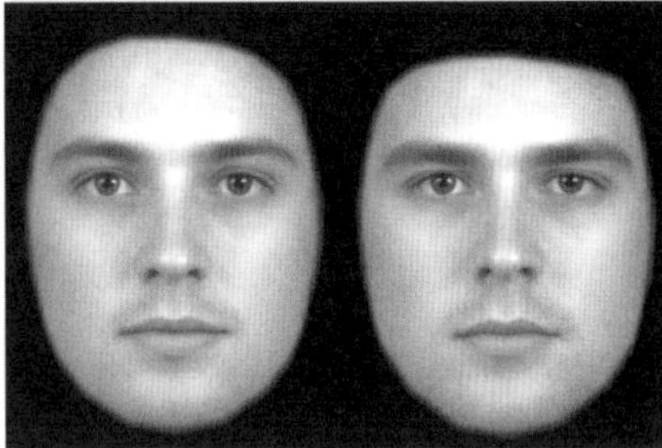

FIGURE 12.15 Facial attractiveness: averaging and sexual dimorphism
The first two images show the effect of merging photographs of 30 Caucasian females (mean age 20.6 years) and 25 Caucasian males (mean age 21.0 years) to create two 'average faces'. Most would agree that the faces appear attractive. The pair of images shows the effect of a 50 per cent feminisation (left) and 50 per cent masculinisation (right) of the composite male face. Females were found to prefer the masculinised male face when their risk of conception was highest.

SOURCE: Reproduced with permission from Penton-Voak, I. S. and Perrett, D. I. (2000) 'Female preference for male faces changes cyclically: further evidence.' *Evolution and Human Behaviour* **21**(1): 39–49.

12.5.6 Female facial preferences: a potential hormone-mediated adaptive design feature

What qualities do females look for in a mating partner? In theory, we might expect a female (in addition to other criteria, such as MHC compatibility) to desire both sets of the following characteristics:

- A partner who is caring, cooperative, honest, loyal and would make a good parent ('dads')
- A partner who has a good immune system (as indicated by symmetry and signs of high testosterone) and who will therefore provide a good set of genes for her children. Note high testosterone can only be supported by males with an efficient immune system since testosterone is an immunosuppressant ('cads').

The problem with finding a male with both these characteristics is that there is conflict between the second and the first set. More specifically, high testosterone may be a reliable indication of a good immune system but it is also implicated in some antisocial qualities such as aggression. But there is now increasing evidence that evolution may have solved this by providing females with what we might call an 'onboard hormone-influenced variable response attractiveness detection unit'. In favour of this notion is work by Penton-Voak and Perrett (2000). They presented five male faces of varying levels of masculinity and femininity in a national UK magazine and asked female readers to choose the most attractive face. Subjects were also asked details of their menstrual cycle. The responses were then ordered into two categories: judgements made at the time of low conception risk (days 0–5 and 15–28 of the ovarian cycle) and high conception risk (the follicular phase of days 6–14). Figure 12.16 shows their findings.

Although there was some ambiguity in the instructions to subjects ('most attractive' could be interpreted as qualities for a long- or short-term sexual partner, for example), the results provide evidence for cyclical changes in female preferences. There seems to be a shift to attend to signs of high testosterone and heritable immunocompetence when conception is most likely.

This study is in line with others demonstrating similar female conditional mate choice strategies. Victor Johnston and colleagues used an ingenious technique to construct a short movie film showing faces gradually morphed from male to female (Figure 12.17). As the frame containing the facial image unrolled, subjects (females) were asked to rate male faces using criteria such as physical attractiveness, intelligence and sensitivity. As predicted, there was a shift in female preference to a more masculine male face in their high-risk phase of the menstrual cycle (Johnston et al., 2001).

If this work is upheld, it seems that ancestral females desired immunocompetent 'cads' during their most fertile phase, and caring 'dads' the rest of the time. This internal unconscious mechanism, coupled with sexual crypsis, ensured that ancestral females could get the best of both worlds.

In this chapter, we have focused on positive mate choice preferences. The other side of the mating coin is rejecting undesirable mates. One type of mate that it is in the interests of both males and females to avoid is someone who is a close genetic relative. This avoidance of inbreeding and incest is the subject of the next chapter.

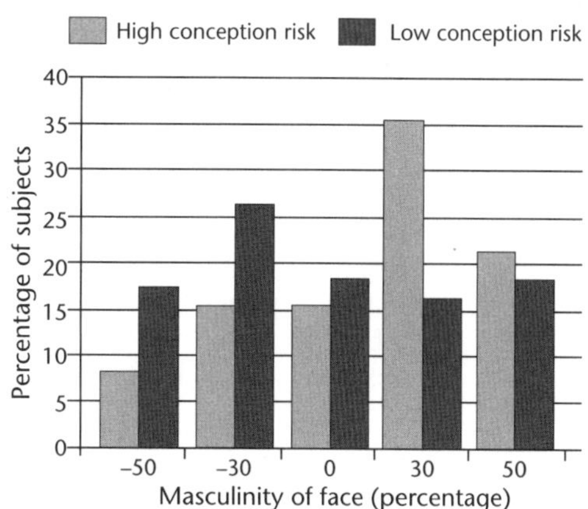

FIGURE 12.16 Percentage of female subjects choosing a face as most attractive in relation to their conception risk and masculinity of chosen face

SOURCE: Redrawn from Penton-Voak, I. S. and Perrett, D. I. (2000) 'Female preference for male faces changes cyclically: further evidence.' *Evolution and Human Behaviour* **21**(1): 39–49.

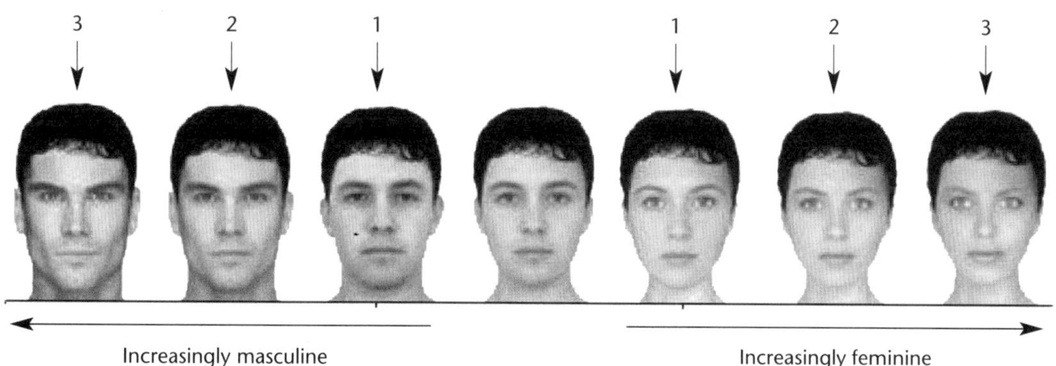

FIGURE 12.17 Typical images used in Johnston et al.'s facial morph movie to assess attractiveness of faces

Image 1 shows composite (averaged) of 16 random male and female faces; 2 shows use of Faceprints software to evolve masculine and feminine versions; 3 shows extreme male and extreme female faces produced by caricaturing masculine and feminine features respectively. The central image is androgynous.

SOURCE: Reproduced with kind permission of Johnston, V. S., Hagel, R. et al. (2001) 'Male facial attractiveness: evidence for hormone-mediated adaptive design.' *Evolution and Human Behaviour* **22**: 251–69.

Summary

- Although males and females share many mate choice criteria, evolutionary considerations would predict a difference between the attributes sought by males and by females in their prospective partners. Females are predicted to look for high-status males who are good providers, whereas males are predicted to look for young, healthy and fertile females who are good child-bearers. Men are also much more likely to seek sex without commitment. There is considerable empirical support for these and related predictions.
- Physical beauty in both males and females is expected to be correlated with signs of reproductive fitness. Evidence is now accumulating that body shapes found to be attractive are those which carry fitness indicators. Facial beauty in both sexes is likely to be related to overall fitness, health and immunocompetence. Faces may convey honest signals about fitness and reproductive value.

Key Words

- **Asymmetry**
- **Body mass index (BMI)**
- **Facial averageness**
- **Fluctuating asymmetry (FA)**
- **Pathogen**
- **Waist to hip ratio (WHR)**

Further reading

Buss, D. M. (1994) *The Evolution of Desire*. New York, HarperCollins.

Based on research from Buss and others. Clearly explains expected sex differences in desire.

Rhodes, G. (2006) 'The evolutionary psychology of facial beauty.' *Annual Review of Psychology* **57**: 199–226.

Good review article.

Rhodes, G. and Zebrowitz, L. A. (eds) (2002) *Facial Attractiveness*, Westport, CT, Ablex Publishing.

Series of recent articles on the subject. Well-organised book and easy to use.

Swami, V. and Furnham, A. (2007) *The Psychology of Physical Attraction*. London, Psychology Press.

Excellent coverage of recent research in this area. Shows how evolutionary psychology can be informed by social psychology and cultural considerations.

Incest Avoidance and the Westermarck Effect

Oh, on these eyes
Shed light no more, ye everlasting skies
That know my sin! I have sinned in birth and breath
I have sinned with Woman. I have sinned with Death.
(*Oedipus Rex*, Sophocles, 428 BC)

The lines above are spoken by Oedipus when he discovers that he has unwittingly married his mother, Jocasta, and has fathered several children with her. Jocasta hangs herself and Oedipus gouges out his own eyes. For the ancient Greeks, as for most human cultures, incest was crime against the natural order and there were strong social rules to prohibit it.

Yet as a problem in biology and cultural anthropology, the formation of rules governing incest and its avoidance seems guaranteed to generate controversy (see, for example, Leavitt, 1990). The reason is probably that so many disciplines and paradigms can claim to have some purchase on the phenomenon and have, accordingly, produced so many competing theories. Moreover, these explanations often invoke core or foundational principles in these disciplines, increasing the odds of what is at stake. Evolutionary biologists see incest avoidance as a biological phenomenon designed to avoid inbreeding reductions in fitness; anthropologists can see it as a cultural phenomenon designed to stabilise the family or form family alliances and so elevate man from a state of nature to culture; Freudians can supply interpretations in terms of the Oedipal stage of infantile sexuality; meanwhile, sociologists often view marriage rules and their variability across cultures as a means of preserving wealth and power. In this chapter, we will consider some of these competing approaches, mainly focusing, of course, on the evidence bearing upon evolutionary explanations.

In discussing **incest** and **inbreeding** it is wise at the onset to distinguish between these two terms. An incestuous relationship is a sexual union between human biological relatives who are more closely related than legislation in that culture permits. What may be incest in one culture (or even a US state) may not be in another. Inbreeding is a more value-free biological term to describe breeding between related individuals. In this discussion, the term 'incest' will imply that humans are the focus; inbreeding will be used to refer to human and non-human animals.

13.1 Early views about inbreeding and the incest taboo

In the late 19th century, it was widely believed that inbreeding, either among humans or within other animals, did not produce any undesirable biological effects. In 1948, the cultural anthropologist Leslie White could still insist that inbreeding does not cause degeneration and that the 'testimony of biologists' was 'conclusive' on this score (White, 1948). The prevailing view from about 1910 to 1960 was that incest avoidance was entirely a cultural restriction with no basis in biology.

FIGURE 13.1 Sigmund Freud (1856–1939) taken in about 1938
Freud is regarded as the father of psychoanalysis. His view was that infants pass through a series of stages that are essential to normal adult functioning. One of these, for boys, was the Oedipal stage. Freud saw the incest taboo as a necessary way of curbing natural sexual desires. Understandably, therefore, he vehemently opposed Westermarck's more Darwinian view.
SOURCE: Library of Congress Prints and Photographic Division, LC-USZ62-72266.

But if the taboo against incest was not biologically based, then what caused it? Many supported the notion that the prohibition of incest was an artificial and hence a supranatural barrier self-imposed by early human family groups to facilitate exogamy. Leslie White typified this view when he noted that 'Co-operation between families cannot be established if parent marries child; and brother, sister' (quoted in Wolf, 2004, p. 7). Incest prohibition, therefore, enabled early families to unite and so began human social evolution. Claude Lévi-Strauss came to a similar conclusion in 1950, when he suggested that the prohibition of incest was a profound step, whereby mankind first renounced animalistic instincts, transcended the natural order and initiated a truly human society as families were united by outbreeding. As Lévi-Strauss said (1956, p. 278):

> If social organization had a beginning, this could only have consisted in the incest prohibition ... it is there, and only there, that we find a passage from nature to culture, from animal to human life.

The theories of Lesley White and Lévi-Strauss are sometimes called 'alliance theories', since they stress the functional role of interfamilial alliances.

In *Totem and Taboo* ([1913] 1950), Freud suggested an alternative function for the taboo (Figure 13.1). He argued that since daughters naturally desired their fathers (the Electra complex) and sons their mothers (the Oedipal complex), then the conflicts resulting from giving free rein to these instincts would make family life impossible. Hence, humans had no alternative but to renounce and repress these desires.

13.2 Westermarck's alternative Darwinian explanation

Although the views of White, Lévi-Strauss and Freud were influential, they were not the only ones in this period. The main rival to these ideas, one which approached the problem from an entirely different Darwinian perspective, was the theory of Edvard Westermarck (1862–1939). Westermarck was born in Helsinki and at the age of 25 taught himself English in order to study the works of Darwin, Morgan and Lubbock in the original language. In 1887, he came to the British Museum in London and studied to produce his doctorial thesis on human marriage. His work first saw publication in 1891 as *The History of Human Marriage*, a work later expanded in 1922 to three volumes (Figure 13.2).

Westermarck suggested that men do not mate with their mothers and sisters because they are disposed not to find them sexually attractive. He proposed that humans (unconsciously) use a simple rule for deciding whether or not another individual is related. If humans avoid mating with individuals they have been reared with during childhood, there is a good chance that they will also avoid mating with close relatives. Children who are raised together will, therefore, experience a sort of negative imprinting against finding each other sexually attractive. At the cultural level, this instinctive disposition transformed itself, according to Westermarck, into a conscious taboo against incest.

Freud firmly rejected Westermarck's ideas since

they threatened to undermine his Oedipal complex, one of the cornerstones of his whole psychoanalytical theory. But other anthropologists and psychologists were also dismissive. There seemed to be several problems:

- The argument, as advanced by James Frazer in his *Totemism and Exogamy* (1910), was that if the inbreeding aversion was a deep human instinct, it would not be necessary to erect a taboo against it. Legal penalties and restrictions were put in force to curb natural antisocial proclivities of humans, such as stealing, violence and adultery; whereas something that was naturally repugnant would not require laws against it.
- It was a common assumption that there was considerable empirical evidence that showed inbreeding to be common among mammals without deleterious effects.
- A belief that since non-human primates (our nearest relatives) showed no aversion to inbreeding, then inbreeding avoidance was not likely to be an innate human disposition.

FIGURE 13.2 Edvard Westermarck (1862–1939) (c.1926)

SOURCE: Reproduced with permission from Åbo Akademis Bildsamlingar.

Perhaps another reason for the neglect of Westermarck's ideas up to about 1963 was the fact that, as a Darwinian, he sought to conjoin the moral imperatives of the taboo with its biological basis; in other words, to show how ethical prescriptions can be integrated with natural science. In this sense, he violated one of the great assumptions of 20th-century ethical thought, namely, that 'is' and 'ought' statements have no logical connection and to attempt to unite them somehow commits what became known as the 'naturalistic fallacy'. Instead, most ethical theorists opted for various forms of transcendental arguments suggesting that morality is something that is rationally derived or intuited and cannot be inferred from the natural world (see Chapter 16). To add to his disfavour in some circles, Westermarck also wrote works offering a relativistic analysis of ethics, arguing that there were no absolute moral truths.

13.3 Testing Westermarck's hypothesis

In many ways, the fate of Westermarck's ideas in the 20th century runs parallel to the broader fortunes of the whole of evolutionary thinking in psychology: an initial upsurge generated by Darwin and his immediate followers, followed by a long eclipse until the late 1960s, and then a resurgence over the last 40 years. By about 1970, it was realised that the **Westermarck effect** may be real after all and deserved careful attention. His primary thesis on incest avoidance can be broken down into a series of linked hypotheses:

1. Inbreeding is injurious to offspring (**inbreeding depression**)
2. Early association inhibits inbreeding by generating an aversion
3. This aversion is an evolutionary adaptation whose function is to reduce the risk of inbreeding depression
4. This aversion manifests itself at the cultural level as an incest taboo (sometimes called the 'representational problem').

We will examine each of these suggestions in turn.

13.3.1 Inbreeding as injurious to offspring (inbreeding depression)

By the early 1960s, evidence was mounting that the human gene pool did contain many recessive and dangerous alleles, consequently inbreeding did carry biological disadvantages. Current thinking suggests that each human carries around two or three lethal recessive alleles. The risks of inbreeding are indicated by a measure called the coefficient of inbreeding (F).

The coefficient of inbreeding

The **coefficient of inbreeding** (F) is the percentage of gene loci at which progeny of inbreeding will be homozygous (that is, the proportion of gene sites on the chromosome where they have inherited two identical copies of any gene) over and above the baseline level of homozygosity of the general population. Another way of saying this is that F is the probability that an individual has both alleles of a gene (that is, a specific locus) identical by descent from the same allele in a common ancestor. As an illustration, we may consider two cousins. The probability that each cousin has an identical specific gene acquired from a common ancestor is 0.125 (see section 2.4.2). If the two cousins mate and produce an offspring, then the chance that the offspring will acquire both copies of this gene is 0.5; so the coefficient of inbreeding is $0.125 \times 0.5 = 0.0625$. Table 13.1 shows some F values which can be seen to relate easily to the genetic relatedness of the parents.

Note that the actual F values may be higher than shown if the whole population has been subject to inbreeding. So for example, the F value for offspring of two siblings might be higher than 0.25 if the parents of the siblings were themselves related in some way. In such cases, a correcting formula can be applied (see Bittles, 2004). This may explain why disorders among children of Pakistani first cousin marriages in the UK are higher than for European first cousins – in Pakistan there is a long tradition of first cousin marriages (see Table 13.2).

Interestingly, Darwin chose inbreeding as his marriage option since he married Emma Wedgwood, his first cousin. As a result, he constantly fretted about the effect this might have on his children, some of whom were rather frail. He even tried, unsuccessfully, through the influence of his local MP and friend John Lubbock, to have first cousin marriages recorded in the 1871 UK census for the benefit of scientific research – reasoning that the number of surviving offspring from first cousin marriages could be used to infer the health of children.

Table 13.1 Coefficients of inbreeding

Sexual relationship	Fraction of genes in common (by common descent) in each parent	Coefficient of inbreeding (F) in resultant progeny
Father–daughter	0.5	0.25
Mother–son	0.5	0.25
Brother–sister	0.5	0.25
Uncle–niece	0.25	0.125
Half-siblings	0.25	0.125
Double first cousins	0.25	0.125
First cousins	0.125	0.0625

NOTE: Double first cousins are where an individual's mother's brother marries that individual's father's sister and produces a child who is then a double first cousin to itself.

Royal families abound with instances of inbreeding. The current monarch of Britain, Elizabeth II, is a third cousin to her husband Prince Philip, since they are both great-great-grandchildren of Queen Victoria. Prince Philip's mother was also inbred ($F = 0.0703$), since her grandmothers were sisters (daughters of Victoria and Albert).

The assessment of the biological consequences of human inbreeding is made difficult by a lack of data, itself a probable reflection of the strong social antipathy to incest as well as the difficulties of establishing true paternity. Finding controls is also difficult, since incestuous relationships often involve confounding variables such as low maternal age, mental abnormalities and low socioeconomic status. Despite these difficulties, it is now quite clear that inbreeding carried health implications for offspring even at the level of first cousin marriages. Table 13.2 shows the results of some recent studies into the effects of inbreeding.

Table 13.2 Scientific studies on effects of inbreeding in humans

Study and reference	Finding
Couples in India and Pakistan (Bittles, 1995)	Higher fertility rates for consanguineous, second cousin marriages (F > 0.0156)
Review (Bittles and Makov, 1988)	Progeny of first cousins' morbidity levels 1–4 per cent higher than unrelated couples
Review of combined data from Europe, Asia, Africa, Middle East and South America (Bittles and Neele, 1994)	Mean excess mortality at first cousin level 4.4 per cent
Five-year study of Pakistani community in Birmingham, UK (Bundey and Alam, 1993)	Rate of lethal or chronic disorders among offspring of first cousin marriages at 10 per cent, rate among non-first cousin offspring 3 per cent. Rate among European first cousin marriage offspring 5 per cent
Study of the highly inbred, isolated Amish community in Lancaster County, Pennsylvania (McKusick et al., 1964; McKusick, 2000)	Rate of Ellis-van Creveld syndrome per live birth 300 in 60,000 compared to US average of 1 in 60,000

The apparent anomaly in Table 13.2, that second cousin and closer marriages have a higher fertility, is resolved when we note that such marriages tend to occur earlier in the woman's life and so her reproductive period is longer. Overall though, we can see that health effects become manifest even at the level of first cousin marriages.

13.3.2 Early association inhibiting inbreeding by generating an aversion

Information on the negative impact on successful pair bonding among humans raised together comes from several sources. One that has a particularly large database (now amounting to the study of over 14,000 marriages) is Arthur Wolf's (1993) study of marriage customs in mainland China and Taiwan. Up until the early 1930s in Taiwan and the mid-1940s in mainland China, cultural practices allowed families to arrange marriages for their sons in one of two ways:

- Major marriage: the son would reach puberty and then marry a young female who would be brought into the house to live with the groom's parents.
- Minor marriage: a family would adopt a very young girl and raise her as a future daughter-in-law until both son and adopted girl reached marriageable age. In Taiwan, these girls are known as *sim-pua* or 'little daughters-in-law'.

It is the practice of minor marriages that provides an invaluable data set for examining any effect of mating inhibition. The children are destined to marry one another, but to measure the success of the marriage, Wolf devised a number of indices that took into account the divorce rate among such marriages as well as the number of children produced. This latter variable was measured by a fertility index defined as:

$$\frac{\text{Births to women aged 15–45}}{\text{Years of marriage between these ages}}$$

Figure 13.3 shows a plot of this statistic for women born between 1890 and 1920 in 85 villages and neighbourhoods. The crucial feature of Figure 13.3 is that it reveals that the older the girl was when she was adopted into the family, the more successful the marriage became in terms of fertility. This is consistent with the Westermarck effect, since the younger the girl at adoption, the longer the co-socialisation with her intended spouse and the longer the time for negative sexually inhibiting imprinting to occur. Compared to major marriages, Wolf found that minor marriages were far less successful in other respects. As he observed:

> To summarise, the divorce rate of minor marriages exceeds that of major marriages by a factor of 2.5 to 1, and the total fertility of major marriages exceeds that of minor marriages by more than 25%. (Wolf, 1993, p. 159)

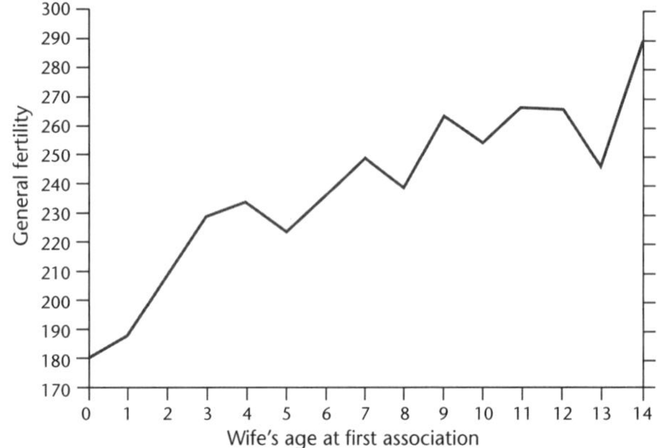

FIGURE 13.3 Fertility of Chinese marriages plotted against age of adoption of girl

The y axis is defined so that a figure of, say, 200 implies a fertility of 0.2 offspring per year of marriage for the woman at age of association on the x axis.

SOURCE: Wolf, A. (2004) Introduction, in A. P. Wolf and W. H. Durham (eds) *Inbreeding, Incest and the Incest Taboo.* Stanford, CA, Stanford University Press, Figure 4.1, p. 80.

Another source of information about the effects on marriage of co-socialisation at a young age comes from Shepher's (1971) study of mate selection among second generation kibbutz adolescents. The kibbutz system studied by Shepher was such that unrelated children from a variety of parents were raised together by trained nurses in heterosexual similar-age peer groups from birth to four years. From six or seven years of age until the age of eighteen, the groups sometimes spanned two years of maximum age difference. The children were exposed to each other constantly and tended to live apart from their parents for most of the day, although they saw their parents in the afternoons. Shepher (1971) looked at the incidence of premarital affairs and then number of marriages among adolescents reared in the same kibbutz. The total number of children and adults in the kibbutz he studied was 65. His findings were stark and can be summarised as follows:

- Not a single case of heterosexual activity between any two native adolescents of the same peer group
- No cases of marriage between any two members of the same peer group
- There was no overt pressure of any kind to refrain from sexual activity
- Of the 2,769 marriages analysed in the whole kibbutz movement, not a single case of true inter-peer group marriage was to be found. Only five sexual couples were found from this number who were in the same peer group before the age of six, but these were never together for more than two years.

Shepher concludes that his findings are in accord with the Westermarck effect and provide evidence for negative imprinting.

We should note, however, that Shepher's work received some sharp criticism when it appeared in book form in 1983. John Hartung (1985a, p. 171) was particularly severe, claiming the work to be 'one of the most sustained and pernicious affronts to empiricism since *Coming of Age in Samoa*'. The essence of Hartung's complaint is his belief that Shepher failed to properly calculate how many marriages to expect between individuals reared in a kibbutz. Much of Shepher's case rests on the fact of no marriages between couples in the same peer group; but, as Hartung notes, girls mature earlier than boys and are unlikely to find partners in the same age group anyway. Furthermore, an examination of Shepher's original data showed that of the 2,769 marriages identified by Shepher, there were 253 between people in the same kibbutz but not, by Shepher's criteria, the same peer group. Significantly, out of those 253 marriages, all but 7 featured a husband who was older than the wife.

More recently, Irene Bevc and Irwin Silverman have conducted studies on sexual activity between siblings. They examined what consequences for sexual behaviour would follow from the separation of siblings during childhood. If the Westermarck effect is well founded, then it could be expected that childhood separation would reduce the inhibition on sexual contacts. In their first study, Bevc and Silverman (1993) found that early separation was positively correlated with later 'mature' post-childhood sexual activity (by which they meant completed or attempted intercourse). In their second, follow-up study, Bevc and Silverman (2000, p. 159) found that, as before, 'early sustained cohabitation between siblings operates as a barrier specific to potentially reproductive acts rather than as a general suppressor of sexual interest'. In other words, separation during childhood did increase the rate of potentially reproductive sex acts (intercourse) in later life compared to non-separated siblings. Their study may also throw light on the debate about the crucial period for 'imprinting'. They found that 15 of the 17 separated sibling pairs in the group who had attempted sexual intercourse were separated for at least a year before either sex reached the age of three. This is consistent with Wolf's suggestion that the critical period is in the region of birth to four years.

13.3.3 Aversion through co-socialisation as an evolutionary adaptation

The evidence so far would tend to offer some support for the idea that the Westermarck effect is an adaptation: it seems to develop in all normal children, whether related or not, and clearly there are costs associated with not having this mechanism in place. But here we can also tackle one of the early objections to Westermarck's ideas, namely, that no such inbreeding avoidance mechanism was to be found in primates. If we were to find one, this would surely count in favour of Westermarck and against the cultural explanations of Freud and Lévi-Strauss, since, presumably, animals do not have the same interests in renouncing their instincts for the good of civilization, nor is it likely that they are able to culturally transmit a Freudian-style taboo.

Inbreeding avoidance in primates

There is now plentiful evidence that primates do, to some degree, avoid inbreeding. One obvious mechanism that achieves this in the wild is sex-biased dispersal. This can be male exogamy (males departing the troop), as in the case of olive baboons, many species of macaque monkey, capuchin monkeys and some lemurs; or female exogamy, as in the case of chimpanzees, spider monkeys and muriquis (South American monkeys). It could be argued that this mechanism is a way of regulating the group size to avoid competition for resources, but this would not explain the sex bias in the individuals that leave.

Reviewing the evidence on primate inbreeding, Ann Pusey (2004) notes, however, that young male chimpanzees have been observed coupling with their mothers and, when they are older, with their daughters. But, as Pusey points out, as the young male advances towards sexual maturity (around nine years old) and becomes capable of siring offspring, the frequency of son–mother copulation falls (Figure 13.4).

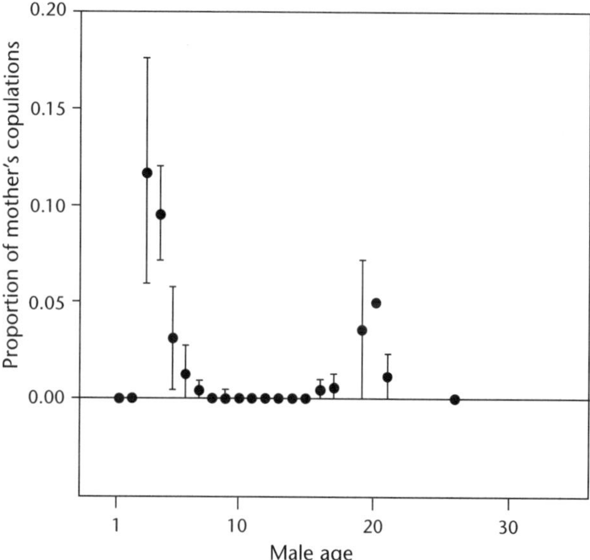

FIGURE 13.4 Copulations between chimpanzee mothers and sons observed in the Gombe National Park, Tanzania

SOURCE: Pusey, A. (2004) Inbreeding avoidance in primates, in A. P. Wolf and W. H. Durham (eds) *Inbreeding, Incest and the Incest Taboo*. Stanford, CA, Stanford University Press, Figure 3.1, p. 69.

Reviewing the literature, Pusey (2004, p. 71) concludes: 'Nonhuman primates provide abundant evidence for an inhibition of sexual behaviour among closely related adults.'

Mechanisms of incest avoidance

Despite the empirical evidence favourable to the Westermarck effect, the mechanism by which it operates is still unclear. Westermarck himself did not speculate on the details of the proximal developmental mechanisms. Wolf (2004) explained the effect of the wife's age on fertility and marital success by using Bowlby's idea of attachment (see Chapter 15). A young female will attach herself to an older member of the family as a source of care and support. In return, this will elicit nurturing feelings from the older member. In this way, attachment brings members of a family closer together emotionally and physically. It is then the job of the Westermarck effect to ensure that this closeness does not become sexual. Wolf (2004) therefore proposes that attachment and the Westermarck effect have evolved together. The younger the wife, the stronger the attachment and the stronger the Westermarck effect in both sexes.

Freudians often feel that evolutionary psychologists misrepresent their ideas and underestimate Freud. In their critique of work by Thornhill (see below), for example, Robert Paul (1991) and David Spain (1991) both point out that Freud did not argue that adults actually desire incestuous mating (a biologically implausible supposition), rather that incestuous erotic love was the first type of sexual love to appear in children and usually began between the ages of four and six and was something that by itself could obviously not lead to mating. In Freudian theory, it is the subsequent formation of the superego at about the age of seven or eight that served to repress these desires. If this happened normally to the maturing child, then the outcome is a voluntary aversion to incest that is indistinguishable (in terms of what is actually manifest) from the Westermarck effect. In this sense, Freudian theory is actually positing an ontogenetic (or developmental in a Tinbergian sense) mechanism consistent with the proximate mechanism that is assumed to lie at the heart of the Westermarck effect. Here is not the place to enter into an evaluation of psychoanalytical theory, but it is worth noting that more empirically verifiable mechanisms may be at hand such as olfactory cues.

Recent experiments on mating preferences in mice and humans have pointed to the possible role of olfaction in kin recognition and mate choice. Wedekind and colleagues found that females preferred the odour of men who differed from them in their major histocompatability (MHC) haplotypes (Wedekind et al., 1995). In later experiments (involving subjects smelling T-shirts worn by volunteers), the pleasantness of the odour as experienced by both males and females was correlated with the degree of MHC dissimilarity between smellers and the wearer of the T-shirt from which the odour was sampled (Wedekind and Furi, 1997). But while this points to a subconscious role for smell in judging attractiveness, it does not by itself support a Westermarck-style inhibitory effect from close association, since the mechanism here might be genetically based and part of a kin recognition system. Experiments by Bateson (1980, 1982) on Japanese quail, however, show that early familiarity inhibits mating, irrespective of relatedness. Experiments on mice involving cross-fostering also show avoidance of mating partners based on familiarity and not just relatedness. Strong evidence in favour of the importance of using familiar smells as a proxy for relatedness comes from experiments whereby male mice can be reared to prefer females even of their own MHC type by raising them with females with different MHC types (Yamazaki et al., 1988). So, at least in mice, it looks like there is a mechanism via olfactory cues consistent with Westermarck-style negative imprinting that disposes mice not to mate with other mice with which they were raised, and that such cues are obtained to some degree by learning rather than a genetic-based recognition of MHC similarity. In a review of the literature, Schneider and Hendrix (2000) think that an olfactory component to the Westermacrk effect in humans is highly likely. They also make the interesting prediction that follows from this that anosmics (people who have lost their sense of smell) should be less inhibited in pursuing sexual contacts with co-socialised individuals. One might expect, for example, a higher percentage of anosmics in people who admit to incest than among the general population.

13.3.4 Inbreeding depression and cultural norms: the representational problem

Variation in laws relating to incest

We now turn to the final hypothesis of the Westermarck effect. Even if we accept that inbreeding in human populations does cause damaging effects, and that individuals develop an inhibition to breeding with those they have been socialised with in their childhood, it is not easy to see how an instinctive, unarticulated aversion can translate itself into a cogently expressed series of norms, rules and laws. The conceptual gap that opens up here is thrown into focus by the consideration that the biological aversion to inbreeding is felt between children who have been co-socialised (whether related or not), whereas the cultural prohibition is directed specifically against kin, with rules beginning to vary as we approach first cousin levels of relatedness. The incest aversion seems to be universally directed at relationships where F is 0.25 (mother–son, father–daughter, brother–sister). At the level of first cousin marriages, practices and customs vary, with some cultures prohibiting such unions and others actively encouraging them (see Table 13.3). Indeed, the whole framework of laws relating to incest seems to be based on a mixture of religious precepts and genetic ideas (see Ottenheimer, 1996).

From aversion to rule

Obviously, from the perspective of the target presented to natural selection, the aversion usually worked to keep close kin away from breeding with each other, despite the occasional 'misfire', as the nascent sexual desires of unrelated children were suppressed. But when and how did humans realise that the aversion could be more effectively channelled against kin marriages only? The psychologist Roger Burton may have provided a way through this impasse. Burton (1973) argued that there probably was a Westermarck-type aversion that reduced the frequency of inbreeding in early human groups. But it was precisely this mechanism that would then have allowed deleterious alleles to accumulate in the heterozygous (that is, unexpressed) condition. When incest did occur, albeit infrequently, the effects would be visible, so that gradually cultures would come to

Table 13.3 A selection of rules governing consanguineous marriages from various authorities

Religious authorities			
Ruling	***Authority***	***Sexual relationship***	***F value of progeny***
Permission required from diocese	Roman Catholic Church	First cousins	0.0625
Permitted	Protestant Church	First cousins	0.0625
Strongly encouraged	Dravidian Hindus of south India	First cousins of type 'a son with mother's brother's daughter'	0.0625
Prohibited	The Bible, Leviticus	Man and his mother, or his daughter	0.25
		Man and his father's brother's wife	0
		Man and his wife's sister	0
		Man and his brother's wife	0
Prohibited	The Koran	Uncle–niece	0.125
Allowed	The Koran	Double first cousins	0.125
Civic authorities			
Ruling	***Authority***	***Sexual relationship***	***F value***
	UK		
Criminal offence	Sexual Offences Act 1956	Sexual intercourse with a woman who a man knows to be his granddaughter, daughter, sister, half-sister or mother	0.125–0.25
Allowed		Relatedness up to and including first cousins	0.0625
	USA		
Marriage illegal	31 states	First cousins	0.0625
Permitted	19 states	First cousins	0.0625
	China		
Prohibited	1981 Marriage Law	First cousins and closer	0.0625

realise that inbreeding among kin produced undesirable consequences. These cultures would then see incest as transgression of the natural and moral order in some way and eventually elaborate a cultural condemnation of it. In early cultures, this prohibition would probably have taken on a religious form, as in the Bible (Leviticus 18) where God is supposed to have dictated to Moses the rules of consanguineous mating. The problem here though is that we now have to explain how cognition (the dawning realisation of the perils of incest) becomes translated into an emotion (revulsion at the idea). In some ways, the original formulation of Westermarck is more consistent. His view was that the moral norms that we articulate are a reflection of our emotional responses; we have, as it were, 'moral emotions'. We explore this below.

Moral condemnation and childhood exposure to brothers and sisters

Such a linkage, between the avoidance of incest at the personal level and the cultural prohibition against it, would tend to suggest that the moral condemnation against incest in third parties is generated by a mechanism the same as, or related to, the onset of personal aversion. From this, it would follow that children who have been raised with siblings of the opposite sex should aver a stronger moral disapproval of incest in third parties than those without opposite sex siblings. If, however, morality is a matter of absorbing cultural norms through socialisation, as cultural constructivists would suggest, then no such relationship should be observed. This theory was put to the test by Lieberman et al. (2003), in their study of the family backgrounds and moral opinions of 186 Californian undergraduates. They looked for a positive correlation between the strength of disapproval of incest and the childhood experience of being raised with a member of the opposite sex. Table 13.4 provides a summary of their findings.

As shown in Table 13.4, they found that the length of co-residence between siblings was positively associated with their moral condemnation of incest more generally. The study took great pains to control for the effects of other variables that might impact on moral sentiments such as family composition and size, sexual orientation and parental attitudes towards sexual behaviour. With all these controls in place, the correlations were positive and significant even when the opposite sex 'sibling' was known not to be a genetic relative.

These results complement the study of Wolf but differ in one important aspect. Wolf suspected that the years from birth to six were those in which the Westermarck effect took place. This study, however, shows that, for males, the degree of moral repugnance continues to rise up to age 18 (see Table 13.4). Lieberman et al. explain this by suggesting that, since inbreeding is more costly for females, the avoidance mechanism swings into play with a low level of socialisation and hence cues concerning relatedness. An inbreeding error is less costly for males, however, and so they delay judgement, not wishing to ignore a potential sexual partner.

Table 13.4 Moral aversion to third party incest in relation to childhood experience

Relationship of cohabitees	Years of co-residence	Correlation between years of co-residence and degree of 'moral wrongness' expressed in relation to third party incest	Significance of correlation P
Males with sisters	0–10, 0–18	0.29, 0.40	0.05, 0.01
Females with brothers	0–10, 0–18	0.23, 0.23	0.05, 0.05
Individual with opposite sex who is not a genetic sibling (that is, not related)	0–18	0.61	0.014

SOURCE: Data from Lieberman, D., Tooby, J. and Cosmides, L. (2003) 'Does morality have a biological basis? An empirical test of the factors governing moral sentiments relating to incest.' *Proceedings of the Royal Society of London*. Series B 270: 819–26.

The unequal costs of incest

The higher costs to females associated with inbreeding would predict, as noted, a more sensitive avoidance mechanism. Support for this notion comes from a similar study to that of Lieberman et al., but carried out independently by Daniel Fessler and David Navarrete (2004) using a data set of 250 undergraduates aged 18–39. They asked their subjects to express their attitude to incest using an 11-point Likert-type scale. In summary, they found that:

- Females raised with male siblings reported a higher aversion to incest than females raised without male siblings
- Males raised with female siblings reported a higher aversion to incest than males raised without female siblings
- Females reported a greater sibling incest aversion overall (see Table 13.5)
- The effect did not depend on the relatedness of co-socialised individuals and was the same with full and step-siblings.

As Fessler and Navarrete (2004) note, their study does not rule out the possibility that families with opposite sex siblings might be exposed to more intense parental instruction about proper sexual behaviour than those with siblings of the same sex, precisely to rule out the possibility of incest.

Table 13.5 Male and female aversion to incest

Male incest aversion score (± SD)	Female incest aversion score (± SD)	't' statistic (two-tailed)	P
4.63 ± 2.14	5.33 ± 1.98	1.44	0.05

SOURCE: Data from Fessler, D. M. T. and Navarrete, C. D. (2004) 'Third-party attitudes towards sibling incest: evidence for Westermarck's hypotheses.' *Evolution and Human Behaviour* **25**: 277–94.

13.4 Keeping it in the family: incest, paternity confidence and social stratification

The anthropologist Nancy Thornhill (1991) provides a different slant on the formation of incest rules. She argues that to focus on rules against biologically damaging inbreeding (that is, where $r > 0.25$) is to miss a major point, since, although such rules obviously exist, such inbreeding is unlikely anyway (due to the Westermarck effect) and marriage rules may be designed for other reasons. She suggests that most incest rules are targeted at people who are not closely biologically related, such as cousins, or only related by marriage, such as a man and his brother's wife; rules defined, therefore, more by social than biological relationships. In support of this, she offers the curious fact that of the 129 societies in her sample with relevant information, only 44 per cent have rules governing incest within the nuclear biological family (where $r > 0.25$), while 88 per cent had rules regulating marriage with other categories of kin such as cousins or in-laws (where $r < 0.25$).

Thornhill (1991) advances several hypotheses to account for these rules governing distant and **affinal** relatives, two of which are:

1. Rules are made to ensure reliability of paternity
2. Rules are made by socially powerful individuals to maintain their privileged position and prevent the accumulation of wealth and power (through family alliances forged by marriage) in subordinates.

From the first hypothesis, Thornhill predicted that rules concerning incest should be more prevalent in societies with patrilocal residence (that is, where a wife moves to a husband's family) than ones with **matrilocal** residence (that is, where a husband moves to the wife's family). The basis of this prediction rests on two considerations: first, patrilocal residence is strongly associated with patrilineal inheritance and it is to be expected that men leaving wealth to their wife's sons will be especially concerned to also ensure that they are their own sons; second, patrilocal residence is also associated with polygyny, implying that a man may have frequent interaction with biologically unrelated women such as the wives of brothers and fathers. These predictions were supported by a chi-squared analysis. Figure 13.5 provides a graphic illustration of the findings.

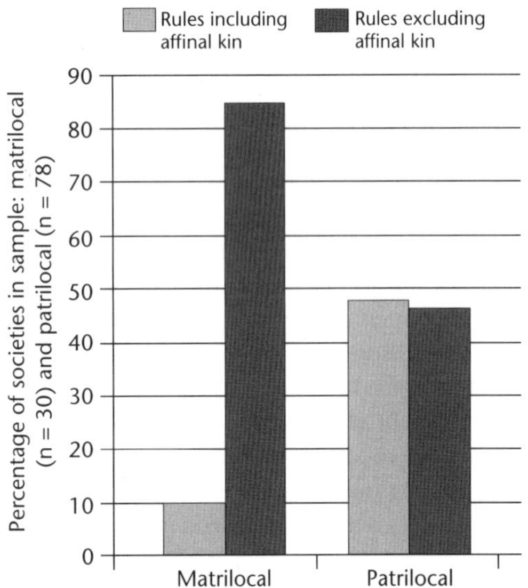

FIGURE 13.5 The presence or absence of marriage rules including affinal kin (in-laws and so on) according to patrilocal or matrilocal residence

NOTE: Yates corrected $\chi^2 = 13.1$, $p < 0.001$.

SOURCE: Plotted from data in Thornhill, N. W. (1991) 'An evolutionary analysis of rules regulating human inbreeding and marriage.' *Behavioural and Brain Sciences* **14**: 247–93.

Thornhill's data also tends to run counter to the family harmony types of explanation in anthropology. According to these views (heavily influenced by Freud), sex within the family is highly disruptive and so rules exist to preserve family structures. However, if this were the case, we would expect rules against both biological kin ($r > 0.25$) and affinal kin to exist side by side, since sexual liaisons between, say, fathers and daughters-in-law should be just as disruptive as kin-based sexual contacts. The data, however, show that there is no significant association between the two types of rule (see Table 13.6).

To test the second hypothesis concerning the concentration of wealth and power, Thornhill (1991) reasoned that rulers in highly stratified societies would seek to maintain their powerful positions by taking measures to keep wealth within their own family circle, while enforcing rules to ensure subordinate families do not accumulate enough wealth to threaten the ruling elite in their family networks. This translates into the expectation that, in highly stratified societies, the rulers will attempt to apply rules against incestuous marriage to others but not themselves. Table 13.7 shows how data supported this expectation.

Table 13.6 Testing for an association between rules against biological and affinal incest

	With affinal rules	Without affinal rules
Biological incest rules	16	18
Without biological incest rules	17	21

NOTE: No association is seen ($\chi^2 = 0$, $p = 0.99$), suggesting that rules against affinal incest are not associated with rules against biological incest.

SOURCE: Data taken from Thornhill, N. W. (1991) 'An evolutionary analysis of rules regulating human inbreeding and marriage.' *Behavioural and Brain Sciences* **14**: 247–93.

Table 13.7 The application of rules concerning marriage in stratified and non-stratified societies

Stratification of society	Rules applied equitably	Rules applied inequitably
Moderate to extreme (n = 20)	0	20
None to slight (n = 35)	25	10

SOURCE: Data from Thornhill, N. W. (1991) 'An evolutionary analysis of rules regulating human inbreeding and marriage.' *Behavioural and Brain Sciences* **14**: 247–93.

In the work of Thornhill on wealth and stratification, we can see how evolutionary and sociopolitical explanations converge. Sociopolitical explanations of cultural phenomena stress the roles of power, prestige and wealth accumulation in the construction of rules that serve specific interests. Evolutionary biology can also call upon these factors while realising that, ultimately, these must also be associated with measures of reproductive fitness. The problem now of course is deciding which mode of explanation to prefer since both types are consistent with the data. A common criticism of this application of evolutionary theory is that it makes no sense to suggest that this concern with wealth is an evolved adaptation, since the whole phenomenon of wealth above subsistence needs is a relatively recent phenomenon made possible by the Neolithic revolution (see Hyland, 1991). Early hunter-gatherers had no wealth to be fussy about. The most plausible response to this, as Thornhill herself suggests, is that there may be an evolved adaptation to be concerned with status, since status, however culturally defined, is likely to be a good predictor of reproductive success. The analysis presumes therefore that humans are alert to environmental cues of status determination.

13.5 Incest and morality

What is also significant about the Westermarck effect is that it provides an intriguing case study into how natural dispositions that have evolved for functional reasons end up as a culturally articulated, transmitted and enforced set of moral codes, or at least rules that some people in a society expect others to adhere to. Although the studies by Liberman et al. (2003) and Fessler and Navarrete (2004) provide support for the relationship between experience and expressions of moral disgust, there remains the ticklish question of why prohibitions should exist for behaviours that no one is supposed to be motivated to perform. We noted Burton's account earlier, but Westermarck himself suggested that such taboos arise as a consequence of our ability to experience the actions of others as if they were our own. Fessler and Navarrete (2004) term this propensity 'egocentric empathy'. Such responses are indeed found when we experience disgust at the thought or sight of somebody eating a foodstuff that we find strongly unpalatable. In Chapters 6 and 8, we observed how mirror neurons might serve this function, although in this case, it is the prospect or thought of incest, rather than the sight of it, which is needed to evoke the aversion.

More generally, the proposed connection between morality and biology places Westermarck in a long line of thinkers who have argued for a naturalistic, as opposed to transcendental, approach to ethics (see Chapter 17). In the naturalists' camp, we have Aristotle, Aquinas, Hume, Adam Smith, Darwin and, more recently, E.O. Wilson and Arnhart. In the transcendentalist tradition – those who think that ethic norms lie outside nature in some way and are either intuited by our ethical sense or derived by an exercise of reason – are to be found Hobbes, Kant, Freud, J. S. Mill and Rawls. If the Westermarck effect continues to gain empirical support, it provides a signpost for how a natural science of ethics can be conceived.

Summary

- Early 19th-views about the prohibition against incest tended to see it as a non-biological, cultural phenomenon. The exception was Westermarck, who postulated that the early co-socialisation of children inhibited sexual desire as a mechanism to avoid inbreeding.
- The Westermarck effect has several components that can be tested empirically. So far, work has tended to suggest: inbreeding does cause a depression in fitness; early co-socialisation tends to inhibit sexual desire; other primates have mechanisms to avoid inbreeding; the proximate mechanism that leads co-socialised humans to avoid mating may be olfactory or even explicable in Freudian terms; and the moral condemnation of incest varies according to the experience of the individual expressing such views.
- An alternative approach to the existence of cultural rules regulating mating and incest avoidance is to see them as ways of ensuring paternity certainty and the concentration of wealth and power.
- The Westermarck effect provides an interesting way in which biological fitness and moral sentiments may be related.

Key Words

- Affinal
- Coefficient of inbreeding (F)
- Inbreeding
- Inbreeding depression
- Incest
- Matrilocal
- Westermarck effect

Further reading

Bevc, I. and Silverman, I. (1993) 'Early proximity and intimacy between siblings and incestuous behaviour: a test of the Westermarck hypothesis.' *Ethology and Sociobiology* **14**: 171–81.

A research paper that looks at actual cases of incestuous behaviour.

Fessler, D. M. T. and Navarrete, C. D. (2004) 'Third-party attitudes towards sibling incest: evidence for Westermarck's hypotheses.' *Evolution and Human Behaviour* **25**: 277–94.

Research paper that examines how people react to the thought of incest in relation to their own childhood experiences.

Freud, S. (1950) *Totem and Taboo*. New York, W.W. Norton.

Famous work where Freud outlines his view of the incest taboo.

Lieberman, D., Tooby, J. and Cosmides, L. (2003) 'Does morality have a biological basis? An empirical test of the factors governing moral sentiments relating to incest.' *Proceedings of the Royal Society of London* Series B 270: 819–26.

A research paper looking at our moral sentiments in relation to family structures.

Wolf, A. P. (1993) 'Westermarck redivivus.' *Annual Review of Anthropology* **22**: 157–75.

A paper where Wolf argues that there is strong evidence supporting Westermarck's ideas.

Wolf, A. P. and Durham, W. (eds) (2005) *Inbreeding, Incest and the Incest Taboo*. Stanford, CA, Stanford University Press.

A collection of useful articles on the Westermarck effect in the light of recent research.

The Disordered Mind

Mental Disorders: Some Theoretical Approaches

O the mind, mind has mountains; cliffs of fall
Frightful, sheer, no-man-fathomed. Hold them cheap
May who ne'er hung there.
(Gerard Manley Hopkins (1844–89) 'No worst there is none', *Poems* 1918)

Gerard Manley Hopkins knew about the mountains of the mind from personal experience. He was one of the finest and most original poets in Victorian Britain, yet in his short and tortured life, his genius and creativity went unrecognised. Today he would be recognised as suffering from major (unipolar) depression. The image of the suffering, slightly abnormal, creative artist has become something of a stereotype, but it is true that many creative people – writers, composers, artists – have suffered from clinical depression in their lives. Examples include Virginia Woolf, Vincent van Gogh, Honoré de Balzac, F. Scott Fitzgerald, Ernest Hemingway, the brothers Henry and William James, Mark Twain, Mary Shelley and Tchaikovsky (see Jamison, 1994).

But mental disorders are not confined to the illustrious. Current definitions of disorders imply that a surprisingly high number of people are suffering from a mental affliction of one form or another. As Comer (2001) pointed out, if we take a random sample of 100 adults in the USA, then 19 will have a significant anxiety disorder, 10 will suffer from profound depression sometime in their lives, 5 will display a personality disorder, 11 will abuse alcohol or some other drug, and 1 will have schizophrenia. Against these seemingly confident statistics must be balanced the very real problem of establishing exactly what constitutes a mental disorder.

14.1 Problems of taxonomy and definition

Pinning down a set of reliable and objective criteria for establishing mental disorders has turned out to be a political and philosophical minefield and many textbooks on the subject open with a discussion of the problems with and limitations of current methods (see, for example, Eysenck, 2004). Traditional definitions of abnormal behaviour usually use criteria such as statistical infrequency, deviations from what would normally be expected from the individual, behaviour that violates moral standards of a culture, and behaviour that causes distress to the sufferer. However, all these criteria have their problems. To take two examples: first, the moral standards of a culture may themselves be questionable; and second, traits and behaviours that lie at the extreme ends of a normal distribution, such as high intelligence, creativity or sporting skill, are not usually thought of as abnormal.

14.1.1 DSM systems

In Britain and America, one of the most commonly used manuals of criteria for abnormal behaviour is the *Diagnostic and Statistical Manual of Mental Disorders* (DSM) published by the American Psychiatric Association. This manual places particular emphasis on the extent of personal suffering experienced by the victim and the impairment of normal life that this causes. The first version of the DSM was published in 1952, and the manual has been through various revisions up to the current version, DSM-IV, published in 1994. DSM-IV describes over 200 mental disorders and a comparison with DSM-I and II reveals a shift from theoretically guided definitions to a more descriptive basis. In one sense, this is a pragmatic move and one to be welcomed: psychology is characterised by a plurality of theories and internal disagreements and the early theories used in DSM-I and II are now regarded as seriously flawed. But in another sense, this atheoretical approach is not conducive to an evolutionary understanding. As Murphy and Stich (2000) note, the **DSM system** takes no account of the underlying mental structures or design features of the brain.

Another problem with the DSM-IV of special relevance to the prospects for an evolutionary approach is that it is a system of categorical classification, that is, patients either have or do not have a mental disorder. Yet basic biology tells us that traits often have dimensional variation. Individuals could have varying degrees of anxiety, for example, and it is just a matter of semantics when the level of the trait is labelled a disorder.

14.1.2 Sociological considerations

The whole process of defining abnormality has also faced a sociological critique. One perspective has been to argue that mental disorders are social constructions – labels used by society (like the terms 'criminal' or 'prostitute') to exclude and stigmatise those whose behaviour is contrary to the norms of that culture (Szasz, 1974). It has also been argued that the variety of criteria and value-laden language used to describe disorders leads to the conclusion that the whole concept of mental abnormality has only a weak scientific foundation (see Lilienfield and Marino, 1999, for a critique of evolutionary foundations, for example).

Changes in the classification of disorders lend some support to this social constructivist perspective and the history of medicine and psychiatry provides numerous examples of the way social bias has impinged on the definition of what is normal and abnormal behaviour and hence the labelling of disorders. In 1851, for example, the Louisiana surgeon and psychologist Samuel A. Cartwright described a number of diseases prevalent among the black population of the southern states. One of these was 'drapetomania' or the disease that caused slaves to run away.

More recently, in 1968, in the DSM-II, homosexuality was classified as a mental disorder, but, significantly (in the light of the rise of the gay rights movement), the latest DSM-IV (1994) no longer refers to homosexuality as a disorder but leaves open an unspecified category of 'distress about one's sexual orientation'.

Box 14.1 Defining mental disorders

Despite the numerous scientific and philosophical difficulties in defining what counts as mental disorder, most classificatory schemes involve some of the following features (see Rosenham and Seligman, 1989; Comer, 2001):

- *Suffering, danger and discomfort:* Individuals with an abnormality will often report that they are suffering. They may pose a danger to themselves or others. Observers may also report discomfort in the presence of such people (they may stare or stand too close, for example).
- *Deviance:* People with an abnormality may display feelings, thoughts and behaviour that deviate markedly from society's implicit or explicit rules of behaviour.
- *Dysfunction and maladaptiveness:* The behaviour of those with a disorder is maladaptive: it prevents an individual from functioning effectively and achieving normal life goals.
- *Irrationality and incomprehensibility:* Abnormal behaviour is often incomprehensible to others who fail to see why anyone would behave in that way.

14.2 Conceptual pluralism in psychology

Before examining evolutionary models of mental disorders, it is worth looking at what McGuire and Troisi (1998) have called the 'state of conceptual pluralism in psychiatry' (a state that applies to psychology as a whole). Valenstein (1998) memorably described the history of psychiatry in the postwar era in terms of two phases: phase one, roughly 1945–60, placed a strong emphasis on psychoanalysis ('blame the mother'); and phase two, from 1960 onwards, stressed the importance of neurotransmitters ('blame the brain'). With the rise of cognitive behavioural therapy (CBT) in recent years, we might be tempted to add that this represents a new phase of 'blame yourself'. The blame the mother period caused much unnecessary heartache and guilt. In 1948, for example, Frieda Fromm-Reichman argued that schizophrenics owed their condition to the coolness and rejection of their 'schizophrenogenic' mothers. Similarly, one of Freud's heirs, Bruno Bettelheim, blamed 'refrigerator mothers' for the onset of autism in their children. Both these conditions are now realised to have a strong genetic component.

Today there exist a variety of paradigms purporting to account for and help treat mental disorders. Table 14.1 outlines some of these.

Table 14.1 Some models of mental disorders

Model	Features
Biomedical	Biomedical models emphasise the importance of genetic and physiological factors in causation. Some physiological system may be malfunctioning or there is a genetic disposition to such disorders in the family. Other causes may be infectious agents (bacteria or viruses). Such disorders typically understood using this perspective are Fragile X, Rett, Williams, Prader-Willi, and Angelman syndromes. Another is phenylketonuria (PKU), a form of mental retardation now reliably established as a product of a genetically determined enzyme deficiency and treated with a diet especially low in the amino acid phenylalanine. This approach would focus on brain biochemistry, particularly the role of neurotransmitters such as serotonin. Therapy resulting from this perspective is often called 'somatic therapy' and involves interventions such as psychosurgery, drugs (such as benzodiazepines used to treat anxiety) and electroconvulsive therapy
Psychoanalytical or **psychodynamic**	The psychoanalytical perspective, closely associated with the work of Freud, tends to view psychological events as antecedent to physiological disturbances. This perspective views disorders as resulting from internal (intrapsychic) conflicts and tensions. It is thought that some of these may result from unresolved issues arising from family relationships. Freud stressed how the part of the mind (the id) concerned with basic instincts such as sex was often in conflict with the conscience (the superego). It was the job of the ego (the rational, self-conscious part of the mind) to reconcile these conflicts by employing defence mechanisms such as denial or repression. Freud argued that mental disorders have their origin in childhood experiences. One does not have to be a Freudian, however, to appreciate that childhood development may influence later adult behaviour. Other psychodynamic approaches emerged after Freud's death in 1939. One with a strong following today is object relations theory, which corrects the excessive bias Freud showed towards sexual factors and childhood experiences. Psychodynamic therapy aims to enable the patient to gain access to repressed feelings and memories to gain insight into their origins and cause
Behavioural	The behavioural model is based on the ideas of John Watson and his followers. The original emphasis was on behaviour learned by classical conditioning (learning by association) and operant conditioning. Later, observational learning (individuals observing others and imitating their behaviour) was added. In this model, mental disorders are maladaptive forms of behaviour that have been learned through classical or operant conditioning or observational learning. This approach focuses on symptoms rather than underlying causes or feelings. Although it has been difficult to substantiate the claim that phobias (for example) are associated with earlier conditioning, this approach has led to behaviour therapy, whereby some anxieties and phobias can be eliminated by a process of extinction. The related aversion therapy can be used to treat such conditions as alcoholism by linking the sight of alcohol with the administration of drugs that cause nausea

Table 14.1 cont'd

Model	Features
Humanistic	The **humanistic** model was developed by Carl Rogers and Abraham Maslow in the 1950s. It stresses the importance of personal responsibility and free will in the process of individuals striving for self-actualisation. From this model emerged client- (or person-) centred therapy, which aims to help patients develop an honest assessment of themselves and increase their self-esteem. This is achieved through the therapist having unconditional positive regard for the patient and developing a good empathetic understanding
Cognitive	The central assumption in the cognitive model is that people suffering from mental disorders have irrational and disturbed beliefs and ideas. Its origin is particularly associated with the work of Albert Ellis (1962) and Aaron Beck (1976). Beck applied his ideas to depression and anxiety disorders by suggesting that people suffering from these conditions had maladaptive thought patterns. He argued that, in the depressive state, patients overgeneralised from negative experiences to view themselves as inadequate and the world as impossibly difficult. Similarly, he suggested that anxiety-ridden patients overestimated the possibility of harmful events and underestimated their ability to cope or the likelihood of someone else helping them. Later, Beck and Clark (1988) suggested that patients of this sort had latent maladaptive schemas that led to their 'cognitive vulnerability' towards such conditions in the face of environmental triggers such as unpleasant events or threats. This model, with its emphasis on latent mental schemas and inherent information-processing circuits, aligns it in part with the approach of evolutionary psychology. Cognitive therapy aims to enable patients to monitor their own thought processes and feelings and correct biases and distortions
Evolutionary	The evolutionary model is perhaps the most recent and employs a variety of approaches. Some researchers suggest that conditions labelled as disorders may simply be extreme forms of naturally occurring traits. Another approach is to look at disorders as mechanisms with adaptive value: the onset of the disorder may have once functioned to actually improve the reproductive fitness of the sufferer by prompting a change in behaviour. The model can also incorporate developmental disruption to the maturation of evolved mechanisms. A common ploy is to see mental problems (like physiological problems) in terms of design compromises or as products of a difference between the EEA and the society that the mind now finds itself in. The rest of this chapter is devoted to this approach

FIGURE 14.1 ***The Scream*, from a lithograph by Edvard Munch (1863–1944) made in 1895**
The Scream is a powerful portrayal of the angst and alienation many people feel sometimes in their lives. Although he lived to be 81, Munch himself suffered from a number of anxieties. He felt a dread when he had to cross an open square and, like van Gogh and Gauguin, was convinced that he was persecuted by others. He suffered a complete nervous breakdown in 1908 and spent six months in a clinic in Copenhagen.

So what approaches do clinicians and psychotherapists actually use? In one survey, 33 per cent of American therapists used solely psychodynamic approaches, 5 per cent only behavioural therapy, 3 per cent client-centred (humanistic) therapy, 5 per cent cognitive therapy and 38 per cent identified themselves as using an eclectic approach – combining elements of the various models (Prochaska and Norcross, 1994). It is fair to say that as far as therapy goes, the evolutionary approach is in its infancy and in the short term promises to supply explanatory as opposed to therapeutic benefits.

In evaluating this plethora of models, on the positive side we might say that this conceptual pluralism could reflect the complex multicausal nature of mental disorders. On the other hand, it may reflect competing professional interests and the historical development of psychology and psychiatry and be a sign of conceptual confusion and immaturity – what Thomas Kuhn called the 'pre-paradigmatic state of a science', characterised by a lack of unity and agreement on foundational principles.

14.3 Evolutionary classifications of mental disorders

The evolutionary approach to mental illness is characterised by a variety of attempts to classify and explain the disorders in terms of the fundamental architecture of the brain as laid down by natural and sexual selection. But even this commitment to a single overarching theory has sustained a wide variety of approaches. We will start with some general schema and then proceed to examine subtheories in more detail in the next section.

For the evolutionist, one of the most promising definitions of a mental disorder comes from Wakefield's harmful dysfunction (HD) analysis:

> A disorder exists when the failure of a person's internal mechanism to perform their functions as designed by nature impinges harmfully on the person's well-being as defined by social values and meanings. (Wakefield, 1992, p. 373)

It is a definition that deftly combines biological and social dimensions, but one that leaves us with the problem of identifying what natural mechanisms exist. One particular advantage of this definition is that it provides clarification on what counts as a disorder. This HD approach would seem to work quite well for some disorders such as autism and schizophrenia (see Chapter 15) but faces a more difficult task with disorders such as substance abuse and personality disorders. There is also the problem that the normal functioning of internal mechanisms can cause distress. If we examine, for example, delusional disorder, jealous type (or 'morbid jealousy'), it may be that such a condition impinges on the sufferer's well-being but it could be the operation of an internal mechanism just as nature intended, that is, functioning to ensure mate fidelity (see Easton et al., 2006). Henriques (2002) provides an interesting appraisal of the HD concept.

Jaak Panksepp (2006) has argued that too often evolutionary psychology has employed the term 'modules' without much concern about their neurobiological basis. His approach is to stress how basic mental systems (**affective** and cognitive) are shared across mammalian species. He cautions against thinking of disorders in terms of 'syndromes', which are merely descriptive phrases without any necessary biological correlates. He uses the term 'endophenotypes' for these basic brain functions that underpin normal and hence abnormal psychology. Applying these ideas to emotional disorders, Panksepp suggests that there are seven basic emotional systems that give rise to emergent emotions and hence are implicated in emotional disorders, as shown in Table 14.2.

Panksepp is surely right to insist that mental disorders should be related to the underlying systems that are malfunctioning. The problem is that such systems are notoriously difficult to probe and, in the case of affective systems, are overlaid by more recent layers of cortico-cognitive information-processing, making their investigation difficult. Until more information accrues from brain imaging, neurophysiology and animal models, such approaches must remain speculative.

Since its inception, evolutionary psychology has stressed that the cognitive and emotional systems we now possess were designed to cope with problems very different from those of the modern world. This historical consciousness adds a new dimension to the question of what counts as a disorder. Figure 14.2 shows a useful schema devised by Crawford (1998b) for conceptualising the relationship between adaptations laid down in the human brain long ago and pathologies. Notice that the two components refer to fitness levels and the experience of well-being respectively.

Table 14.2 Panksepp's postulated relationship between basic emotional systems (endophenotypes), commonly experienced emotions and major psychiatric disorders

Basic emotional system or endophenotype	Emergent emotions	Related emotional disorders
SEEKING: A system that generates a host of investigatory and exploratory activities to gather resources and information from the world	Interest Frustration Craving	Obsessive compulsive disorder (OCD) Paranoid schizophrenia Addictive personalities
RAGE: Like other sub-neocortical systems, the higher cortico-cognitive areas of the brain can inhibit, guide and regulate anger	Anger Irritability Contempt	Aggression Psychopathic tendencies Personality disorders
FEAR: Circuitry to keep individuals away from danger or prime their response to it. There are 'high roads' of cognitive circuitry and 'low roads' of more rapid thalamic responses. Typical circuit is the link between the amygdala and the periaqueductal grey of the midbrain	Simple anxiety Worry Psychic trauma	Generalised anxiety disorders Phobias Post-traumatic stress disorder variants
PANIC: Typical example may be separation distress	Sadness Guilt/shame Shyness	Pathological grief Depression Agoraphobia
PLAY: The urge to play is common in mammals. Allows individuals to practise emotional responses and learn limits of behaviour	Joy and glee Happy playfulness	Mania Attention deficit hyperactivity disorder (ADHD)
LUST: Erotic desire that fully emerges at puberty stimulated by gonadal hormones	Erotic feelings	Fetishes
CARE: Hormonally primed urges such as the maternal instinct	Nurturance Attraction	Dependency disorders Attachment disorders

SOURCE: Adapted from Panksepp, J. (2006) 'Emotional endophenotypes in evolutionary psychiatry.' *Progress in Neuro-Psychopharmacology and Biological Psychiatry* **30**: 774–84.

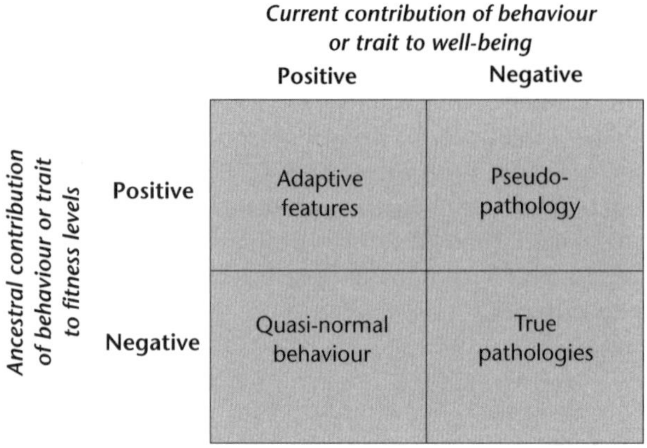

	Current contribution of behaviour or trait to well-being: **Positive**	**Negative**
Ancestral contribution of behaviour or trait to fitness levels: **Positive**	Adaptive features	Pseudo-pathology
Negative	Quasi-normal behaviour	True pathologies

For key, see next page.

FIGURE 14.2 Mental pathologies and fitness

SOURCE: Adapted from Crawford: wwwsfu.ca/faculty/Crawford, accessed 5 July, 2006 and Crawford, C. (1998b) Environments and adaptations: then and now, in Crawford, C. and Krebs, D. L. (eds) *Handbook of Evolutionary Psychology*. Mahwah, NJ, Lawrence Erlbaum, Table 9.1, p. 283.

Key to Figure 14.2

True pathologies: Behaviour that had deleterious consequences for humans in ancestral environments and still reduces fitness today. Could arise from malfunctions of otherwise adaptive mechanisms, spontaneous mutations not yet eliminated by natural selection, or the fitness cost of other adaptations. Examples include PKU, autism, Korsakoff's syndrome, incest, Huntingdon's chorea.

Adaptive: Behaviours, modules, traits that enhanced the fitness of our ancestors and still do in current environments. Examples include theory of mind, kin-directed altruism, reciprocal altruism, gossip, outbreeding, incest avoidance (as conditioned by the Westermarck effect, for example), beauty (mate value) recognition devices and so on.

Pseudopathologies: These conditions fall under the 'out of Eden' hypothesis (see section 14.4.6), that is, behaviour that was once fitness enhancing but no longer increases fitness in current environments. Often it is the social environment that has changed so much that such behaviours now have a negative effect on fitness. Examples include morbid male jealousy, high preferences for food high in calories, sugar and salt, exchange of resources for sex (paying for sex would once have increased male fitness but modern prostitution does not). A more disturbing example might be psychopathology. There is the possibility that this 'disorder' is in fact a frequency dependent strategy that served the fitness ends of a minority of ancestral humans (see Chapter 15).

Quasi-normal behaviours: These behaviours would once have probably detracted from ancestral fitness levels but as levels of cultural sophistication increased, they became tolerated or even thought desirable for social and ethical reasons. Examples include adoption of unrelated children, true altruism, mate fidelity in the face of temptation, deliberate reduction of family size for ecological reasons.

14.4 Evolutionary accounts

In this section, we examine several ways in which evolutionary thinking can help to account for the origins and persistence of mental disorders.

14.4.1 Genetically based disorders: mutations

Most medical conditions with a strong genetic aetiology are explained by random mutations. In any gene pool, mutations will spontaneously arise, most of which will eventually be removed by natural selection. But natural selection takes time to operate and equilibrium will be established between the rate of accumulation of mutations and the rate of removal by selection. Many of these mutations (and they probably lie behind over 1,000 disorders such as albinism, dwarfism, PKU), unless they confer some counterbalancing advantage, will be on their way out. The selection against genes that have damaging effects late in life will be weak and the frequency of such genes in the population high. It is expected that late-acting deleterious alleles will accumulate and cause senescence. An example of this is Huntington's chorea, which operates in middle age. Selection against deleterious alleles may also be weak if, when they segregate, they show linkage to other highly important genes.

14.4.2 Defence mechanisms

Nesse and Williams (1995, p. 230) criticise most current psychiatry for putting the cart before the horse, for 'trying to find the flaws that cause the disease without understanding normal functions of the mechanisms'. As an analogy, they consider coughing. The traditional psychiatric approach, parodied by these authors, would be to describe and catalogue coughing, study the neural mechanisms at work when coughing takes place and identify the cough control centre in the brain. Psychiatrists would then proceed to observe how certain substances, like codeine, suppress coughing, leading to speculations that coughing may be the result of a lack of natural codeine-like substances in the body. This obviously flawed line of reasoning is sometimes called the 'treatment aetiology fallacy'. Amid all this, how much clearer it becomes when we know that coughing is a natural defensive reaction of the body to expel foreign matter from the lungs, the oesophagus or the mouth. The analogy reinforces the point that the best way to understand the mind is to establish the purposes for which it was designed.

Organisms face a constant barrage of threats throughout their lives and natural selection has put in place a series of adaptive responses to deal with these.

Under this heading, we can identify two possible types of disorder: disorders that are the normal operation of defence mechanisms that have conventionally been labelled as disorders since they are unpleasant; and disorders that result from malfunctioning or dysregulated defence mechanisms. Examples of the former might include fever, pain, anxiety, jealousy and pain in general; and examples of the latter, autoimmune diseases and anxiety disorders. If the former category is suspected, then we can predict that the form of the defence response should match its function. For example, anxiety should bring about changes that are appropriate to deal with the threat. In this case, the evidence is positive: increased heart rate, glucose metabolism, clotting, sweating and breathing all change in line with a fight or flight response. We can also predict that fears should be related to actual fears in ancestral environments. Fear of high places should not be expressed in creatures for which high places pose no threat.

Experiences such as coughing, nausea, vomiting, diarrhoea and pain in general are both unpleasant and usually functional, and it is their very averseness that motivates us to avoid situations that cause such responses (Williams and Nesse, 1991). More generally, of course, this is precisely why natural selection designed the circuitry behind classical and operant conditioning. In operant conditioning, behaviours associated with a reward become more frequent, while those followed by punishment or painful consequences become less so. Many other defensive responses are products of classical conditioning. We can easily develop an aversion to cues that once preceded a series of painful experiences, such as the sight or smell of food or drink that we may have consumed to excess or that once caused illness.

Some defence mechanisms are innate. Of these some are general (or domain non-specific) such as the stress response to a whole variety of situations. Others fall under the heading of domain-specific mechanisms, in that they have been selected to respond appropriately to specific types of threat. Table 14.3 shows a few of these.

Table 14.3 Defensive responses to threats

Threatening situation	Defensive response
Infection	Expulsive defences such as spitting, vomiting, coughing, diarrhoea, sneezing, rhinorhea. Responses to combat infectious agents such as inflammation, fever
Loss of mate's fidelity	Jealousy
Loss of resources	Worry, suspicion
Loss of friends	Anger, anxiety, guilt and other social emotions specific to cause and context

SOURCE: Adapted from Nesse, R. M. (2005) 'Natural selection and the regulation of defenses: a signal detection analysis of the smoke detector principle.' *Evolution and Human Behaviour* **26**: 88–105, p. 26.

14.4.3 The smoke detector principle

Defences involve costs, however, and as organisms travelled along their phylogenetic trajectory, natural selection had a tricky problem to solve: how to balance the costs of the defence against the costs of the threats. Typically, the organism has to decide if the signal for a threat (a sound or moving shadow, for example) is real or simply the product of background 'noise'. Figure 14.3 shows a simple decision-making matrix indicating the costs of different outcomes in a simple scenario.

	Harm actually present (signal plus noise)	Harm absent (noise only)
Defensive response activated	True positive cost (CD)	False alarm cost (CD)
No response	False negative (harm cost) CH	Correct rejection cost (zero)

FIGURE 14.3 Decision matrix (costs and benefits) faced with a signal of a potential threat

At a very simple level, it would be pointless if the cost of the defence (CD) were to be greater than the cost of the harm (CH). Even if CD < CH, taking evasive or defensive action when the signal is not reliable is a waste of resources. But missing a true positive could bring harm or death. We can model the situation by considering the probability p(H) that the signal is indicative of a real threat. An optimal system would only express the defensive reaction when:

$$CD < p(H) \times CH.$$

Or its equivalent:

$$p(H) > CD/CH.$$

As Nesse (2005) points out, this can yield surprising results. If the metabolic costs of a panic-flight reaction are, say, 300 kcals (CD) and the damage that results from an unexpected attack amounts to 300,000 kcals (CH), then CD/CH = 0.001. For p(H) to be larger than this means that the probability of a signal being real only has to be greater than 1 in 1,000 for it to be worth activating a defensive response. Put another way, from 1,000 episodes of an actual response, just fewer than 999 will be false alarms and yet the system is still cost-effective.

The logic that generates this expectation of a high frequency of false alarms is sometimes called the **smoke detector principle** (Nesse, 2005) (or adaptive conservatism hypothesis). In a typical household fitted with a smoke alarm, the alarm will go off in response to a whole range of false signals (burning toast, smoking cooking fat and so on); it is an irritating diversion but one that is worth the cost, given the implications of not attending to a true threat that may appear only very infrequently. The animal kingdom provides numerous biological examples of this principle in action. Birds, for example, will often take flight at the slightest signal.

The equation above can also be applied to situations where the cost (CD) is an advantage not won. In the context of human mating behaviour, the reproductive success of the male is more greatly enhanced by separate multiple matings than that of the female. It should follow that males should be particularly alert to signals of sexual intent from females. So much so that the error rate from ambiguous cues (that is, reading sexual interest when none is there) can be quite high (see Haselton and Buss, 2000).

The simple equation above only tells us the threshold for p(H). It is in the interests of any organism, of course, to avoid paying unnecessary false alarm costs and to use information from the environment to gauge p(H) as accurately as possible. This is the basis of the clinical treatment known as behavioural exposure therapy. Individuals with simple **phobias** can be cured by repeated exposure to cues that set off phobias. Typically, the patient is introduced to the cue (the sight of a spider, for example) in a relaxed setting and counselled not to flee or panic. Usually, the initial surge in anxiety reduces. Gradually, over time, the patient unlearns the association between the sight of the spider and the need to panic and flee (see Marks and Tobena, 1990). Conversely, every occasion the response is set off and the threat turns out to be real, so the threshold of signal to noise lowers such that it pays to regard lower levels of signal above noise as posing a real danger. Nesse (2005, p. 101) thinks that panic disorder may result from the repeated firing of alarm signals and panic response: a 'runaway positive feedback process acting on an adjustable defence threshold'. In such cases, a panic attack from some sort of internal dysregulation (neurochemical or cognitive) is interpreted as a signal of real danger; panic sets in which serves to reinforce the association between the internal signal and the adverse reaction.

14.4.4 Preparedness theory

Phobias

An important point to note is that fears and phobias are not randomly distributed. Very few people, for example, have phobias about pencil sharpeners but many more have phobias about small animals such as rats and spiders. One approach to explaining this is the **preparedness theory** of Seligman (1971). This states that the non-random distribution of fears is caused by an evolutionary predisposition (preparedness) that operates during fear conditioning. The salient point here is that we are not born with innate fears, but we are born to more readily associate harm with some stimuli than others. More specifically, the hypothesis suggests that we become readily conditioned to fear objects that were once harmful to us in the EEA, but are more resistant to fears developing due to associations with harmless objects and events. Adverse experiences provide the trigger but adapted predispositions channel its effect. In a sense, this hypothesis is a mixture of adaptationist thinking with a Pavlovian conditioning interpretation. Lumsden and Wilson (1982, p. 2) emphasised this point when they said:

> It is a remarkable fact that phobias are easily evoked by many of the greatest dangers of mankind's ancient environment, including closed spaces, heights, thunderstorms, running water, snakes and spiders. Of equal significance, phobias are rarely evoked by the greatest dangers of modern technological society, such as guns, knives, automobiles, and electric sockets.

Preparedness theory suggests that we are born with specialised circuits that bias learning and the fear response. Ohman and Mineka (2001) make a persuasive case for the existence of a fear module with its origins in the distant past of our mammalian ancestry. They propose that the module (like many modules suggested by evolutionary psychology) has four basic characteristics:

- *selectivity:* the module attends to stimuli that were once indicative of threats in the past
- *automaticity:* the module is likely to have its origins in primitive-brained ancestors and so may not be under conscious control
- *encapsulation:* because of its ancient origin, the module may be impermeable to influences from other modules
- *specialised neural circuitry:* the module is likely to lie in subcortical dedicated circuitry.

On the last point, recent work suggests that fear and fear learning is organised around the amygdala, a subcortical structure shared with other mammals (see Chapter 8; also Fendt and Fanselow, 1999 for a review).

These predictions and others related to this whole area of study have received support from three lines of enquiry: experiments on humans investigating classical conditioning, experiments on the fear reaction in rhesus monkeys, and experiments on humans exploring illusory correlations.

Classical conditioning experiments

Studies on the laboratory conditioning of humans to 'fear-relevant' stimuli, such as snakes and spiders, and 'fear-irrelevant' stimuli, such as pictures of houses or flowers, began in the 1970s (see Ohman, 1979 for a review of this early work). Typical procedures involved conditioning human volunteers by presenting an image (snake, spider, house, flower and so on) accompanied by an uncomfortable electric shock. The strength of the conditioning was then measured by skin conductance responses. Subjects reliably developed a conditioned response to these images: fear was elicited by the image alone after repeated association with the electric shock. Initial findings suggested that it was easier to condition subjects to display a fear response to fear-relevant stimuli (snakes and spiders) than to neutral images (house, flowers).

If such fear-relevant stimuli are tapping into innate circuitry of an ancient origin, it could be expected that the dissociation (or 'extinction') of the response should take longer with fear-relevant stimuli that fear-irrelevant ones. That is, the skin conductance response to the repeated presentation of the image with no further shocks should decrease more quickly with images of houses and flowers than with snakes. There is considerable experimental evidence supporting just this prediction (see Ohman and Mineka, 2001).

One objection to these fear conditioning and extinction experiments is that the impact of the fear-relevant stimuli comes from cultural (that is, ontogenetically derived) information about the hazards posed by various objects and is not phylogenetically mediated – culture tells us at an early age that snakes are dangerous and flowers are benign. To circumvent this problem, Hugdahl and Karker (1981) conditioned humans to phylogenetically fear-relevant stimuli (snake images), ontogenetically fear-relevant stimuli (electric sockets) and neutral geometric shapes. All three groups showed reliable conditioning to these stimuli in the acquisition phase. But in the second phase (dissociation), the snake and spider conditioning effects showed much greater resistance to extinction than electric sockets or geometric shapes. Similar results were reported by Cook et al.

(1986) where enhanced resistance to extinction was noted for conditioning to snakes compared to guns.

Experiments on rhesus monkeys

In the 1980s, it was noted that monkeys who were born in captivity and had spent their entire lives in the Wisconsin primate laboratory in the USA showed no apparent fear of snakes. Yet monkeys of the same specie in the wild showed an intense fear of snakes bordering on phobia. Cook and Mineka (1987) reasoned that the phobic reaction of wild rhesus monkeys was probably vicarious (that is, acquired by watching other monkeys), since few monkeys would have survived to associate snakes with pain, due to the prevalence of poisonous snakes in their natural environment in India. A number of ingenious experiments have supported this and the related idea that monkeys are selective in their associations. Lab-reared monkeys watching the fearful reactions of wild monkeys to snakes (both real and toy) quickly acquired the fearful reaction themselves (Cook et al., 1986). In later experiments, lab-reared monkeys were shown wild monkeys behaving fearfully to simultaneously presented images of flowers and snakes without any cues as to which of these images was causing fear in the wild monkey. In follow-up trials on the lab-reared monkeys, they showed fear to snake images but not to flower images (Cook and Mineka, 1987).

Illusory correlations

Evidence supporting these predictions comes from work on illusory correlations. In these experiments, mild electric shocks are administered to volunteer subjects when images of objects are shown. In a study by Tomarken et al. (1989), several objects were used including images of what were supposed to represent historically real risks such as a snakes and spiders, and neutral images such as flowers or mushrooms. The study used two groups of people: those with a pre-existing expressed fear of the fear-relevant stimulus (snakes or spiders) and those who expressed no pre-existing fear. The linkage between the pain from the shock and the image was random but averaged 1:3 for each object, such that for every three images of the snake, flower or mushroom, a painful shock was felt; the other two times, either nothing was felt or an audible tone was heard. When the volunteers (high fear and low fear subjects) were asked afterwards to estimate the percentage of time a pain followed an image of a snake, it was consistently overestimated at 42–52 per cent, compared to the actual association of 33 per cent (see Figure 14.4) (Tomarken et al., 1989). Similar biases have been shown with blood injury stimuli and shock (Pury and Mineka, 1997) and socially relevant fear stimuli such as angry or disgusted facial expressions and aversive outcomes (see review in Ohman and Mineka, 2001).

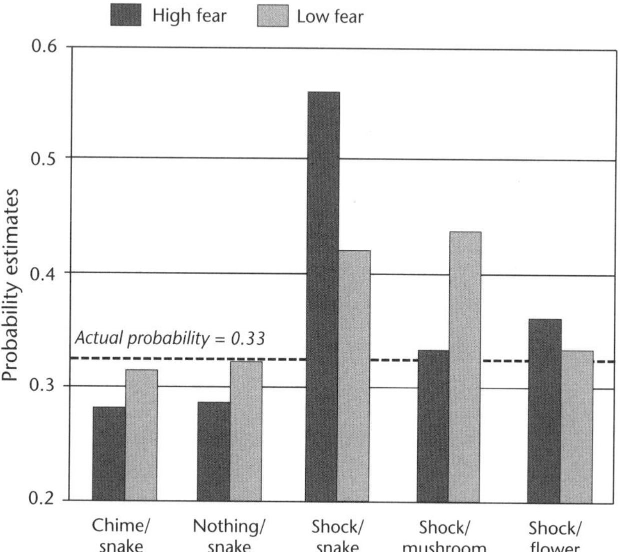

FIGURE 14.4 Covariation bias in the estimation of the conditional probability of outcomes
Note that both groups of participants overestimated the covariation of snake images and shock. The actual probability for all covaried stimuli was 0.33.
SOURCE: Adapted from Ohman, A. and Mineka, S. (2001) 'Fears, phobias, and preparedness: toward an evolved module of fear and fear learning.' *Psychological Review* **108**(3): 483–522.

According to the DSM-IV, one of the diagnostic criteria for phobias is that adult and adolescent sufferers know that the fear they are experiencing is unreasonable and disproportionate to the situation. It is as if, therefore, cognitive control is unavailable. This dissociation is predicted by preparedness theory, which postulates the non-cognitive nature of prepared selective associations, and by a modular view of fear, which postulates automaticity and encapsulation. To test these suggestions, Ohman and Soares (1994) developed a technique to investigate if the fear response was elicited when the images of the fearful object were presented subliminally. The procedure involved showing an image of the object for about 30 milliseconds followed by a masking image of longer duration. The subjects were not able to report the nature of the fast image and so in effect were not conscious of it. Figure 14.5 shows how, for subjects who already feared snakes or spiders, the fear response (as measured by skin conductance) was still elicited.

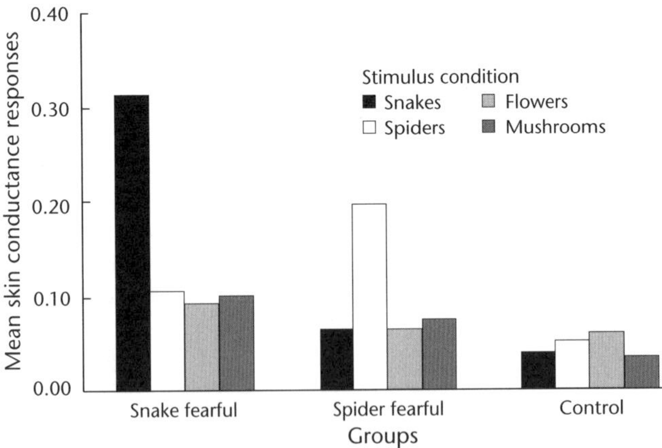

FIGURE 14.5 Fear reactions to subliminally presented images

SOURCE: Adapted from Ohman, A. and Soares, J. J. F. (1994) 'Unconscious anxiety: phobic responses to masked stimuli.' *Journal of Abnormal Psychology* **103**: 231–40, p. 236.

One problem with this whole line of research is identifying what counts as an ancestral hazard (sometimes referred to in the literature as 'phylogenetically relevant stimuli'). Deciding on what were real risks to hominins half a million years ago necessarily involves a good deal of conjecture. We have to construct the past environment of these early humans and speculate on likely predators and sources of injury. Nor is this environment of evolutionary adaptedness a single environment. We may spend a lot of time reconstructing the environment of the African savannah only to find that humans migrated out of Africa and then faced a whole new range of environments including the snowy wastes of a Europe passing through various ice ages. Furthermore, as we saw in Chapter 8, part of the fear reaction (which must surely be linked with phobias) is a fast response that bypasses the neocortex and is likely to be associated with old parts of the brain – in which case we need to look at conditions faced by primates long before the genus *Homo* arose.

The study of Tomarken et al. (1989) noted earlier can be revisited in the light of this difficulty. The positive interpretation of this experiment is that humans can be more readily conditioned to associate pain with images of snakes and spiders than flowers or mushrooms because the former were once hazardous to our ancestors and we are prepared to accept conditioning to them readily. There have been critics of this approach. Delprato (1980, p. 89), for example, argued that:

> Considering the fact that approximately 100 species of poisonous mushrooms have been identified in the USA alone ... it is reasonable to suspect that mushrooms have posed a greater threat to the survival of the human species than have snakes and spiders combined.

A response to this might be that the evolved aversion mechanism to poisonous food stuffs was not the fear module but taste.

Another critique is that some fears that we suppose to be ancestral may not have been that hazardous. Many people show a fear of spiders, but to what extent did spiders represent a significant threat to early humans? True, there are some poisonous spiders but they account for only about 0.1 per cent of the 35,000 documented species. It is hard to believe that these small creatures posed such a threat to our existence that we are now genetically programmed to become easily sensitised to painful associations with them.

There is also evidence that some phobias do arise in relation to objects and situations with little apparent relationship to past environments. Table 14.4 shows the results of a large survey of 20,000 subjects based on interviews and questionnaires administered in the USA. The second column shows the percentage of subjects expressing an 'unreasonable fear' and the third column the percentage that would meet the criterion of having a formal phobia diagnosis. Only the four more common phobias are shown. The results are interesting in that they show that unreasonable fears are widespread, with 1 in 6 people (18.5 per cent) meeting the criteria for a formal phobia of something. The problem for the evolutionary paradigm is that whereas it is possible to weave a story explaining how bugs, snakes, mice and bats may have once been real hazards through direct attack or spreading disease, it is harder to relate public transport to the plains of the African savannah – although one could argue that fear of public transport is really a version of agoraphobia or fear of strangers. A more realistic view would see such phobias as a mixture of evolutionary preparedness (where phylogenetically relevant stimuli are highly effective in initiating the fear response) and individual learning experiences, bearing in mind that the outcome of individual learning and experience is itself an expression of an evolutionary derived conditioning mechanism (see Ohman and Mineka, 2001).

Other studies also show that phobias can develop in relation to recent (in evolutionary terms) scenarios. Kuch et al. (1994) report that 38 per cent of survivors of road accidents develop a phobia related to driving a car. In another survey of common fears, Kirkpatrick (1984) found the surprising result that of the 133 stimuli categories used, women rated fear of roller-coasters the most important. For men, the fear of being punished by God was ranked the highest. Such findings are challenging to even the most ingenious evolutionary psychologist.

Table 14.4 Selection of results of an epidemiological study of phobias on sample of 20,000 American subjects

Stimulus	Percentage of subjects expressing unreasonable fear	Percentage of subjects classifiable as having a phobia
Bugs, mice, snakes, bats	22.4	6.1
Heights	18.2	4.7
Water	12.5	3.3
Public transport	10.5	3.2
Any phobia	60.2	18.5

SOURCE: Data extracted from Chapman, T. (1997) The epidemiology of fears and phobias, in G. C. L. Davey (ed.) *Phobias: A Handbook of Theory, Research and Treatment*. Chichester, John Wiley & Sons, Table 20.3.

Phobias may represent extreme forms of perfectly natural and useful emotions. Possibly, anxiety and a series of mild phobias represent the price we have to pay for ensuring that our genes stay clear of danger and survive long enough to replicate themselves. The fact that some people suffer more than others could then be explained by the normal distribution of these traits, as explored in the next section.

14.4.5 The trait variation argument

If we consider such traits as height, weight, intelligence, extraversion and introversion, it is obvious that people lie along a spectrum of quantitative measures of these variables. In such cases, the population as a whole usually describes a normal distribution, with most people in the middle and fewer people at each extreme. Such bell-shaped or Gaussian curves are found when numerous independent (such as genetic and/or environmental) factors contribute to the final size of the measured quantity. In biological contexts, when the number of genes involved in the expression of a trait is high, there is more scope for mutational variation. Under these conditions (unlike a simpler trait involving just a few genes), significant genetic variation can be maintained even in the face of strong selection. This variation could allow the possibility of some individuals existing at either end of the distribution. An obvious example is height. Height brings advantages but only up to a point. Extreme tallness (achieved by a significant minority of men) can lead to stature-related health problems. Applying this reasoning to the case of the affective system (emotion-generating mechanisms) means that some individuals could be oversensitive and suffer from, say, excessive anxiety, and others undersensitive and be foolhardy or excessive risk takers.

14.4.6 Out of Eden hypothesis

Over the past 10,000 years, roughly since the invention of agriculture, humans have so transformed their way of life that for the majority of the world's six billion people, conditions today are now vastly different to those in the EEA. Could this transformation be the root cause of several psychiatric disorders? This view, sometimes called the 'out of Eden' hypothesis, 'genome lag' or the mismatch hypothesis, has attracted numerous adherents. The notion has a superficial plausibility: by abandoning the hunter-gatherer lifestyle to which we are genetically suited and moving to live in densely populated cities with huge disparities of wealth and power on show daily, and where humans are sealed off from the natural diurnal and seasonal rhythms of life, we have set our genes and culture on a collision course. So possibly our Stone Age genes sit uncomfortably in a space-age culture.

One area that may illustrate this approach is mate choice and estimations of self-worth. In choosing a mate, an individual takes into account (among other factors) the attractiveness (fitness) of a potential partner coupled with some estimation of the sexual appeal or market value of themselves. Hence, in pairing up, humans estimate their own relative attractiveness and personal circumstances before deciding what minimum level of attractiveness will suffice in a partner. Now, in ancestral environments, where humans ranged in groups of about 100–150, such assessments were probably reasonably accurate. There would, for example, be very few extremely attractive or extremely wealthy people. The problem with modern culture is that people are now frequently exposed to images of highly desirable and successful men and women in the form of fashion models, actors and pop stars. The overall effect may be to bias our perception of the true frequency of such people in our social group. Men may become dissatisfied with their partners, thinking they have settled for someone too low down on the scale of market value; women may underrate their own attractiveness and take drastic actions, such as cosmetic surgery or crash dieting, to improve their appearance (Buss, 1994). In short, our ability to beam images of beautiful people around the globe may lead many to experience low self-esteem. Here is an example, then, where psychic distress may be caused by clever technologies and our 21st-century mode of existence.

So, are our Stone Age genes and 21st-century modes of life out of alignment in other ways? The problem with this general theory is that despite abandoning the EEA, humans are thriving as never before. Compare the global population of *Homo sapiens* at six billion with that of our nearest relative, the common chimp (*Pan troglodytes*) at a few hundred thousand. Moreover, some psychiatric disorders, such as schizophrenia, seem to be present in all societies and are even found among the few genuine hunter-gatherer cultures left. It seems difficult, therefore, to maintain that schizophrenia is a problem of modern lifestyles.

It is easy to draw sharp contrasts between ancestral and modern environments by choosing features that have drastically changed, such as population densities, jet travel, computer technology and so on. However, such a comparison is pointless unless we specify the nature of the adaptations that are supposed to be out of place as a result. We may live surrounded by the comforts of modern technology but the fundamental patterns of life go on: couples meet and have babies; people make friends and enemies, argue and make up; we gossip avidly about each other; and have intense relationships with family members. Despite the dire prognostications of some science fiction writers that humans will become slaves to machines, looking at much modern technology and its uses it is clear that the contrary is more often the case: we have shaped our technology around our natures. Computers would much prefer to operate using instructions given in zeros and ones, yet we force them to communicate with us using little pictures and a clumsy mouse to point at them. We often visit or live near to our relatives, dwelling places are often designed for the nuclear family, we work in groups with hierarchies – all these features are probably not far removed from the ancestral condition. As the old song goes: 'The fundamental things apply as time goes by'.

Crawford (1998a) advises that we should assume a basic similarity between ancient and contemporary environments with respect to particular adaptations, unless there are signs of stress and malfunction in humans, the behaviour is rare in the majority of cultures, or unusual reproductive consequences are observed.

The Neolithic revolution

If we are looking for an occasion when we most radically departed from the hunter-gatherer lifestyle that has characterised our social life for 90 per cent of our time on this planet, then the first and perhaps the most important has to be the Neolithic revolution. During the Neolithic or 'New Stone Age' period, humans moved from a position of foraging and hunting for food on a daily basis in small groups to a settled mode of existence involving the cultivation of crops and the domestication of animals. This technological revolution, which occurred about 10,000 years ago, made city life possible and current civilisation is based on it. If anything should have been a shock to our hunter-gatherer genes, then this should be it.

Although humans did very well in population terms out of the Neolithic revolution, we may still be prey to disorders that kick in after the age of reproduction. Examples include reproductive cancers in women. Compared to their hunter-gatherer ancestors, modern women experience earlier menarche and later first birth. This increases the exposure of reproductive tissues to oestrogenic hormones and in turn cell proliferation; and highly proliferating cells have a higher risk of cancer.

Changes in dietary habits since Palaeolithic times have been enormous. Modern humans in developed industrial nations get about 55 per cent of their calories from cereals, milk, milk products (for example cheese), sugars and alcohols. For hunter-gatherers, this figure (with the exception of some natural sugars) would be near to zero. Most of the calories consumed by hunter-gatherers would have come from fruits, vegetables, nuts and honey (Eaton and Cordain, 1997). The hunter-gatherer diet would also have been much lower in saturated fats. Meat from wild game animals has about 4 g of fat per 100 g of meat, compared to typical supermarket meat from domesticated animals of 20 g per 100 g. Modern humans also consume far more salt than our ancestors. Apart from populations living near the coast, the majority of our forebears lived in areas where salt was scarce. Consequently, we have an appetite for salt, which, now it is available cheaply, leads us to overconsume it. We need to add to this picture the fact that most humans in developed nations are far more sedentary than their Palaeolithic ancestors. Diseases linked to this high fat, high sugar and high salt modern diet and sedentary lifestyle include conditions such as breast and colon cancer (high fat, low fibre), hypertension (high salt) and coronary heart disease (high fat).

Modern humans also suffer from eating disorders where they consume too much (bulimia) or too little food (anorexia). Eating disorders are relatively rare in traditional cultures. It is likely that natural selection put in place strong measures to avoid starvation but weaker ones to guard against overeating. The ready availability of cheap food in modern society has led to a near epidemic of obesity in some cultures. Eating disorders often begin with efforts to lose weight. Nesse (1999) has argued that serious dieting sets off an adaptive response to consume more food leading to a cycle of binge eating and ultimately lack of control.

Another example of a massive change brought about by culture on our everyday life is the schooling of children. Humans have been speaking language for probably about 100,000 years and are biologically primed to learn languages quickly and efficiently while young. But reading and writing are skills that only emerged over the past 10,000 years, have no natural basis in brain architecture and, accordingly, are skills only mastered after careful schooling, with tedious repetition and external sources of motivation. So asking children to sit quietly in a classroom to master these arts is asking a lot of young primates whose natural inclinations are to be active, playful and exploratory. With this in mind, we should consider the prospect that children diagnosed with ADHD may simply be those youngsters who are naturally more playful than average. Estimates of the incidence of ADHD vary from 1.7 to 16 per cent of children in the USA and UK (Goldman et al., 1998). The fact that boys are affected more than girls, in a ratio of about 3:1, may be significant in view of the well-established fact that males naturally take more risks than girls.

The stress response: then and now

Natural selection provided our ancestors (and by descent ourselves) with a coordinated system for responding to stress, which we call the fight or flight response (see Chapter 8). In the face of a perceived physical threat (approach of a predator or aggressive signals from an unfriendly rival), the nervous system primes a variety of organs and releases a cascade of hormones to prepare the organism to fight or flee: blood is directed to the large muscles; the heart rate increases; pupils dilate to gather more information; non-urgent systems such as digestion and the immune response are put on hold; stress hormones such as cortisol and adrenaline are released; stored glucose is mobilised (glucogenesis); and adipose tissue is broken down to release fatty acids. This is a great system for putting up a spirited physical fight or running away at speed. The problem for modern humans is that this stress response is activated by stressors for which running away or fighting are not options. Instead, we seethe with frustration in traffic jams or nurse egos bruised by nasty office politics, and all those glucose and fatty acid molecules are not utilised as nature intended. This by itself can give rise to circulatory and cardiac health problems. The ill effects are compounded by the reduction in the efficiency of the immune system that everyday stresses bring, making us more susceptible to the pathogens that are in plentiful supply in modern, high-density population environments.

Mental health

Statistics on the incidence of mental health lend some support for the mismatch theory. For those Americans born before 1905, only about 1 per cent suffered an episode of major depression before they reached the age of 75. For modern Americans born after 1955, about 6 per cent were reported to have had an episode of major depression before they reached the age of 30 (Meyer and Deitsch, 1996). Although we should treat such figures with caution (since they are subject to trends and fashions in diagnosis and incidence of self-reporting), they are pointers to the fact that the comfortable life, abundant food and extended lifespan we have enjoyed over the past 150 years do not seem to have delivered commensurate benefits to our mental health. Happiness seems to be increasing more slowly than GNP. There are numerous possible causes of this. Our working week of 40 hours may be longer than that needed to sustain life in the Old Stone Age, leaving less time for relaxation and leisure. We also live in groups that are huge compared to our ancestors. We probably see or encounter more strangers or non-kin in a week than a hunter-gatherer would have in a lifetime. We are surrounded by people who have no kin affiliation to us and are not so concerned about our welfare as our kin would be.

14.4.7 Design trade-offs

We have already met the idea that natural selection often involved trade-offs. The large brain of human babies implies greater infant dependency and raises the mortality risks of childbirth (see Chapter 5). Post-meiotic cells may provide another example. It is a well-known fact that most brain cells do not divide and reproduce after birth. The number of brain cells you have now is about the same as you were born with. Brains probably work better with non-dividing neurons but leave humans at risk of irreversible damage from diseases such as Alzheimer's. The smoke detector principle introduced earlier also falls into this category. It makes adaptive sense to attend to slight signals of danger, even though we may seem to be overanxious.

Genes that enhance fitness in one way may bring about disorders in other ways. This is sometimes called the **pleiotropy** hypothesis. Pleiotropy refers to the well-documented genetic fact that a gene or set of genes can have more than one effect. People prone to manic depression, for example, may also be highly creative and attractive to the opposite sex (see Chapter 15).

Linked with this approach is the argument that the same genes that cause an apparently fitness-reducing condition in some people give rise to effects in others that increase their fertility. This is an argument that has been applied to explain the possible genetic basis of male homosexuality (although this is no longer classified as a disorder). A gene for male homosexuality could thrive in the gene pool if, when found in the sisters of homosexuals, it increases their fertility.

There are a number of well-established cases of maladaptive genes surviving against apparent odds in the human gene pool. A simple change to the base sequence on our DNA is known to cause the distressing condition of sickle-cell anaemia. The substitution of one amino acid (glutamine becomes valine) in haemoglobin causes an alteration in the shape of red blood cells – they appear sickle-shaped – and a reduction in the oxygen-carrying capacity of the blood. The sickle-shaped cells produced when the defective gene is present on both chromosomes (that is, the chromosomes are homozygous) are quickly broken down by the body, blood does not flow smoothly and parts of the body are deprived of oxygen. The physical symptoms range from anaemia, physical weakness, damage to major organs, brain damage and heart failure. There is no cure for the condition, which causes the death of about 100,000 people worldwide each year. Sickle-cell anaemia is by far the most common inherited disorder among African-Americans and affects 1 in 500 of all African-American children born in the USA. Its high frequency in the population, and the fact that natural selection has not eliminated it (many of those suffering die before they can reproduce), is probably due to the fact that, in Africa, possession of one copy of the sickle-cell gene confers some resistance to malaria. If Hb is taken to be the normal haemoglobin gene and Hbs the sickle-cell gene, then people who inherit both sickle-cell genes, that is, one from their father and one from their mother, and who are therefore Hbs Hbs, suffer from sickle-cell anaemia and die young. People who inherit only one copy of the sickle-cell gene and are Hb Hbs are said to have sickle-cell trait and only some of the red blood cells are oddly shaped. It is in this latter condition that the gene gives an advantage in protecting against malaria, since the malarial parasite *(Plasmodium)* cannot complete its life cycle in the mutant cells. It is the prevalence of malaria in African countries that explains why this apparently maladaptive set of genes survived in the gene pool and is now found among African-Americans.

Just how small an advantage is needed to ensure that genes that are maladaptive when an individual carrier has copies from both parents remain in the gene pool can also be seen in the condition of cystic fibrosis. A child with **cystic fibrosis** is born when the relevant genes are homozygous in the recessive state, that is, it has two copies of the defective gene, one from each parent. If it has only one copy, then it is said to be a **carrier**. People who are carriers live perfectly normal and healthy lives, never realising they are carriers until they mate with another carrier. About 1 in 25 Caucasians are thought to be carriers of the recessive allele for cystic fibrosis. The chance of two carriers meeting is thus about $(1/25)^2 =$ 1 in 625 or 0.0016. The chance of a child from a union of these parents having both recessive alleles and hence displaying the condition is one-quarter of 0.0016 = 0.0004, or 1 in 2,500. Hence, about 1 in 2,500 of Caucasian children are born with cystic fibrosis (see section 1.5). It is thought that the heterozygous condition only needs to carry a 2.3 per cent advantage compared to non-carriers for the recessive allele to persist indefinitely (see Strachan and Read, 1996).

The trade-off type of explanation has also been plausibly applied to depressive illnesses and schizophrenia (see case studies in Chapter 15).

In this chapter, we have sketched out an intellectual toolkit that evolutionary thinking provides for the examination of mental disorders. The range of concepts at the disposal of the evolutionary psychiatrist includes viewing mental problems as:

- Ailments due to mutations not yet eliminated from the gene pool, or due to alleles that carry some counterbalancing advantage
- Apparent disorders that are really the response of defence systems that appear oversensitive but, despite causing distress, are actually cost-effective in the long run (the smoke detector principle)
- An inbuilt preparedness to acquire phobias to stimuli that were phylogenetically significant (snakes, spiders and so on)
- Trait variation – the disorder lies at the extreme end of a natural variation
- Disorders due to a mismatch between genes selected in an EEA now vastly different to modern environments
- Disorders that are design compromises.

In the next chapter, we apply this toolkit to a number of specific case studies.

Summary

- The classification of mental disorders is a controversial topic. A robust definition of a disorder will need to include biological and sociological components.
- Current psychiatry suffers from a state of conceptual pluralism, with a number of different approaches to identifying and treating disorders. These approaches include cognitive, behavioural, humanistic, biomedical, psychodynamic and evolutionary.
- The evolutionary approach invokes a whole set of ideas to explain mental disorders. They are united by a focus on the natural functions of the brain as designed by natural selection.

Key Words

- **Affective**
- **Carrier**
- **Cystic fibrosis**
- **DSM system**
- **Humanistic**
- **Phobia**
- **Pleiotropy**
- **Preparedness theory**
- **Psychodynamic**
- **Smoke detector principle**

Further reading

Andreasen, N. C. (1987) 'Creativity and mental illness prevalence rates in writers and their 1st degree relatives.' *American Journal of Psychiatry* **144**: 1288–92.

An interesting article that examines the high frequency of mental illness among creative people.

Jamison, K. R. (1994) *Touched With Fire: Manic Depressive Illness and the Artistic Temperament.* New York, Free Press.

An authoritative and readable work. Examines the lives of artists and writers over three centuries and the link between creativity and bipolar disorder.

McGuire, M. and Troisi, A. (1998) *Darwinian Psychiatry.* Oxford, Oxford University Press.

An excellent overview of various Darwinian approaches to this subject and especially the 'out of Eden' hypothesis.

Murphy, D. and Stich, S. (2000) Darwin in the madhouse, in P. Caruthers and A. Chamberlain (eds) *Evolution and the Human Mind.* Cambridge, Cambridge University Press.

A chapter that examines several evolutionary approaches but emphasises the application of the idea of modularity. It offers a severe criticism of DSM-IV methodology.

Nesse, M. and Williams, C. (1995) *Evolution and Healing: The New Science of Darwinian Medicine.* London, Weidenfield & Nicolson.

A general work that examines a whole range of disease states (mental and physical) from an evolutionary standpoint. A useful book to start exploring this whole area.

Wakefield, J. C. (1992) 'The concept of mental disorder: on the boundary between biological facts and social values.' *American Psychologist* **47**: 3733–88.

An article that advances a useful definition of mental disorders.

Mental Disorders: Some Case Studies

This is the Hour of Lead –
Remembered, if outlived,
As Freezing persons, recollect the Snow –
First – Chill – then Stupor – then the letting go –
(Emily Dickinson, 1830–86, 'After great pain, a formal feeling comes')

In this chapter, we will examine a number of specific mental disorders in the light of evolutionary thinking. The illnesses covered are depression, psychopathy, schizophrenia and autism. They are chosen since each illustrates some specific evolutionary approaches.

15.1 Depression

15.1.1 Types of depression and their incidence

The term 'depression' describes a wide range of emotional states from low mood and sadness to clinical or suicidal depression. Sadness is a natural human emotion and is usually taken as a sign of normal functioning so long as there is some obvious cause. But clinical depression is more severe and long-lasting, often without any obvious cause for its onset. The classification of depression in the DSM-IV falls under 'mood disorders'. There are several varieties: major depressive disorder (unipolar), bipolar and dysthymic disorder. Table 15.1 gives some summary information on these states.

The term 'unipolar' refers to the fact that in people who suffer from this condition there is one abnormal state: that of depression with all its associated symptoms. It is sometimes thought that there are two discrete categories of **unipolar depression**: reactive and endogenous. Reactive depression is a response to some painful event such as the loss of a loved one, redundancy at work, or an instance of personal failure. Endogenous depression arises from inside the person and is usually more serious. These terms are not used in the DSM but instead we find major depressive disorder (MDD), which is severe but short-lived and dysthymic disorder, which is less acute but may last for much longer.

As the term 'bipolar' indicates, in this condition there are two states often called 'mania' and 'depression'. People afflicted by this condition (many of whom in history have been talented and creative people) experience violent mood swings from mania (typified by frantic activity, irritability and recklessness to a degree that sometimes necessitates hospitalisation) to depression itself.

There is evidence of some genetic basis to **bipolar depression** but less so for unipolar depression. Early studies on twins by Price (1968) found that monozygotic twins (MZ) were much more likely to both suffer from manic depression if either one of them did than were non-identical or dizygotic (DZ) twins. Significantly, this was observed even if the monozygotic twins were reared apart (see also Table 15.3).

As the figures in Table 15.2 show, unipolar depression is quite common, indeed it is responsible for a significant proportion of the global burden of all diseases. Table 15.2 shows a ranking of what are known as 'disability adjusted life years' (DALYs) using data from surveys up to the year 2001. The DALYs are calculated from loss of life due to the condition and years of life spent suffering with the condition adjusted for the severity of the condition. They provide a measure of the importance of various conditions to the global totals for human suffering in terms of years lost of human life per year.

Table 15.1 Clinical names for some depressive states

Depressive condition	Incidence	Symptoms (some or all)
Major depressive disorder (old name: unipolar)	Varies from country to country, but in Western countries about 10 per cent of men and 20 per cent of women become depressed sometime in their lives. Typical age of onset about 40. Risk of suicide much lower than for bipolar disorder. Some famous sufferers include Winston Churchill, Kurt Cobain, Judy Garland, Marilyn Monroe, Sylvia Plath	Sad depressed mood, loss of interest in usual activities, sleep disturbance, tiredness, loss of energy, negative self-concept, weight loss or gain
Bipolar disorder (manic depression) DSM-IV distinguished between bipolar I and bipolar II, where the mania in bipolar II is of lower intensity than bipolar I. Not all agree with this separation	Typically begins in adolescence or early adulthood and continues throughout life. Lifetime risk of disease about 1 per cent for men and women. Suicide rate about 15 per cent of those that suffer from disease. Famous sufferers include Virginia Woolf, John Ruskin, Tchaikovsky, Tennyson, Coleridge	Depressive symptoms as above, coupled with mania: elation, talkativeness, unjustified high self-esteem. Sufferer will swing from depressed to manic state with normal mood periods in between. The frequency of change varies from person to person. Tends to run in families, indicating a genetic basis. Sometimes responds to lithium treatment. Mild symptoms of bipolar disorder for at least two years with person not without symptoms for more than two months sometimes diagnosed as cyclomythia
Dysthymic disorder, a depressive sadness	More common than major depressive disorder but less severe	A milder form of unipolar disorder lasting for at least two years with more days affected than not. Includes periods of poor appetite or overeating, insomnia or hypersomnia, low energy or fatigue, low self-esteem, poor concentration or difficulty making decisions, feelings of hopelessness

Table 15.2 The top 10 leading causes of the global burden of disease (2001)

	Cause or condition	DALYs (millions of years)	Percentage of total DALYs
1	Perinatal conditions (low birth weight, birth trauma and so on)	90.48	5.9
2	Lower respiratory infections	85.92	5.6
3	Ischemic heart disease	84.27	5.5
4	Cerebrovascular disease	72.02	4.7
5	HIV/AIDS	71.46	4.7
6	Diarrhoeal diseases	59.14	3.9
7	*Unipolar depressive disorders*	*51.84*	*3.4*
8	Malaria	39.97	2.6
9	Chronic obstructive pulmonary disease	38.74	2.5
10	Tuberculosis	36.09	2.3

SOURCE: Adapted from Table 3.14, http://www.dcp2.org/page/main/Data.html, accessed 2 August, 2006; Lopez, A. D., Mathers, C. D., Ezzati, M. et al. (2006) *The Global Burden of Disease and Risk Factors,* DCPP Publications.

Table 15.2 serves to reinforce the point that childbirth is still an enormously risky event. It is noteworthy, however, that depression ranks seventh in terms of a scale of global suffering, above more headline-grabbing causes of human suffering such as cancers and road traffic accidents that do not even figure in the top 10.

15.1.2 Depression as an adaptive strategy

There is some evidence that susceptibility to depression has a genetic basis. Table 15.3 shows a selection of concordance data for unipolar and bipolar depressive disorders. It is clear that bipolar disorder seems to have a stronger genetic causation than unipolar disorder.

Table 15.3 Concordance values for depressive disorders showing data from several sources

Relationship	Concordance values
Major depressive disorder	
Monozygotic twins	59 per cent (Bertelsen et al., 1977) 46 per cent (McGuffin et al., 1996)
Dizygotic twins	30 per cent (Bertelsen et al., 1977) 20 per cent (McGuffin et al.)
Bipolar disorder	
Monozygotic twins	80 per cent (Bertelsen et al., 1977) 40 per cent (Craddock and Jones,1999)
MZ twins reared together	69 per cent (Rush et al., 1991)
MZ twins reared apart	67 per cent (Rush et al., 1991)
Dizygotic twins	16 per cent (Bertelsen et al., 1977) 5–10 per cent (Craddock and Jones, 1999) 13 per cent (Rush et al., 1991)

The genetic basis at least of bipolar disorder does not solve the question as to whether depression is an adaptive trait or not. The tendency to depression could be a result of trait variation: some individuals carry a burden of genes at the extreme end of the spectrum of normal variation that disposes them to become depressive.

It could also be that bipolar depression results from an unfortunate combination of mutant genes that would tend to run in families – genes that are malfunctional but have not yet been selected out. This seems unlikely, however, since bipolar disorder affects about 1 per cent of people and substantially lowers their fitness. Suicide rates from those suffering from this disorder are about 15 per cent. In view of the relatively high frequency of the disorder and the strong selection against it, it would seem unlikely to be a product of a recent mutation. The problem then becomes identifying the advantage that must come with this disorder to balance out the deleterious consequences. Several candidates have been proposed. There have been reports, for example, that successful creative people have high rates of manic depressive illness (Jamison, 1994); that mood disorders have a higher than average frequency in writers and their relatives (Andreasen, 1987); and that the relatives of people suffering from the condition are more creative (Richards et al. 1988).

The route from these positive features to higher fitness might then be through sexual selection: creative people are attractive to the opposite sex. This model (essentially an argument based on the concept of pleiotropy) is further explored in the context of schizophrenia (see below).

Development and attachment

Depression could also be the result of disrupted developmental processes. Indeed, many approaches to mental disorders (psychoanalytical, behavioural and psychosocial) stress the importance of early life experiences. Such explanations do not fall outside an evolutionary framework. At birth, the brain arrives with mental hardware and neural circuitry preformed to respond to stimuli in specific ways. Different environments will produce different people and adverse environments may distort the development of the brain away from the norm. The biologist Ernst Mayr used the term 'open programs' to describe how environmental conditions shape the tendencies in the genotypes to produce the phenotype. For example, humans are almost certainly programmed to bond affectionately with their mother. If this process is interrupted, abnormal psychological growth may result. In this sense, then, evolutionary theory provides the ultimate causal framework, whereas psychoanalytical, behavioural and psychosocial models provide the proximate hypotheses (McGuire et al., 1997).

A good example of this evolutionary informed ontogenetic approach is the work of the British doctor and psychoanalyst John Bowlby on **attachment** theory. In the 1930s, Bowlby worked as a psychiatrist at a centre for troubled youngsters in London. Here he was struck by the blank emotional response of many of the youths he met. Then he noticed a pattern: children who showed a lack of emotional response, and typically were in trouble with the authorities, tended to be those who had been separated from their mothers at an early age.

At the time of Bowlby's original insights, the study of human emotions and relationships was dominated by Freudian psychoanalysis. The study of non-human animals, particularly in America, was also heavily influenced by the behaviourism of Watson, Skinner and their followers. Bowlby felt that both Freudianism and behaviourism were not adequate to the task of explaining his observations on attachment. Fortuitously, Bowlby then came across the work of two men, the Cambridge comparative psychologist Robert Hinde and the Austrian ethologist Konrad Lorenz. Both were sensitive to the importance of evolutionary theory in explaining animal behaviour. This was just the encouragement that Bowlby needed and he went on to formulate his attachment theory within an evolutionary framework.

Bowlby (1969) proposed that all primates are born programmed to form close emotional ties with nearby adults such as mothers. Bowlby argued that how a growing child conceived of social relationships depended on this early infant experience and disruptions to this crucial early process of attachment could bring about disorders later in life. Bowlby suggested that the process of forming an attachment was adaptive for both parents and offspring. Babies, he argued, are born programmed to attach to the adult that gives them care early in life, this usually being the parent who provides the baby with a source of protection and nurture. Bowlby argued that just as the newborn baby is attracted to the smiles and face of its mother, so the faces of babies and their voices serve to stimulate and release the nurturing response of the mother.

According to Bowlby (1980), the attachment bond had to form before the age of two and a half years or else it would be difficult for an attachment to form after this. Should attachment not take place, then permanent emotional damage would result, leading to problems with relationships, sadness and depression later in life.

Bowlby's work has had an enormous impact on thinking about the relationship between parents and children in the first few years of life. His work has been criticised, extended and modified, and other possible explanations for the effects he observed have been offered. However, the central tenets of attachment theory have entered mainstream psychological thinking and the consensus seems to be that attachment has an important role to play in the long-term mental health of the growing child. Where the theory is weak is in explaining why children differ so much in their resilience to neglect.

15.1.3 Depression as a means of seeking help and conserving resources

Depression may serve adaptive needs by signalling the depressed state to others in the hope of eliciting help. It may also serve to conserve energy and direct resources to solve immediate problems. This latter notion is the conservation-withdrawal strategy coined by Engel and Schmale (1972) who first applied the term to the 'despair' pattern of behaviour seen in a lost infant monkey. Both these ideas have problems, however. The depressed state may not bring help and could serve to alienate friends. The conservation-withdrawal strategy implies that depression should cease with the passage of time, but this is sometimes not the case (Nettle, 2004).

Postpartum depression

Another idea is that depression may serve as either a 'cry for help' or a way of cutting losses. Hagen (1999), for example, has argued that postpartum (or postnatal) depression (PPD) is an adaptive mechanism, whereby mothers signal their need for greater investment from kin and their partner by presenting themselves as unable to cope. In addition, Hagen suggests that PPD is also a strategy to incline mothers to cease investing in offspring that may not be viable, that is, may not reach reproductive age themselves. Factors that may signal to the mother that care should be withdrawn from the child or that extra support should be sought include:

1. Insufficient investment from the father.
2. Problems with the pregnancy or the infant, suggesting that the offspring may have low viability.
3. A hostile environment inauspicious for raising offspring.
4. Large opportunity costs, in other words, investment in the offspring could be more profitably directed towards other activities such as mother's survival or existing offspring.

Figure 15.1 shows data plotted from a similar study by Field et al. (1985), whereby questions addressed to the expectant mother during her third trimester of pregnancy showed that answers revealing a lack of support or commitment to the child were strongly predictive of later PPD.

Watson and Andrews (2002) have suggested a more general model, which would include PPD as a special case. In their view, the depressive state serves two functions:

1. Diversion of energy away from normal social and physical activities to enable the sufferer to concentrate on the problem that caused the depression. This is sometimes called the 'social rumination function'.
2. A signal of the need for more investment from the depressive's kin and social network.

In this view, depression serves a similar function to pain and is therefore an adaptation. As Nettle (2004) points out, there are problems with this approach:

- The analogy with pain may be misleading, since all normal humans experience pain but there is no evidence that all humans have the capacity to become clinically depressed.

Prepartum questions

1. Are you single or separated?
2. Do you have marital problems?
3. Do you often feel that your husband (boyfriend) does not love you?
4. Was your pregnancy unplanned (accidental)?
5. Do you more or less regret that you are pregnant?
6. Can you honestly say at this time you really do not desire to have a child?
7. Did you become very depressed or extremely nervous in the period following the birth of your last child?

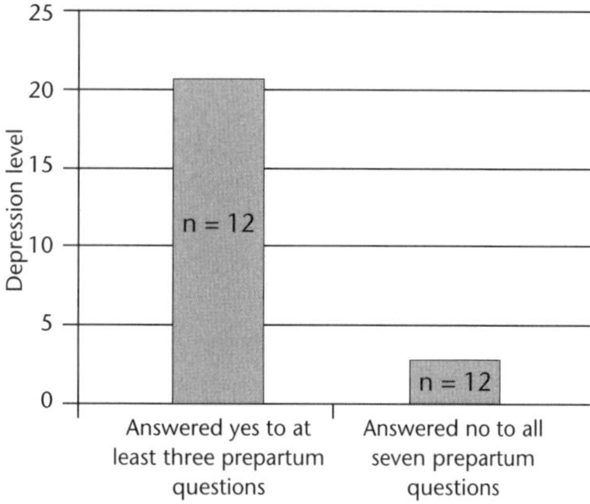

FIGURE 15.1 Conditions associated with postpartum depression

SOURCE: Reproduced from Hagen, E. H. (1999) 'The functions of post-partum depression.' *Evolution and Human Behaviour* **20**: 325–59, Figure 1; data from Field, T., Sandburg, S., Garcia, R. et al. (1985) 'Pregnancy problems, postpartum depression, and early mother–infant interactions.' *Developmental Psychology* **21**(1): 1152–6.

- The heritable variation in true adaptations is usually very small. This is because natural selection fixes adaptations in species and quickly uses up most of the heritable variation, leaving only residual variation around the adaptively optimal peak. In such cases, then, we would expect such variation as is observed to be largely due to environmental influences. As a very simple example, most humans have five fingers at birth and any variation (people with fewer than five fingers) tends to be due to environmental influences such as a surgeon's knife, or an unfortunate accident. In this case, the heritability will be low (it will not be zero since some people are born with six fingers). Yet the evidence shows that depression does run in families. The odds ratio for first degree relatives is 2.84 (with 2.31–3.49 giving 95 per cent confidence intervals) (Sullivan et al., 2000). The only way for such genetic heritability to exist is through mutations that have yet to be eliminated, gene linkage or by such genes conferring some other advantage.

- It is not at all obvious that depression does enhance the abilities of those afflicted to solve social and life problems. There is evidence, for example, that depressives fare worse than normal people in their execution of interpersonal skills (see Nettle, 2004).
- Although there is evidence that supports the prediction that PPD elicits greater investment from social allies (Hagen, 2002), there is also plenty of evidence to suggest that more general depressive symptoms exacerbate social problems, leading to marital failure, hostility and rejection. There may be some truth in the old adage that 'laugh and the whole world laughs with you, cry and you cry alone'.

15.1.4 Depression as a bargaining strategy

Edward Hagen (2003) sees postpartum depression as a special case of his more general theory: that depression is a form of bargaining in which the sufferer indicates a willingness to impose costs on the rest of the group unless he or she is helped to overcome a particular problem. This could explain why depression is associated with that ultimate fitness-reducing act: suicide. The threat of suicide is the prospect that an individual will withdraw their efforts entirely from the group, and so it behoves the group to help. Some, but not all, suicide attempts are fatal, but averaged over many generations, it is possible that the benefits gained to genes related to suicide from actual help given outweigh the losses in the cases where suicide was successful.

The bargaining model also attempts to explain the sex bias in depression (women suffer at roughly twice the rate of men). It is possible that depression was a more effective strategy for women in the EEA than men for the following reasons:

1. Physical power was a less effective strategy for women than men in resolving conflicts.
2. Female exogamy (or patrilocality) is the modal pattern for human groups, meaning that women were more likely than men to find themselves among non-kin and hence conflict situations.
3. Whereas powerful men would seek to control female sexuality (the limiting resource in human mating), females could bargain with their most powerful weapon: their reproductive and child-care investments.

Table 15.4 shows Hagen's view of how the symptoms of depression may serve a bargaining function.

Table 15.4 Selected symptoms of a major depressive disorder and their possible function according to the bargaining model of depression

Symptoms (taken from DSM-IV)	Putative bargaining function
Sad or depressed	Signal to others of state
Loss of interest in virtually all activities	Reduction of investment in oneself and others
Hypersomnia	Reduce productivity
Fatigue	Reduce productivity
Feelings of worthlessness	Communication to others of feeling undervalued
Recurrent thoughts of death	Threat to reduce (totally) future productivity

SOURCE: Adapted from Hagen, E. H. (2003) The bargaining model of depression, in P. Hammerstein (ed.) *Genetic and Cultural Evolution of Cooperation*. Cambridge, MA, MIT Press, Table 6.1, p. 101.

15.1.5 Social competition hypothesis

The social competition hypothesis of depression is particularly associated with the work of John Price and colleagues. Price et al. (1997) adapted concepts from the study of the competitive encounters between non-human animals as found in behavioural ecology and applied them to human contexts. Their view is that depression is part of an 'involuntary subordinate strategy' with three main functions:

1. A mechanism to prevent individuals from making a 'come back' by inhibiting aggression towards

superior rivals and creating a 'subjective sense of inadequacy' (Price et al., 1999, p. 242).
2. A means of communicating submission and 'no threat' signals to rivals.
3. A means of facilitating the acceptance of defeat and promoting behaviour leading to termination of the conflict followed by reconciliation.

The essential idea behind the pathological state is that if 3 becomes blocked for some reason, then 1 and 2 become intense and prolonged, leading to depressive illness. Price et al. contend that the social competition hypothesis is consistent with the cognitive distortions of depression. Hence, the world of the depressive is coloured by a pessimistic attitude to the likelihood of success and a sense that the conventional attributes of a successful life such as rank and ownership are of little value. This model does make an interesting prediction: depression should reduce the tendency of the depressive to be aggressive or competitive to those of higher social rank but not necessarily to those of lower social standing (see Price et al., 1997).

15.2 Psychopathology

The common view of disorders (although challenged by some of the adaptive hypotheses already considered) is that they are pathological. But there remains the possibility that some people, conventionally labelled as suffering from some disorder, are behaving in adaptively efficient ways just as nature intended. In this case, such disorders would deserve the label **pseudopathologies** in the sense of Figure 14.2. One candidate for this category is the condition known as **psychopathy** or 'sociopathy' (in the psychiatric literature, the usage of these terms is not always consistent (see Mealey, 1997), but here the term 'psychopathy' will be taken to include sociopathy). Psychopaths are said to possess an 'antisocial personality' disorder. This is described in DSM-IV as:

> A pervasive pattern of, and disregard for, the rights of others that begins in childhood or early adolescence and continues into adulthood ... This pattern has also been referred to as psychopathy, sociopathy, or dysocial personality disorder. (American Psychiatric Association, 1994, p. 645)

Psychopaths are renowned for displaying some of the following characteristics:

- *Antisocial personality and behaviour:* They typically fail to show empathy or remorse and do not appreciate the consequences of their actions. They tend to be cold-hearted, selfish, insensitive, egocentric, callous and aggressive.
- *Deceit and manipulation:* Psychopaths often commit crimes that rely upon the trust and cooperation of others such as bigamy and fraud.
- *Emotional make-up:* Psychopaths have shallow emotions, are impulsive and crave excitement.

A couple of examples will suffice to show these traits in action. The psychopath Kenneth Taylor (a dentist) battered his wife to death in 1984 and then was surprised that no one could sympathise with him for his loss; Jack Abbott killed a waiter who had asked him to leave a restaurant but then denied he had done anything wrong since it was a clean wound and the victim was worthless (see Pitchford, 2001).

The prevalence of antisocial personality disorder is about 3 per cent in males and 1 per cent in females, but using the more strict criteria for psychopathy developed by Hare (1993), the rate of psychopathy is estimated to be about 1 per cent. Hare also estimates that about 20 per cent of male and female prisoners are probably psychopaths.

We will consider here three possible causes of psychopathy (two of which illustrate the idea that such people may not have a disorder at all):

1. psychopathy as psychopathology
2. psychopathy as a single, genetically based strategy
3. psychopathy as a product of genetically variable and environmentally contingent strategies.

15.2.1 Psychopathy as psychopathology

This is probably the traditional view and is based on the idea that psychopaths are impaired in some way. In other words, one or more mental functions (or modules) have been damaged by injury, have been distorted during development in an adverse environment, are the product of some genetic malfunction, or have been impaired by some other cause. This view is the one on which the evolutionary approach probably has the least to say.

15.2.2 Evidence for psychopathy as design

A variety of empirical evidence lends support to the idea that psychopathy is not an aberration (that is, pathological in the sense of category 1 above) but a functionally designed lifestyle with a genetic basis. There is evidence, for example, that criminality and psychopathy share common heritable factors; so that children of psychopaths who are adopted away from their parents show a tendency towards criminality; similarly, psychopathy is found with a significantly higher frequency than chance would suggest among children of criminals who have been adopted away from their natural parents (Moffitt, 1987; Cadoret et al., 1990).

As noted earlier, the defining characteristics of psychopathy appear to be a whole cluster of attributes such as irresponsibility, selfishness, callousness, promiscuity, aggression and antisocial behaviour. It is not easy to see how such a combination could result from pathological development faults. If, however, these attributes are part of a minority adaptive strategy 'designed' by natural selection, it is equally not immediately obvious how this could be tested. Martin Lalumière and colleagues, however, have suggested that measuring developmental instability might provide clues. The logic here is that if psychopathy is associated with measures of disrupted development, such as pregnancy difficulties, stress during delivery, prolonged labour and low birth weight, this would support the contention that psychopathy is a consequence of developmental damage to an individual phenotype and not a normal expression of their genotype. As Lalumière et al. (2001) note, such developmental problems are known to be associated with other mental disorders, such as autism, anxiety disorders, and mental retardation. If, on the other hand, psychopaths seem to have experienced normal development, this adds weight to the view that they are behaving according to their natural design. The authors investigated a sample of 800 male offenders (of which 152 were psychopaths according to the 'psychopathy checklist-revised' (PC-R) system designed by Robert Hare) from a maximum-security psychiatric hospital. Figure 15.2 shows how psychopaths and non-psychopaths compare in terms of the obstetrical problems they experienced at birth and in infancy. The obstetrical problems scale was devised using a checklist including such items as fetal distress during labour, neurological impairment in infancy, toxaemia, prolonged labour and so on.

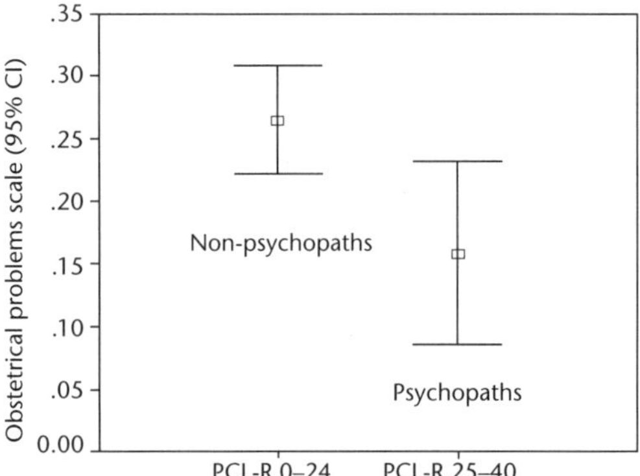

FIGURE 15.2 A comparison of the incidence of obstetrical problems in psychopathic and non-psychopathic criminal offenders

SOURCE: Lalumière, M. L., Harris, G. T. et al. (2001) 'Psychopathy and developmental instability.' *Evolution and Human Behaviour* **22**: 75–92.

As Figure 15.2 shows, the psychopaths suffered fewer obstetrical problems than the non-psychopaths and the difference was very significant ($P < 0.005$). In the same study, the authors also investigated the fluctuating asymmetry (FA) of three groups: non-offenders, non-psychopathic offenders and psychopathic offenders. FA is instructive since it is fairly well established that a high FA is a strong indicator of developmental instability and that a low FA is associated with high phenotypic quality (see for example Manning and Pickup, 1998). Here the finding was less straightforward: the FA of the psychopathic group was lower than the non-psychopathic offenders but higher than non-offenders.

15.2.3 Psychopathy as a genetically based and adaptively functional single strategy

If psychopathology is a designed feature of a minority of humans, we need to establish what advantages it brings. One model draws upon the idea of mixed evolutionary stable strategies that might result in a population of people both competing and cooperating. In the context of psychopathy, this model suggests that psychopaths are playing a successful single evolutionary strategy, possible only as a minority behavioural pattern in a much larger group of prosocial individuals. A number of people have worked on this notion (see Harpending and Sobus, 1987; Mealey, 1997). Lalumière and Seto (1998) argued that psychopaths were behaving like cheaters or defectors in prisoners' dilemma scenarios. They suggested that such a strategy would have been successful in the EEA if several conditions held:

- Most members of the group were cooperators
- It was possible to move from one group to another
- Detecting defectors involved costs.

These three conditions are intuitively obvious. The psychopathic strategy only works by exploiting cooperators and is disastrous if played against itself. Psychopaths must frequently move on in case their reputation alerts group members to their tactics. Finally, psychopaths rely upon the fact that it is time-consuming to detect them; if no costs were involved, then constant checks and precautions could be put in place against cheating. If this model is correct, then all cultures are predicted to have a fairly common baseline frequency of psychopaths who thrive as a minority exploiting the rest of us.

15.2.4 Psychopathy as an environmentally contingent strategy

This model was suggested by Linda Mealey. She called the condition 'secondary psychopathy' to distinguish it from 'primary psychopathy' (category 2 above). In this account, individuals differ genetically in the way they develop strategies in response to different environments. The idea is that certain environments promote secondary psychopathy in some individuals but not in others (Mealey, 1997). Primary sociopaths are postulated to carry genes that design them from birth to pursue a life strategy of cheating and manipulation and the condition is therefore a **frequency dependent** adaptation. Secondary psychopaths, however, follow a facultative cheating strategy. The trigger for secondary psychopaths could be risk factors encountered in their upbringing such as low economic status, inconsistent discipline, punishment from elders rather than rewards, and a high number of siblings. We should note, however, that most psychopaths (primary psychopaths if we accept Mealey's distinction) appear very at ease with themselves. Often they can be intelligent and charming, with no evidence that they faced adverse or harsh experiences in their childhood (Hare, 1993).

The model has received some criticism (see Crusio, 1995) but has the advantage of making specific predictions about how the incidence of psychopathy varied with cultural conditions. Primary sociopathy is an obligate strategy whose frequency should not change much across cultural boundaries. Secondary sociopathy, on the other hand, since it is triggered by and exploits specific conditions, should vary in frequency. It is predicted that virtually all upper-class sociopaths will be primary, since the privileged conditions such people experience in their development would not be expected to initiate secondary sociopathy. Secondary sociopathy is predicted to be less heritable than primary and so, as the incidence rises in a culture, any differences between male and female frequencies should reduce. This is because the effect of any sex-linked genes (and any other genes) contributing to secondary sociopathy will be less, since an increased number of secondary sociopaths is due to a greater influence of the environment as a whole.

15.3 Schizophrenia

Schizophrenia was first delineated at the start of the 20th century by Emil Kraepelin and Eugen Bleuler. It was Bleuler who called the disorder 'schizophrenia' meaning a fragmenting (schizo = split) of the mind (phren). Schizophrenia appears to be an ancient condition. A minimum date for its first occurrence must be at least 60,000 years ago, since schizophrenia has been observed in Australian Aboriginals and this group became isolated from the rest of humankind at about this time. Furthermore, schizophrenia occurs in all cultures at about the same frequency (about 1 per cent) and so was probably well established in human evolution before genetically isolated enclaves were formed (see Polimeni and Reiss, 2003).

Schizophrenia is a common and very serious disorder, causing great suffering to those afflicted. Most people who develop the disorder are unable to work, do not marry or have children, and about 10 per cent eventually commit suicide. Unlike simpler disorders such as depression, the symptoms of schizophrenia are manifest in several mental processes: speech, thought, behaviour and emotion. According to DSM-IV, the criteria for schizophrenia include such things as delusions, hallucinations, disorganised speech, lack of emotion, lack of motivation, speaking very little or uninformatively, and social and/or occupational dysfunctions. Most schizophrenic hallucinations consist of voices, leading to the suggestion that schizophrenics are confusing their own inner speech and thoughts for external voices (see Crow below).

Despite these commonalities, it should be noted, however, that schizophrenia is also a heterogeneous condition with varying age at onset, responsiveness to drugs and clinical course, leading some to question the very validity of the condition as a diagnostic category (Byne et al., 1999).

15.3.1 Biology and genetics of schizophrenia

As a genetic disorder (and the evidence is plentiful), schizophrenia is complex and does not follow a simple Mendelian pattern of inheritance. Gottesman (1991) summarised the data from about 40 twin studies. He found that the concordance rate for a twin is 48 per cent if he or she had an identical twin and 17 per cent if the individual had a fraternal (dizygotic twin). Having parents or siblings who were schizophrenic also increased the concordance levels above background (see Table 15.5).

We should note that no studies have shown concordance levels of 100 per cent. This indicates that although there may be a genetic component to schizophrenia, environmental influences are also at work. Both the onset of schizophrenia and the course that it takes may be strongly influenced by the social dynamic of family life.

Table 15.5 Concordance rates for the development of schizophrenia

Relationship	Concordance rates (that is, probability of individual developing schizophrenia if relative (shown left) is afflicted)
Monozygotic twins	0.48
Dizygotic twins	0.17
Sibling	0.09
Both parents	0.45
One parent	0.15
Random individual (background rate)	0.01

SOURCE: Data from Gottesman, I. I. (1991) *Schizophrenia Genesis: The Origins of Madness*. New York, W. H. Freeman.

Another useful concept that assists the study of the heritability of psychiatric disorders is that of relative risk (lambda_R), which is defined as:

$$\text{Lambda}_R = \frac{\text{Risk or incidence of disorder among relatives of those affected}}{\text{Risk or incidence in general population}}$$

When the relatives in question are siblings (brothers or sisters), then the term becomes lambda_S. So lambda_S becomes:

$$\text{Lambda}_S = \frac{\text{Risk to an individual if his/her brother or sister is affected}}{\text{Risk or incidence in general population}}$$

Table 15.6 shows some lambda_S values for a variety of disorders.

Diabetes, although obviously not a mental disorder, is included to illustrate the fact that such studies can help to establish if the condition has a heritable component. In this case, we note that type I diabetes is more strongly inherited than type II.

Table 15.6 Approximate risk ratios for various disorders

Disorder	Approximate risk ratio (lambda$_S$)
Autism	>75
Type I diabetes	20
Schizophrenia	10
Bipolar disorder	10
Panic disorder	5–10
Type II diabetes	3.5
Phobic disorders	3
Major depressive disorder	2–3

SOURCE: Data from Smoller, J. W. and Tsuang, M. T. (1998) 'Panic and phobic anxiety: defining phenotypes for genetic studies.' *American Journal of Psychiatry* **155**(9): 1152–62.

We should treat these statistics with caution. The fact that you are ten times more likely than average to develop schizophrenia if you have a brother or sister with the condition does not unequivocally point to a genetic foundation: it could be that you are more likely because you have been exposed to similar conditions to your brother or sister. Another alternative is that you carry genes that dispose you to react in certain ways to environmental stimuli. Table 15.5 above shows that you have a 48 per cent chance of contracting the disease if you have an identical twin with the condition. This leaves a 52 per cent chance of not developing schizophrenia, yet your genes in this monozygotic condition are identical to those of your brother or sister.

The undoubted importance of genetic factors, however, is further supported by adoption studies. One approach is to look at the children of schizophrenic parents who were given up for adoption. Tienari (1991) examined the children of 155 schizophrenic mothers who were given up for adoption compared with 155 adopted children of parents who were not schizophrenic. He found that the rate of schizophrenia was 10.3 per cent for those with schizophrenic mothers compared to 1.1 per cent of those with non-schizophrenic mothers.

Because of these genetic influences, the heritability of schizophrenia is quite high: about 80 per cent, which is slightly higher than that of height. The concordance values in Table 15.5 above imply that many genes must be at work, otherwise the figure for fraternal twins would not be so low compared to identical twins. If it were a single gene, for example, since the chances of your fraternal twin also possessing the same gene would be 50 per cent, then we would expect the concordance to be roughly 50 per cent of that of identical twins (see Tsuang et al., 2001 for a review).

15.3.2 Evolutionary explanations of schizophrenia

The strong evidence that there are genes disposing some people to schizophrenia raises the usual question as to why these genes persist in the gene pool and have not been selected against. The seemingly ancient origin of the disease, its strongly negative impact on fitness and its fairly prevalent occurrence suggest that we are not simply looking at some recent mutation event. There have been efforts therefore to identify counterbalancing advantages to the condition.

Advantages to individuals

In evolutionary terms, advantages can operate on individuals, kin (through kin selection) and, more rarely, on groups. The relatively high prevalence and genetic basis to schizophrenia alongside substantial fecundity disadvantages for individual patients is called the 'central paradox of schizophrenia' – a concept first formulated in these terms by Julian Huxley and colleagues. Huxley et al. (1964) postulated that a genetic advantage, specifically better resistance to wound shock and stress, existed in those who carried the gene but did not experience its pathological expression. There has, however, been little evidence forthcoming for the advantages they proposed and it now seems unlikely that the advantages they described should be so unrelated to the dysfunction. But their concept of the 'central paradox of schizophrenia' remains. Allen and Sarich (1988) have postulated that the genes responsible for schizophrenia give advantages to individuals living in groups. They suggest that overt schizophrenia would be maladaptive but that some asymptomatic heterozygotes lacking full symptoms might still possess useful individualistic qualities that enable them to resist, as they say, the 'shared biases and misconceptions of the group'. In other words, such individuals might flourish by exploiting original and creative niches unavailable to conformists.

More recently, Brad Folley and Sohee Park (2005) found that schizophrenics do better on certain creativity tasks. At the anecdotal level, John Nash, a mathematician who was awarded the Nobel prize for economics and whose life was made the subject of the film and book *A Beautiful Mind*, developed schizophrenia in his thirties. Daniel Nettle and Helen Clegg (2006) also examined the possibility that some schizotypal traits such as increased creativity may be attractive to the opposite sex. They surveyed a sample of 425 poets, visual artists and other adults. They found that active artists had a higher number of sexual partners than those without artistic ambitions. But this was indirect evidence only, since the survey was not directed at schizophrenics as such.

It is also possible that schizophrenia results from mutations on a whole array of genes. Each variation may be valuable and advantageous in itself but when all the gene variants come together in one individual, the brain is hopelessly distorted. Randolph Nesse (2004) refers to this as the 'cliff edge effect'. Unlike the normal distribution of traits and their fitness effects, the effect of schizogenic traits increases fitness up to a critical point when it suddenly falls.

Advantages to kin

There is also evidence (admittedly much of it not yet rigorously tested) that schizophrenics have other family members who are highly creative. Bertrand Russell had a son and granddaughter who were schizophrenic, and Albert Einstein's son by his first wife developed schizophrenia. It may be, then, that schizophrenia-inducing genes reduce fertility drastically in the homozygous condition, where schizophrenia is fully present, but give advantages to carriers (that is, relatives of the afflicted) in the heterozygous condition. In support of this, Karlsson (1974) did find that relatives of schizophrenic patients achieved superior academic success. Even Galton, one of the founders of eugenics, noticed this effect in the 19th century, when he noted: 'I have been surprised at finding how often insanity has appeared among the near relatives of exceptionally able men' (quoted in Ridley, 2003, p. 122).

15.3.3 The group-splitting hypothesis

Stevens and Price (1996) have proposed a novel hypothesis for the persistence of schizophrenia. They suggest that schizoid personalities could have acted in the past to perform the valuable function of dividing human groups when they become too large.

Humans are, without doubt, social primates. Our inherited biological capacity for complex language, shared by no other animal, shows that group living has been a fundamental component of human evolution. Scientists differ over dating the origin of human language, but if we take 100,000 years ago as a conservative estimate, then it is hard to imagine what humans used language for if not to facilitate group living. We may say that language indicates group living, although group living does not always lead to language. Our very psychology seems geared up for group dynamics. When strangers assemble, they readily form groups, and our moral behaviour seems to distinguish between the in-group and the out-group – a disposition all too easily exploited for tribal conflict and warfare.

Group living carries benefits and drawbacks and all groups have their optimum size. As the size of a group increases, so more eyes and brains become available to watch out for predators and find food. However, beyond a certain size, characteristic of the species and the environment in which it lives, negative effects begin to outweigh the advantages. Food may be more easily spotted but then there are more mouths to feed, requiring a greater range of foraging and travel and hence increased exposure to predators. Social tensions within the group will also increase. It follows that as a group grows in size through reproduction, or the movement in of outsiders, there comes a point when the optimum group size is exceeded and fission will increase the fitness of each individual. Stevens and Price (1996) estimate that the critical size of early hunter-gatherer groups may have been around 40–60. Dunbar (1996b) puts this figure somewhere between 100 and 150. Whatever the group size when fission occurs, the argument of Stevens and Price is that groups could be persuaded to split by a charismatic leader, who would promise benefits to those who followed him (or her). They maintain that it is the symptoms of a schizoid personality, such as cognitive dissonance, mood changes, bizarre beliefs, hallucinations and delusions of grandeur, which would have induced others, already discontented by the conditions prevailing in a group beginning to exceed its optimum size, to follow and form a new community. They suggest that charismatic leaders such as Hitler and David Koresh (leader of the Branch Davidian sect near Waco, Texas) may have suffered from schizophrenia.

The problem with this hypothesis is that it is difficult to test. It could be possible to examine if cult leaders show signs of a schizoid personality but extremely difficult to show that such personality types formed a significant part of our evolutionary history. Furthermore, most people suffering from schizophrenia are not organised enough in their thought processes to form a new group. Stevens and Price (1996) are also aware that suggestions that changes in gene frequencies can favour groups are unpopular now in evolutionary biology. Group selection is a controversial topic and it looks like conditions only rarely observed are required for it to happen at all.

15.3.4 The brain lateralisation and language development hypothesis

A number of theories have been proposed suggesting that schizophrenia may be a disadvantageous byproduct of human brain evolution. One of the most intriguing theories of recent years has been that of the Oxford psychiatrist Tim Crow. Crow (2000) has developed a remarkable theory that links the speciation of *Homo sapiens* with brain lateralisation and psychoses such as schizophrenia. Crow's theory builds upon and explains the facts that brain lateralisation (see Chapter 8) and handedness are not found in other primates to a large degree, and neither is language. Furthermore, there is some evidence that cerebral asymmetry is a predictor of cognitive ability, that is, individuals who have higher brain symmetry and display less handedness tend to score lower in cognitive tests. Crow's overall argument is quite complex but in summary form is as follows:

- Sometime around 23 million years ago, a 3.5 mB block of genetic information translocated from the X to the Y chromosome. This sequence was later (date as yet not established) paracentrically inverted on the Y chromosome short arm in the speciation of *Homo sapiens*.
- Such genes present in a homologous form on X and Y chromosomes are responsible for asymmetric brain development. A candidate region is Xq21.3/Yp11.1.
- Brain lateralisation following these genetic events made language possible, leading to the current normal state of the left hemisphere specialising in semantics (meaning) and phonetics (sounds) and the right in identifying emotions from the tone of the speaker.
- According to Crow (2000), schizophrenia is the failure in the balance of activities shared between these hemispheres; more specifically, a failure to establish dominance of the phonological sequence in one hemisphere. This leads to a failure in 'indexicality', that is, an inability of the sufferer to distinguish his own thoughts from the speech output he generates and the speech input he receives from other people.
- In essence, schizophrenia is the price humans pay for language.

There is a lot of evidence consistent with this theory. Schizophrenics do have lower lateralisation in handedness (their preference for one hand over the other is less pronounced than normal), and mixed handedness is also associated with other disorders such as dyslexia and autism. Crow (2000) also points out that there is evidence that studies on the brains of schizophrenics reveal that there is diminished asymmetry.

The problem for this hypothesis then becomes to explain why such deviations from normal asymmetric development exist. There are two alternatives. One is essentially the trait variation argument: that schizophrenia is an extreme form of variation of the genes that code for brain asymmetrical development. The other is the pleiotropy argument: that schizoid genes confer, or once conferred, some other advantage. Possibly, for example, the symptoms were less damaging in the past when cultural life was less complex. Crow's view is that schizophrenia is an extreme form of a natural variation. Those who wish to retain schizophrenia as a separate and identifiable category will naturally resist this.

15.4 Autism

The term **autism** was first employed by Leo Kanner in 1943 to describe some unusual children he saw in his clinic in Baltimore. By an uncanny coincidence, the Austrian paediatrician Hans Asperger independently discovered this condition and gave it the same name. The word 'autism' comes from the Greek *autos* meaning self, and was chosen to reflect the fact that the children seemed cut off from the social world and resided in a world of their own (Kanner, 1943). It is now recognised as a condition that affects 4 or 5 in every 10,000 children and seems to have a genetic basis, with estimates of concordance rates varying from 0 to 24 per cent for DZ twins and 73 to 90 per cent for MZ twins, making this one of the most strongly heritable of all psychiatric conditions (see Steffenberg et al., 1989; Bailey et al., 1995). About 75 per cent of children with autism are also mentally retarded, although the other 25 per cent have IQs that fall within the normal range for their age. The term **Asperger's syndrome** is now used for highly functioning autistics, that is, people with autistic traits in the normal IQ range. Some autistic individuals (sometimes called 'autistic savants') show outstanding talents in non-social areas such as artistic ability, musical talent, facility with languages or feats of mental arithmetic. The film *Rain Man* was praised as a sensitive and accurate portrayal by Dustin Hoffman of an autistic young man (in real life, one Raymond Babbitt) with special talents.

According to DSM-IV, autism involves three categories of symptoms:

1. Qualitative impairment to formation of social relationships, including lack of eye contact and poor facial recognition.
2. Qualitative impairment in communication skills, such as limited use of speech to maintain a conversation, little grasp of metaphor.
3. Restricted, repetitive and stereotypical patterns of behaviour, such as obsessive adherence to routines, inability to cope with change, fixation on detail.

15.4.1 Autism and mind blindness

In recent years, much research into autism has concentrated on the apparent inability of autistics to appreciate the mental perspective of others. To understand the importance of this, it is worth examining for a moment the way normal people function in social gatherings. If you reflect upon occasions when you are interacting socially, you will quickly realise that you relate to people not usually on how they behave but what you think they are thinking about. There is now compelling evidence that autistics are unable to realise that people have ideas and beliefs different to their own, they suffer from mind blindness, or, as it is often phrased, they do not have a 'theory of mind'.

Theory of mind

The term 'theory of mind' was first introduced in an influential paper on chimpanzees by Premack and Woodruff (1978). They argued that chimps, although lacking linguistic ability, could still detect intentions and desires in other chimps. The concept was quickly taken up by child psychologists and it is fair to say that subsequently the idea has found more support in its application to humans than to non-human primates (see Heyes, 1998). This ability is also sometimes referred to as mind-reading, mentalising, folk psychology and the intentional stance.

The theory of mind mechanism (or, as some would prefer, 'module'), often abbreviated to ToMM, was first proposed by Alan Leslie in 1987 as a system existing in the brains of normal humans for inferring the mental states of others. The essence of the mechanism is to enable individuals to form beliefs about the beliefs of others. To illustrate this point, Leslie (1987) used the distinction between primary and secondary representations. Primary representations are beliefs about the world such as 'there is a crocodile in that stream'; secondary representations are beliefs such as 'Tony thinks there is a crocodile in that stream'. Notice that here the primary representation could be false, while the secondary one is correct. Essentially, the theory of mind involves second-order intentionality: thoughts of the form 'I know that you know'. Leslie has also used the term 'M-representations' to describe this ability.

Despite the origins of the concept in primate research, Baron-Cohen thinks that there is now little evidence to suggest the great apes have a fully fledged theory of mind, in the sense of being able to ascribe goal states (of the form 'X desires Y') and epistemic states (of the form 'X thinks that Y thinks'). He believes that there is evidence that humans possessed this ability at least as long as 40,000 years ago, as evidenced by art forms that appeared at this time (Baron-Cohen, 1999). It seems likely that the theory of mind originated alongside language and perhaps with the speciation event that gave rise to *Homo sapiens* some 200,000 years ago.

Onset of theory of mind

A standard way to investigate the functioning and ontogenesis of theory of mind in children is to use various false belief tasks, as pioneered by Wimmer and Perner (1983). An influential variant of this test was developed by Baron-Cohen and colleagues (1985) to compare the functioning of theory of mind in normal, Down syndrome and autistic children. The children were presented with the story of two dolls, Sally and Ann, one of whom (Ann) hides a marble while the other (Sally) is out of the room. Sally returns and the child is asked by the experimenter where Sally will look for the marble (see Figure 15.3 for an explanation of the procedure). To succeed in this task, the children have to realise that Sally can hold a false belief: a belief that is different to their own (since they know the marble is in Ann's box) and a belief that, in this case, is not true. Two control questions were also asked: 'Where is the marble really?' ('reality question'), and 'Where was the marble in the beginning?' ('memory question').

Table 15.7 Results from the Sally–Ann false belief test

	Autistic children (n = 20)	Down syndrome children (n = 14)	'Normal' children (n = 27)
Naming question	100	100	100
Reality question	100	100	100
Memory question	100	100	100
Belief question	20	86	85

The researchers found that most autistic children were incapable of passing the false belief test, even though they could remember the names of the dolls and where the marble really lay (Table 15.7).

A simpler false belief task also routinely employed is the 'Smartie test' (Smarties are small colourful sweets sold in distinctive cardboard tubes). The tube

FIGURE 15.3 The Sally–Ann false belief test
The experimenter introduces the names of the dolls to the child. The child is later asked to recall the doll's names (the 'naming question'). Sally hides a marble in her basket and then leaves the room. While she is away, Ann takes the marble out of Sally's basket and puts it in her own box. Sally returns and the child is asked the question (the 'belief question'): 'Where will Sally look for her marble?' The correct response is to point to or name Sally's basket. This correct answer shows that the child is capable of knowing that Sally will believe the marble to be where in fact it is not. The incorrect response is to point to or name Ann's box.

SOURCE: Redrawn after Baron-Cohen, S., Leslie, A. M. and Frith, U. (1985) 'Does the autistic child have a theory of mind?' *Cognition* **21**: 37–46.

is opened in front of the child and shown to contain pencils and not the expected Smarties. The tube is closed and the child is then asked what another observer would think is inside the tube. Children under about four years and autistic children usually answer 'pencils', consistent with the interpretation that they do not have a fully developed theory of mind and so are incapable of distinguishing between their own beliefs and what another is likely to know.

Baron-Cohen (1997) has suggested that several modular mechanisms are involved in the full articulation of mind-reading:

- *an intentionality detector:* a component that is tuned to moving objects and decides if their movement has been caused by an intentional agent
- *an eye-direction detector:* a component that detects the direction of gaze of another person and infers that if the other person's eyes are directed at an object, then that person must see that object
- *a shared attention mechanism* (SAM): a component that decides if the self and the other are looking at the same object or event
- *a theory of mind mechanism* (ToMM): a component that is activated by inputs from the SAM and enables the individual to form 'M-representations' of the form 'agent–attitude–proposition', for example Sally (agent) thinks (attitude) the marble is in the box (proposition).

Baron-Cohen believes that one group of autistic individuals lacks a functioning SAM and ToMM, and another group lacks a functioning ToMM.

The neurosciences are just beginning to locate these suggested components of a theory of mind. It looks like that part of the visual system called the superior temporal sulcus is responsible for detecting motion and head and eye direction; and an area bordering the anterior cingulated and medial frontal cortex is a likely site of the mechanism that represents the mental state of self and others (Griffin and Baron-Cohen, 2002).

More recently, Baron-Cohen (2006) has incorporated the importance of the theory of mind into a theory that proposes that autism is an extreme form of the hypersystemising brain. The idea here is that humans have two main ways of predicting the outcome of changing events. One is to assume that the change is 'agentive', meaning that some intentional consciousness is at work, with goals and desires not dissimilar to the observing subject. In this case, empathy can be brought to bear to mentally intuit what the other agent might be thinking and planning. This requires the exercise of the theory of mind. The other way is to assume the change is non-agentive but that some system (natural or mechanical bound by the laws of nature) is at work. In such cases, the cognitive apparatus brought to bear looks for patterns and law-like behaviour. Typically, understanding what has caused a child to become upset would require the exercise of empathy, while diagnosing a fault in an automobile would require the exercise of system-based thinking (no matter how strong the tendency to think the vehicle has a mind of its own). Baron-Cohen suggests that differences between male and female cognitive styles result from differing genetic influences on the systematising ability. Males are typically more systematising than females and females better at empathising (see Geary, 1998, for a review of evidence that supports this). Baron-Cohen (2006) postulates that autistic children result from a union of parents who both carry highly systematising genes. This explains, he maintains, why autistic people prefer rigid routines and are most comfortable with situations that are highly lawful and regular, unlike the complex interaction and events that characterise the social world.

The work of Baron-Cohen on autism can be interpreted in one or both of two models of mental disorders. It is possible that autistic people lack a theory of mind due to some damage to one of the several modules involved in its functioning. This could be due to impairment at birth, a genetic defect or an adverse environment that disrupted the normal development of the module. The ideas also fall under the trait variation argument. Genes for systematising exist in all people but the trait has a high degree of variance. In this view, autistic people have a chance combination of highly systematising genes and so lie at one end of a spectrum of abilities. The genes themselves and the ability they design (systematising) are highly valuable in giving humans the ability to predict and hence control events and so have been maintained in the gene pool. As Baron-Cohen (2006, p. 871) suggests: 'it is likely that the genes for increased systematizing have made remarkable contributions to human history'.

15.4.2 Autism and genomic imprinting

Christopher Badcock and Bernard Crepsi (2006) have advanced an intriguing theory of autism that extends Baron-Cohen's idea of the extreme male brain and provides a basis for its origin in the concept of genomic imprinting. To understand this idea, we need to consider the different, and sometimes competing, interests of genes for the same function that we inherit from our fathers and mothers (see also section 6.2.8). On autosomal chromosomes we all carry two copies of all alleles: if we consider genes that help to build the limbic system, for example, then we possess genes that direct the construction of this from both parents. But, as Haig (2000) pointed out, the fitness effects of the expression of these genes may be different for maternally and paternally derived versions. Because of the possibility that the mother will have children by more than one father, the siblings of any individual could be less related to each other than each of them is related to their mother. Consequently, a father's genes in any offspring will have less interest in the welfare of their siblings than their mother will, and as a result, they will be better served by genes that promote selfish behaviour than those that are maternally derived. In addition, a father's genes can be expected to benefit more from behaviour of his offspring that seeks more resources from the mother than is optimum from the mother's perspective, since paternally derived genes are less interested in future offspring of the mother than is the mother herself. Similarly, paternally derived genes will be less altruistic towards siblings than the mother would prefer. Whereas the genes from any single father may be present in one offspring but not others, the mother's genes for brain function (the case in point) will be present in all her offspring and so can be expected to benefit more from cooperative behaviour than those from the father. The crucial point, then, is that whereas maternally and paternally derived genes all code for the same areas of the brain (essentially the whole brain), paternal genes will benefit more than maternal ones from the actions of the brain such as the hypothalamus, the amygdala and other parts of the limbic system that are concerned with instinctive drives, self-centred appetites and emotions. In contrast, the fitness of maternally derived genes stands to gain more from behaviour that promotes reciprocal social interactions. Badcock and Crespi speculate that these latter behavioural traits are directed by the frontal lobes of the brain and suggest that such maternal genes will be favoured by (and so favour) an empathetic style of cognition and one that benefits from social learning from a mother keen to promote cooperation among her offspring. As these authors observe, from this perspective, the brain behaves like a 'social placenta' – an organ that links into wider social networks to extract resources for the individual to flourish.

The fact that genes from the maternal and paternal lines benefit differently from the operation of different parts of the brain suggests that we might expect that paternal alleles for the frontal lobes will be imprinted (that is, switched off to reduce their effect) compared to paternal alleles for the more impulsive and egocentric parts of the brain; conversely, maternally derived genes for these instinctive drive areas will be imprinted compared to the maternal genes for social cognition and empathy. In this view, autism could be caused by alterations (mutations) in **imprinted genes** or genes associated with imprinted genes (some genes regulate others) and so could result from an impairment of the maternal brain (with a normal or enhanced paternal brain), or enhancement of the paternal brain (with a normal or deficient maternal brain). There is also the possibility that autism could result from some developmental disruption to the evolved balance between the maternal and paternal brain.

If this view is correct, we should view autism as part of a spectrum of human cognitive diversity rather than a simple categorical disorder or disability. It also raises the intriguing possibility that schizophrenia is located at the other end of the spectrum and is related to a strongly maternally biased style of perception and cognition (Badcock, 2004).

15.4.3 Autism and mirror neurons

In the late 1990s, two groups of researchers, Vilayanur Ramachandran and Lindsay Oberman (2006) at the University of California, San Diego, and Andrew Whitten's group (Williams et al., 2001) at the University of St Andrew in Scotland, both suggested that the symptoms of autism (lack of empathy, language deficits and poor imitation skills) are just what would be expected if mirror neurons were dysfunctional. Ramachandran and Oberman came up with an ingenious experiment using an electroencephalograph (EEG) to test if mirror neurons are impaired in autistic individuals.

An EEG measures electrical activity within the brain. It is non-invasive and relies upon electrodes strapped to the scalp to measure electrical activity within the skull. Brain waves recorded by an EEG can be broken down into different categories. One interesting set, called mu waves, falls within the 81–3 Hz frequency range. In the 1950s, Gian Emilio Chatrian showed that these waves are suppressed when a person makes a voluntary muscle movement such as opening and closing the hand. Later studies showed that this suppression also occurred when a subject watches someone else perform the same act. This pattern of behaviour strongly suggests that mirror neurons are at work.

To investigate the operation of these mirror neurons in autistic children, Ramachandran and Oberman (2006) compared the mu wares of 10 high-functioning autistic children with 10 age- and gender-matched controls. The control group, as expected, displayed suppression of mu waves when they moved their hands or watched videos of a moving hand, but the autistic subjects only displayed suppression of mu waves when they moved their own hands.

This work clearly points towards an aetiology of autism that is related to dysfunctional mirror neurons. But some of the other symptoms of autism, such as repetitive motions, hypersensitivity, aversion to certain sounds and avoidance of eye contact, are less easily related to the behaviour of mirror neurons alone. Ramachandran and Oberman (2006) suggest that there may also be problems with connections between the visual cortex and the amygdalae, causing the amygdalae to trigger inappropriate bodily responses to visual information.

Reconciling these different views about autism – theory of mind deficits, cognitive spectrum ideas, imprinted genes and mirror neurons – may not be easy or possible. Baron-Cohen's view that autism results from malfunctioning theory of mind modules is a major advance but in essence it is merely clarifying the symptoms rather than offering a causal explanation. The work on mirror neurons look highly promising, since it offers a neural basis for the operation (and malfunction) of theory of mind circuitry or modules. If, on the other hand, autism is part of a cognitive spectrum and does have its roots in dysfunctional mirror neurons, then one might expect a variation in the sensitivity and behaviour of mirror neurons within the population. If this perspective is correct, possibly males, on average, typically have a less sensitive mirror neuron system than females.

Summary

- Depression is a common condition and can be described as unipolar and bipolar. There seems to be strong genetic component to bipolar depression. Depression may be an adaptive response when the sufferer is in difficulty: it may serve as a bargaining tool or as a means of signalling a desire for more resources.
- There is some evidence that psychopathy (sometimes called sociopathy) may not be a dysfunction at all. It may be the operation of normally functioning mechanisms that exist in a minority of people. As such, it could be a frequency dependent strategy to extract resources from a much larger population of 'normal' individuals by cheating, without the usual feelings of guilt and remorse.
- There seems to be a genetic component to schizophrenia, hence many theories seek to look for what advantages such a debilitating condition could bring. There is some evidence that relatives of schizophrenics (and some schizophren-

ics themselves) are gifted. One theory sees the prevalence of schizophrenia as the price humans pay for the evolution of language.

- Autism is a strongly heritable condition. A common view is that autistic individuals lack a properly functioning theory of mind. This may be due to some damage to the development of this part of the brain or as part of a natural spread of traits. In the latter case, autistic individuals have the extreme form of the 'male brain' and see the world in terms of patterns and routines rather than agencies acting.
- Recent work suggests a role for mirror neurons in the cause of autism. Autistic individuals, among other things, may have dysfunctional mirror neurons.

Key Words

- **Asperger's syndrome**
- **Attachment**
- **Autism**
- **Bipolar depression**
- **Frequency dependent**
- **Imprinted genes**
- **Pseudopathologies**
- **Psychopathy**
- **Schizophrenia**
- **Unipolar depression**

Further reading

Allen, J. S. and Sarich, V. M. (1988) 'Schizophrenia in evolutionary perspective.' *Perspectives in Biology and Medicine* **32**: 132–53.

Although now dated, this article gives a useful review of some of the early attempts to link schizophrenia to evolutionary theory.

Badcock, C. and Crepsi, B. (2006) 'Imbalanced genomic imprinting in brain development: an evolutionary basis for the aetiology of autism.' *Journal of Evolutionary Biology* **19**(4): 1007–32

A fascinating, carefully argued and detailed paper reviewing the evdience for genomic imprinting as a cause of autism.

Baron-Cohen, S. (1997) How to build a baby that can read minds: cognitive mechanisms in mindreading, in S. Baron-Cohen (ed.) *The Maladapted Mind.* Hove, Psychology Press.

Here Baron-Cohen advances his argument that autism results from a malfunctioning of several linked modules.

Baron-Cohen, S., Leslie, A. M. and Frith, U. (1985) 'Does the autistic child have a theory of mind?' *Cognition* **21**: 37–46.

A classic paper where the link between autism and a lack of theory of mind is proposed.

Bowlby, J. (1980) *Attachment and Loss*, vol 3. *Loss, Sadness and Depression.* New York, Basic Books.

An important book where Bowlby states his influential theory of attachment.

Crow, T. J. (2000) 'Schizophrenia as the price that *Homo sapiens* pays for language: a resolution of the central paradox in the origin of the species.' *Brain Research Reviews* **34**: 118–29.

A radical and controversial paper linking language to schizophrenia.

Wider Contexts

The Evolution of Culture: Genes and Memes

CHAPTER 16

The greatest enterprise of the mind has always been and always will be the attempted linkage of the sciences and the humanities.
(E. O. Wilson, 1998)

It is a truism to observe that the major human achievements of the past 10,000 years, which distinguish us so markedly from the rest of the animal kingdom, are the results of cultural and not biological evolution; our genes are roughly the same as our Palaeolithic ancestors, whereas our culture has changed by an incredible degree. Such is the disparity in these changes, and such is the complexity of cultural change, that it has proved difficult to establish the relationship between cultural evolution and genetic evolution. The subject is complicated and has stimulated a variety of models, some employing highly complex mathematics, designed to address the problem. In this brief chapter, we will sketch out some lines of approach.

16.1 Modelling culture

If evolutionary psychology is able to say anything about human culture, we need a working definition of what culture is. The problem is that the word culture has a range of connotations. To the historian of art and literature, it may signify something that is morally improving or uplifting. To an anthropologist, it may mean the fabric of beliefs and values common to a society. A biologist would identify it as something that is passed down by social learning. The definition of culture adopted here will be that **culture** is information (knowledge, ideas, beliefs and values) that is distinct from that stored in the human genome and which is socially transmitted. Note that this does not a priori rule out any evolutionary purchase on culture. The fine details of culture are obviously not represented in the genome, but the type of culture we generate and the functions that it serves could, in principle, be directed by our genes. It is of course an extravagant claim and one not to be swallowed lightly. Neither is this a definition that would be accepted by everyone. Darwinian anthropologists such as Irons take a more phenomenological view and see culture in terms of behaviour and not mental constructs. The logic of this position is that mental constructs such as customs and beliefs only have biological significance to the extent that they influence behaviour (Irons, 1979). The definition here, however, is more tolerant of a range of evolutionary models that have been developed to explain culture. In any case, behaviour can also be transmitted through social learning.

A number of key questions are raised when we examine the relationship of evolution to culture. Three crucial ones are:

1. How does culture interact with brain hardware?
2. If we accept (using the definition above) that cultures are transmitted and inherited by social learning, then what exactly is transmitted? What are the units of cultural inheritance and what processes are involved in cultural transmission and evolution?

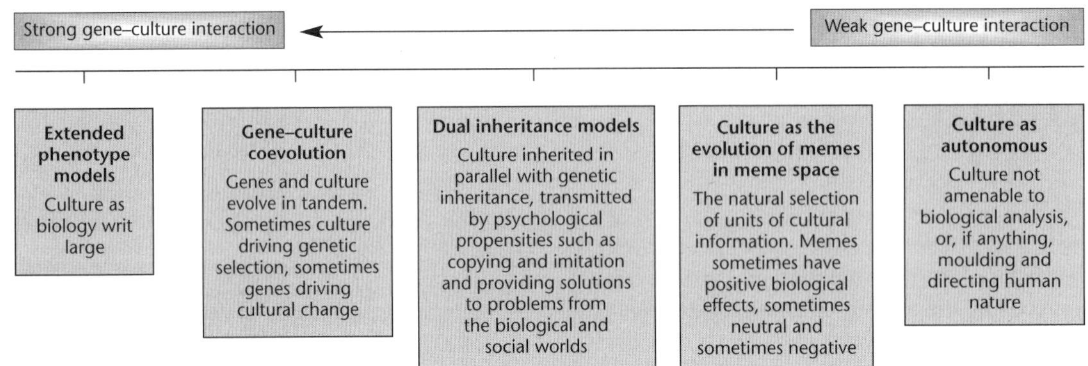

FIGURE 16.1 A spectrum of models of gene–culture interactions

3. If cultural evolution is linked in some way to biological evolution, what is the relationship between cultural evolution and biological evolution?

A number of models have been proposed to tackle these problems and, at the risk of oversimplifying some complex work, we will reduce these to five categories and discuss each in turn. The five categories can be represented as different positions on a spectrum, as shown in Figure 16.1, with the models that stress the autonomy of culture to the right and those proposing a strong linkage between cultural evolution and genetic evolution on the left. Apart from the idea that culture is entirely autonomous (far right of Figure 16.1), the other models have significant areas of overlap and the differences between them often amount to stressing one process or feature above another.

16.2 Culture as autonomous

This model starts from the observation that, although very little genetic change has occurred over the past 10,000 years, culture has changed enormously. Similarly, it takes seriously the paradox that cultural practices and behaviour vary enormously across the globe, despite the fact that humans are genetically very similar (modern genetics shows that the human gene pool is far less diverse than, say, chimpanzees). These considerations might suggest that there is no linkage between biological and cultural evolution, and that culture has its own laws of development, which are at present only dimly understood and probably best studied by the humanities and the social sciences. In this case, the evolutionary biologist retires from the field gracefully or perhaps makes the passing observation that cultural evolution seems to be Lamarckian, in that the achievements of one generation can be passed on to the next. The very definition of culture adopted earlier asserts that culture does not reside in the genome. Many cultural changes are fitness neutral, changes in fashions and styles, for example, are probably best approached sociologically using cultural theory rather than through biology. Some cultural practices are maladaptive: chastity among monks and nuns is hardly designed to promote fitness. So how can culture reflect universals of human nature?

Advocates of this view (characterised by Tooby and Cosmides (1992) and the standard social science model; see Chapters 1 and 4) often suggest that culture is also a primary determinant of human nature and behaviour: we are shaped by nurture and so our behaviour reflects the cultural norms around us. This approach was typified by the French sociologist Emile Durkheim (1858–1917), when he tried to dissociate social facts such as crime and suicide from individual psychology. According to his famous dictum, social facts can only be explained by other social facts.

But this model is probably too dismissive of the power of evolutionary thinking. Culture is after all the product of human minds that have been shaped by selection. Humans seem to be thriving as never before (in terms of population numbers at least) within complex cultures and so cultures do seem to promote some fitness gains. It has already been noted that some features of what is often regarded as the province of culture such as marriage contracts and inheritance patterns (see section 9.4) can be tackled using evolutionary concepts. One of the most telling criticisms of this approach is that it confuses the description of the phenomenon with the very thing to be explained. Noting that there is a cultural difference between two groups of genetically similar human populations does not logically imply that there is a force of culture at work.

Furthermore, many have pointed to the troubling circularity in Durkheim's view that social facts can only be explained by other social facts. On this basis, it is difficult to see how social facts ever arise. It is obvious that experience shapes our behaviour but it does so in non-random ways. Experience has to act upon a biological substrate (the brain or body) and the way the brain reacts is primarily a product of biological hardware rather than sociological software. As an analogy, consider some bodily responses to experience. When Caucasians are exposed to the sun, their skin turns brown; this is an inherent protective mechanism laid down by natural selection so that the body responds in a fitness-promoting way. Similarly, if we exercise our muscles, they grow larger, not smaller; this again makes perfect adaptive sense: if the body senses that muscles are in constant use, it stimulates their growth to facilitate whatever activity the phenotype has become involved in. The crucial point is that the effects of experience are facultative responses. This is seen most clearly in conditioned responses. The way we learn from experience through conditioning and association is inherently designed to help us to cope with our local (phylogenetically relevant) ecologies and make appropriate behavioural adjustments in the light of experience. Hence, as we noted in section 1.4.4 on the demise of behaviourism, organisms have inherent learning biases and tendencies that exist before the onset of experiences.

In further support of these objections, we may recall how humans can make quite rapid and subtle adjustments in their behaviour in relation to environmental and cultural cues. In section 11.3, for example, we examined evidence that showed that a woman's sexual behaviour is influenced in predictable ways by the sex ratio in the local culture. Similarly, in section 12.4, we saw how judgements of attractiveness depended on local norms and the significance of body weight and shape in cultures of differing degrees of affluence.

But Durkheim was partially right: social facts can explain social facts but only by incorporating biological facts. As an illustration, consider the correlation examined in section 9.4 between levels of promiscuity in a culture (potentially social facts) and the prevalence of matrilineal inheritance (other social facts). It was noted that high levels of promiscuity in some way seem to promote the practice of matrilineal inheritance (wealth of father left to sister's children). But this linkage would remain a mere association if we did not understand how the genetic interests of the father are served by this. Without this intermediary set of biological facts and theories, we cannot understand why promiscuity should promote matrilineality and not some other set of social facts such as crime or male infanticide.

16.3 Cultural evolution as the natural selection of memes

Of all the models, this is the most revolutionary. It suggests that the world of ideas (the characteristic products of culture) evolve in a Darwinian fashion but not necessarily through the medium of physical objects such as genes. To appreciate the logic of the model, consider the minimum set of conditions for some sort of Darwinian selection and so evolution to operate:

1. There exist in the world entities capable of self-replication.
2. The process of replication is not perfect, errors are made and the next copy may not perfectly resemble its template.
3. The number of copies of entities that can be made depends on the structure of the entities and hence their manner of interaction with the world outside.
4. As a result of the finite nature of resources, operating spaces and so on, these entities experience differential reproductive success.

From these four minimal conditions, we should be

able to witness Darwinian evolution. It is easy to appreciate that the entities above may not be strands of DNA. It is conceivable that, on other planets, the molecular basis of replication may be completely different. It is more of a shock to realise that the entities may not need to be physical at all; they may, in short, be ideas existing in and moving between brains.

Dawkins (1989) was not the first to note this insight but he articulated it forcefully in selectionist terms and coined the term **meme** to describe elements of thought or culture that replicate in human brains. As an analogy, the meme idea works surprisingly well. Memes move from brain to brain like parasites from host to host. We can catch them vertically from our parents, as in the case of rules of behaviour inculcated in childhood, or horizontally from each other, as in the case of peer pressure or conformity to fashion (Box 16.1). Some memes are truly parasitic, in the sense of damaging the survival chances of the host or the host's genes. Chastity, celibacy and self-sacrifice for noble ideals are all memes that damage their host's biological success. But this need not concern memes if memes survive: if self-sacrifice is held up (probably by a linked gene) as a laudable act, then others will fall under the sway of the meme, and the meme will survive. Many memes, however, are mutualistic, in that they assist their replication by ensuring the well-being of the host. Examples here include elementary rules of hygiene, methods of fashioning tools, avoiding disease and so on. The incest taboo may be a case of genes and memes directed to the same end if, as it seems, the Westermarck effect is based on a genetic developmental programme. The taboo becomes the meme that reinforces the genetically based mechanism of avoiding incest.

Box 16.1 The spread of memes

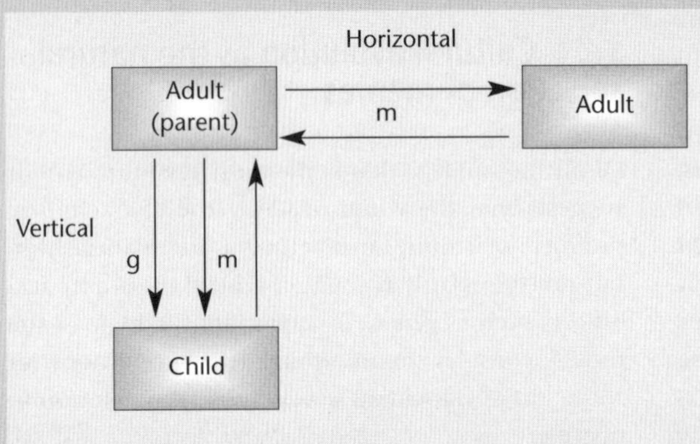

Memes (m) can be passed vertically in the case of parent to offspring or horizontally in the case of one organism to another.

Vertical: When passed vertically down the generations, memes can accompany genes (g). In early traditional cultures, there was probably a large degree of synergy between genes and memes and this may persist into modern times. The linked memes in Catholicism that restrict birth control and also insist that offspring are raised in the faith, for example, have the dual effect of increasing the spread of memes and the genes of those professing the memes. In these cases, sociobiological and memetic explanations yield the same results.

Horizontal: In horizontal transmission, genes do not accompany memes and memes may, from a biological point of view, be fitness reducing. The meme that suggests a career is more important than children reduces biological fitness but it may nevertheless spread through imitation. A fanatical devotion to chastity may be highly successful, in that biological energies are diverted into meme replication (chastity) rather than gene replication.

The important question to raise is whether the meme is an ingenious and amusing analogy or whether it provides a serious set of testable hypotheses that may really help us to understand the evolution of culture. In *The Selfish Gene*, Dawkins (1989) is in earnest when he explores the potential of memetics and uses the model to good (and characteristically provocative) effect in explaining the spread and maintenance of religious beliefs. To any secular rationalist, there forever looms the formidable problem of explaining the fact that, throughout history, the vast majority of people have subscribed to a set of religious ideas that:

- are inconsistent with other equally fervently held systems
- require believers to accept a suspension of the natural laws of the universe
- place a great strain on their biological inclinations and drives
- are held in spite of contradictory, little or, at best, ambiguous empirical support.

To Dawkins, such beliefs represent the invasion of minds by memes and people who are victims of these memes he calls 'memeoids'. Memes for some particularly unlikely tenet of belief may move around, linked with other memes that help their survival such as the meme for the virtue of faith. If we define faith as 'belief despite the lack of evidence', then the 'virtue of faith' meme suggests that faith is in itself a good thing to have. People of 'little faith' are castigated as deficient in some way. Most religious systems have a 'virtue of faith' meme as part of their meme complex.

The concept of the meme is similar to that of the 'culturegen' introduced by Lumsden and Wilson (1981). As these authors note, some culturegens could have positive biological fitness benefits such as those associated with food taboos and food preparation techniques. Viewing certain animals as unclean may have a sound basis in hygiene if such animals carry parasites that can infect humans. Alternatively, certain techniques of preparing food may help to protect against parasites. It is probably no accident, for example, that food in hot countries, where ambient temperatures are such to allow bacteria to multiply rapidly on foodstuffs, is often highly spiced. Many spices (for example garlic and chilli) are known to have antimicrobial properties. In support of this thinking, Sherman and Billing (1999) conducted an analysis of 4,578 meat-based recipes from 36 countries and found that as average annual temperatures increased, so did the proportions of indigenous recipes using antibacterial spices. As a follow up, Sherman and Hash (2001) predicted that the rise in spice use frequency with local temperature should be less marked for vegetable dishes, since the cells of dead animals are less protected against, and more conductive to, the growth of bacteria and fungi than the cells of dead plant matter. In a survey of 2,129 vegetable dishes in 36 countries, they found that although spice use did increase with ambient temperature, this was far less rapid than for meat dishes, and, furthermore, by every measure, vegetable-based recipes were less spicy than equivalent meat-based ones.

Susan Blackmore is one psychologist who has high hopes for the new science of memetics. In her book *The Meme Machine* (1999), she attempts to show how memes could account for the explosion in brain size that occurred about 2.5 million years ago. To follow her argument, we must adopt the concept of a meme as that which is subject to imitation. Once hominins started to imitate each other, a new replicator was born: the meme. Memes are passed on by imitation. An early hominin such as *Homo habilis* might observe another fashioning a tool in a particularly effective way, or enjoying and apparently thriving on a new foodstuff. Imitation would then bring distinct rewards at the level of both genes and memes. The genes of the imitator, and hence any genetic disposition to imitate, would be passed on if the activity conferred some survival advantage. By a parallel process, the memes would thrive, since more people are now practising whatever it was that enhanced survivability. At this stage of our evolution, genes and memes were probably closely linked in terms of the advantages they gave to the biological body.

Once imitation enhances genetic and memetic fitness, then further selection pressures are set up. Those individuals who are good imitators and, crucially, imitate other successful imitators will do better. In addition, those who choose mates who are good imitators will also leave more viable offspring. Blackmore suggests that this combination of natural and sexual selection drives up brain size by a process of positive feedback. As brains become selected for imitating, then language emerges more or less inevitably, since language is one way that memes can be propagated and obtained. As Blackmore (1999, p. 119) observes:

> I suggest that the human brain is an example of memes forcing genes to build ever better and better meme-spreading devices.

Memes are unlikely to exist alone. The meme for

altruism, for example, is linked with other memes that confer memetic advantage. A combination of memes Blackmore calls 'memeplexes'. Religious systems, as suggested by Dawkins and others, are complex memeplexes. They have strict rules of adherence, and employ enhancers and threats. An enhancer encourages the spread of the meme by an appeal to its virtues, such as 'blessed are the merciful: for they shall obtain mercy' (Bible, Matthew, 5: 7). Advertising slogans are full of memes connected with enhancers: 'a Mars a day helps you work, rest and play' is successful in that it promises benefits from the activity (eating a Mars chocolate bar) as well as using rhythm and rhyme that fit in well with our neural circuits. Some memes carry threats. One of the many biblical memes to renounce worldly pleasures, for example, is sanctioned by a divine threat: 'if any man love the world, the love of the Father is not in him' (1 John, 2: 15).

Blackmore makes the radical suggestion that the concept of the self is a memeplex – or 'selfplex' as she calls it. She suggests that the 'I' acts as a sort of protective coat around other memes. By entertaining an 'I', we feel more strongly bound to the memes that go along with it and hence we fight for their survival. Yet, ultimately, the self has no centre:

> That is, I suggest, why we all live our lives as a lie, and sometimes a desperately unhappy and confused lie. The memes have made us do it – because a 'self' aids their replication. (Blackmore, 1999, p. 234)

There are of course problems with the meme theory and there may be a circularity in some of its attempts to explain the propagation of memes. The concepts of 'self' and 'I' can only protect other memes if they carry some force in themselves. The power associated with the self needs further explication. The analogy with DNA is also not perfect: natural selection works because inheritance is discrete and is not blended. If any new change in the genome simply blended with other genes, novelty would be quickly lost and evolution would grind to a halt. But it is not at all clear that memes act like discrete units. They seem to blend and merge in ways quite unlike the information on DNA. Whereas genes have a universal language, there seems little prospect of a uniform meme language in brains. Gene mutation is random and undirected, whereas changes to memes are goal directed. Perhaps, however, we should not expect all replicators to behave like DNA.

Another serious problem is understanding the conditions by which some memes are copied well and others are ignored. In the case of biological features such as camouflage, it is easy to see why such genes have been selected. With memes, it is not at all obvious in purely memetic terms why, say, the meme for flared trousers should have increased in frequency in the early 1970s, suffered a setback in the 1980s and then revived (partially) in the late 1990s. Such phenomena may become clearer when we understand imitation biases, a subject that forms part of the model considered in the section below.

16.4 Dual inheritance theories

According to **dual inheritance theory**, there are two types of process that can generate design: natural (and sexual) selection acting upon genes, and a broader variety of selective processes acting upon culturally transmitted variants. One of the most influential theories of dual inheritance comes from Richerson and Boyd (2005). According to their ideas, both the psychological basis of culture (naturally selected and genetically based learning mechanisms and biases) and the pool of transmitted ideas are adaptations to solve the problems of a rapidly changing environment that early hominins faced. They argue that the way humans absorb, select and modify culture at the individual level leads to adaptive cultural evolution at the population level. In this view, culture becomes a changing reservoir of useful (but sometime maladaptive) knowledge and skills passed between humans in tandem with genetic evolution. This model is similar to that of memetics, but more precise in the way it envisages how culture is shaped and is inherited.

16.4.1 Imitation and bias

Culture, as defined earlier, is transmitted by imitation and social learning. Indeed, human infants (as any parent will testify) are excellent mimics – far exceeding the capacity of other infant primates. But, as

Rogers (1989) pointed out, there is something of a paradox at the heart of cultural transmission by imitation. Imitation is useful because it provides a rapid and cheap way of acquiring valuable cultural knowledge and skills – a process certainly faster than trying to continually figure out new solutions to life's little problems. But as the number of imitators rises (favoured by their fitness gains from copying others), so the fewer there are who are genuinely creative. There will come a point when imitators will start imitating each other and culture will stagnate and fitness levels fall as new problems are not solved. It may look as if we have reached an impasse. Mathematical models of this process show that the system reaches equilibrium when the cost of knowledge production to innovators equals the cost of being wrong to imitators (who are imitating out-of-date solutions). In such cases, the population ends up with a mixture of innovators and copiers in which (and this is the surprising point) both types have fitness levels the same as a population of innovators (Boyd and Richerson, 1995). The problem now is obvious: how can culture, which is supposed to deliver benefits by cheap and fast transmission of ecologically and socially useful knowledge, ever evolve if this process is no better than learning afresh?

Richerson and Boyd (2005) suggest that culture could escalate if innovators are given special privileges by imitators to compensate them for the costs they endure in devising original solutions. In addition, imitators would be best served by being selective: specifically, they should imitate the most successful in local populations, since the most successful must be doing something right (either imitating wisely themselves or discovering new tricks) to acquire the status they have. A moment's reflection on contemporary culture shows that this is exactly what we do: genuinely creative innovators become cultural icons that we reward handsomely and imitate. Hence the considerable sums of money paid by advertising executives to celebrities to endorse a product: they know that this is an effective way to change our behaviour. The simple rule 'imitate the successful' is one of several learning heuristics that can assist the assimilation, transmission and evolution of culture (Box 16.2).

Box 16.2 Fast and frugal heuristics for acquiring a culture

In the tradition of Gigerenzer's work on cognitive heuristics (see Chapter 7), Richerson and Boyd (2005) identify several heuristics that individuals may use to reduce the cost and improve the effectiveness of social learning. The use of heuristics means that individuals do not have to be exceptionally clever at creating cultural solutions: simple heuristics can, like the force of natural selection, produce clever adaptations when exercised by thousands of individuals over long periods of time. Two of these are the conformist bias and the prestige bias.

Conformist bias: This is a frequency dependent bias since it is the frequency of the behaviour and not the content or the person performing it that excites attention and emulation. In the conformist bias, the strategy is to imitate the most frequent behaviour. This bias may often yield the best solution since the most advantageous form of behaviour is often likely to be the commonest. There is a mass of evidence from the social psychology literature to show that people do indeed have a strong tendency to adopt the views and beliefs of the majority (see Myers, 1993).

Prestige bias: This is a model-based bias, since it is the characteristics of the person performing the behaviour (the model) rather than its content or frequency which is of interest. The logic behind this heuristic is that locally successful and prestigious people must be behaving in ways that have contributed to their success; therefore, it will pay to do what they do.

16.4.2 Changing environments and social learning

In the model of Boyd and Richerson (1995), cumulative cultural adaptations will be favoured when a number of conditions are found. One of these is a large spatial variability in environmental conditions, such as when we move around the earth's surface and climatic zones. In such situations, organisms with similar genes (that is, members of the same species) can exploit spatial variation in environmental conditions without relying upon genetic variation. This is precisely what we observe in the case of humans who have colonised most of the globe. There are vast differences, for example, between the environments inhabited by the Inuit people living in the Artic regions of Alaska and Greenland and the Maasai in the tropics of Kenya. Interestingly, there are biological differences between the two groups that partially facilitate this. The Maasai are darker-skinned, taller and slimmer than the Inuit. The shorter, more rounded bodies of the Inuit help to conserve body heat, while the slim bodies of the Maasai help them to lose heat. But these differences notwithstanding, the main reason the two populations are able to survive is due to the knowledge embodied in their cultures, which is fine-tuned to local ecologies and provides cultural solutions to living in these difficult conditions. This cultural knowledge would be passed on by parents and other group members. There is an old African saying that 'It takes a village to raise a child', meaning that child-rearing is something shared by the broader community (a process known as 'alloparenting') and not just the biological parents. It is this spatial variation of environmental conditions that humans encountered and were able to master (eventually) as they spread out from Africa about 200,000 years ago.

Another condition favouring cultural adaptation is that environments must change over time at about the right rate. If environments change very slowly, then organic evolution (a slow process in humans since we have long generation times) can keep track and ensure that we remain biologically adapted. If environments change too rapidly, then cultural knowledge quickly becomes out of date and redundant. In such cases, individual learning will be favoured over sociocultural learning.

Boyd and Richerson (1995) think that the last half of the Pleistocene provided patterns of climate variation that were just right for the evolution of culture. They extend their idea to explain why complex culture as a solution to survival only appeared once: the late Pleistocene was the first time in earth's history that a large-brained organism (the hominins) faced climate change at the crucial level of variability. In section 5.2.2, we noted how several cooling episodes over the past five million years may have facilitated the evolution of bipedal hominins. On top of these cooling periods were fluctuations caused by Milankovitch cycles of roughly 23,000, 41,000 and 100,000 years. These cycles by themselves are unlikely to lead to adaptations for cultural learning, since they are long enough for species to cope with by changing their range or by organic evolution. However, the recent availability of high-resolution data for climatic changes over the past 80,000 years from Greenland ice cores has revealed a pattern of rapid short-scale fluctuations. It now looks likely that between about 100,000 to 10,000 years ago, the earth's climate oscillated from glacial to nearly interglacial conditions every 1,000 years or so, with some fluctuations happening over only 100 years. Such rapid changes would have had major effects on local vegetable and animal populations as well as the type of clothing and shelter needed every few generations. The hypothesis of Boyd and Richerson (1995) is that to cope with such changes, cultural transmission became advantageous. Cultural evolution could give rise to complex adaptations much faster than genetic evolution and was therefore ideally suited to variable Pleistocene environments.

16.4.3 Maladaptive cultural variants

Once a system of cultural evolution (and specifically the learning mechanisms that drive it) is in place, there are costs to be had. Many problems that humans face, such as how many wives (or husbands) to take or how family life should be organised, have low 'trialability', that is, we do not have time to try out alternatives. Rather, we have to rely upon tradition and custom in the hope that they will provide tested and time-honoured solutions. But the downside of this is a tendency towards credulity: we may accept some beliefs and values that are not at all fitness enhancing. Indeed, there is no reason to expect genes and memes (to borrow a term from a related model) to be in total accord. Just as sexual and natural selection can pull in opposite directions, so too might cultural and genetic evolution. The interesting question is how far cultural or memetic fitness can evolve against the grain of biological fitness. The answer is probably a significant distance. Cultural entities have a distinct advantage over genes in that they can multiply and spread very rapidly both vertically and horizontally (that is, within a peer group). If their transmission rate is high enough, they could, in principle, considerably reduce the fitness of the biological bodies that are their hosts. The situation is analogous to the outbreak of an epidemic carried by a virulent microbe: populations can be reduced drastically until those that are left are the immune ones, or until so many are killed that the microbe can no longer be carried to new hosts.

Maladaptive culture may be generated by precisely those heuristics that serve us well over the long term. In the case of prestige bias, for example, people may emulate role models who are famous but are exemplary in the wrong ways, such as being too thin or prone to drug use. Another natural bias that is a candidate for the generation of cultural errors is the confirmation bias: a tendency whereby we tend to look for evidence that confirms our beliefs, rather than, as good Popperians should, look for instances whereby our beliefs would be refuted (see Chapter 7). Pascal Boyer (1994) argues that the widespread acceptance of supernatural beliefs is fostered by 'abductive reasoning'. Abductive reasoning (sometimes known by philosophers as the 'fallacy of affirming the consequent') means accepting a premise as correct if the implications of the premise are observed. Imagine you have a belief in the healing powers of a local deity. You pray and get better. Was your premise (that is, the existence and benevolence of your god) correct? Not necessarily: people get better anyway.

16.4.4 Culture and life history

Once climatic variability (over space or time) places a premium on culturally assisted behavioural flexibility, it becomes important for juvenile humans to have sufficient time to acquire cultural skills and knowledge to function effectively in their groups. This would select for an extended juvenile period and so a delay in reaching reproductive age; humans would become increasingly K selected (see Chapter 3).

Humans have a long juvenile period compared to other primates. The average age that girls experience their first menstrual period in modern cultures is between 12.5 and 13.5 years. This age is known to be dependent on environmental factors such as food intake and has been declining in Europe over the past 150 years. In 1860, for example, it stood at about 16.5 years and by 1960 had reduced to 13.5 years. Using these data and studies of traditional cultures, it seems likely that our ancestors in the Old Stone Age would have fully reached reproductive age at about 18–20 years (Kaplan et al., 2000). This figure seemed to have increased in the hominin lineage and then declined recently under the influence of cultural factors. Table 16.1 shows estimates by Bogin (1999) of the period of reproductive immaturity for some of our ancestor species based on fossil evidence on bone sizes and dental development.

Table 16.1 Estimates of the juvenile period of selected hominins

Species	Age at reproductive maturity
Australopithecines	12
Homo habilis	12–13
Homo erectus	14–15
Early *Homo sapiens*	Late teens, early twenties

This prolonged immaturity of humans in terms of physical growth is shared, to a degree, by some other primates. In most farm animals, for example, puberty is reached when they attain about 30 per cent of their adult weight, whereas the figure is about 60 per cent for humans and chimps. But despite this similarity, in terms of developmental timing, the pre-reproductive period in humans is much longer than in other primates. Significantly, its increase in the *Homo* lineage took place alongside a threefold increase in brain volume since *A. afarensis*. It is plausible, then, that this long period of development was needed to enable young humans to acquire the social, physical and cognitive skills to function as an adult in demanding ecological and social environments. It is also noteworthy that this long period of immaturity runs alongside a much lower infant mortality in humans compared to our nearest relatives. Child mortality is about 50 per cent for children in modern hunter-gatherer cultures compared to between 60 and 90 per cent for other primates (Lancaster and Lancaster, 1983).

Humans seem to exist at the extreme end of a tendency observed in other primates, namely, that the size of the adult brain is related to the length of the juvenile period (Figure 16.2).

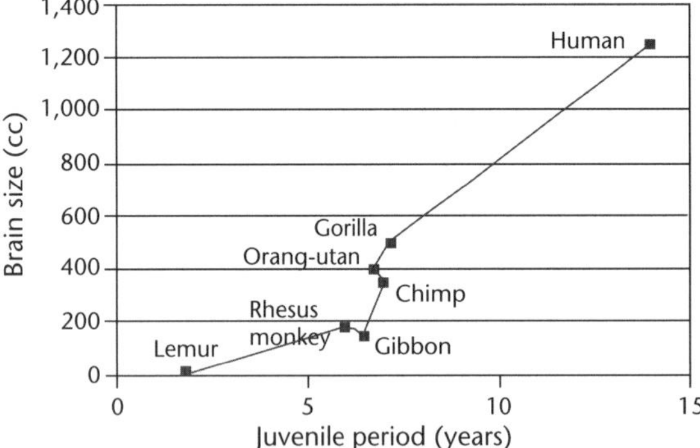

FIGURE 16.2 Brain size in relation to the length of juvenile period for selected primates

SOURCE: Redrawn from data in Bonner, J. T. (1988) *The Evolution of Culture in Animals*. Princeton, NJ, Princeton University Press, p. 207.

If the extended juvenile period in humans is related to the need to acquire cultural adaptations, then this is an interesting case of how cultural change (in this case, the wholesale value of culture) can drive genetic change (delayed sexual maturity). If this turns out to be correct, this provides an example of gene–culture coevolution, a model considered in the next section.

16.5 Gene–culture coevolution

In this view, culture has some autonomy from genes but is constrained by an 'elastic' leash. As the metaphor suggests, cultural change can apply a force that pulls along genetic change, but cultural change cannot drift too far away from the interests of genes before it is pulled back. This approach has spawned some highly complex mathematical models. In early models by Lumsden and Wilson (1981), culture is depicted as the outcome of the behaviour of individuals but in situations where the behaviour of individuals is itself influenced by the existing culture and individual 'epigenetic' rules of development. We can imagine a growing individual as bombarded by bits of culture or 'culturgens' (an early word for memes). Genetically based rules of development will influence a child to accept some culturgens and reject others. Genes and culture have therefore shaped the final behaviour of the adult, and, moreover, he or she then contributes in the re-creation of culture. Cultural and genetic change can take place together because culture sets the environment that may alter gene frequencies for genes that describe the epigenetic rule of development. In this manner, culture and genes roll on together.

There are numerous variations in approach to this subject but two examples should serve to illustrate the general line of thinking – yam cultivation and lactose intolerance (Feldman and Laland, 1996). In West Africa, people often cut down trees to cultivate yams. In these areas, heavy rainfall then results in pools of stagnant water exposed to the air that are ideal breeding grounds for mosquitoes. This cultural practice seems to have set up a selective pressure favouring resistance to malaria. Consequently, in these areas, the frequency of the sickle-cell anaemia mutant (see section 14.4.7) that confers some resistance to malaria is higher than would otherwise be expected. In short, a cultural practice has changed gene frequencies.

Lactose intolerance is more complicated. The naturally occurring sugar found in milk is called lactose and can be digested so long as the body is able to produce the enzyme lactase. Most mammals stop producing lactase at the time of weaning when offspring stop taking their mother's milk. Thereafter the young remain intolerant of milk sugar. This makes good biological sense: no point going to the energetic expense of synthesising lactase beyond the time when it will be useful. The milk of cattle and other animals has probably been an important component of the diet of some human populations for about 6,000 years. The prevalence of dairy farming among some human groups may have led to a selection pressure for genes allowing the absorption of lactose beyond childhood weaning. It is unlikely that early hunter-gatherers were able to continue to synthesise into adulthood the enzyme lactase that allows the digestion of milk. Although in the West we tend to assume that milk is a natural part of the diet, the majority of adults in the world lack the enzyme needed to digest lactose. If this latter group drink milk, the lactose is fermented by bacteria in the gut rather than broken down by lactase, leading to flatulence and diarrhoea. The fact that this was not known until the 1960s shows how the study of nutrition was culturally biased.

There are populations, particularly in Africa and Asia, that do exhibit this typically mammalian pattern of adult lactose malabsorption (LM). There are others, however, especially people of northern European and Scandinavian descent, who exhibit lactose persistence, meaning that they can drink milk from cattle and other animals throughout their life.

To understand this distribution of genetic differences, we need to look at the role milk products played in selection pressures. One answer is that malnourishment in some environments would have helped the spread of genes that would assist in the absorption of lactose and so enable individuals to survive. This is plausible, but there are probably additional factors (Durham, 1991).

It is known that lactose, like vitamin D, also helps the absorption of calcium in the gut. Vitamin D is produced in the body by the effect of ultraviolet radiation on the skin. Vitamin D deficiency, and hence poor calcium absorption, is a serious risk for people living in high latitudes where the intensity of sunlight is low. For these people, lactose absorption had a double advantage in providing calories and essential calcium. This example neatly illustrates how a cultural change – the invention of agriculture and the domestication of animals – can bring about a genetic shift in human populations.

This culture-historical hypothesis has now received considerable support (for example Simoons, 2001). Figure 16.3 shows the distribution of lactose malabsorption (LM).

If we consider populations that are lactose tolerant (LA), then a question arises as to why dairying was practised in these areas and not others. Gabrielle Bloom and Paul Sherman (2005) investigated this by compiling a massive database on LM and LA frequencies, temperature, climate and the incidence of cattle diseases. They found that LM was negatively correlated with latitude and positively with temperature, that is, LM decreased with distance away from the equator and increased with temperature. There was also a positive correlation between LM and the number of cattle diseases found in the area. Their results suggest that LA increases with latitude, since conditions away from the equator favour the keeping of cattle.

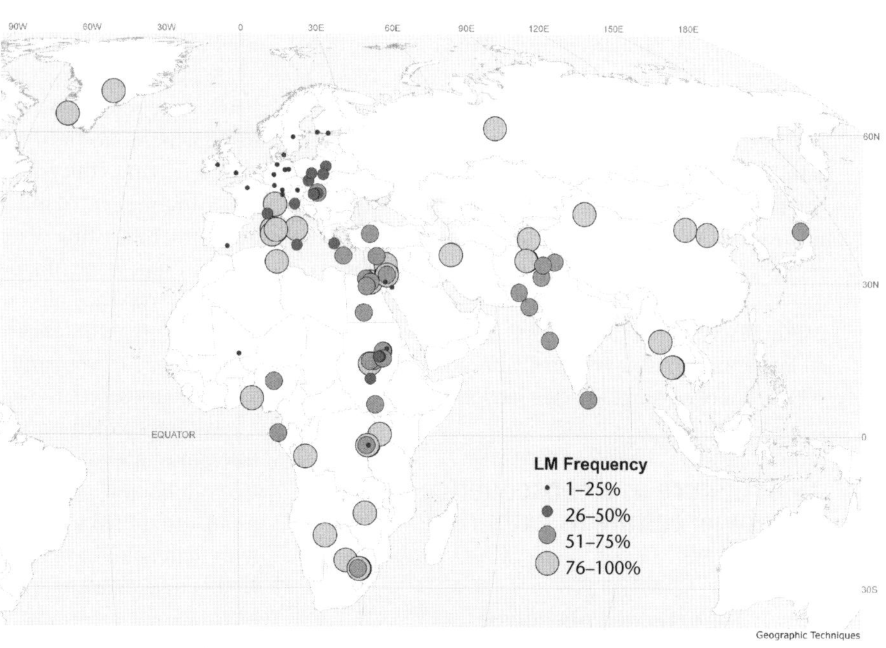

FIGURE 16.3 Global distribution of adult LM in a sample of 91 populations

Lactose malabsorption (LM) is associated with areas unfavourable to the rearing of cattle.

SOURCE: Bloom, G. and Sherman, P. W. (2005) 'Dairying barriers affect the distribution of lactose malabsorption.' *Evolution and Human Behaviour* **26**(4): 301–13, Figure 2, p. 306.

Both the prevalence of sickle-cell anaemia in Africa and lactose tolerance can also be seen as varieties of the **Baldwin effect** (Chapter 1). Essentially, in the case of lactose tolerance, an environment that favoured animal husbandry led to an advantage to those people who where initially flexible enough to digest at least some milk products. This enabled the survival of those populations who then experienced a steady selective pressure for those who were genetically disposed to tolerate lactose. Although Lumsden and Wilson (1981) called this a leash, a waltz might be a better analogy, since leash suggests that genes have the ultimate controlling hands. Or to borrow a metaphor of Richerson and Boyd (2005), if there is a leash, then the dog on the other end is pretty smart and has a mind of its own.

There are several more ancient examples where we can observe cultural change bringing about significant genetic change: hominin physique, jaw size and language. Modern *Homo sapiens* are generally less robust than some earlier species. Paleoanthropologists have argued that one reason for this might be the cultural evolution of projectiles as hunting weapons. Once an animal can be killed from a distance by throwing a spear or shaped stone, there is less need for a strong (and energetically expensive) physique. Jaw size and tooth size could have fallen in the hominin lineage following the use of fire to cook meat. Cooking breaks down the fibres of meat and reduces the need for large jaws to tear and chew meat. The evolution of language (whatever the initial cause) placed a selective pressure on the vocal tract of humans, such that it eventually became modified to produce spoken language (see section 6.3).

16.6 Culture as a consequence of genotype: culture as extended phenotype

16.6.1 The extended phenotype

The notion of the extended phenotype was introduced by Richard Dawkins (1982) in a book of that title. The subtitle of the book, *The Long Reach of the Gene*, encapsulates the essential idea that the effect of genes can extend outside the vehicles that carry them. Dawkins realised that it was somewhat arbitrary to limit the effect of genes to protein synthesis and behaviour. Genes can bring about changes in the world outside the bodies that contain them. Some obvious examples are the mounds that termites build to control their environment, the dams built by beavers, and the bowers built by bower birds (see also section 9.5.1).

It would be foolish to suggest that the precise details of culture are described by the human genome, but it is possible with some plausibility to suggest that genes themselves give rise to development processes that predispose humans to develop certain cultural forms. This seems to be the position argued by Wilson in his pioneering book *Sociobiology: The New Synthesis* (Wilson, 1975), where he suggested that cultural forms also confer a genetic survival value. Religious systems, for example, often prescribe correct forms of behaviour in relation to food, cooperation and sex. Incest taboos may reflect an instinctive mechanism designed to avoid homozygosity for recessive alleles. Ethical norms may be a way of our genes persuading us to cooperate to serve their best interests (see Chapter 17).

Examples of how culture may reflect biological needs are discussed by Gary Cziko (1995) as part of a more general account of the power of selectionist thinking. One concerns the farming of rice on the small Indonesian island of Bali. The planting, irrigation and harvesting of rice are not organised according to a secular or functional timetable but rather a religious one. The Balinese follow a religious calendar of activity in accordance with a belief in Dewi Sri, the rice goddess. The outcome, however, is that this religious practice serves to maintain high yields of rice.

We have noted earlier the stance adopted by the discipline of evolutionary psychology to the effect that our bodies and minds should be adapted to solve problems of the Pleistocene Epoch, and not the problems of modern life except in so far as they are similar. In this framework, culture is to be seen as a response of universal features of human nature to transient environments. Tooby and Cosmides (1992) have articulated this model. They argue that culture can be divided into three components: metaculture, evoked culture and adopted (or epidemiological) culture:

- *Metaculture* represents those universal components of culture that are the product of natural selection, such as language, grief over the loss of loved ones, association with kin and so on.

- *Evoked cultures* are formed when different environments act upon features of human nature. The outcome will be different types of culture according to different local environmental conditions.
- *Adopted or epidemiological culture* is when one culture spreads to another in a manner analogous to the spread of a disease.

They argue that sociologists have paid too much attention to epidemiological culture compared to the other more biologically based forms.

A good example of evoked culture in this sense is the practice of food sharing. Anthropologists have found that the degree to which individuals share food in a culture is related to the variance in food supply. Among the Ache tribe of Paraguay, meat obtained from hunting is highly variable in its supply. The chances of a hunter returning with a kill are only about 60 per cent. Correspondingly, within the Ache, meat is shared equitably among the tribe: the kill is passed to a 'distributor' who allocates portions to families according to their size. In the same tribe, however, food obtained by gathering is of low variance – it can be obtained fairly reliably, given effort and persistence. Significantly, among these people, whereas meat is distributed communally, gathered food is shared only within the kin group. The evolutionary logic to this is straightforward. With a high variance food such as meat, which is difficult to store, there is little point in gorging oneself or one's family on the products of a day's hunt. Once immediate needs have been met, there is little extra benefit to be gained. There is a strong advantage, however, to sharing surpluses among the whole group, since hunting involves some degree of luck and the widespread sharing of meat ensures that the hunter and his family will be fed if he is unlucky at some future date. With gathered food, one reason for a lack of success would be laziness not luck; consequently, the benefits of reciprocal altruism are not so compelling. This analysis is supported by Elizabeth Cashden in her studies on the San people in the Kalahari desert. Cashden (1989) found that some San tribes are more egalitarian than others. The Gana San tend to hoard food and are reluctant to share it outside family circles, whereas the !Kung San show a much higher degree of food sharing. The reason may be that the variability of food supply is high for the !Kung San and low for the Gana San, triggering and evoking two different but evolved psychological mechanisms.

16.6.2 Culture as sexual display

In the film *The Dead Poets' Society*, Robin Williams plays an inspirational English teacher, Keating, who takes over a new class. One exchange with his pupils is interesting:

> KEATING: Now, language was developed for one endeavour, and that is? Mr Anderson? Come on! Are you a man or an amoeba?
> NEIL: Uh, to communicate.
> KEATING: No! To woo women.

This is not so far from the approach elaborated by the evolutionist Geoffrey Miller (2000), who argues that we should view the arts in terms of sexual selection theory. For Miller, the generation of art is part of courtship display, a demonstration of cognitive fitness. Just as the peacock displays its extravagant and gaudy tail to attract peahens, and male bower birds build complex bowers from twigs and colourful feathers to attract females, so humans (especially males) display their intellectual and manual skills through artful displays: telling complex stories, producing paintings, composing sonatas, making artefacts, drafting love sonnets and so on. In further support of his theory, Miller notes how creativity (measured in terms of, say, output of novels, songs, sales of music and so on) peaks at the period of peak fertility of young men. And, in a comment that is bound to infuriate many (especially women), he notes that:

> Males produce about an order of magnitude more art, music, literature ... than women, and they produce it mostly in young adulthood. This suggests that ... the production of art, music, and literature functions primarily as a courtship display. (Miller, 1998, p. 119)

The essential idea of Miller is that artistic displays make men more attractive to women. In support of this, there is plenty of anecdotal but suggestive evidence of the large number of sexual partners enjoyed by musicians, artists and writers. Creativity, it seems, is sexy. Miller provides some quantitative support of his hypothesis in an analysis of jazz musicians (Figure 16.4). The distribution of output in terms of number of records against age has a fairly narrow profile: most music is produced by men aged between 20 and 40, the very age when they are investing heavily in mating effort (Miller, 1999a). Jazz music is seen as a form of sexual display, since it peaks with young men and men greatly outnumber women in terms of its production – a sure case of men blowing their own trumpets.

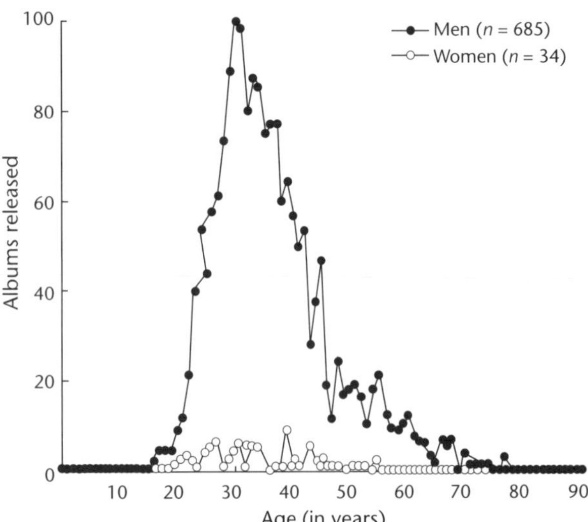

FIGURE 16.4 Albums of jazz music produced by males and females
Output of jazz albums as a function of age and sex of the principal musician/composer, reflecting a random sample of 1,892 albums between the 1940s and 1980s in the USA and Britain.

SOURCE: Miller, G. F. (1999a) Sexual selection for cultural displays, in R. Dunbar, C. Knight and C. Power (eds) *The Evolution of Culture*. Edinburgh, Edinburgh University Press.

The hypothesis is still rather tentative and controversial and certainly it is unable to explain the form that artistic creativity takes. In its favour, we might note that many works of art and creative displays are made by young men and that male pop stars and other cultural icons are extremely attractive to the opposite sex. Perhaps Picasso was aware of this function of art when he said that he 'painted with his penis'. He was speaking metaphorically (we hope) but it is significant that both financially and sexually Picasso was very successful. On the other hand, modern culture has many examples of young females enjoying the cultural output of other females (for example cult female authors such as J. K. Rowling, or pop stars). Moreover, it is easy to find exceptions to the idea of art as a young man's sexual display, such as the remarkable efflorescence of poetry by Thomas Hardy when in his seventies, or the late music of an aged Vaughan Williams. But Miller's ideas remain intriguing. He regards much of human culture as 'wasteful sexual signalling' that began with language, art, music, humour and clothing but was then closely followed by religion, philosophy and literature (Miller, 1999b).

This approach has also been applied to ancient cultural artefacts. The conventional view of stone tools is that they were objects made by hominins in the Old Stone Age for mundane tasks such as hunting, animal butchery, digging and cutting wood and other materials. But in an adventurous paper, Marek Kohn and Steven Mithen (1999) have assembled evidence to support their argument that stone hand axes (see Figure 16.5) manufactured as early as 1.4 million years ago are best seen as products of sexual display. They agree that many stone tools (such as arrowheads and scrapers) did have a utilitarian function but argue that hand axes served other purposes. In support of their case, they note that the production of hand axes presents a number of puzzling features, such as their huge numbers in the archaeological record and the high degree of symmetry and precision to which they are crafted, which archaeologists find difficult to explain. Table 16.2 presents these puzzles and shows how Kohn and Mithen use sexual selection theory to explain them.

The various models discussed above have many features in common. But there is the feeling that human culture is still a highly elusive and complex phenomenon.

Dennett (1995) speaks of Darwinism as a 'universal acid' that passes through everything, leaving behind 'sounder versions of our most important ideas'. In the next chapter, we examine how evolutionary thinking can possibly offer a sounder version of another cultural phenomenon that has long attracted a bewildering variety of theories: our ethical and moral sensibilities.

Table 16.2 The production of hand axes in the light of sexual selection theory

Hand axe puzzle	Explanation by sexual selection theory
Why are they found in such large numbers at individual sites? Large numbers are often found showing no signs of wear	The production of hand axes was by males and used as a signal of cognitive and manual skill. But to avoid displays of cheating (a male could simply display an axe made by another), females had to witness the production of the axes
Why was so much time invested in the production of such axes when stones, worked to a lesser degree, are suitable for tasks such as butchery and cutting?	Axes served as displays of intelligence, skill, health and aesthetic appreciation, and so needed to be more refined than for utilitarian tasks
Why do hand axes show a high degree of symmetry?	Humans have a perceptual bias in favour of symmetry. Symmetry seems to be one criterion used to assess mate value (see section 12.5). This perceptual bias was exploited by (male) axe manufacturers to please the onlooker
How can we explain the existence of giant hand axes such as found at Furze Platt and Shrub Hill in England, which are too large and unwieldy for practical use?	This is a case of social display – the giant hand axes were never intended to be used but rather to show the skill of the fabricator

SOURCE: Based on ideas in Kohn, M. and Mithen, S. (1999) 'Handaxes: products of sexual selection.' *Antiquity* **73**: 518–26.

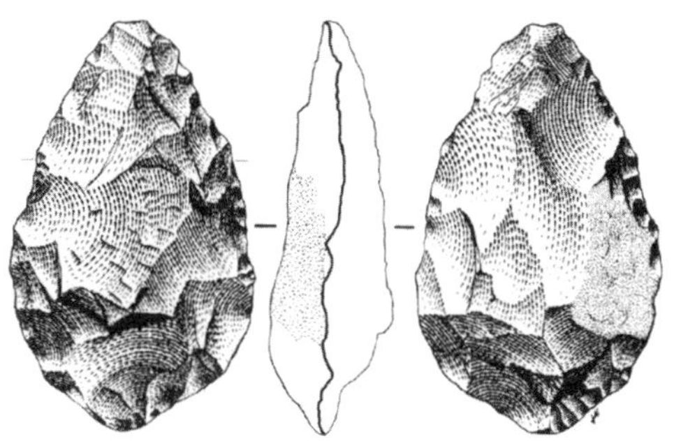

Figure 16.5 Typical Acheulean hand axe from the Douro valley, Zamora province in Spain
The axe is about 12 cm in length and shows signs of careful working to achieve a high degree of symmetry.

SOURCE: Image courtesy of Jose-Manuel Benito Alvarez.

Summary

- There are at least five models of how genes and culture could interact. Culture could be totally autonomous (biologists remain silent), culture could be an expression of the long reach of the human genotype, culture and genes could coevolve together in a dynamic interaction, both genes and culture are inherited and subject to their own selection forces (dual inheritance models) or culture could be the natural selection of the memes.
- The dividing lines between these models are not always clear and they do not necessarily provide exclusive explanations of cultural phenomena. Although all these models are in their infancy, several have already provided illuminating and plausible accounts of cultural change.

Key Words

- Baldwin effect
- Culture
- Dual inheritance theory
- Memes

Further reading

Barkow, J. H., Cosmides, L., and Tooby, J. (1992) *The Adapted Mind.* Oxford, Oxford University Press.

A manifesto for evolutionary psychology. See especially Chapter 1 on psychology and culture, and Chapter 18 on culture.

Blackmore, S. (1999) *The Meme Machine.* Oxford, Oxford University Press.

Bold claims are made for the power of the memes. A provocative work with an interesting analysis of the self.

Dunbar, R. I. M. and Barret, L. (2007) *The Oxford Handbook of Evolutionary Psychology*. Oxford, Oxford University Press.

A substantial work containing many useful chapters in this whole field. Section VII covers recent ideas on cultural evolution from several authorities.

Richerson, P. J. and Boyd, R. (2005) *Not by Genes Alone.* Chicago, University of Chicago Press.

Readable and interesting account of dual inheritance theory. The treatment of religion is especially interesting.

Ethics

Nature, Mr Allnut, is what we were put in this world to rise above.
(Katherine Hepburn, as Rosie Thayer, in *The African Queen*)

Ethics as we understand it is an illusion fobbed off on us by our genes to get us to co-operate.
(Wilson and Ruse, 1985)

The status of moral values has been the subject of almost endless dispute. One tradition suggests they may be thoughts in the mind of God; another that they are eternal truths to be reached by rational reflection (much like the truths of mathematics); another that they are social conventions; and another that they reflect enduring features of human nature. The relationship of evolutionary biology to this question has had a troublesome history. Early efforts to draw moral guidance from the process of evolution, such as the recommendations of the social Darwinists, met with a critical response from moral philosophers. The effect of this has been that whole generations of moral philosophers have given the biological sciences a wide berth and consequently often remain poorly informed about recent advances in evolutionary thought and the neurosciences. On the other hand, scientists are beginning to obtain a good grasp of the evolution of the moral sentiments and the centres of the brain associated with emotion and motivation, but have been fearful of committing the naturalistic fallacy (see below) and so have fought shy of extrapolating their findings to ethical questions. After all, no one wants to be seen to be committing an elementary logical blunder. But in recent times this has begun to change. In this chapter, we examine the prospects for placing ethics on a Darwinian base.

17.1 Does Darwinism signal the end of ethics?

17.1.1 The challenge of evolution

To many religious fundamentalists and even secular dualists (that is, those who think that humans are composed of body and mind and that these are different entities; see Figure 17.1 below), Darwinism is to be distrusted because it is thought to undermine the basis for moral belief. The argument runs that since Darwinism relocates humanity in the natural world as just another species with no privileged status, then there is nothing outside ourselves on which to ground objective moral truths. In fact, looking to a nature 'red in tooth and claw' (as Tennyson said), our quaint moral codes seem to be an aberration. This was the general moral alarm generated by evolutionary thinking in Darwin's day and it still persists: Darwin, by placing humans in nature and denying final causes and teleology – the idea that nature proceeds according to some plan or purpose – has stripped humanity of its moral compass and cast us adrift.

When such fears, historical and contemporary, are analysed, they can usually be seen to be based on a series of readings of the implications of Darwinism for the moral life. We can grade these in order of increasing philosophical sophistication roughly as follows:

1. Darwinism erodes belief in the existence of God and since God is needed as a basis for standards of

right and wrong, so the absence of God implies no common standards.

2. By supplying a completely naturalistic account of the mind, cognition, emotion and so on, Darwinism completes the materialist paradigm and so destroys the idea that humans have a free will. Without a free will, humans cannot choose morally correct or incorrect behaviour; furthermore, we have thereby lost the ground for praising the good and punishing the bad, since people who commit such behaviour have no choice.
3. By showing that traditional moral virtues such as altruism may be a product of kin selection or habitual game theory strategies, their status as morally praiseworthy actions is compromised. After Darwin, altruism becomes enlightened genetic self-interest and morality therefore collapses into a series of fitness-enhancing rules. In such a world, genuine moral behaviour is impossible. The worry here is that empiricism will simply chart our preferences but not be able to provide a basis for choosing among them. Moreover, as part of the scientific destruction of metaphysics, Darwinism makes the search for objective moral truths futile. What were once taken to be eternal verities are merely subjective, or at best species-typical, codes and feelings. As E. O. Wilson noted in the formative days of sociobiology:

   > In the first chapter of this book I argued that ethical philosophers intuit the deontological canons of morality by consulting the emotive centres of their own hypothalamic-limbic system. (Wilson, 1975, p. 563)

   But, it is argued, the consequence of this is that, without an external authority, anything goes and we slide into moral relativism.
4. Darwinism, as a science, is only able to provide facts about the world not values to adhere to. Because of the ought/is dichotomy, we cannot derive prescriptive statements from descriptive ones. However compelling a picture of the human condition Darwinism supplies, we cannot derive statements about how we *ought* to behave because 'oughts' belong to the realm of values, and values cannot be inferred from facts.

So, after Darwin, is humanity left without a moral compass? We will tackle each of the concerns above.

17.1.2 Existence of God and a basis for objective moral standards

Whether or not you believe in a god may be linked to your views about Darwin, but that is not our primary concern here. For many, Darwinism did indeed sweep away the last vestiges of a theocentric world view. Others, however, have managed to reconcile their religious feelings with the acceptance of organic evolution. Given that the focus here is on morality, it is more productive to examine the idea that a belief in a god is necessary for a belief in objective moral standards.

This question has a long history and was addressed in classical times by Socrates in the dialogue written by Plato called *The Euthyphro* (*c.*350 BC). In this work, Socrates asks his pupil Euthyphro: 'Is something right because God commands it or does God command it because it is right?' The latter view, that by definition what God commands is good, is sometimes called the 'divine command theory of ethics' or 'voluntarism'. It is this view that would seem to be threatened by Darwin, since if Darwinism challenges a belief in God and if goodness is that which God commands, then we have lost our source of goodness and our grounds for good and evil have disappeared.

The Euthyphro question has been the subject of philosophical enquiry ever since it was raised. If we accept the divine command theory (a position taken by Descartes, Calvin and Luther), it would follow that if we were to discover that God actually commanded what we now and mistakenly believe to be cruel and unjust, such as genocide on the basis of ethnicity, then we would be forced to adjust our moral sentiments and accept this as good. The consistent voluntarist would have to reply 'so be it'. Most people, however, would feel deeply concerned about this and think that some mistake must have been made. It will not do for the voluntarist, however, to simply say that God would never command such things because He is good. If God is good, then goodness is an attribute we can attach to him; an attribute that can be expressed as God does good things, that He recognises the good and puts it into practice. If this is the case, then the voluntarist must give up his position that whatever God wills is good simply because God wills it.

If we respond to the Euthyphro question, as Plato does, by accepting the first alternative, that God commands things that are right in themselves, then we have tacitly accepted that goodness is a property we can ascribe to God but also to other things. Thus God would forbid the torture of innocent children for pleasure because it is wrong, He recognises it is wrong and exhorts us not to do it. It is not wrong simply because He forbids it, otherwise we have the problem with the divine command theory outlined above. In this scenario, if goodness can be separated from God and exists independently of Him, then any doubts about the existence of God brought about by Darwinism, or any other branch of science, are not really relevant to the question of objective moral standards.

Few theologians and philosophers now accept the divine command view of ethics. If we follow their lead and reject it, then believers and non-believers alike have no grounds to fear that Darwinism will destroy any concept of goodness.

17.1.3 Freedom of the will

The traditional moralist might have two related concerns here. One is that Darwinism, as part of the materialist paradigm, forms part of a completely naturalistic explanation of human actions and such an explanation, de facto, allows no room for choice or free will. Since morality is intimately linked to the capacity to make choices (we do not, for example, impute moral intentions to earthworms), then a world where choice is unavailable has no moral dimension. The second related concern is that if humans are not responsible for their actions, then praise and chastisement – the carrot and stick of moral systems – are deprived of their authority except as positive and negative stimuli for learning.

Although these concerns were raised long before the advent of Darwinian psychology, it does seem that the success of a gene-centred view of natural selection and the demonstration of the adaptive significance of human behaviour have once again raised fears about our moral autonomy. Commentators such as Steven Rose and Stephen Jay Gould, both forceful critics of evolutionary psychology, have raised these concerns. Rose, for example, in a scathing review of Wilson's *On Human Nature*, wrote:

> Although he does not go as far as Richard Dawkins (*The Selfish Gene*) in proposing sex-linked genes for 'philandering', for Wilson human males have a genetic tendency towards polygamy, females towards constancy (don't blame your mates for sleeping around ladies, its not their fault they are genetically programmed). Genetic determinism constantly creeps in at the back door. (Rose, 1978, p. 45)

In a similar vein, Gould notes with concern that:

> If we are programmed to be what we are, then these traits are ineluctable. We may, at best, channel them, but we cannot change them either by will, education, or culture. (Gould, 1978, p. 238)

The objections of Rose and Gould are, in fact, easily dismissed. Janet Richards, a British philosopher, picks through such arguments and shows convincingly how they assume premises that are simply not valid (Richards, 2000). The first error is the idea that we are genetically programmed. Genes, in concert with environmental influences, build bodies and neural circuits that dispose people to behave in certain ways. We are not, however, constrained to behave only in those ways. Complex organisms such as humans are capable of decision-making, taking into account a whole range of sometimes competing emotions and factors. Evolutionary psychology tells us about the functional origin of emotions, drives and decision-making algorithms but not how compelling any of them are.

Linked with this whole approach is the worry that by explaining the basis of what we have labelled 'moral action' in terms of natural processes, we have somehow destroyed the moral component. The problem of altruism (taken here to mean occasions when individuals sacrifice their own interests to advance those of others) is a case in point. The first problem posed by altruism (see also Chapters 2 and 9) is the fact that it undeniably does exist: people do the most self-sacrificing things that are difficult to explain in terms of kin selection, mutualism or reciprocal altruism. Yet at the genetic level, it is probably true to say that altruistic genes cannot survive for very long, for this would require the existence of genes that increase the frequency in the gene pool of other genes while decreasing their own frequency. The appearance of such genes would be a short-lived phenomenon at best. The way through the conundrum is that there is no simple one-to-one correspondence between genes and behaviour. Genes can only serve their purposes by building cells, organs and bodies. In the case of humans, they built complex organisms with an emotional-cognition system that, on average, in the past did a good job in ensuring the survival of the genes responsible. It is true that many of the dispositions we have are towards kin or in the expectation that favours will be returned, but natural selection also delivered up genuine emotions such as sympathy and concern for others that had the effect of promoting the general welfare of the group. In the end, this proved a good breeding ground for these genes and was better than a social world limited to dog eat dog and the incessant competitive pursuit of self-interest. The outcome of mutualism and reciprocity (see Chapter 9) meant that lots of gains were made through cooperative interactions and the feelings that motivated them. The crucial point is that just because such emotions did serve the interests of the genes that lay behind them, this in no way invalidates the existence of the emotion and the action. Explaining the origin of a phenomenon still leaves the phenomenon intact (Keats' complaint that Newtonian optics had spoilt the beauty of the rainbow by explaining its origin in refraction was a foolish romantic posture; see Cartwright and Baker, 2005). Furthermore, the fact that such dispositions evolved to increase the frequency of the genes responsible does not mean that this is what the dispositions will always do.

The essential point here is that there is not a simple correspondence between the objectives of our multiplying genes and our own welfare as we now experience it. The realisation that pain was designed as a way of protecting genes from destruction does little (even for the most enthusiastic Darwinian) to lessen its unpleasantness. Similarly, altruism, compassion, sympathy, gratitude, trust and concern for others are real emotions directed away from the self and serve the common good in ways that we can label as morally approvable, even though they were brought into being by selfish replicators. To suppose that an explanation of something destroys its status is the fallacy of 'nothing but ism'. The Elgin Marbles (sculptures from the Parthenon Frieze now housed in the British Museum) may be made of calcium carbonate but they are still outstanding works of art. As Richards (2000, p. 178) notes:

> An explanation of what something is in other terms is not enough to show that it is not a real case of that something.

In general terms, we could say that while a reductionist explanation may be a necessary condition to undermine the status of the thing it explains, it is not a sufficient one. Newtonian optics can account for the formation and position of the rainbow but rainbows remain, apparently hanging there in the sky (although since Newton, we may be inclined to be more sceptical about their divine origins). Similarly, Darwinian reasoning helps us to understand the origins of altruism, but does not cause it to vanish.

Those who subscribe to this worry have more disciplines to fear than just evolutionary psychology, for if the concern is that revealing the causes of behaviour patterns places them beyond personal control, then this same concern also applies to environmentalism. Furthermore, it is not at all obvious that behaviours that unfold as a partial consequence of some genetic programme are harder to modify than those induced by environmental conditioning. Hence the cogency of the phrase coined by the Jesuit missionary Francis Xavier (1506–52): 'give me a child until he is seven and I will give you the man'. It follows that those who are worried about the genetic determinant of behaviour as subversive to the notion of conscious self-control and moral responsibility are really worried about determinism as a whole, be it genetic, environmental or some combination. Clearly, the issue resolves to the more general one of free will, determinism and responsibility.

It is usually thought that free will is incompatible with materialism: if the universe is made up of matter and energy obeying fixed laws, then it is both deterministic and determined and so free choice is an impossibility. This is a huge philosophical question and will only be explored briefly here, but deep enough I hope to allay the fears outlined above. It is instructive to compare this supposed denial of free will within scientific naturalism or materialism (of which Darwinism is an essential part) with what might be expected from an alternative paradigm such as that of **dualism**. The long tradition of dualism from Plato, through Christianity and Descartes still bears heavily upon thought. Most people find it natural to think of themselves as a physical shell (the body) containing something (soul, mind, psyche, spirit) that is non-physical. This view is sometimes called the 'ghost in the machine' – the idea that an immaterial agency haunts a physical shell. It is a view that is deeply misleading and at odds with the life sciences, but it is the philosophical implications that reward exploring. Figure 17.1 shows how the various interpretations of Darwin stand in relation to the idea of mind–body duality.

Let us suppose that the dualistic world described above (the ghost in the machine argument) is non-deterministic. If it were, then we are back to the earlier objection that free choice is impossible. Janet Richards (2000) explores this scenario with her telling example of a man facing the temptation of adultery. In a predetermined world, he is unworthy of praise for resisting or condemnation for succumbing since he had no choice – his decision was made for him by whatever circuitry he carried in his brain. How does the potential philanderer fare in a dualistic world? If we accept that not all things are determined, then things can just happen without a cause. Now, the decision of the man was either predetermined by the

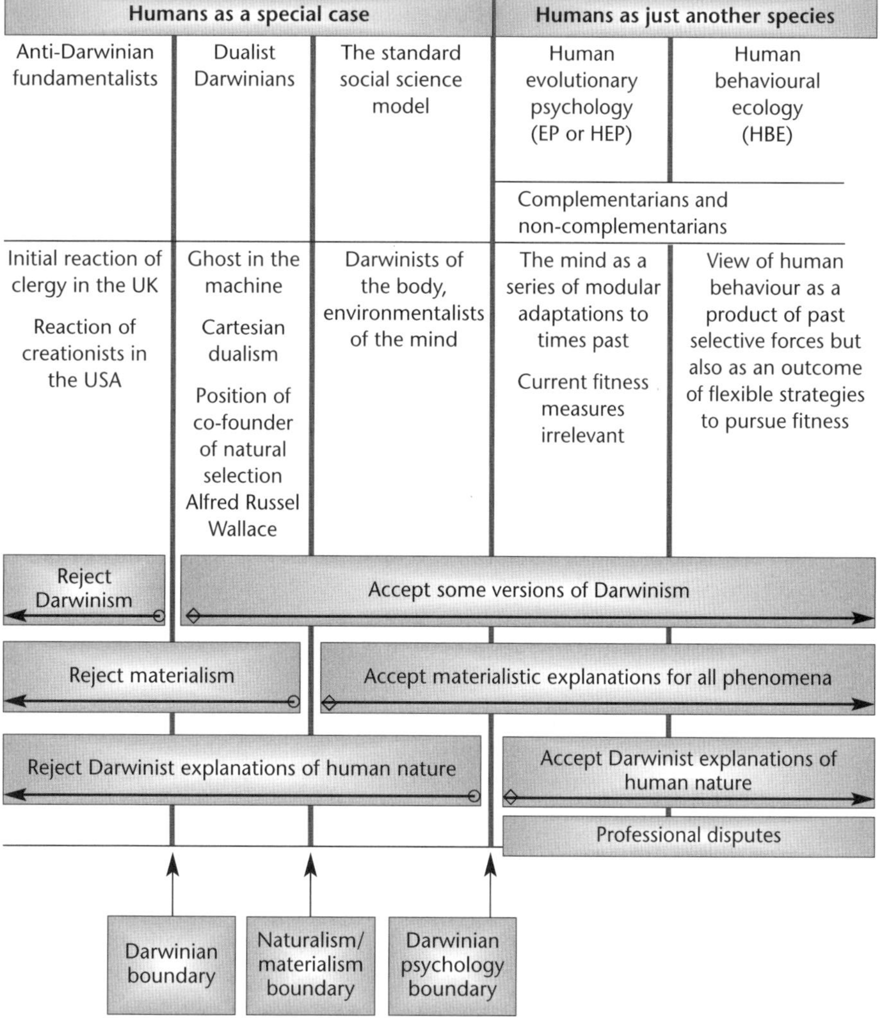

FIGURE 17.1 A spectrum of responses to Darwin expressed as a series of boundaries
Left of the Darwinian boundary are those who reject Darwin outright; to the right, stand those who accept some version of Darwinism. Left of the naturalism–materialism boundary stand those who reject the application of materialism to all phenomena; to the right, stand those who accept a materialist explanation for all phenomena. Left of the Darwinian psychology boundary stand those who think Darwinism has little to say about human nature; to the right are those who (in one of at least two disciplines) accept Darwinian explanations for the human mind and behaviour.

SOURCE: Developed from an idea in Richards, J. R. (2000) *Human Nature after Darwin*. London, Routledge, p. 54.

laws of nature or it suddenly took place without cause. In the former, the man is not responsible, for reasons already discussed. But in the latter case (indeterminacy), if his decision or behaviour appeared spontaneously, then he is equally incapable or merits blame or praise since 'he' did not cause it. As Richards (2000, p. 140) says:

> The trouble with trying to bring non-determinism into all this is that uncaused events are, by definition, things for which nobody and nothing can be responsible.

If free will has thereby been demonstrated to be incompatible with determinism and indeterminism, it seems that we have ruled it out of existence altogether or that we have an imperfect conception of what 'free will' actually means. A long philosophical digression on this would be out of place here but it is worth considering Richards' suggestion that we need to distinguish between two types of responsibility. One is 'ordinary' responsibility, which means being of sound mind and not under the influence of drugs or disturbed in some way; the other is 'ultimate' responsibility, which is something to do with a chain of causation. As our understanding of genetics, development and enculturation becomes more sophisticated, so we will achieve a better understanding of the forces that make us what we are. It does not follow, however, that praise and blame for moral and immoral acts respectively become unnecessary. Instead, they both serve as essential ingredients to inject into the decision-making of all of us. They become 'instrumentally necessary for the maintenance of social order' (Richards, 2000, p. 151).

17.1.4 Transcendentalism, empiricism and the slide to relativism

The origin and authority of our moral standards lies at the heart of perhaps the two most essential questions in metaethics: 'If ethical norms lie outside the natural world, where are they and how can we recognise them?'; and 'If they are just products of our biology, what authority should they have and why should we trust them?'

Wilson (1998, p. 238) summed up the vast literature on this topic as follows:

> Centuries of debate on the origin of ethics come down to this: Either ethical precepts such as justice and human rights are independent of human experience or else they are human inventions.

In other words, one is either a transcendentalist or an empiricist. By 1998, Wilson was arguing even more passionately that a transcendental approach to ethics must give way:

> The choice between **transcendentalism** and empiricism will be the coming century's version of the struggle for men's souls. Moral reasoning will either remain centred in idioms of theology and philosophy, where it is now, or will shift towards science-based material analysis. (Wilson, 1998, p. 240)

A Darwinist will, of course, take the empiricist's route, but it is sometimes feared that this will only end in moral **relativism** ('anything goes and all systems are equal'). The argument usually goes something like this: Darwinism destroys our belief in objective moral standards (since evolution has no direction and there is no external authority) and since there are no objective standards, we should respect all conventions equally and allow cultures to follow their norms. A hint of this line of reasoning (although not its conclusion) was given by Darwin himself when he wrote that if human society were like that of bees then:

> There can hardly be any doubt that our unmarried females would, like the worker bees, think it a sacred duty to kill their brothers, and mothers would strive to kill their fertile daughters; and no one would think of interfering. (Darwin [1871] 1981, p. 73)

A Darwinian might respond to the accusation of moral relativism in a number of ways. The first is to point out that the absence of objective standards outside our species-typical inclinations does not mean there are no standards. We might wish simply to emphasise the original source of our moral convictions: the human hypothalamic-limbic system. In other words, accept that morality is a facet of our nature and since there is a universal human nature (laid out for our inspection by Darwinism), so there are universal canons of morality that apply to our species (accepting that evolution may have made us prone to some cross-cultural and intersexual variation).

Another response might be to recognise that there are perfectly valid refutations of relativism and that consequently no philosophically minded Darwinian would wish to advocate this position. One argument is that it is self-contradictory: if there are no absolute standards, then the idea that we 'should' respect all standards is fallacious, since 'should' is an exhortation made impossible by the original premise of no objective standards. Any relativist (on the grounds that all systems are equal) criticising one system for attacking another is surreptitiously privileging its own position (that is, relativism). This is known as 'pragmatic self-refutation'. Relativism is also incoherent in practice. If we take the case of a culture adopting a relativist stance to values within it, then relativist law makers would have to allow laws respecting pro-abortionists and anti-abortionists. Yet they are contradictory – one cannot have both.

17.1.5 The fact–value dichotomy

This is perhaps the most general and telling complaint against a Darwinian approach to ethics. The essence of the objection is the view that whereas Darwinism probably has a great deal to say about how we derived those emotional and intellectual convictions that form the basis of moral thinking and sentiment (a valuable exercise in its own right), it can never tell us what is the correct course of action or whether our intuitively guided behaviour is 'really right'. We may be able to show that moral values are indeed the goals of our adaptations, but how do we tell if such values really are morally correct?

Shortly after the emergence of sociobiology, Kitcher (1985, p. 434) dismissed it as inadequate for this task:

> With its emphasis on the dictates of neural systems that have allegedly been fashioned to maximize the inclusive fitness of individuals who possess them, pop sociobiological ethics lacks any theory of the resolution of conflicts.

Recently, the philosopher Janet Richards (2000, p. 252) restated this inevitable separation of facts and values:

> A Darwinian understanding of evolved characteristics cannot on its own offer any guide to what one ought to be doing. It can give us the information that may be relevant to the achieving of our ends, whatever they are, but it cannot specify the ends ... In a Darwinian world values remain obdurately separate from the facts.

It is easy to see that there must be some sort of separation between facts and values, between the 'is' and the 'ought', but how these two should be divided is still the subject of much dispute. An example from one unfortunate episode in mixing evolutionary facts and social values will illustrate the perils of mixing these two domains.

In the 19th century, a movement began called **social Darwinism** (see Chapters 1 and 18). Put very simply, social Darwinists tried to read off from the operation of the natural world a moral and political message about how societies should be structured and what were the obligations of the state to the needy. The conclusions reached from this dubious process were usually that since nature had progressed and evolved without any intervention, so the obligations of the state towards the weak were minimal: schemes to help the poor and feeble would only make matters worse by enabling them to breed and weaken the race.

Nowadays, the term 'social Darwinist' is one of abuse. Denouncing someone as a social Darwinist is often thought to be a sufficiently crushing argument in itself. But why exactly is social Darwinism an untenable exercise? Spencer's phrase 'survival of the fittest' has become a catch phrase for those who advocate the virtues of free competition. There may indeed be virtues, but Darwinism, to the disappointment of any contemporary would-be social Darwinists, must remain silent on the issue. At one level, it is not at all clear that nature runs strictly along 'red in tooth and claw' lines anyway: animal groups show plenty of signs of cooperation, and even vampire bats share a meal with their needy brethren. If we look at some taxa, such as the ants, competition between individuals seems entirely suspended in favour of caring and sharing for the common good. If we wish to model human society on the natural world, it is difficult to know which group of organisms we should consider: the message from, for example, ants, bats and dandelions will be entirely different.

We could acknowledge the fact that nature (as far as we can tell) is not regulated by some external conscious agency and that indeed the purposeless process of natural and sexual selection has led to such complex organisms as ourselves. But does it follow that society should also be left to the unregulated outcome of the effects of individuals all pursuing their selfish ends? The answer is no. To believe otherwise is to make a huge and invalid leap of logic. The way in which humans want their social world to operate is a matter of values; biology is no more reliable a guide to what values we should hold than, for example, chemistry or astronomy. The suggestion that one can infer values from descriptive facts is now known as the 'naturalistic fallacy', and it is often claimed that it was David Hume in his *Treatise of Human Nature* ([1739] 1964) who exposed this fallacy – although, as we shall see, Hume had his sights on a different goal.

Returning to the logic of social Darwinism, we can show that the reasoning is fallacious, but we need to do better than simply to evoke Hume. What the social Darwinist does is to confuse the consequences of natural processes with their value. If fierce unbridled competition got us to our present state, there is no obvious reason why it should still serve our ends. Suppose, for example, that one could demonstrate that periodic famines on a global scale or massive doses of gamma rays from solar flares were instrumental in the course of the evolution that led us to *Homo sapiens*. I doubt if even the more ardent social Darwinist would suggest that famine and ionising radiation are to be welcomed as a means by which we can improve the human stock. Men, on average, are taller than women; this does not imply that they ought to be.

The social Darwinist is also guilty of smuggling teleology in through the back door. Social Darwinism should be really called social Spencerism, in that it was Herbert Spencer rather than Darwin who kept the ideas of progression in his system of thought. The abyss into which Darwin stared was always too much for Spencer, who clung to a belief in steady evolution towards perfection. The essential point here is that it is the very purposelessness of the natural world that makes it a doubly unreliable guide. Natural selection does not make organisms better in any absolute way: it merely rewards reproductive success. There is no progress measured on an absolute scale but merely change. The whole thing is not going anywhere.

The invocation of Hume's law – the impossibility of deriving the 'ought' from the 'is' – is often thought to be sufficient to deal the death blow to ethical reasoning that seeks a basis in the natural order. But we should be careful before abandoning a possibly fruitful line of enquiry. At some stage, the determined Darwinian will want to give a naturalistic account of value and morality and this, in the absence of any transcendental notions of goodness, will presumably have to be based on a factual account of the natural world. The prospects for this enterprise are examined in the next section.

17.2 Prospects for a naturalistic ethics

17.2.1 The naturalistic fallacy

Despite its legendary reputation for thwarting any attempt to link values with natural facts, the naturalistic fallacy is not as straightforward as might be supposed. In his *Treatise of Human Nature*, Hume wrote some passages that have been taken as proof that one cannot derive an 'ought' from an 'is' and that, accordingly, the world of facts and values must for ever remain separate. There is now a growing consensus, however, that this is far from what Hume intended (see Walter, 2006). Hume's main point (see Box 17.1 below) was that moral sentiments are already facts (the 'is') of human nature – they exist as passions in the mind of humans. Hume's argument was that he could not see how reason alone could enable the connection between the 'is' and 'ought' to be made. Hume himself was frustrated that people misunderstood his argument. In 1752, he published a work designed to clarify his position called *An Enquiry Concerning the Principles of Morals*, in which he stated:

> The hypothesis we embrace is plain. It maintains that morality is determined by sentiment. It defines virtue to be whatever mental action or quality give to a spec-

> tator the pleasing sentiment of approbation; and vice its contrary. (quoted in Walter, 2006, p. 36)

So, far from wishing to place ethics beyond the reach of naturalism, Hume stands with the empiricists in opposition to the idea that morality must somehow be the process of grasping transcendental truths through the exercise of pure reason or other means. For Hume, moral truths are inherent facts about human nature.

The term **naturalistic fallacy** is not to be found in Hume's works but was coined by the British philosopher G. E. Moore in his *Principia Ethica* of 1903. Although Moore's reasoning and the fallacy itself are often linked with Hume, Moore does not refer to Hume at all. Moore was keen to insist that moral value or 'goodness' was not a quality that can be read off from external objects. In his view, goodness was like 'yellowness', a subjective experience that was not reducible to anything else. To say, for example, that pleasure was good (as the utilitarians were forced to do) only made sense if goodness was a separate quality. Moore wanted to show that goodness was a non-natural property – a 'simple, indefinable, unanalysable object of thought' (quoted in Curry, 2005, p. 161). Moore is generally credited with derailing the project to link evolution and ethics because of his successful attack on Spencer's attempt to view evolution as progressive, tending towards good ends. Moore's attack on social Darwinism was correct, in that we cannot simply say that what is natural is good: within the whole project of naturalism, everything is natural and if this implied that everything is good, then the very meaning of good (together with moral problems) would disappear. But Moore's debunking of social Darwinism does not damage the central plank of the Darwinian approach, which is that the passions are natural and some natural passions are moral in that they promote the common good.

Alex Walter is keen to challenge the shadow cast by the putative naturalistic fallacy over this area and suggests that we should also be wary of the 'anti-naturalistic fallacy'. He defines this as follows:

> We must recognize that while not all natural facts are relevant to ethical or moral discourse, all facts that are relevant to ethical and moral discourse will nonetheless be natural facts. (Walter, 2006, p. 35)

This is a statement of belief with which the Darwinist can concur: ethical reasoning must somehow take into account the natural facts of the world (and, by implication, human nature) because there are no other facts to be had. At some level, values and facts must be related. It is, for example, the factual nature of the human condition that enables us to express what human wants are and what are good things for humans. We value a society that allows couples to have children, for example, because this is allowing freedom of expression to our biological nature. This is the approach adopted by Wilson, who thinks crying 'naturalistic fallacy' in such debates is an invalid objection:

> The posing of the naturalistic fallacy is itself a fallacy. For if ought is not is, what is? To translate is into ought makes sense if we attend to the objective meaning of ethical precepts. They are very unlikely to be ethereal messages outside humanity awaiting revelation. (Wilson, 1998, p. 250)

For Wilson, 'the ought' is sought not in the transcendental realm but by looking at the public will:

> Ought is just shorthand for one kind of factual statement, a word that denotes what society first chose (or was coerced) to do, and then codifies ... Ought is the product of material process. The solution points the way to an objective grasp of the origin of ethics. (Wilson, 1998, p. 251)

The problem here is that this explains the origins of the compulsion to behave (that is, the experience of 'ought') but does not help in formulating a new code or answering moral questions, for example, 'Ought we to ban human cloning?'

Box 17.1 David Hume (1711–76)

FIGURE 17.2 David Hume
Hume's views on the nature of emotions and human morality have much in common with contemporary evolutionary psychologists.
SOURCE: from a steel engraving by Holl after Smith, published by C. Knight, 1850.

The central feature of Hume's philosophy is the demonstration that factual knowledge must be based on sensory experiences that are processes and understood by inferences and habits of the mind. To the 21st-century reader, this will seem uncontroversial enough, but at the time, it ran counter to a strong tradition in Western thought that knowledge could be obtained through the exercise of reason and faith alone. He applied this perspective to morality and concluded that morality is not something that could be derived from the exercise of reason alone. If we witness an immoral act, we do not 'see' the immorality out there as we see objects. Rather, we find it immoral because of a 'sentiment of disapprobation' that arises in us. One might possibly object that we feel the disapproval because the action we witnessed broke some rationally derived 'rule of right' or moral code that we have previously established, but here Hume demonstrates that we tend to justify rules by reference to our moral sense and not the other way round. The fundamental point about Hume's approach is that he argues that morality arises from the sentiments, passions and feelings that all humans share.

Hume anticipated some of the thinking later to be found in evolutionary psychology. He thought affection for children to be natural and that affection for relatives lessened as the relatedness decreased. He also wrote on the inequalities in parental certainty and thought this could explain why chastity was valued by men as a virtue in women. On the topic of more general reciprocity between non-kin, however, he thought this was not a natural moral sentiment and had to be enforced by artificial devices such as promises.

17.2.2 A Darwinian updating of Hume

Many Darwinists (for example Curry, 2005) look to Hume for inspiration as to how ethics can be reconciled with evolution. One of the benefits of Darwinising Hume is that it helps to clarify what we mean by the term 'value' and provides an account of how values can exist in a world of facts. In Hume's view, human values are products of the passions, and passions include things such as love, hate, anger, malice, clemency and generosity. These passions or sentiments are inherent in human nature and those that promote the common good we call the 'moral passions'. In the Darwinian world, animals have adaptations that enable them to pursue their goals, such as moving towards food, moving away from predators, seeking mates, fighting off competitors. This list of goals becomes the organism's values and the desires for these goals (and other mechanisms involved in their pursuit) are the adaptations laid down by natural selection. In the case of humans, our values are the goals of our adapted sentiments and moral values are those that serve the common good.

In this view, the question as to whether moral values are subjective (as in the tradition of emotivism) or objective (as in realism) receives an interesting answer. Moral values are subjective, in that they reside inside our heads, and objective, in that they are common to all (normally functioning) humans. They are as objective as, say, colour vision or the peacock's tail (Curry, 2005).

This perspective suggests that morality is a set of procedures to help humans, in the face of limited resources, limited sympathies, personal diversity and competitive instincts, to secure the fruits of cooperation and arrange their equitable distribution. Morality is the name we give to those emotions and inclinations that steer us away from the temptation to cheat and reap immediate selfish rewards towards cooperative behaviour that helps us to reap the benefits of mutualism, reciprocal altruism and prisoners' dilemma cooperative interactions. Emotions such as sympathy, empathy and compassion enable us to experience the perspective of others and bind communities together. The crucial point is that whereas the original evolution of these tendencies took place according to the cold and ruthless logic of natural selection – they were selected for if they gave an advantage to the genes of those possessing them – we are now left enriched by these feelings that spill over into a whole range of contexts, enabling us to feel empathy with and pity for the suffering even when we 'know' cognitively that any help we give is unlikely to be returned even by indirect means. An analogy for this might be sexual attraction. Experiencing sexual urges for someone highly attractive was once an efficient way of ensuring that your genes paired up with a high-quality complementary set. But now we can experience the urge and enjoy the behaviour it directs even when we are told that the other person is infertile (through contraception, for example).

As we have discussed in previous chapters, psychological traits germane to moral theory that we are likely to possess from the study of the behaviour of humans and other animals are:

- Taking care of family and kin
- Sympathy for others
- Reciprocation
- Punishing cheats.

The essential importance of reciprocity in delivering mutual gains (of which the prisoners' dilemma is but one model; see Chapter 9) is illustrated by the importance that moral philosophers have attached to the emotions and feelings that are associated with it, such as trust, commitment, punishment, guilt and forgiveness.

This realisation that the roots of moral thinking and action must somehow lie deep in our natures explains the cogency of some of the objections often hurled against **utilitarianism**. Within mainstream moral philosophy, utilitarianism remains one of the most accessible and coherent single creeds. It is of undoubted use and its principles are widely applied in such things as cost–benefit analysis and the allocation of medical resources. Although there are various versions of utilitarianism, the underlying logic is that the correct action is that which tends to promote the greatest good (pleasure or happiness) to the greatest number. It is a valuable philosophy but it is not foolproof. As an illustration, we can consider the argument William Godwin (husband of Mary Wollstonecraft) advanced in his *Political Justice* (1794). Godwin thought that morality (contra Hume) must be a product of reason. Imagine, he said, a burning building containing a famous author producing works of benefit to all mankind (we could update this argument by imagining a scientist on the verge of a cure for cancer) and a chambermaid who could be your wife or mother. You can only save one person, so who do you save? Godwin thought the answer was unquestionably the benefactor of mankind, since in the future, he or she would produce the greatest good. Moreover, it mattered not one jot if the chambermaid was related to you; as he said: 'What magic is there in the pronoun "my" to overturn the decision of impartial truth?' Most modern readers will not find the decision so easy – feeling that a wife or mother (or to update Godwin further, a husband, partner or father) surely has some pull on our obligations. Post Darwin we can answer Godwin: the magic in that 'my' is the product of hard-wired sensibilities shaped by millennia of kin selection that define our humanity and will always frustrate attempts to reduce morality to a calculus of pleasure and pain.

Since Godwin's backfiring thought experiment, the use of hypothetical moral dilemmas has been a common way to explore the psychology of moral reasoning. In such studies, participants are given a scenario and asked what outcome is most desirable. One commonly explored dilemma is the 'trolley problem'. A participant is told to imagine a runaway trolley hurtling along the track. A switch is available that would send the trolley down a sidetrack. The dilemma arises because there are various combinations of people and animals at the end of each track that will be killed if the switch is used or not. The dilemma explores what type of lives take precedence over others. Studies (mostly using westernised university students as participants) have shown that the responses of participants vary in systematic and predicable ways (Petrinovich et al., 1993). Typically, options that spare the greater number of individuals are favoured, as are those that favour kin and friends, and those that favour humans over animals. Also, social monsters (psychopaths) were readily sacrificed. Perhaps surprisingly, the personal characteristics, beliefs and religious affiliations of the participants had relatively little effect. It is tempting to conclude that this pattern of preferences reflects a general pattern of human moral intuitions. Favouring kin obviously relates to kin selection and favouring friends and sacrificing social monsters would clearly serve to promote reciprocal altruism. It is possible, however, that such responses reflect a Judeo-Christian upbringing rather than an evolved human nature.

In an effort to test the wider generalisability of these findings, O'Neill and Petrinovich (1998) explored the responses of Taiwanese university students to similar dilemmas. Very few of these students had experienced an upbringing influenced by Judeo-Christian teachings and the majority had experiences of the very different teachings of Eastern religions (Taoism and Buddhism). US students were used as a control group. The researchers found that although the ethics and teachings of Western and Eastern religions were very different, as were the social norms of each culture, the reactions to the dilemmas were similar. For all groups, it was thought better to take action (that is, move the switch) than do nothing; and preferable to save kin and friends compared to strangers. Again, humans were preferred over other animals. The authors argue that this is support for a universal, evolved moral response.

17.2.3 The behavioural ecology of morality

If we argue that morality is an expression of a set of common human adaptations, then we run straight into a tricky problem: ethical norms and moral values vary greatly across the globe. This could imply that they cannot simply be an expression of a universal human nature. But this would be to misunderstand what a universal human nature means. Humans are born with a development programme with enough similarities to enable all people to be classified as the same species. The developmental programme does not have infinite plasticity but has been designed to respond in adaptive ways to local physical and social environments in the same way as other genetically guided adaptations do. In relation to morality, a Darwinian approach would lead us to have the following expectations (testable by empirical means):

- Morality (like other attributes) may have a heritable component with a distribution of tendencies in a population.
- Research on primates may yield homologous systems in place for moral feeling and action.
- There are likely to exist sex differences in morality such as the moral weight attached to chastity and promiscuity (see section 11.3 on sexual strategies) and affections towards kin. If moral values are the goals of our adaptations, then, as we have already noted, men and women have different psychological adaptations. This point actually reinforces one of the arguments within feminism: there are female values and perspectives different from those of men that can be used to critique male assumptions, values and institutions.
- Since people of different ages face different adaptive challenges, so moral values and standards are likely to vary over the life cycle. An example might be a mother's attitude to abortion. Evolutionary reasoning would predict that a women's tolerance of the acceptability of abortion may decrease with her age. Evidence (see Chapter 10) supports this at least in terms of the practice of abortion.
- Moral reasoning and the internalisation of codes of conduct is likely to vary with local conditions. A good example of this concerns marriage rules and paternity confidence (see Chapter 13). Conditions likely to affect moral reasoning and societal norms are likely to be the age and sex profiles of the population, levels of inequality and resource availability.

- The move into a 21st-century environment very different from that in which we evolved may require a rational retooling of our moral convictions. The customs, practices and convictions that served the common good in the Old Stone Age may no longer do so today.

17.2.4 Game theory and moral philosophy

A promising line of research in the effort to naturalise ethics is the linkage between game theory and moral reasoning. Some interesting parallels emerge when we revisit the prisoners' dilemma in the light of some traditional ethics theories. Figure 17.3 shows a restatement of a typical prisoners' dilemma payoff matrix.

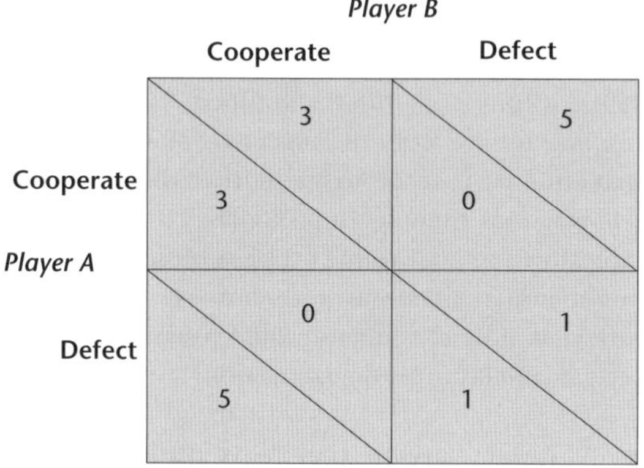

FIGURE 17.3 Typical payoff values for a prisoners' dilemma scenario

As we have observed, the problem posed by prisoners' dilemma situations is how to get individuals to cooperate for their long-term advantage. Rapoport and Chammah (1965) introduced the idea of two kinds of rationality: individual and collective. Individual rationality is that which maximises the return to the individual; collective rationality is that which, if each follows, makes each better off than if each had simply used individual rationality. We have observed earlier (see Chapter 8) how it is likely that our emotional reactions in such situations serve this function of enforcing cooperation and reciprocity for the good of ourselves.

To enforce collective rationality, some have thought that what is needed is some central authority to ensure that both players cooperate for their own good. This was the position advocated by Thomas Hobbes in his famous work *Leviathan* (1651). Hobbes thought that in the natural state men were so egoistic and competitive that a restraining central authority (either a monarch or an assembly) was needed to ensure that civil society could flourish. Sadly, this does not really solve the game's dilemma, since the addition of an enforcer and a corresponding punishment regime changes the nature of the game by introducing a whole new set of payoffs (if there is a punishment for defection, then the rewards of defection have changed according to the probability of being found out). Also, the approach of Hobbes does not really help us to work out how biology solved the prisoners' dilemma, since in nature we can detect no Leviathan gently coercing organisms to cooperate. In addition, Hobbes' view of human life without government as 'solitary, poor, nasty, brutish and short' does not really accord with what we know about early human groups and hunter-gatherer societies.

It is instructive to consider the ethical system of another great philosopher, Emmanuel Kant (1724–1804), in the light of game theory. Kant's approach to ethics in his *Critique of Practical Reason* (1788) had the ambition of attempting to find a universal basis for morality that is not dependent on subjective experience. He believed that morality cannot simply reside in human desires and inclinations since these vary between people. So Kant looks to the application of reason to establish universal moral laws since reason is peculiar to humans and shared by all. He then investigates what the rationally derived moral law might look like and applies the test of consistency. He summarises his argument as follows:

> I am never to act otherwise than so that I could also will that my maxim should become a universal law. (Kant, 1785)

Bertrand Russell responded to this by saying that it was really just a 'pompous' way of saying that we should 'do unto others as you would have them do unto you'.

Russell is perhaps a little unfair, in that Kant also thought that such maxims (or **categorical imperatives** as he called them) should also avoid logical self-contradiction. Hence the idea that 'I may sometimes make a promise that I will not keep' is wrong, since if everyone were to act according to this rule, then a promise would never be a promise and the rule would necessarily destroy itself. In essence, Kant's view was that morality flowed from the consideration of rational beings possessing immortal souls. We have a duty to follow the precepts that our reason reveals to us; similarly, we should distrust the emotional feelings that cloud our judgement when we face moral problems.

With game theory in mind, we can restate the categorical imperative as 'choose a strategy which, if chosen by everyone else, would lead to the best outcome from your point of view than any other'. This is consistent with Kant's insistence on treating people as ends and not means and his exhortation to only follow maxims that you can will to be a universal law without self-contradiction. We can see that 'cooperate' fulfils this criterion but 'defect' does not. If everyone defected, you would be worse off than if everyone cooperated. Kant's categorical imperative is really a version of the golden rule: 'do unto others as you would have them do unto you' (Bible, Luke, 6: 31). Kant thought that he had arrived at this through reason and that human beings as rational agents had a duty to follow its dictates. Indeed, Kant was so optimistic about the power of reason that he thought it was the answer to organising social affairs even for people who are inherently bad:

> The problem of organising a state, however hard it may seem, can be solved even for a race of devils, if only they are intelligent. (Kant, quoted in Barash, 2003, p. 132)

The problem is that Kant left open the question about what motive there is to cooperate. Individual rationality would dictate that we cheat if we can get away with it, and to cooperate we need virtues and these arise from affective not cognitive states.

Interestingly, a rule utilitarian would also come to the same conclusion that 'cooperate' is the best move to adopt. Utilitarians recommend acting so as to maximise happiness for the greatest number of people. Act utilitarians argue that each act should be considered in terms of its pros (creation of happiness units or hedons) and cons (creation of unhappiness units or dolons). For both to cooperate yields a total of six units of benefit, higher than any other combination. Rule utilitarians argue that it is not always feasible to weigh up every action so we choose actions which, if chosen as a rule by everyone, would lead to the greatest happiness. It is clear that 'cooperate' would be the rule that utilitarians would advocate following.

17.2.5 Moral development

It is obvious that there is a developmental dimension to moral behaviour and reasoning: children behave and think differently from adults. One of the most influential accounts of moral development came from the work of Lawrence Kohlberg carried out through longitudinal studies on young people in the 1950s (Kohlberg, 1984). Kohlberg was initially inspired by the work of Piaget (1932) and the idea that children pass through a series of discrete cognitive stages as they develop. Kohlberg's method was typically to present young people with scenarios that posed moral dilemmas and then ask through an interview how the subject would respond. He drew several conclusions from his work:

- As people develop, they move through increasingly sophisticated levels of moral reasoning. Six stages can be identified (two at each level) (Table 17.1)
- The acquisition of a higher stage of moral reasoning supersedes and displaces those below
- The higher the stage reached, the more likely the subject is to behave according to this level.

Kohlberg's work and that of his followers has generated a vast literature well outside the scope of this work to assess. In brief, however, a few criticisms (see Krebs, 2000) commonly levelled are:

- It is a rationalistic approach to morality that assumes people use reason (and not, say, intuition) to solve moral problems. Even if rational solutions can be provided, it may be that this is post hoc rationalisation after the moral 'passions' have done their work in pointing to an answer.

Table 17.1 Kohlberg's stages of moral development and some possible evolutionary parallels

Kohlberg's level and stage	Evolutionary derived parallels
Level 1: Preconventional (what's in it for me?)	
Stage 1: Obedience to authority; avoiding punishment	Deference to authority; submission to more dominant individuals; individuals maximising their own immediate somatic benefits and minimising losses. Operation at this stage sets up dominance hierarchies: defer to those more powerful, oppress those less powerful
Stage 2: Self-interest; return favours, getting even, simple exchanges	Fulfilment of obligations in direct reciprocity: you scratch my back and I'll scratch yours; mutualism within or between species; tit for tat strategies
Level 2: Conventional morality	
Stage 3: Conforming to what is expected; maintaining reputation; caring for others; exercising the virtues of trust, loyalty and gratitude	Indirect reciprocity: A scratches the back of B who scratches the back of C who scratches the back of A. In human groups, indirect reciprocity maintained by concern for reputation and punishing cheats
Stage 4: Maintaining social systems; obeying social laws; contributing to the moral order in society	
Level 3: Post-conventional (principled conscience)	
Stage 5: Fulfilling social obligations; pursuing the greatest good for the greatest number; concern with the welfare of all	Difficult to find non-human examples of this. Humans can rationalise at this level but few may actually behave at this level with real-life everyday problems. This level invokes calculated rational judgement with little emotional input. Difficult to see this as a natural adaptation. Possibly a cultural acquisition that is followed with difficulty
Stage 6: Treating people as ends in themselves; respect for the equality of human rights; following 'universal principles of justice'	

SOURCE: Based on ideas in Krebs, D. L. (2005) The evolution of morality, in D. Buss (ed.) *The Handbook of Evolutionary Psychology.* Hoboken, NJ, John Wiley & Sons.

- It is androcentric and does not reflect the responses and feelings of women. Weight is lent to this criticism by the fact that Kohlberg's initial work was done on young men.
- There are differences between the way people actually behave and the way they suggest answers to hypothetical problems.
- The structural framework is employed by people more flexibly in practice than Kohlberg supposed and people in real-life situations often operate at different levels according to context.

Despite these criticisms, there does seem to be considerable agreement that the six stages Kohlberg postulated do represent a useful classification system for levels of moral reasoning. James Rest, for example, has devised a 'defining issues test' based on Kohlberg's ideas that is still widely used in moral psychology (see Rest, 1975).

The psychologist Denis Krebs has analysed Kohlberg's ideas through an evolutionary lens. He concludes that people rarely put into practice moral conclusions reached through Kohlberg's stages 5 and 6. In addition, people move up and down between levels. Even sophisticated adults, for example, will sometimes solve moral dilemmas by simply deferring to authority (stage 1 in Kohlberg's schema). On the positive side, Krebs has seen plenty of evidence that naturally evolved behaviour, such as kin-directed altruism, mutualism and direct and indirect reciprocity, do coincide with Kohlberg's stages 1, 2, and 3. Individuals who have identified with their society and have made an investment into it will also operate at stage 4. Krebs thinks that stages 5 and 6 are different and may be recent cultural artefacts; he even suggest that there is 'no evidence that people from non-industrialized societies make stage 5 or stage 6 moral judgements' (Krebs, 2005, p. 768), suggesting that such mechanisms are not adaptations since they do not develop regularly and routinely.

Table 17.1 draws upon the ideas of Krebs to show how Kohlberg's stages might (or might not) have evolutionary derived parallels in human cognitive and affective systems. Much work remains to be done in this area. We need to ascertain whether people actually do progress through the stages Kohlberg proposed or whether these are post hoc rationalisations that can be given by subjects once their moral intuitions have given them a solution. What will be crucial to ascertain is whether or not these higher stages really do have any adaptive component (in the intuitive sense of 'fairness' and 'justice', for example). Krebs' judgement that no non-industrialised societies operate at this level seems a little harsh and evidence from historic societies would be worth examining.

17.2.6 But is it right?

You may have noticed that we have been skirting around a central question: can this view of morality provide a sufficient basis for judging if an action is morally right?

We have been dealing with at least two layers to morality. One is the phenomenon of moral behaviour, of which a good Darwinian may be able to give a plausible account, in other words, why people erect rules and choose to live by them and how such rules relate (or otherwise) to fitness gains in any given environment. Another layer is the question of whether such codes and rules are right. A large number of people, from T. H. Huxley onwards, have argued passionately that ethics transcends nature and have despaired at any attempt to draw ethical premises from evolutionary thought. A modern exponent of this view is the Harvard biologist Stephen Jay Gould. Gould (1998, p. 21), who has done much to expose the sexist and racist bias in some attempts to capture human nature, is of the opinion that 'evolution in general (and the theory of natural selection in particular) cannot legitimately buttress any particular moral or social philosophy'.

In this whole debate, there does seem to be an entanglement of different issues. One concerns the ontological status of moral values or where the good resides; another concerns how to justify (in this light of their ontological status) the 'rightness' of these values; and another is how we use knowledge of this status to solve moral problems and resolve conflicts. The problem for the Darwinian approach is that it is not clear how to answer the second and third concerns from knowledge of the answer to the first. Darwinians are clear that the ontological status of the moral sense resides in those sensibilities (sense of fairness, right conduct and justice) that natural selection supplied us with. So here is the crux of the matter: in justifying the rightness of these, we have to stop somewhere, otherwise we can keep asking how the justification argument (whatever it is) is also justified and so on ad infinitum (Curry, 2005). Theologians might stop at divine commands, Kantians stop at rules that can be generalised without contradiction. Relativists stop at the idea that such values are social conventions outside any transcultural justification. Utilitarians stop at the equation of pleasure or happiness equals the good. Darwinians stop at psychology: our psychological adaptations incline us to recognise something as good or bad. We also need to add that the sensations of 'rightness' gained their force according to whether those dispositions solved conflict problems to the betterment of all players. In other words, the only 'rightness' that resides outside biology is the criterion of delivering up maximum utility in social interactions.

17.3 Applied ethics: moral problems

So what does it mean to say that action X is wrong? In this new view, it could mean that doing X is not the way to reach agreed ends. We might wish to argue, for example, that the death penalty for adultery is morally incorrect since it is not the best way to ensure a happy marriage and a stable family. Or we might mean that X is not an accurate reflection of natural moral sentiments. Notice here that we have to use the word 'moral'. Killing a rival for some minor infringement might be *naturally* tempting for some people but it is hardy a morally correct course of action since it would not promote the general good.

This is the view shared by Larry Arnhart (2005, p. 208), when he writes:

> Correct moral judgements are factual judgements about the species typical pattern of moral sentiments in specified circumstances.

This brings us to another problem: if moral reasoning is the existence of decision rules inherent in our information-processing capacities (guided and lent force by the emotions) that are adaptations to solve collective action problems, why do moral problems (often called dilemmas) arise? There are at least three reasons (see also Curry, 2005):

1. *Conflict of values:* Our values reflect our goals and like other goals that an organism has (such as to fight or flee, or reproduce or save resources) these goals may often be in opposition. In the case of our moral sympathies, there are conflicts between forgiveness and punishment, for example.
2. *Conflict with others of different values:* It is to be fully expected that values will vary according to age, sex and location.
3. *Novel environments:* Our values were forged to ensure our well-being in the Old Stone Age. Many aspects of modern life throw up dilemmas and polylemmas that are difficult because such anciently derived values are difficult to apply in modern situations. This may be the case with the debate over the morality of cloning and abortion, for example. Another example concerns global warming, where mutual gains (industrial development) cause future problems on a timescale that we find hard to factor into our decision-making or too abstract to comprehend with our very human-oriented value systems.

In this latter point lies the hope of not being restricted to facing the 21st century with a Stone Age morality. We can use our reason to formulate game theoretic and contractarian approaches to investigate how benefits can be maximised and spread, while relying on our innate moral sensibilities to drive the enterprise with some urgency and give moral force to its conclusions. We saw that with the prisoners' dilemma, individually rational action led to an overall detriment (the 'tragedy of the commons'). In this sense, morality served the function of preventing failures in individual rationality. The reverse of this is also likely to be true: rationality can help to prevent the failures of morality.

Summary

- The acceptance of a Darwinian approach to morality does not imply that morally responsible behaviour is impossible.
- The treatment of ethics has traditionally been dominated by transcendentalist and rationalist approaches. Darwinism seeks to understand ethical reasoning and moral behaviour empirically, relating our moral sense to natural adaptations for group living.
- Much headway in the naturalistic approach to ethics can be made by combining the insights of Darwin and Hume.
- Game theory is a useful conceptual schema for understanding moral interactions and moral dilemmas. There is evidence that we have natural dispositions to solve prisoners' dilemma-type situations in mutually beneficial ways.
- Some parallels can be drawn between the stage of moral development as identified by developmental psychologists and the ways in which organisms solve group living problems as identified by behavioural ecologists and evolutionary psychologists.
- On the subject of metaethics, Darwinists recognise no source of absolute authority and 'rightness' outside the evolved human mind.

Key Words

- Categorical imperative
- Dualism
- Naturalistic fallacy
- Relativism
- Social Darwinism
- Transcendentalism
- Utilitarianism

Further reading

Arnhart, L. (1998) *Darwinian Natural Right: The Biological Ethics of Human Nature.* New York, State University of New York Press.

Arnhart makes a brave attempt to link facts and values and show that biology can provide a sound basis for ethical thought.

Ridley, M. (1996) *The Origins of Virtue.* London, Viking.

Discusses game theory and human cooperation. Interesting and controversial application of game theory to politics and environmental issues.

Wilson, E. O. (1998) *Consilience: The Unity of Knowledge.* New York, Knopf.

A powerful and broad-ranging work in which Wilson airs his conviction that all knowledge must be linked together. Wilson attempts to show that ethics can be understood in evolutionary terms.

Wright, R. (1994) *The Moral Animal.* London, Little, Brown.

A readable work that explores the implications of Darwinism for understanding our moral sentiments and the problems of modern social life. Wright makes ingenious use of Darwin's own life to illustrate his arguments.

Epilogue: The Use and Abuse of Evolutionary Theory

Those who forget history are condemned to repeat it.
(George Santayana, 1863–1952)

We pass lightly over sentiments like those of Santayana these days, and we do so at our peril. For there is one area more than any other where such thoughts should still strike us with full force, and that is in the history of attempts to define human nature. Nothing could be more important, yet in no other area has science been so betrayed. Good ideas have been neglected and bad ones pursued to tragic ends. In this field in particular, scientists have an obligation to the past and the future to be aware of the history of their discipline and the social ramifications of their ideas.

The impact of Darwinism on other disciplines has been enormous: philosophy, theology, psychology, anthropology, literature, politics – the list could be continued – have all been radically altered by the import of evolutionary ideas. Given that the theory deals with fundamental questions about the human condition, this is not altogether surprising. The results, however, have not always been welcomed and, in the sphere of politics in particular, the consequences have been messy. There have been unwarranted extensions from facts to values, and some odious political philosophies have sought their imprimatur from evolutionary thought. In this final chapter, we will examine the interaction between scientific ideas and their social contexts. In particular, we will attempt to separate readings of Darwin that fall foul of empirical or logical errors from those which are legitimate. The first part of the chapter describes the way in which evolutionary ideas have been subject to a variety of political interpretations and extrapolations over the past 150 years. The sections that follow then tackle some of the philosophical issues raised.

18.1 Evolution and politics: a chequered history

When evolutionary theories of human development and origins began to emerge in the early 19th century, they were attacked by the political and religious establishments as being radical, destabilising and a threat to the social order. The evolutionary views of Lamarck were met with contempt and dismissed as atheistic, revolutionary and subversive. Nothing better illustrates the hostility to evolutionary ideas among conservatives than the reaction to a book written by the Edinburgh publisher and amateur naturalist Robert Chambers. The book was called *The Vestiges of Natural Creation* and was published anonymously in 1844. In fluent prose, mixing religious speculation and scientific facts, Chambers gave expression to the idea that life had evolved and that species were mutable. The Anglican establishment came down hard on Chambers. Adam Sedgewick, a Cambridge don and former tutor of Darwin, called it a 'filthy abortion' that would sink man into a condition of depravity and poison the wellsprings of morality. One of the reasons why Darwin delayed publication until 1859, despite the fact that he had the essential mechanism of natural selection to hand by about 1838, was that, as a respectable and prosperous middle-class Whig, he feared the use to which it would be put by radical agitators such as the atheists and Chartists who were clamouring for reform. As Desmond and Moore (1991, p. 321) remark in their masterly study of Darwin's life:

> Anglican dons believed that God actively sustained the natural and social hierarchies from on high. Destroy this overruling Providence, deny this supernatural sanction of the status quo, introduce a levelling evolution, and civilisation would collapse.

When Darwin's *On the Origin of Species* finally came out in 1859, there had been a sea change in British life. Despite Darwin's own anxieties on the eve of publication, wealthy entrepreneurial Britain received his ideas gladly. After the publication of *The Origin*, there grew up a movement known as social Darwinism. In fact, much of the thinking contained in this movement can be found in the writings of Herbert Spencer before 1859, and the movement could more deservedly be called social Spencerism; the association with Darwin has, however, stuck. It was, in fact, an assortment of ideas rather than a fully coordinated political philosophy, but the basic premise was that evolutionary biology could teach a political lesson.

Since, as biology has shown, struggle, competition and survival of the fittest are natural phenomena that have operated to shape well-adapted and complex organisms such as ourselves, this is clearly how the social world should be organised. The natural world had operated to weed out the weak and feeble, there had been no support from any central authority, yet naked competition between individuals pursuing their own ends had indubitably led to progress. To the social Darwinists, the political message was clear: colonialism, imperialism, laissez-faire capitalism, disparities of wealth and social inequalities were all to be justified and encouraged. One of the leading social Darwinists in America was William Graham Sumner (1840–1910), a professor at Yale University. For Sumner, any redistribution of wealth from rich to poor favoured the survival of the unfittest and destroyed liberty:

> Let it be understood that we cannot go outside this alternative: liberty, inequality, survival of the fittest; not liberty, equality, survival of the unfittest. The former carries society forward and favours all its best members; the latter carries society downwards and favours all its worst members. (quoted in Oldroyd, 1980, p. 215)

Darwin himself was not immune to the ever-present temptation to mix social and biological concepts, when he observed in a letter that 'the more civilised so-called Caucasian races have beaten the Turkish hollow in the struggle for existence' (Darwin, 1881). But if the capitalists and their apologists drew succour from Darwin, so did the Communists. In a letter of 1861, Marx wrote that 'Darwin's book is very important and it suits me well that it supports the class struggle in history from the point of view of natural science' (quoted in Oldroyd, 1980, p. 233).

It is easy to see why social Darwinism appealed to industrialists, entrepreneurs and those who had gained or stood to gain from the operation of the free market. Its additional appeal for Marx was that it eliminated teleology and design from nature. Marx saw that evolution could be used to undermine his ideological enemy – organised religion. Ironically for contemporary Marxists, Darwinism has proved to be a double-edged sword. Marx's own views on human nature were ambiguous, but most Marxists have adopted the view that human nature is plastic, in the sense that 'being determines consciousness'.

The political affiliations of another group that drew inspiration from Darwinian ideas, the eugenics movement, are harder to define. Eugenics is often treated as a subset of social Darwinism but is in fact dissimilar in both motivation and policies. The eugenics movement in Britain began with the work of Francis Galton (1822–1911). Galton, who was Darwin's half-cousin, adopted a strong hereditarian position and argued that there was a correlation between a person's social standing and his or her genetic constitution (Figure 18.1). As early as 1865, Galton had tried to sway public opinion to his view that the upper classes should breed more and the lower classes less, but with little effect. However, the eugenics movement flourished in Edwardian Britain, where the social strains between rich and poor and the effects of international competition were beginning to tell (MacKenzie, 1976). The general worry was that if the lower classes were breeding faster than the upper, a general lowering of the genetic stock of Britain would result.

FIGURE 18.1 **Francis Galton (1822–1911)**
Francis Galton was Darwin's half-cousin and a Victorian statistician, anthropologist, inventor and polymath. He rejected the inheritance of acquired characteristics and attempted to measure the inheritance of traits such as intelligence in human groups. He coined the term 'eugenics' as well as the phrase 'nature versus nurture'.
SOURCE: Engraving from *The Popular Science Monthly*, May 1886.

On the eve of the First World War, the eugenics movement flourished on both sides of the Atlantic. The first International Congress of Eugenics, held in London in 1912, had Winston Churchill as the English vice-president, with Charles W. Eliot, the president of Harvard University, as the American vice-president. Eugenics societies included some distinguished geneticists as well as the likes of socialists such as Beatrice and Sidney Webb. In Britain, eugenics ideas appealed particularly to the professional middle classes. They preyed upon middle-class fears of a rising working-class population and a concern among the establishment with the poor medical condition of working-class recruits for the Boer War. It was particularly attractive to the professional middle classes and intellectuals, since it suggested that experts and meritocrats like themselves should play a role in an efficient, state-organised society (MacKenzie, 1976). In America, Galton's recommendations on selective breeding were not taken seriously until Lamarckism was discredited among biologists and social scientists (Degler, 1991). In a Lamarckian framework, if the environment worked upon individuals and the modifications thereby induced could be inherited, the main hope for social progress lay in improvements to social conditions. Once the inheritance of acquired characters is removed as a scientific possibility, selective breeding becomes a serious option for improving the race.

One of the most prominent eugenicists in America was Charles Davenport. Davenport held positions at the universities of Harvard and Chicago before becoming director of the Eugenics Record Office at Cold Spring Harbor, New York. Davenport and his workers initially adopted the Mendelian assumption that each human trait was the work of one gene. They then traced the genealogical path of traits such as criminality, artistic skill and intellectual ability. Their warning to the nation about the effects of uncontrolled breeding is exemplified by their analysis of the Jukes family. Davenport examined the burden on society brought about by the offspring of one Margaret Jukes, a harlot and mother of criminals. He concluded that as a result of her protoplasm multiplying and spreading through the generations, the state treasury was worse off to the tune of $1.25 million in the 75 years up to 1877 (Richards, 1987; Degler, 1991). As the eugenicists saw it, one way to stem the march of degenerate protoplasm was to restrict immigration into the USA of those racial types who were expected to belong to inferior stock.

In the UK, the movement was attractive to Fabian socialists who believed in state intervention to cure the inefficiencies of an unplanned economy. For this reason, it could be better described as socialist Darwinism rather than social Darwinism. By today's standards, the eugenicists made some outrageous proposals. There were suggestions, for example, that the long-term unemployed should be discouraged from breeding since they obviously carried inferior genes. Major Leonard Darwin, the fourth son of Charles Darwin, in his book *Eugenic Reform*, was strongly opposed to the advancement of scholarships to bright children from the lower classes. His reasoning was that once such children were promoted by their educational attainments to the class above, their fertility would decrease, whereas if they were left as they were, they would probably have more children, so their gifted genes would be more likely to propagate. In addition, Major Darwin argued, the existence of scholarships would worry the parents of children already in the higher social classes, since they would now face more competition, and this would further reduce their already low fertility. Looking back, these ideas appear comic, but in other countries they led to extreme and tragic consequences. In the 1920s, 24 American states passed sterilisation laws and, by the mid-1930s, about 20,000 Americans had been sterilised against their will in an effort to stamp out inferior genes.

By the 1930s, natural scientists in Britain and America were realising that the early deliberations of the eugenicists were based on faulty assumptions about the nature of inheritance. Most traits were not simply the product of single genes as had been supposed. Features such as intelligence, moral rectitude and personality were, if they had any genetic basis, the consequence of the action of many genes in concert with environmental influences. Consequently, it was extremely difficult to predict the outcome of any given union of parents. Even enthusiasts for negative eugenics realised that there were formidable problems. If a genetic abnormality caused an abnormality in the homozygous condition, heterozygous carriers could go undetected. It was not at all clear to the eugenicists what could be done about carriers.

By the late 1930s, scepticism over the viability of eugenic principles among Western biologists and social scientists turned to revulsion as it became clear to what depths the Nazis had sunk in their application of eugenic ideas. It is known that, while in prison, Hitler imbibed the ideas of eugenics from *The Principles of Heredity and Race Hygiene* by Eugene Fisher. In the hands of Hitler, the eugenic ideal of improving the national stock had become twisted to a concern with racial purity – the enhancement of the Aryan race and the outlawing of mixed-race marriages between Aryans and the supposedly inferior Jews, Eastern Europeans and blacks. When the Nazis came to power in 1933, they set about the systematic forced sterilisation of schizophrenics, epileptics and the congenitally feeble-minded. Deformed or retarded children were sent to killing facilities, an estimated 5,000 dying in this way. Seventy thousand mentally ill adults were also targeted and put to death (Steen, 1996). The horrific culmination of this reasoning was the Holocaust and the extermination of about six million Jews, homosexuals and others deemed unfit.

While support for biological accounts of human nature ebbed away in the 1930s and 40s in Britain and America, Konrad Lorenz in Vienna was developing his theories of instinct and was laying the foundations of ethology. The reception of Lorenz's ideas in the English-speaking world was heavily influenced by an essay review of his work written by the biological psychologist Daniel Lehrman (1953), and it seems likely that Lehrman's appraisal was influenced by the apparent sympathy shown by Lorenz for Nazi ideology. Others were similarly concerned. When Lorenz won his Nobel prize in 1973, for example, Simon Wiesenthal, the head of the Jewish Documentation Centre in Vienna, wrote to him to suggest that he should refuse it (Durant, 1981).

The precise linkage between Lorenz's science and his early attraction to the Nazi cause is not straightforward. There is no doubt that Lorenz initially showed sympathy for the Nazi ideal. He joined the Nazi Party after the annexation of Austria by Germany (the Anschluss) and wrote articles for *Der Biologie*, a journal with explicit Nazi connections. In some aspects of his thinking, he displayed typical Nazi fears, such as the belief that urban man had unwisely acted to suspend the cleansing force of natural selection and consequently faced biological deterioration (Kalikow, 1983). The assertion that Nazi **ideology** moulded his scientific approach is, however, debatable. Lorenz might have developed his theory of instincts and the assertion that human aggression has an innate basis with or without the rise of the Nazis, and he certainly continued to develop these ideas after the war. More generally, it would also be wrong to overstate the connection between scientific Darwinism (as opposed to a more nebulous commitment to evolutionist thinking) and Nazi ideology. There is little evidence, for example, that the Nazis approved of the idea that the Aryan race evolved from ape-like ancestors roaming the African plains.

The Nazis drew upon other intellectual traditions such as Hegel's philosophy of the state, the concept of superman found in Nietzsche and a romantic and racist attachment to the idea of the German *Volk*.

Marxists also distorted evolutionary theory for political purposes. In revolutionary Russia, Darwinian theories of natural selection never really took hold. The revolutionaries regarded natural selection as tainted with capitalist notions of competition. It was in this ideological climate that a quack geneticist called Trofim Lysenko (1898–1976) managed to steal the show. In place of natural selection, Lysenko asserted the mechanism of Lamarckism. Mendelian genetics was denounced as 'bourgeois', and its adherents were forced to recant or were exiled to Siberia to reconsider their position. Lysenko claimed that his philosophy and methods would bring about improvements in the Russian grain harvest. The effects were, however, disastrous. Only in the 1950s, after Stalin's death, did the science of genetics recover in Russia.

18.1.1 Race, IQ and intelligence

In the USA in the 1930s, support for eugenics ideas faded. In psychology also, the theory of instincts, which always seemed to have a slender empirical base, was gradually abandoned. A supposed link between race and intelligence was, however, more difficult to sever. One reason was that, by the 1920s, plenty of experimental data had accumulated to suggest that there were biological differences in the innate mental capacities of different racial groups. Looking back, we can only groan at the crudity of the tests used and the mentality that lay behind their use. However, the sheer momentum built up by the process of intelligence testing, linked to immigration control, helps to explain the seeming paradox that as behaviourism was abandoning the concept of instincts, the idea of a link between race, biology and ability remained and was not really expunged from psychology until much later.

The greatest challenge to the validity of intelligence testing between racial groups came from Otto Klineberg, a psychologist at Columbia University. Klineberg became professionally acquainted with Boas and dedicated his book *Race Differences*, published in 1935, to him. Klineberg systematically and methodically set about testing suggestions of Nordic superiority and black inferiority. His results showed that, once background and upbringing were controlled for, the data were far more ambiguous. Any differences that did remain Klineberg attempted to explain through environmental influences. Despite Klineberg's well-intentioned and well-researched efforts, the connection asserted to exist between heredity and intelligence was weakened but never totally severed. Despite a preference for cultural and behaviourist explanations in the social sciences, anthropology and psychology in the 1940s, 50s and 60s, the issue of race and intelligence arose periodically to trouble the academic community (Harwood, 1976, 1977).

18.1.2 A poisoned chalice

Against this background, it is hardly surprising that biological theories of human behaviour are treated with great suspicion. It is a natural tendency of ideologies to seek support and corroboration from other fields of thought, and Darwinism has always provided an axe for virtually anyone to grind. Looking back, it now seems almost inevitable that when evolutionary explanations of human behaviour began to emerge in the 1970s, some hostility would result. Thus, in Wilson's *Sociobiology: The New Synthesis*, published in 1975, 95 per cent of the subject matter was concerned with animal behaviour and only 5 per cent with humans, yet the outcry was vociferous and at times hysterical. Referring to the discipline of sociobiology, the Science as Ideology Group of the British Society for Social Responsibility in Science wrote that:

> Sociobiology arrives at a time when there are wide-ranging challenges to the existing social order being made ... It is of course racist and sexist and classist, imperialist and authoritarian too. (BSSRS, 1976, p. 348)

Attacks like this were common throughout the late 1970s and the early 1980s. The general thrust was that, by identifying universal human traits that were genetic rather than cultural, evolutionary theory applied to humans was guilty of propagating an oppressive **biological determinism**. Rose et al. (1985, p. 30) saw the universities as having a special role in this process:

> Thus, universities serve as propagators and legitimators of the ideology of biological determinism. If biological determinism is a weapon in the struggle between classes, then the universities are weapons factories, and their teaching and research facilities are the engineers, designers, and production workers.

Such ideological reactions were common in the UK and America from those left of centre. In the 1990s, the debate over the application of evolutionary ideas to human nature thankfully shifted somewhat from the zone of class conflict, and ideas began to be considered more for their scientific merits than their position in an ideological battleground.

The awkward and uncomfortable history of the political uptake of biology, the cranky and outrageous ideas that have resulted, and the hysterical reactions to even moderate attempts to probe the biological basis of human nature should not blind us to the fact that there are serious and important issues at stake. Nor will it do simply to assert that scientists have no moral responsibility outside their subject area. Evolutionary thinkers have a duty of care in this area, and we must carefully examine the tangled relationship between evolutionary thinking and moral, social and political thought as objectively as possible. In the next few sections, we will look at just a few of the specific issues that arise from the implications of Darwinism.

18.2 The eugenics movement

Eugenics represents the other side of the coin from social Darwinism (see Chapter 17). Rather than let nature take its course, the eugenicists wanted to intervene to put it right. Eugenicists were concerned that the processes operating in urban societies were such that people producing the most offspring resided in lower socioeconomic groups and were therefore genetically inferior to those in the higher strata of society. Needless to say, those promulgating the idea regarded themselves as being genetically superior. The remedy for the eugenicists lay not in competition and laissez-faire – since civilisation for ethical reasons had already accepted the burden of helping the weak – but in active measures to encourage the spread of good genes (positive eugenics) and discourage the spread of weak ones (negative eugenics). The whole eugenics programme was so fraught with scientific, ethical and practical difficulties that no one today would seriously advocate the sort of measures proposed in the first half of the 20th century. In fact, any hint of sympathy for eugenics ideas in the UK is regarded as a blight on the career of a politician.

The fertility of different social groups is of course an empirical matter and could be settled by statistical means. It is also possible that certain genes are more frequently found in certain social groups than others. Problems start hereafter, for who is to define what genes are desirable and worth increasing in frequency? As if this were not enough, there is then the practical problem of how the state could alter gene frequencies without an unacceptable infringement of other human values, or whether the state even has a responsibility to its gene pool that overrides its responsibility to the welfare of individuals and the preservation of individual freedoms.

We must be alert to eugenics issues since gene technologies are increasingly delivering into our hands powerful tools to screen individuals for genetic defects. Prenatal screening enables doctors to assess whether the fetus is genetically defective for a wide range of conditions. If, for example, parents choose to terminate a pregnancy because of the condition of cystic fibrosis, this involves the judgement that a child with cystic fibrosis is too great a burden to justify his or her birth. Some have argued that this is a form of eugenics through the back door. The comparison with eugenics thinking is not, however, strictly accurate in these cases. Certainly, the motivation of the parents is not to eliminate the genes from the human population: their concerns are about the suffering of the child and the burden to the family. In fact, it is extremely difficult to alter gene frequencies by such procedures.

18.3 Evolutionary biology and sexism

When sociobiology emerged in the 1970s, it was quickly denounced as sexist. 'Sexism' is in fact quite a complex term that needs to be unpacked carefully to examine this accusation. Evolutionary thinking could be sexist in the sense that concepts from socially constructed gender roles are transported into the natural world and have a distorting effect. To speak of the 'queen bee', for example, is really a metaphor that, if taken too literally, could give a very misleading effect of what actually happens in the hive, where, if anything, the queen seems to be controlled by her 'workers'. There are particular problems here in describing the sexual behaviour of animals, it being only too easy to transpose concepts from the social world to the biological and then back again. What may seem to be a dominant and resourceful male reigning over his harem could be a group of females with their own social bonds clubbing together to choose the best-looking male.

Another example concerns the way in which we give an account of how some species of ants make 'slaves' of others. In human slavery, members of the same species are violently coerced into labouring for others, yet the application of the word 'slave' to ants may be misleading. When ants make 'slaves', they capture immature members of a different species. The captured individuals then mature in the nest of their captors and perform housekeeping tasks, apparently without coercion. Perhaps a better metaphor here would be that of domestication.

The reception of scientific ideas can also be influenced by views on the social roles of males and females. In the patriarchal climate of Victorian Britain, where women were denied effective political power, Darwin's view that the female could, through her power of choice, exert an effect on the male was received with much scepticism. One might postulate then that ideas from evolutionary thinking are accepted by the scientific community if they conform to contemporary social expectations.

Both these points have epistemological and political dimensions and need to be considered carefully. There seems to be no doubt that, in constructing knowledge of the world, scientists employ metaphors that betray a social origin, and that such metaphors may therefore condition a particular image of reality. Knowledge is rarely, if ever, value neutral. It is produced by people with specific social, personal or professional interests. Even in the most abstruse field, someone decides that something is worth knowing about, and that invokes a value commitment. The important question here has to be whether or not our view of external reality is so distorted by this process that our image of the world is entirely a social construction. The answer, we think, has to be no. Unless you are a thorough-going relativist with respect to scientific knowledge (in which case you probably would not have read this far anyway), it has to be acknowledged that the world is not plastic enough to sustain any interpretation. Moreover, the checks and balances built into the methodology of modern science ensure that false images will eventually be exposed.

In a detailed paper, the Californian primatologist Craig Stanford (1998) has suggested that interpretations of the behaviours and social systems of chimpanzees *(Pan troglodytes)* and bonobos *(Pan paniscus)* have been heavily influenced by contemporary gender roles and images of men and women. Put briefly, chimps have been interpreted in masculine terms and bonobos in feminine ones. Stanford may be right, and the paper exists in the literature for all to judge. The fact, however, that we have the intellectual tools to realise potential bias shows that we can transcend our prejudices. The very fact that we realise that analogies such as queen or harem or slave-making are simply that – imprecise metaphors – shows that we are not imprisoned by them. More generally, the very phrase 'natural selection' is a metaphorical extension of the way in which humans select, but no respectable biologist really believes that some conscious agency is at work doing the selecting.

One has to concede that particular lines of enquiry may be socially conditioned. At a trivial level, the funding arrangements of science will always ensure that social priorities enter into the direction of scientific research, for example. The crucial point, however, is that the results of the enquiry process are scrutinised by standard scientific procedures. In short, the outcome of the research is not logically predetermined by the motivating factors. To suppose that it is entails committing what Popper called the 'genetic fallacy' (nothing this time to do with genes): the belief that the origin of ideas impacts on their truth value.

Another line of attack on evolutionary accounts of human nature is that it is sexist because it points to innate differences between the sexes. The concern here stems from the belief that to suggest differences in gender-specific behavioural dispositions means that these dispositions are fixed and therefore incapable of moderation, and the genetic basis of behaviour can be used to legitimate social roles. The first point to note here is that there are obvious physical differences between men and women that have a strong genetic basis. Men cannot bear babies or lactate. In relation to height and musculature, there are of course environmental influences, and girls who are well fed and nourished may grow to be larger than boys who are malnourished, but, on average, when raised in similar conditions, men are slightly taller and more muscular than women. Girls, on the other hand, mature physically and emotionally faster than boys. These are not sexist statements in the sense of denigrating one sex or the other, nor are they sexist in that they entail distortions, deliberate or otherwise, of the world; they are descriptive statements about human development. If they are sexist, so too are large portions of the sciences of anatomy and physiology. The facts could be used for sexist purposes, but that needs to be tackled on a different level and in no way challenges the data themselves.

I suspect that more concern is expressed over the supposed sexist implication of evolutionary accounts of behaviour than over physique because behaviour is what defines us as human. In terms of our bodies (apart from our extra-large brains), we are very similar to the great apes, but in behavioural terms, we have a sophisticated culture that apes lack. The fear that a biology of mind destroys our humanity runs deep. For many, a belief in the autonomy of the mind and its susceptibility to beneficial moulding by culture represents the last raft of refuge from scientific attacks on the uniqueness of the human species. Copernicus and Darwin (and some would say Freud) effectively sank any claims that humans are the chosen species occupying a special place in creation. Those who have such worries should take heart: in a meaningful sense, evolutionary theory confirms that we are unique. It just adds the timely reminder that all species are unique.

It has been argued throughout this book that there are fundamental differences in the behavioural characteristics of human males and females. This should not worry us; instead, we should celebrate and take delight in the fact. Aristotle Onassis spoke up for about half the human species when he said that: 'If women didn't exist, all the money in the world would have no meaning.' It would be surprising in the extreme if genes conditioned our physique and the structure of our brains but stopped short at wiring us up for behaviour and handed it entirely over to culture.

Evolutionary biology has of course been used to provide corroborating evidence in the legitimisation of the social roles of men and women. The arguments usually amount to the idea that certain contemporary roles are more suited to one sex or the other because of the ancestral division of labour that became encoded in our genes. It might be thought that this approach could provide valuable information in, for example, job selection, but in fact, beyond surrogate motherhood or wet nursing, for which we could reasonably rule out men, any information we have on the evolutionary basis for sex differences is useless in this respect. In height and physique, for example, men and women are not strongly dimorphic. To use sex as a guide to these qualities would be useless, given the overlap of the spread of values in male and female populations. Even strength is a quality increasingly less useful in a society in which muscle power is increasingly displaced by mental agility. The fact remains that virtually all modern social roles can be performed by both men and women, so sex alone is not a reliable criterion for assessing suitability for a particular role.

Other such extrapolations often fall prey to the naturalistic fallacy. To say, for example, that women or men should perform some tasks because they did so in the hunter-gatherer stage of evolution is to leap from facts to values. Western society has fortunately seen the sense of all this, and discrimination based on sex, in theory at least, is largely outlawed.

If anything, evolutionary accounts of human sexuality provide a strong antidote to sexism. There is no room in biology for the suggestion that one sex is in some way superior; the concept simply has no meaning. In sexual reproduction, each sex inherits half its genome from its mother and half from its father. Whatever we think of the genes that meiotic shuffling has given us, we must give blame and thanks equally.

Racism and sexism are not eliminated, however, by simply attacking a biology that suggests there are biological differences. If this were the case, then presumably we would have to countenance the unequal treatment of races and sexes if real differences were discovered. In fact, there are real differences, of course, but the crucial point is that the interests of people are given equality of consideration for the furtherance of their (sometimes differing) interests. We might agree, for example, with the right of a woman to have an abortion; but this makes no sense when applied to a man.

18.4 Evolutionary biology and racism

Racists have often turned to biology for support. Even before the evolutionary thinking of the 19th century, racists, particularly in the USA, used a mixture of biology and religion to justify their exploitation of African natives. It was both bad theology and bad biology. Following the advent of Darwinian thinking, the exponents of racism had to shift their ground but, unsurprisingly, came to similar conclusions as before: that some races were higher or more developed than others. This view crept into medicine. Down syndrome, a problem caused by an error in the chromosomal inheritance of a child, was called 'mongolism' by its Victorian discoverer, John Langdon-Down. To him, it seemed an appropriate term; sufferers from this condition had slipped a few places in the evolutionary hierarchy to resemble a race lower than the Europeans – the Mongols.

In the main, modern evolutionary thought and the science of genetics are destructive of racist ideas. It turns out that the concept of race is not a particularly useful one for the biologist. It was realised long ago that all races belong to the same species, *Homo sapiens*. (Given the fact that racism is a problem in our culture, one can only shudder at what the world would be like had another *Homo* species survived into the present epoch.) If we start with, for example, skin colour as a criterion for dividing people into groups, it transpires that only about 10 genes out of a total of at least 50,000 on the human genome are responsible for skin colour. We might then look for correlations between skin colour genes and others. When we do, patterns in the distribution of one set of genes are not matched by distributions in others. The human races are remarkably homogeneous, possibly because of our relatively recent origin. Most of the genetic diversity between individuals occurs because they are individuals rather than because they are members of the same race. Put another way, most of the world's genetic diversity is found in any one race you choose. On the whole, the evolutionary approach to human behaviour is concerned with human universals – cross-cultural features that unite the different groups of the world and reveal our common evolutionary ancestry. The mental modules or Darwinian algorithms to which evolutionary psychologists refer were laid down before races differentiated.

It follows that the concerns of the eugenicists over the heritability of various traits is not of particular concern to the evolutionary theorist. The concept of heritability describes the percentage of variation between individuals that results from inheritance. Assuming for the moment that IQ has some validity, the heritability of this feature is a measure of the extent to which differences between individuals are attributable to genetics or environment. If we say that IQ has a heritability of 50 per cent, this means that half the variation in IQ between, for example, two people is caused by genetic influences and half by environment. A heritability of 100 per cent would imply that all the difference between individuals is caused by genes, and 0 per cent would imply that any difference is entirely due to upbringing.

Now, in studying human nature from a Darwinian angle, we are dealing with low heritabilities. The premise is that all humans have mental hardware that predisposes them to behave in ways that are adaptively similar. This mental hardware is laid down by the genes, but the variance is small. As an analogy, consider the number of lungs (two) possessed by most people. The heritability of this is near zero: nearly all people are born with two lungs. If we examine people who have only one lung, it will usually be found to be a product of the environment – usually the surgeon's knife. The possession of two lungs is an inherited trait (very adaptive) but with low heritability. A feature such as eye colour will have nearly 100 per cent heritability, differences between people being almost entirely the result of genetic influences: the environment does not shape eye colour. This raises another point: features with low heritability tend to be more interesting. Heritability itself is not a good guide to establish whether something is under genetic control.

But racism exists and is in need of an explanation as well as a cure. We must consider the slightly frightening prospect that racism has some adaptive function. To offer an explanation is of course not to condone the behaviour. If we explain racism sociologically, for example, which is commonly done, this neither supports racism nor excuses it. Nor, crucially, does it undermine the sociological approach. If there is a biological basis to racism, it is something that we must face squarely.

Numerous attempts have been made to investigate whether the roots of racism lie in biology. One promising line of research stems from the inclusive fitness theory of Hamilton: inclusive fitness is increased if individuals are nice to those who bear copies of the same genes. Some have seen this as a basis for ethnicity: by distinguishing between those most likely to share copies of your genes and those less likely to do so, it is possible to distribute cooperative behaviour more effectively. Reviewing the evidence, however, Silverman (1987) concluded that most intergroup conflicts that have taken place are within an ethnic grouping rather than between such groups, and that such conflicts are explicable in terms of resource competition. Silverman also concluded that if racism were a fundamental part of the human psyche, any small gains to inclusive fitness that were gained by racial discriminations would be outweighed by a loss of the ability to form cooperative coalitions between groups as conditions changed. Thus, racism at this simple level would not be adaptive.

What may have been adaptive is the post hoc rationalisation that accompanies group conflict. As we have argued in Chapter 17, morality may be an adaptive device to ensure that we cooperate in conditions where it is favoured by fitness gains. If, however, circumstances favour the exploitation of a former ally or cheating on obligations, racism could be a device to guard us from the full illogicality of our moral position. In this sense, racism may have acquired a function as a consequence rather than a cause of intergroup strife. Silverman's arguments are plausible and in a sense give reason for hope. Our self-deceptions are often fragile and open to elimination by education.

An exposé of the political abuse of evolutionary ideas may suggest that there is some sort of natural and therefore suspicious affinity between Darwinism and unpalatable political philosophies. Here, we should be wary. Scientific ideas can be taken up into pre-existing debates without inspiring them, and indeed with little logical connection to them. Sexism, racism, militarism and imperialism, for example, all existed before the Darwinian revolution and will probably persist for a long time thereafter. Looking again at the eugenics movement in America, it can be seen that it drew particular inspiration from the new science of Mendelian genetics, yet the early Mendelians rejected the notion that natural selection had been important in evolution. For the eugenicists, a belief in the necessity of artificial selection was not based on any acceptance of the power of natural selection. When it comes to notions of racial superiority, it was, as Bowler (1982) notes, Lamarckism that was more easily incorporated into attempts to construct a racial hierarchy with Europeans at the top.

In summary, this section should have demonstrated that Darwinism provides no ammunition for the eugenicist and little comfort for the racist. It does, however, provide an essential ingredient in our search for self-knowledge, and this theme is explored in the next section.

18.5 The limits of nature

18.5.1 Reductionism and determinism

In some circles, particularly where politically correct thought control is in operation, the word **reductionism** has a strong pejorative tone. It carries the implication that one is doing damage to a complex topic by reducing it to its component parts. So is evolutionary reasoning reductionist, and is this a bad thing?

In a sense, all the natural sciences are reductionist in that they subscribe, albeit in most cases indirectly, to ontological statements. In modern science, it is the accepted view that the building blocks of the universe are matter (in the form of particles) and energy, or quarks or strings or whatever the latest theory suggests. No serious biologists believe that, when they study life, they are studying some immaterial force, just as psychologists do not think that the human psyche consists of some non-physical entity that inhabits our brains. In this way, biology and evolutionary theory are reductionist. It is the case, however, that thinking at the level of fundamental particles is not particularly useful for the evolutionist.

Evolutionary thought is no more or no less reductionist than, for example, geology, meteorology or economics, but its concepts and levels of explanation lie on a different plane from that of particle physics. It must be said that all science is reductionist, in that it reduces the bewildering variety of phenomena to the operation of fewer essential laws, principles or theories. This is the case for the evolutionary theorist as it is for the social scientist. Any subject that attempts to explain a set of facts by a larger set of principles, which only have meaning and application in relation to those facts, has missed the point of what to explain means.

The real danger and threat of reductionism arise from what might be called 'methodological imperialism' – the view that only one type of explanation counts. Consider eating an apple. Physiology will tell you how it is digested, and environmental chemistry will tell you what happens to the carbon dioxide you breathe out as a result. Evolutionary theory will tell you why you find it sweet: you cannot manufacture your own vitamin C so you must obtain it from foodstuffs; consequently, there was a selective advantage long ago to finding the taste of fruits sweet. Try feeding an apple to domestic cats, animals that can manufacture their own vitamin C, to observe the contrast. Evolutionary theory also tells us why plants go to the trouble of packaging their seeds in a nutritious and expensive coat, but it does not have much to say on how the apple got into your hand, how it came to be in the shop or why the farmer bothered to grow it. Here, it must gracefully give ground to economics and perhaps human geography. Human knowledge is essentially pluralistic. If physicists ever reach the Holy Grail of a super-theory of everything, there will still be employment for literary critics, historians, economists, archaeologists and so on. Each domain of experience must call upon levels of explanation and notions of causality appropriate to that domain. Quantum gravity is of no use in explaining the fall of a man who jumps from a tower block.

Determinism is another concept that has had a bad press. The fear of determinism is related again to an anthropocentric view that humans tower over the rest of the animal world, aloof in their self-consciousness and confident in their power of free will and rational thought. Again, however, all science is based on the notion that events are caused. The whole of scientific progress is the story of how mysterious phenomena are brought within the fold of causal explanation. Social scientists, feminists and Marxists all assume that events and behaviours are determined: they do not just happen. The real fear is of course that biological determinism is somehow limiting, an affront to human dignity, but humans have always been deterministically limited. Science, and biology in particular, clarifies and even helps to overcome those limits. Anatomy and physics tell us that we cannot fly unaided; biology tells us we cannot breathe underwater or on the moon. Men cannot have babies, nor can women father them.

Yet to explain, reduce and identify limits need not be limiting. With assistance, we can fly at twice the speed of sound five miles up, land on the moon and breathe under water. And to those who point to the fate of Icarus, we should note that he knew too little science rather than too much.

18.5.2 The perfectibility of man

There is an age-old philosophical debate that goes back to the time of the Greeks concerning the origin of human vices and virtues. In the modern period, the debate was sharply defined by Hobbes and Rousseau. Hobbes, writing in England in the 1650s after the chaos of a civil war, argued that, in the natural state, the life of man was 'nasty, brutish and short'. Left to his own devices, man would live in a squalid state of perpetual struggle and conflict. The solution for Hobbes was for the state to impose order from above to curb the excesses of human nature. At the other end of the debating spectrum lies Jean-Jacques Rousseau. In his *Discourse on Inequality*, published in 1755, Rousseau argued that humans are by nature basically virtuous but are everywhere corrupted by civilisation. Rousseau gave Europeans the image of the noble savage living in a state of bliss before the arrival of civilisation. Rousseau's arguments were in part polemical and designed to expose the decadence in French culture, but his picture of the noble savage stuck and was profoundly influential. Ever since the time of Rousseau, weary Europeans have sought examples of the blissful and guiltless lives that Rousseau described.

The reality has, however, never really matched up

to the expectations, but on one occasion it looked as though Rousseau's vision had been found. In 1925, Margaret Mead went to the Polynesian island of Samoa to study the life of the islanders. Mead spent just five months among the islanders before returning to New York (Figure 18.2). Her subsequent account in *Coming of Age in Samoa*, published in 1928, was a seminal work. Mead claimed to have discovered a culture living in a state of grace, free from sexual jealousy or adolescent angst. Violence was extremely rare, and young people enjoyed a guilt-free, promiscuous lifestyle. Mead became a major celebrity; her books were bestsellers and became required reading for generations of undergraduates. She even had a crater on the planet Venus named after her.

FIGURE 18.2 Margaret Mead (1901–78)
In her influential book, *Coming of Age in Samoa*, Mead argued that the sexual mores of young people are determined by their culture. Derek Freeman has, however, questioned the validity of Mead's work and has claimed that Mead was the victim of a hoax by her subjects.
SOURCE: Library of Congress Prints and Photographs Division, Washington, DC, 20540.

Unfortunately, Mead was duped. At the onset of her career, she was strongly influenced by the anthropologist Franz Boas, who, appalled at the eugenic thinking that he encountered in his native Germany, propounded a culturalist view of human nature. Mead imbibed this, and her work was a product of her own expectations coupled with faulty data collection. Her errors were exposed by Derek Freeman (1996), who, like Mead, spent time (five years) among the Samoans but who came to an entirely different conclusion. Mead had constructed her account of the carefree love lives of the Samoans from the reports of just two adolescent girls, Fa'apua'a and Fofoa. When Freeman interviewed the girls, by then old ladies, he heard how, in a state of embarrassment about Mead's questioning of their sex lives, they had made up fantastical stories of free love. So it was that a whole view of human nature in social anthropology was based on a prank by two young women.

18.6 So human an animal

18.6.1 Fine intentions

It is easy to see why the left and the liberal intelligentsia should be so attracted to environmentalist conceptions of human nature. For a start, right of centre ideologies have often looked to a static human nature to support their claims. At a deeper level, however, there lies the often-unquestioned assumption that if human vices are the product of social circumstances, then by changing the circumstances, we can change human nature – for the better of course – and the perfectibility of man is at hand. Similarly, feminists have often argued that the unequal distribution of power between the sexes, the differences in historical cultural achievements between men and women, gender stereotypes and the 'glass ceiling' are products not of biological differences between the sexes but of socialisation in a patriarchal society. Change the society and we can change the roles.

Such thinking seems to have lain at the heart of the environmentalist programme of Boas. As a Jew, Boas found the anti-Semitism in Germany in the 1870s and 80s both discouraging and alarming. He foresaw a career path strewn with obstacles and disappointments merely as a result of his own racial identity. In contrast, Boas saw an America (before the proliferation of eugenic ideas and restrictive immigration policies) beckoning, with an outlook that stressed equality of opportunity and intellectual freedom.

Boas almost single-handedly swung American anthropology away from explanations based on inherited mental traits towards cultural relativism. The transformation in anthropology was mirrored in the lives of individual social scientists. Carl Kelsey, who was a sociologist at the University of Pennsylvania, is a particularly interesting example. In his early career, Kelsey embraced Lamarckism and regarded the race problem in America as a product of the inherent differences between blacks and whites brought about by the exposure of thousands of generations to radically different environments. The downfall of Lamarckism that led some scientists to turn to eugenics as a method of effecting national improvement led others such as Kelsey to move in the opposite direction. If, as Boas had shown, nurture was instrumental in shaping character, Kelsey reasoned that social progress could be achieved by improving environmental conditions. Such a procedure had the additional merit of being faster than either selective breeding or waiting for the inheritance of acquired characteristics. Within this framework, Darwinism was an irrelevance. Some psychologists were quite open in their commitment to a science that was in keeping with liberal values. One such was Thomas Garth of the University of Texas, who in 1921 laid down a rule for students who were set upon examining racial differences:

> In no case may we interpret an action as the outcome of the exercise of an inferior psychical faculty if it can be interpreted as the outcome of the exercise of one which stands higher on the psychological scale, but is hindered by lack of training. (quoted in Degler, 1991, p. 190)

The rule is an amusing and ironic allusion to the canon laid down by Morgan 26 years earlier (see Chapter 1).

There is at present much controversy surrounding the proper application of Darwinism to human psychology. The journalist Andrew Brown draws a picture of two main warring camps, which he calls the Gouldians and the Dawkinsians, the former being sceptical and the latter optimistic about the whole project. The Gouldians include the late Stephen Jay Gould, Steven Rose and Richard Lewontin. Among the Dawkinsians, we find Richard Dawkins, the late John Maynard Smith, Daniel Dennett and Helena Cronin. If one wishes to examine the extra-scientific commitments that lie behind the rival philosophies of these groups, one may discern among the latter a distrust and rejection of the pretensions of organised religion. This is coupled with a belief that a science of human nature can guide social action and policy. As Brown (1999) notes, however, it is the background of the Gouldians that is especially interesting. Of the three names noted, all are Jewish and all vaguely Marxist. Sociologically, this may be significant: Jewishness would give a strong motivation for being suspicious of any attempt to draw up a biological profile of human nature, and Marxism too, at least in the 20th century, became largely committed to a philosophy of biological egalitarianism and the belief that social existence determines human nature.

18.6.2 Retrieving our humanity

The history of ideas tells us that Darwinism is not the property of any single political ideology. It is a scientific view of nature that can be used to inform political discussions but one that does not translate easily into simple political remedies. It is simply misguided to imagine that the scientific enterprise of examining the evolutionary roots of human behaviour is somehow impugned by the errors of the past. In the coming years, skill will be needed to sift the legitimate from the spurious applications of Darwinism. We already factor a knowledge of human nature into our social systems in a myriad of ways. Consider the undeniable and biological propensity for humans to fall asleep. This is not something we learn – we are born with this tendency – but modern society relies upon the ability and willingness of some individuals to work through the night. A knowledge of biology tells us that there is a price to pay in terms of performance and fatigue, and elementary psychology tells us that we may need inducements to persuade people to work through 'unnatural' hours. But it can be done. Biology is not destiny, but it can provide a useful contour map.

This is the approach taken by the Australian philosopher Peter Singer (1998), who argues that Darwinism informs us of the price that we may have to pay to achieve desirable social goals. Uninformed state attempts to make socialist man have failed because they ignored human nature (Figure 18.3). For Singer, some aspects of human nature show little or no variation across cultures and must consequently be taken account of in any social engineering.

Singer's list includes concern for kin, an ability to enter into reciprocal relationships with non-kin, hierarchy and rank, and some traditional gender differences. To ignore these is, according to Singer, to risk disaster. The abolition of hierarchy in the name of equality, as attempted in the French and Russian revolutions, for example, has all too often simply led to a new hierarchy. This, for Singer, is not an argument in favour of the status quo. The political reformer, like a good craftsman, should have a knowledge of the material with which he or she works. The trick is to work with the grain rather than against it.

A set of deeper problems arises with the view that the promise of a humane society lies only within an acceptance of the opinion that our humanity is culturally determined and defined. Supposing we could structure a society to shape people in the way in which we desire. Who draws up the blueprint for *Homo perfectus*? From where do we draw our notions of what constitutes ideal man? Reason by itself is not enough. Reason needs motives to act; it needs the will, beliefs, goals, ideals, something to serve, in other words, human nature. Reason cannot lift itself up by its own bootlaces. Perfectly rational man would be a monster.

The blank slate approach to human nature that is still unquestioned in some branches of the social sciences would, if it were taken seriously, be a tyrant's licence to manipulate. Liberals would have to stand back powerless and impotent as a tyrannical state moulded its people into instruments of whatever crazy ideology was in fashion. There would be no basis for any objection since this would have been jettisoned when biology was thrown out: if human nature is anything that a society structures it to be, there is nothing to be abused.

Associated with the view that human nature is culturally determined is the philosophy of cultural relativism. If there is no fixed nature, there can be no single way of life conducive to its expression or fulfilment. Consequently, there is no judgemental moral high ground. Cultures in which the limbs of criminal offenders are severed, mixed-race marriages are forbidden and females undergo genital mutilation must be contemplated in silence. As the French philosopher Finkielkraut (1988, p. 104) said: 'God is dead but the Volksgeist is strong.'

We should remember that the Enlightenment project of progress through reason, science and the intellectual challenging of authority delivered human freedom precisely at the expense of culture. To resurrect culture as the new authority risks all that we have gained and threatens to tip us into a state of intellectual bankruptcy and moral free fall. Fortunately for anyone so inflicted, Darwinism is the best antidote around to the fashionable fallacies of postmodernism.

In the coming decades of the 21st century, more pieces of the human jigsaw will be put in place, with advances in the human genome project and neurobiology, and with Darwinism cutting ever deeper into the human psyche. One of the major challenges ahead will be knowing how to conduct research in this area wisely and ethically, and, just as importantly, how to integrate scientific findings into social life with intellectual and moral rigour.

FIGURE 18.3 A child chipping away at the Berlin Wall
Decades of Communism failed to mould human nature to eschew the freedoms and private property of the West. Political and ethical systems fail when they ignore the biological dimension to our common humanity.

Looking back over the 20th century, historians will probably see a struggle for the ownership of human nature. They will note how many scientists and intellectuals, sometimes for the best of motives, allowed it to be snatched away by the social sciences and cultural relativists. At last, evolutionary thinking has made a strong bridgehead in psychology; and as Blaise Pascal noted: 'If the earth moves, a decree from Rome cannot stop it.'

Summary

- Human biology has often been a battleground where competing ideologies have struggled to secure scientific support for political actions. The whole procedure has usually involved errors of fact and logic.
- The evolutionary approach to human nature is largely divorced from eugenic concerns and the attempts of some psychologists to search for inherited racial differences.
- The plasticity of human nature has often been asserted from an ideological belief that the good society is more consistent with a science showing that human behaviour is a product of culture than one showing it to be understandable in evolutionary terms. A Darwinian approach to human nature is not only consistent with the search for the good society, but is also in fact a prerequisite for it.

Key Words

- **Biological determinism**
- **Ideology**
- **Reductionism**

Further reading

Alocock, J. (2001) *The Triumph of Sociobiology*. Oxford, Oxford University Press.

Impassioned account of why sociobiology got it right.

Browne, K. (1998) *Women at Work*. London, Phoenix.

A book that will infuriate many. Browne gives an evolutionary analysis of why gender roles are as they are.

Gould, S. J. (1981) *The Mismeasure of Man*. London, Penguin.

A sensitive work that shows the folly of politically motivated science, in this case craniology and IQ measurements.

Moore, J. and Desmond, A. (1991) *Darwin*. London, Michael Joseph.

Lively, extremely well-written biography of Darwin. Contains little about the science of evolution but provides a penetrating analysis of the social context of Darwin's ideas.

Pinker, S. (2003) *The Blank Slate*. London, Penguin.

Exposé of the origins and follies of the blank slate view of the mind.

Ruse, M. (1993) *The Darwinian Paradigm*. London, Routledge.

Ruse has written numerous works on the philosophical and political dimensions of Darwinism. This collection of essays is a good introduction to his thoughts.

Segerstrale, U. C. O. (2000) *Defenders of the Truth: The Battle for Science in the Sociobiology Debate and Beyond*. Oxford, Oxford University Press.

Detailed account of some of the ideological battles associated with the rise of sociobiology and evolutionary psychology.

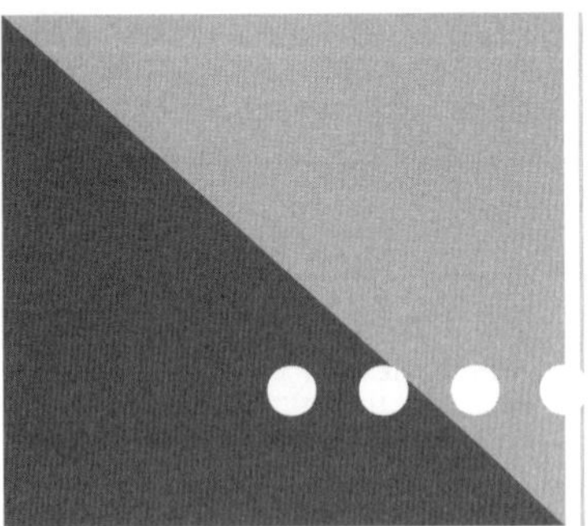

Glossary

Acheulean (also spelt Acheulian). A name given to a style of tool manufacture that began about 1.5 million years ago and lasted until about 100,000 years ago. A typical Acheulean tool is the bifacial hand axe, often associated with *Homo erectus*.

Adaptation. A feature of an organism that has been shaped by natural selection such that it enhances the fitness of its possessor. Adaptation can also refer to the process by which the differential survival of genes moulds a particular trait so that it now appears designed for some particular survival-related purpose.

Adaptive significance. The way in which the existence of a physical or behavioural feature can be related to the function it served and may continue to serve in helping an animal to survive and reproduce.

Affective. In psychology, this means pertaining to emotion as opposed to cognition. Often used in the context of affective disorders that are disturbances of mood or emotion.

Affinal. Individuals related through marriage and not through genetic descent.

Allele. A particular form or variant of a single gene that exists at a given locus on the genome of an individual. There may be many forms of alleles within a population of one species. A simple example is eye colour, alleles for which exist in two forms: blue or brown. Each gene occupies a specific position on the chromosome called the locus. At each locus, a human possesses two alleles, one inherited from the father and one from the mother. An allele is therefore a sequence of nucleotides on the DNA molecule.

Allometry. The relationship between the size of an organism as measured by, for example, length, volume or body mass, and the size of a single feature such as brain size. The relationship can often be expressed by mathematical allometric functions or graphs showing allometric lines.

Allopatric speciation. Sometimes known as geographic speciation. A process by which two populations of the same species become geographically isolated and in time form two groups that can no longer interbreed and hence two or more new species.

Altricial. Refers to offspring that are born in a state of helplessness, requiring constant care and attention. Humans are an altricial species and so are many types of birds.

Altruism. Self-sacrificing behaviour, whereby one individual sacrifices some component of its reproductive value for another individual. Self-sacrifice for a relative is termed kin altruism. What may appear as help or self-sacrifice may be enlightened self-interest, as in mutualism, or reciprocal altruism, where a favour is given in the expectation of some return eventually. Sacrifice at the level of the phenotype can be interpreted in terms of the 'self-interest' of the genes involved.

Amino acid. The molecular building blocks of proteins. There are 20 main amino acids in the proteins found in organisms. The particular sequence of amino acids in a protein determines its properties and is itself related to base sequences on DNA.

Amygdala. A small almond-shaped structure found in both hemispheres of the brain. The amygdalae form part of the limbic system and are involved in the processing of emotions.

Androgens. Male sex hormones such as testosterone.

Anisogamy. A situation where the gametes from sexually reproducing species are of different sizes. Males produce small, highly mobile gametes in large numbers; females produce fewer and larger eggs.

Archaic *Homo sapiens*. A term once given to a species of hominin that seemed intermediary between *Homo erectus* and *Homo sapiens*.

***Ardipithecus ramidus*.** A very early member of the hominin genus (and hence an ancestor of humans). Lived in Africa about 4.4 million years ago.

Asexual reproduction. Production of offspring without sexual fertilisation of eggs (see **parthenogenesis**).

Asperger's syndrome. Applied to people with autistic traits in the normal IQ range.

Asymmetry. A measure of the departure from symmetry of features that could in principle be symmetrical. Asymmetry is thought to be increased by poor conditions and stress.

Attachment. When applied to humans, this refers to the emotional bonding of one individual to another, usually a parent and child.

John Bowlby developed attachment theory in a series of seminal papers published in 1958.

Australopithecines. The earliest hominids that appeared on the African plains about four million years ago.

Autism. A disorder describing people who have difficulty with language and social relationships. One recent suggestion is that autistic individuals lack a complete theory of mind.

Autonomic nervous system (ANS). A division of the nervous system involving important processes such as heart rate, breathing, digestion and sweating that are not normally under conscious control.

Autosome. Any chromosome other than the sex chromosomes.

Baldwin effect. A process (named after James Baldwin) whereby behavioural changes initiated by the phenotype can ultimately bring about genetic change. One example is the evolution of lactose tolerance.

Base rate fallacy. An error of reasoning such that an individual fails to take account of the incidence (base rate) of some factor in the wider population.

Base. DNA consists of a phosphate–sugar backbone with attached nitrogenous bases. In DNA, a base can either be cytosine (C), guanine (G), adenine (A) or thymine (T). In RNA, uracil (U) is substituted for thymine. The precise sequence of bases on DNA serves as a set of instructions for building the cell.

Behaviourism. The school of psychology largely founded by Watson that suggests that observable behaviour should be the subject matter of psychology.

Biological determinism. A belief (often referred to pejoratively) that behaviour is caused by the biology of an individual such as physiology or genetics.

Bipolar depression. Sometimes called bipolar affective disorder. A clinical condition whereby the sufferer oscillates between periods of mania and depression.

Body mass index (BMI). A measurement of the relative weight of an individual taking into account height. It is calculated as weight/height^2.

Bounded rationality. A term that suggests people make decisions using limited and non-complete information; in contradistinction to classical rationality, which assumes that all factors are taken into account.

Broca's area. A part of the brain involved in speech production and language comprehension. The area is named after the 19th-physician Paul Broca.

Carrier. An individual who is heterozygous for an inherited trait (often a disorder), but does not display the symptoms of the disorder.

Categorical imperative. A concept central to the whole moral philosophy of Immanuel Kant. Kant thought such imperatives, by their very nature, to be rules we have to follow, irrespective of our desires and circumstances. To arrive at such imperatives, Kant devised a number of formulations. The most commonly encountered is the idea that we should act only according to maxims that can be willed to become universal laws without self-contradiction. Applying these formulations, Kant ended up with rules such as 'it is wrong to lie and steal'. The idea that it is permissible to steal, for example, is ruled out, since the very concept of stealing presupposes the existence of private property – if stealing were permitted, then the whole notion of private property would disappear; in other words, the statement 'it is permissible to steal' runs into a self-contradiction when universalised.

Central dogma. The idea that the information flow from DNA through RNA and to the proteins of cells that make up an individual is one way and irreversible.

Central nervous system (CNS). The largest part of the human nervous system. It consists of the brain and the spinal cord. It is fundamental to the control of behaviour.

Cerebral cortex. An area of the brain resembling a folded sheet of grey tissue that covers the rest of the brain. In humans, it is associated with 'higher functions' such as speech, language and reasoning.

Cerebral hemisphere. The brain is divided into two cerebral hemispheres. The right side tends to specialise in visuo-spatial functions and emotional processing; the left side is specialised for reasoning and language.

Chromosome. Structures in the nucleus of a cell that house DNA. Chromosomes contain DNA and proteins bound to it. Chromosomes become visible to the optical microscope during meiosis and mitosis.

Cladistics. A means of producing a phylogenetic classification. In cladistics, a group shares a more recent common ancestor than members of a different group.

Codon. Three nucleotides along DNA that specify one amino acid.

Coefficient of inbreeding (F). The probability that a person with two identical copies of a gene at a specific locus on their genome acquired both copies from a single ancestor.

Coefficient of relatedness (r). The r value between two individuals is the probability that an allele chosen at random from one individual will also be present in another individual. Can also be thought of as the proportion of the total genome present in one individual present in another as a result of common ancestry.

Corpus callosum. A bundle of nerve

fibres joining each cerebral hemisphere.

Crossing over. During meiosis, chromosomes in a diploid cell may exchange segments of DNA. It ensures a recombination of genetic material. The process results in highly variable gametes.

Cuckoldry. A cuckold is a married man whose wife (unknown to him) is mating with another partner. In biology, cuckoldry is the term given to a mating strategy, whereby the female obtains genetic material from one male but resources from another.

Culture. An evolutionary account of culture is one of the most difficult problems facing Darwinism. As yet there is no consensus on the definition of the term. In common parlance, culture is usually taken to mean the knowledge, belief systems, art, morals and customs acquired by individuals as members of a society. More recently, evolutionists have attempted to remove behaviour from this definition, leaving behind the view that culture comprises socially transmitted information.

Cystic fibrosis. A disease that occurs in people who possess two copies of a particular recessive allele, individuals with only one copy being carriers. The condition results in the secretion of abnormally thick mucus, increased sweat electrolytes and autonomic nervous system overactivity throughout the body.

Demographic transition. A phase in the economic development of cultures such that both mortality and fertility fall, often leading to families that produce children at or below the replacement rate (two children per couple). This phase has already been reached by most advanced industrialised nations such as those of Europe and the USA.

Diploid. A diploid cell is one that possesses two sets of chromosomes; one set is obtained from the mother and one from the father. Humans are diploid organisms. Compare with **haploid**.

Dizygotic. Refers to twins that result from two independent fertilisations of two eggs by two sperm and are therefore non-identical. See also **zygotic** and **monozygotic**.

DNA (deoxyribonucleic acid). The molecule that contains the information needed to build cells and control inheritance.

Domain specific mental modules. The idea that the human mind contains specialised modules for dealing with specific problems, as opposed to general intelligence capabilities or domain unspecific, problem-solving mechanisms.

Dominance hierarchy. The ranking of individuals in a social group. The hierarchy is usually established by aggression and conflict but once established it can be used to settle conflict issues without fighting.

Dominant allele. An allele that is fully expressed in the phenotype. An allele A is said to be dominant if the phenotype in the heterozygotic condition Aa is the same as that in the homozygotic condition AA. In this situation, allele a is said to be recessive. Alleles can be dominant, recessive or partly dominant.

DSM system. A means of classifying mental disorders according to the *Diagnostic and Statistical Manual of Mental Disorders* published by the American Psychiatric Association. It first appeared in 1952. The current edition is referred to as DSM-IV. A revision is planned for 2012. It is widely used by health professionals, although it frequently receives criticism about the scientific validity of the categories it uses.

Dual inheritance theory. A theory that proposes that both culture and biology evolve by separate but interacting processes. The two realms can interact and affect the behaviour of humans. In these models, genes can influence the course of cultural evolution but cultural change can also bring about genetic change (see the **Baldwin effect**). In the models of Boyd and Richerson, culture enables humans to adapt to rapidly changing environments by acting as a biologically useful source of information, customs and norms.

Dualism. The belief that humans have two aspects to their lives, physical and non-physical, matter and mind, or body and soul.

Empiricism. The belief that all knowledge is or should be based on experience.

Encephalisation. The enlargement of the brain (relative to body size) over the course of evolution.

Encephalisation quotient. A relative measure of brain size that compares actual brain size with expected brain size for a species of a given body weight.

Environment of evolutionary adaptedness (EEA). A concept highly favoured by evolutionary psychologists. That period in human evolution (over 30,000 years ago) during which the mind and body plans of humans were shaped and laid down by natural selection to solve survival problems operating then.

Environmentalism. In the science of behaviour, environmentalism is the belief that social and cultural factors are paramount in determining (human) behaviour.

Epistemology. The branch of philosophy that studies the nature, acquisition and limitations of knowledge and belief.

Ethology. A branch of biology dealing with the natural behaviour of animals.

Eugenics. A largely discredited set of beliefs that advocates selective breeding among humans to remove undesirable qualities and enhance the frequency of desirable genes.

Eusocial. A term used to describe highly socialised societies, such as found among ants and bees, where some individuals forego reproduction to assist the reproductive efforts of other members of the social group.

Evolutionary stable strategy (ESS). A set of rules of behaviour that once adopted by members of a group is resistant to replacement by an alternative strategy.

Exogamy. The practice of mating with someone outside the natal group, usually by a migration to the group of the other partner. Exogamy facilitates outbreeding. The opposite to exogamy is endogamy.

Extrapair copulation. Mating by a member of one sex with another outside what appears to be the stable pair bond in a supposed monogamous relationship.

Facial averageness. By photographic and computer-assisted techniques, the 'average' appearance of a set of faces can be created. Averaged faces tend to be regarded as attractive, a finding that has spawned a number of adaptive explanations, but they may not be optimally attractive.

Fitness. The term, crucially important to evolutionary theory, continues to elude a precise and universally agreed definition. Fitness can be measured by the number of offspring that an individual leaves relative to other individuals of the same species. Direct fitness (sometimes called Darwinian fitness) can be thought of as being proportional to the number of genes contributed to the next generation by production of direct offspring. Indirect fitness is proportional to the number of genes appearing in the next generation by an individual helping kin that also carry those genes. Inclusive fitness is the sum of direct and indirect fitness.

Fixed action pattern. An innate or instinctive pattern of behaviour that is highly stereotyped and stimulated by some simple stimulus.

Fluctuating asymmetry (FA). FA refers to bilateral characters (that is, one feature on each side of the body) for which the population mean of asymmetry (right measure minus left measure) is zero and the variability about the mean is nearly normal. In individuals, the degree and direction of asymmetry varies (hence fluctuating).

Founder effect. If a new group of organisms is formed from a few in a larger population, the new group is likely to have less genetic variation and have an average genotype that may be shifted in some direction even though the shift was not the result of natural selection.

Frequency dependent (selection or strategy). A process by which the strategy that is best for an organism and is naturally selected is dependent on the strategies pursued by other competing organisms.

Function. Sometimes used as shorthand for the adaptive value of a behavioural trait. The word has experienced an ambiguous usage in the human sciences. In the early years of the 20th century, a school of functionalism grew up in psychology and sociology including William James (1842–1910) and John Dewey (1859–1952). Eventually, the evolutionary adaptive significance of behaviours to individuals was lost within this movement. In psychology, functionalism became concerned with how individuals became adapted or adjusted in their own lives rather than how traits that had been selected over time became manifest. In sociology and anthropology, functionalists looked at how current behaviours of social practices contributed to the stability of the current order. Bronislaw Malinowski (1884–1942), in his pioneering study of kinship among the Trobriand islanders, for example, tried to show how kinship served the social order as a whole rather than the genetic fitness of individuals. The functional approach also used analogies between organs of the body and parts of a social system. The situation is further confused by a modern school of functionalism, a merger of cognitive science and artificial intelligence, which uses the word in terms analogous to mathematical functions where the mind is interpreted in terms of computer-like inputs and outputs.

Game theory. A mathematical approach to establishing what behaviour is fitness maximising by taking into account the payoffs of particular strategies in the light of how other members of the group behave.

Gamete. A sex cell. Gametes are said to be haploid in that they only contain one copy of any chromosome. A gamete can be an egg or a sperm.

Gene. A unit of hereditary information made up of specific nucleotide sequences in DNA.

Gene pool. The entire set of alleles present in a population.

Genetic drift. A change in the frequency of alleles in a population due to chance alone (as opposed to selection).

Genome. The entire set of genes carried by an organism.

Genomic imprinting. A mechanism where cells express either the maternal or paternal allele of a gene but not both.

Genotype. The term can be used in two senses: loosely, as the genetic constitution of an individual, or as the types of allele found at a locus on the genome.

Genus. In classification, the genus is the taxonomic category (taxon) above the level of a species but below that of a family. Hominids (*Homo sapiens*, *Homo erectus* and so on), for example, belong to the genus *Homo*.

Good genes. An approach to sexual selection that suggests individuals choose mates according to the fitness potential of their genome.

Grandparental solicitude. The tendency of grandparents to invest energy and resources in the upbringing and welfare of their grandchildren.

Green beard effect. A term used by Richard Dawkins to describe the largely theoretical possibility that individuals may grow and display some phenotypic marker (the 'green beard') to indicate to others of the same species the presence of common genes. The idea is that such markers may allow individuals to direct their altruistic efforts more effectively.

Grooming. Ostensibly the cleansing of skin, fur or feathers of an animal by itself or, more significantly, by another of the same species. The function of grooming may lie deeper than that of simple hygiene and could be an indication of the formation of alliances and the resolution of conflict.

Group selection. Selection that operates between groups rather than individuals. The notion was attacked and shunned in the 1970s but may be making a comeback in studies on human evolution.

Haemophilia. A sex-linked genetic disorder expressed in human males and characterised by excessive bleeding following injury.

Hamilton's rule. A formula which predicts that individuals will perform altruistic acts to their relatives so long as the genetic benefit exceeds the costs (taking into account the degree of relatedness between the donor and the recipient).

Handicap. Features of an organism that seem at first sight to have a negative impact on fitness. The handicap principle was an idea advanced by Zahavi in 1975 to account for what appear to be maladaptive features, or handicaps, of an organism, such as the long train of a peacock or the huge antlers of deer. Zahavi suggested that these features were honest advertisements of genetic quality, since an animal, usually the male, must be strong in order to grow and bear such a burden.

Haploid. A condition where cells only contain one copy of any chromosome. Some organisms are haploid but in humans only the gametes are haploid.

Herbivore. An organism that eats only plants.

Heritability. The extent to which a difference in a character between individuals in a population is due to inherited differences in the genotype. It is often expressed as a number between zero and 1 that refers to the proportion of variance in a character in a population that is ascribable to inherited genetic differences.

Heterozygous. A state describing individuals that have two different alleles (for example Aa) at a given genetic locus.

Heuristic. A fast and computationally inexpensive means of solving a problem such as a simple rule of thumb procedure.

Hominid. Any member of the family Hominidae, which includes humans, chimpanzees, gorillas and all their extinct ancestors. The term is open to variable interpretation, depending on which classification system is accepted.

Hominin. A member of the subfamily Homininae. Includes extant humans and extinct precursor species following the breakaway from the chimpanzee lineage some seven million years ago. The term is open to variable interpretation depending on which classification system is accepted.

Homology. A feature present in several species and present in their common ancestor.

Homozygous. A state describing individuals that have identical alleles for a given trait at a given locus (AA or aa).

Honest signal. A signal that reliably communicates the quality of an individual in terms of its fitness.

Humanistic. A term applied to a branch of psychology and psychotherapy, both of which grew out of dissatisfaction with behaviourist and psychoanalytic approaches. Humanistic psychology emphasises the holistic nature of humans, human consciousness, and the meanings, values and intentions adopted and sought by human beings. A leading exponent of this approach was Abraham Maslow (1908–70).

Hypothalamus. A structure deep within the forebrain that controls a range of autonomic functions, for example hunger, thirst, body temperature, wakefulness and sleep. Linked to the hormonal system.

Hypothesis. A conjecture set forward as a provisional explanation for a phenomenon.

Ideology. A set of beliefs, values and assumptions that structure understanding and inform policy decisions. An ideology usually supports the political interests of particular groups.

Imprinted genes. Genes that express their functions only when inherited from the mother or father but not both. See **genomic imprinting**.

Inbreeding. Breeding between two individuals who are close genetic relatives. The fitness of offspring produced is sometimes reduced by inbreeding depression.

Inbreeding depression. The phenomenon that descendants of individuals who mate with close relatives tend to be lower in fitness. It can be brought about by a lowered genetic variety or by the fact that an individual may be homozygous for recessive and deleterious alleles.

Incest. Sexual activity between two individuals who are too closely related to be allowed to marry according to local social rules and customs.

Inclusive fitness. Fitness that is measured by the number of copies of one's genes that appear in current or subsequent generations in offspring and non-offspring. Kin-directed altruism, for example, is said to increase inclusive fitness. See **fitness.**

Infanticide. The deliberate killing of an infant shortly after its birth.

Innate releasing mechanism. A hypothetical mechanism or model devised to help explain how an innate response is triggered by a sign stimulus.

Intensionality. A term used to express degrees of self-awareness and the awareness of the mental states of others. First-order intensionality is self-consciousness (I know); second order is the awareness that others may have self-awareness (I know that you know); third order is the knowledge that others may be aware of your thoughts (I know that you know that I know) and so on.

Intersexual selection. A form of selection driven by the exercise of choice by one sex for specific characteristics in a mating partner of the opposite sex.

Intrasexual selection. Competition between members of the same sex (typically males) for access to the opposite sex.

Isogamy. A condition where the gametes from each partner engaged in sexual reproduction are of equal size. Isogamy is common among protists and algae. Higher plants and all animals display anisogamy.

James–Lange theory. A theory that emotional states are responses to the physical reaction of the body to external events, rather than the idea that physiological changes follow emotional states.

K selected. Species that produce small numbers of slow-growing and long-lived offspring that require considerable parental investment. Humans are a K-selected species.

Kin discrimination. The ability of an animal to react differently to other individuals depending on their degree of genetic relatedness.

Kin selection. The suggestion that altruism can evolve because altruistic behaviour favours increases in the gene frequency of the genes responsible. A situation where altruism to relatives is favoured and spreads.

Kluver–Bucy syndrome. A behavioural disorder resulting from a malfunction in both temporal lobes. Named after Heinrich Kluver and Paul Bucy who performed experiments on rhesus monkeys. The condition is linked to a disruption of the normal operation of the amygdalae. In humans, it is characterised by socially inappropriate behaviours, hypersexuality, bulimia and flattened emotions.

Lamarckian inheritance. Shorthand for the inheritance of acquired characteristics. A mechanism (among others) proposed by the French evolutionist Lamarck, whereby it is supposed that characters or modifications to characters acquired by the phenotype can be passed on to offspring through genetic inheritance. The mechanism is now rejected as without foundation.

Lamarckism. Doctrines associated with Jean Baptiste Lamarck. Usually taken to refer to his mechanism of inheritance.

Lateralisation. A term referring to the fact that the human brain is separated by a deep longitudinal fissure into two cerebral hemispheres. The hemispheres appear superficially similar but there is evidence that different functions are carried out in each hemisphere.

Lek. A display site where (usually) males display and females choose a mate.

Life history theory (LHT). A theory that considers how organisms allocate resources to various life processes, such as mating, growth and repair across the lifespan.

Limbic system. Part of the brain located beneath the cerebral cortex. Contains a number of structures, such as the amygdalae and the hippocampus, involved in learning, memory and emotion.

Lineage. A sequence showing how species are descended from one another.

Linkage disequilibrium. A condition where genes travel together in the process of inheritance. The result is that the frequency of a group of linked genes is different to what would be expected from random recombination.

Locus. The particular site where a particular gene is found on DNA.

Machiavellian intelligence. The idea that one of the prime factors leading to the growth of intelligence in primates was the need for an individual to manipulate its social world through a mixture of cunning, deceit and political alliances.

Maternal–fetal conflict. The theory that the fetus and the mother that carries it may have different genetic interests and so engage in a biological tug-of-war over the level of resources passed to the fetus as it develops.

Matrilocal. A social system where women tend to remain in their local group after marriage, while men tend to leave their family group to join that of their wives.

Meiosis. A type of cell division whereby diploid cells produce haploid gametes. The double set of chromosomes in a diploid cell is thereby reduced to a single set in the resulting gametes. Crossing over and recombination occur during meiosis.

Meme. An activity or unit of information that can be passed on by imitation.

Memetics. The scientific study of memes.

Menarche. The first menstrual period or bleeding experienced by a girl. In the USA, the average age of menarche is now about 12 years. It is a key feature of puberty.

Mendelian genetics. A mode of inheritance first described by Gregor Mendel. In diploid species showing Mendelian inheritance, genes are passed to offspring in the same form as they were inherited from the previous generation. At each locus, an individual has two genes (haploid), one inherited from its mother and one from its father.

Menopause. The cessation of monthly ovulation experienced by women usually in their late forties. Women are infertile after the menopause. The adaptive significance of the menopause is probably related to the risks of childbirth and the need to care for existing children.

MHC (Major histocompatibility complex). A set of genes coding for antigens responsible for the rejection of genetically different tissue. The antigens are known as histocompatibility (that is, tissue compatibility) antigens. There are at least 30 histocompatibility gene loci. The genes of the MHC are subject to simple Mendelian inheritance and are co-dominantly expressed; that is, alleles from both parents are equally expressed. Each cell, therefore, in any offspring has maternal and paternal MHC molecules on its surface. The human MHC is found on the short arm of chromosome 6.

Mirror neurons. Neurons directly observed in some primates and thought to exist in humans that fire both when an animal performs an action and when the animal observes that same action performed in a conspecific. Their discovery in the mid-1990s has excited a great deal of interest, since they may lie at the heart of imitation, learning and theory of mind capabilities.

Mitochondrial DNA. A short section of DNA found only in the mitochondria of cells and in humans only inherited from the mother.

Mitochondrial Eve. The name given to the woman who is the most recent matrilineal common ancestor to all humans. The mitochondrial DNA in all humans can be traced back to her.

Mitosis. Part of the process where one cell divides into another identical one with the same number of chromosomes as the original. The growth of an organism is through mitosis. Sexual reproduction requires meiosis.

Modularity. The belief that the human brain consists of a number of discrete problem-solving units, or areas of specialised function (such as theory of mind and face recognition) shaped by natural selection to solve specific problems. See **domain specific mental modules.**

Monism. The belief that the universe is composed of one basic substance. In the context of human behaviour (materialistic), monism would assert that behaviour has physical causes.

Monogamy. The mating of a single male with a single female. In annual monogamy, the bond is dissolved each year and a fresh partner found. In perennial monogamy, the bond lasts for the reproductive life of the organisms.

Monozygotic. Refers to (identical) twins that result from the separation of a single fertilised egg or zygote and are therefore identical in terms of their DNA. See also **zygotic** and **dizygotic**.

Morgan's canon. The assertion that explanations for the behaviour of animals should be kept as simple as possible. Higher mental activities should not be attributed if lower ones will suffice.

Multiregional model. A theory that proposes that *Homo sapiens* arose in various parts of the world from pre-existing ancient stock that had already reached there (such as *Homo erectus*).

Mutation. In modern genetics, a mutation is a heritable change in the base sequences in the DNA of a genome. Most mutations are deleterious.

Mutualism. A symbiotic relationship between individuals of two species such that both partners benefit.

Nativism. The idea that children are born with a basic understanding of some features of the world (such as number, space and time).

Naturalism. The belief that all phenomena can be explained by scientific laws and principles without recourse to the supernatural or entities outside the remit of science.

Naturalistic fallacy. A fallacy of reasoning that supposes it is possible to derive 'ought'-type statements from factual statements. It was introduced by the British philosopher G. E. Moore and then (confusingly) applied by others to the 'is/ought' problem identified by Hume.

Neocortex. The outer layers (about 2–4 mm in thickness) covering the entire cerebral cortex. A late addition to the brain in evolutionary terms. Involved with higher brain functions such as sensory perception, conscious thought and language. A highly folded structure, it comprises about 75 per cent of the brain volume. Contains about 20 per cent of all the neurons of the brain.

Neoteny. The retention of typically juvenile features into adulthood.

Neuron. A cell found in the nervous system (brain and spinal cord).

Oestrus. A period of heightened interest in copulation experienced by female animals.

Ontogeny (coming into being). The development and growth of an organism from a fertilised cell, through the fetus to an adult.

Oogenesis. The formation of female sex cells (ova). See also **spermatogenesis**.

Operant conditioning. A type of learning whereby an action (operant) is carried out more frequently if it is rewarded.

Operational sex ratio. The ratio of sexually receptive males to females in a particular area or over a particular time.

Optimality. The idea that the behaviour of animals will be that which is ideally suited to bring maximum gain for minimum cost or, more precisely, where gain minus cost is maximised. The assumption that all behaviour must be optimal can be misleading.

Orbitofrontal cortex. The portion of the outer layer of the brain that lies just above the eye sockets. Involved in directing socially appropriate responses to emotional events.

Order. A unit of classification above the family and below the class. Humans belong to the order of primates.

Out of Africa hypothesis. The idea that the species *Homo sapiens* arose once in Africa and that members of this species migrated out of Africa about 150,000 years ago to eventually populate the globe.

Ovarian cycle. A cycle of events controlled by hormones in the ovary of mammals that leads to ovulation.

Ovulation. The release of a female sex cell or ovum (plural ova) from an ovarian follicle.

Paradigm. A cluster of ideas and theories that are consistent and form part of a distinct way of understanding the world. Evolutionary psychology can be regarded as a paradigm.

Parasite. An organism in a symbiotic relationship with another (host) such that the parasite gains in fitness at the expense of the host.

Parental investment. Actions that increase the survival chances of one set of offspring but at the expense of the parent procuring more offspring.

Parent–offspring conflict. Disputes between parents and their children over the allocation of resources. Such conflicts are the focus of attention by evolutionary psychologists, in that they may have a basis in the different reproductive interests of the parties concerned. Typically, the best reproductive interests of young parents may be served by allocating resources to future children – a decision that may not be optimum from the point of view of an existing child.

Parthenogenesis. Asexual reproduction. Production of offspring by virgin birth.

Paternity certainty. An expression of the confidence that a male has that he is really the genetic father of any offspring.

Pathogen. A disease-causing organism.

Patrilocal. A social system whereby women leave their family group upon marriage to live with the natal group of their husband (contrasts with **matrilocal**).

Phenotype. The characteristics of an organism as they have been shaped by both the genotype and environmental influences.

Phenotypic plasticity. A debated term. The average value for a phenotypic character between two individuals with identical genomes or two populations with similar gene frequencies may be different because of different environmental influences. Phenotypic plasticity is often used to refer to irreversible change and phenotypic flexibility to refer to reversible change.

Phobia. An irrational and persistent fear of specific objects or situations. The symptoms of the disorder include an unreasonable desire to avoid the object or situation. When the phobia becomes acute, the sufferer is often diagnosed as suffering from an anxiety disorder. Examples of phobias include arachnophobia (fear of snakes), xenophobia (fear of strangers) and agoraphobia (fear of open spaces outside home).

Phylogeny. The branching history of a species showing its relationship to ancestral species. A phylogenetic tree can be used to infer relationships between existing species and their evolutionary history.

Phylum. A taxonomic category above class but below kingdom. Humans belong to the order of primates, the class of mammals and the phylum Chordata.

Placenta. The organ that provides nutrients and oxygen to an embryo.

Pleiotropy. The ability of a single gene to have a number of different effects on the phenotype.

Polyandry. A type of mating system such that a single female mates with more than one male in a given breeding season.

Polygamy. The mating of one member of one sex with more than one member of the other sex. The two varieties are polyandry and polygyny.

Polygynandry. A mating system in which males and females mate with each other in a promiscuous way.

Polygyny. A mating system whereby a single male mates with more than one female in a given breeding season.

Polymorphism. A condition where a population may have more than one allele (at significant frequencies) at a particular locus. The different forms of the alleles may all be adaptive in their own right.

Population. A group of individuals, usually geographically localised and interacting, that all belong to the same species.

Positivism. The belief that science represents the positive and more advanced state of knowledge. Also the supposition that only objects that can be experienced directly form part of the proper process of scientific enquiry. Positivism seeks to repudiate metaphysics.

Precocial. A term applied to species whose young are born relatively mature and mobile from the moment of birth. The opposite of precocial is altricial.

Preparedness theory. A theory that humans are born with biases and propensities to quickly learn specific responses. In psychology, it is often used to explain the basis for some phobias.

Prisoners' dilemma. A model of the problem faced by individuals in knowing how to act to serve their own best interests in the context of uncertain knowledge of how others, who may be rivals, will also act. The analogy is drawn with two prisoners who if they cooperate with each other will receive a lighter sentence.

Prosimians. A diverse group of small-bodied primates. Most species live in Africa. Best-known examples include bushbabies and lemurs. Thought to be more 'primitive' than the simians.

Proximate cause. In behavioural terms, the immediate mechanism or stimuli that initiates or triggers a pattern of behaviour.

Pseudopathology. A category of illness that may result from the operation of a response or strategy that was once adaptive but is no longer so in current environments.

Psychodynamic. A view of the human mind that studies behaviour in terms of the action of motivational forces, desires and drives (hence 'dynamic'). Its application to therapy often draws upon the work of Sigmund Freud, who saw early childhood experiences and repressed emotions as the key to adult behaviour.

Psychopathy. A term closely related to sociopathy. A psychopath is someone who lacks empathy, is impulsive, self-centred and manipulative. The condition does not feature in DSM-IV but is closely related to antisocial personality disorder. The relationship and differences between psychopathy, sociopathy and antisocial personality disorder are the subjects of an ongoing debate.

Recessive allele. See **dominant allele.**

Reciprocity. The donation of assistance in the expectation that favours will be returned at some future date.

Recombination. An event whereby chromosomes cross over and exchange genetic material during meiosis. Recombination tends to break up genes that are linked together.

Reductionism. The attempt to explain a wide range of phenomena by employing a smaller range of concepts and principles that are more basic.

Relativism. The view that there is no privileged knowledge or belief system that can claim supremacy over another. A relativist would argue that the validity of belief systems cannot be decided by universal criteria.

Reptilian brain. A section of the brain labelled by McLean to describe the oldest (in evolutionary terms) part of the brain that deals with basic survival and reproductive functions.

Reverse engineering. A way of thinking about the consequences of evolution. It starts with a contemporary understanding of the function of adaptive behavioural or physiological attributes and tries to infer what problems our ancestors faced to give rise to these adaptive solutions.

RNA (ribonucleic acid). Molecules that act as intermediaries as the hereditary code in DNA is converted into proteins. Messenger RNA (mRNA) is the molecule that carries information from DNA in the nucleus to the sites in the cytoplasm where protein synthesis takes place.

r-selected species. Species that produce many short-lived and quickly maturing offspring that need little parental investment.

Schizophrenia. A psychological disorder characterised by hallucinations, delusional behaviour and disordered thoughts. Thought to be present in some form in about 1 per cent of any population.

Selection. The differential survival of organisms (or genes) in a population as a result of some selective force. Selectionist thinking is the approach that looks for how features of organisms can be interpreted as the result of years of selection acting upon ancestral populations. Directional selection tends to favour an extreme measure of the natural variability in a population and the average measure will gradually move in this direction. Disruptive selection tends to favour more than one phenotype. Stabilising selection tends to favour the mean values currently found and ensures that variation is reduced. See also **sexual selection.**

Sex chromosomes. Chromosomes that determine if an individual is male or female. In humans, a female possesses two X chromosomes and a male one X and one Y chromosome.

Sex ratio. The ratio of males to females at any one time. At birth for humans the ratio is about 1.06.

Sexual crypsis. Sometimes known as cryptic oestrus or concealed ovulation. The suggestion that in human females the time of oestrus is concealed.

Sexual dimorphism. Differences in morphology, physiology or behaviour between the sexes in a single species.

Sexual selection. Selection that takes place as a result of mating behaviour. Intrasexual selection occurs as a result of competition between members of the same sex. Intersexual selection occurs as a result of choices made by one sex for features of another.

Sexual strategies theory (SST). A theory developed by evolutionary psychologists David Buss and David Schmitt, but ultimately deriving from the work of Charles Darwin and Robert Trivers, that proposes that men and women have a variety of context-dependent but evolutionary derived strategies that they enlist when pursuing a mate. One distinctive but controversial feature of this theory is that men and women are supposed to have different strategies for long- and short-term partners.

Siblicide. The killing of a sibling (brother or sister) by a brother or sister.

Smoke detector principle. A term popularised by Nesse to argue that it may be natural to demonstrate higher levels of anxiety than necessary, since the cost of reacting to a false alarm is much less than not reacting to a rarer true alarm.

Social Darwinism. A rather loose collection of ideas and philosophies that took root in the USA and Europe during and after the last quarter of the 19th century. Social Spencerism would be a more accurate description, since, like the British philosopher Herbert Spencer, social Darwinists advocated minimum state control in the economy and free competition.

Speciation. The creation of a new species through the splitting of an existing species into two or more new species.

Species. Using the biological species concept, a species is a set of organisms that possess similar inherited characteristics and crucially have the potential to interbreed to produce fertile offspring.

Sperm competition. Competition between sperm from two or more males that is present in the reproductive tract of the female.

Spermatogenesis. The formation of male sex cells (sperm).

Strategy. A pattern of behaviour or rules guiding behaviour shaped by natural selection to increase the fitness of an animal. A strategy is not taken to be a set of conscious decisions in non-human animals. A strategy may also be flexible, in that different biotic and abiotic factors may trigger different forms of optimising behaviour.

Symbiosis. A relationship between organisms of different species that live in close association and interact. See **mutualism** and **parasite**.

Symmetry. A state where the physical features of an organism on one side of its body are matched in size and shape by those on the other. Symmetry may be an indication of physiological health since stress increases asymmetry. Animals may, therefore, use symmetry as an honest signal of fitness. See **fluctuating asymmetry**.

Taxon (plural taxa). A named group in classification. It may be a species, a genus, a family, an order or other category.

Taxonomy. The theory and practice of classifying organisms.

Teleology. The belief that nature has purposes, that events are shaped by intended outcomes.

Testis (plural testes). Male sex organ that produces sperm and associated hormones.

Theory of mind. The suggestion that an important component of the mental and emotional life of humans and some other primates is the ability to be self-aware and to appreciate that others also have awareness. The theory of mind also implies that an individual is capable of distinguishing between the real intentional or emotional states of others and those that may be feigned. Having a theory of mind is an essential component of Machiavellian intelligence.

Tit for tat. A strategy that can be played by individuals stuck in a prisoners' dilemma. The strategy is to never be the first to defect but to follow defection or cheating by the other player with a retaliation. The tit for tat strategy, on average, works well against a variety of other strategies.

Tragedy of the commons. A term introduced by Garrett Hardin (1968) to describe a system where the logical strategy of individual self-aggrandisement leads to a failure of resources when this strategy is pursued by all.

Transcendentalism. The belief that some knowledge and truths lie outside the plane of ordinary human existence.

Ultimate causation. The explanation for the behaviour of an organism that reveals its adaptive value.

Unipolar depression. Sometimes called major depression or major depressive disorder. It may be a single episode of a severely depressed mood that lasts over two weeks, or recurrent periods that may occur over many years. If the patient experiences episodes of mania or elevated mood, then a diagnosis of bipolar disorder is usually made.

Utilitarianism. A branch of moral philosophy that suggests that the moral value of an action is related to its contribution to the overall utility. When faced with an ethical dilemma, utilitarians would identify the best course of action as that which delivers the maximum happiness (or pleasure) to the greatest number.

Waist to hip ratio. The circumference of the waist divided by that of the hips.

Wernicke's area. A part of the brain involved in language, named after the 19th-century Polish physician

Karl Wernicke. Like Broca's area, it lies in the left hemisphere of the brain. It is connected to Broca's area by the arculate fasciculus.

Westermarck effect. Named after the Finnish anthropologist Edvard Westermarck, a mechanism whereby males and females co-socialised from an early age tend not to develop sexual feelings for each other.

Zygote. A fertilised cell formed by the fusion of two gametes. See **monozygotic** and **dizygotic**.

Further reading

Keller, E. F. and Lloyd, E. A. (eds) *Keywords in Evolutionary Biology*. Cambridge, MA, Harvard University Press.

Many of the terms in the glossary are uncontentious, but some have disputed meanings. The reader is referred to the above title which considers 37 key terms in some depth.

Bibliography

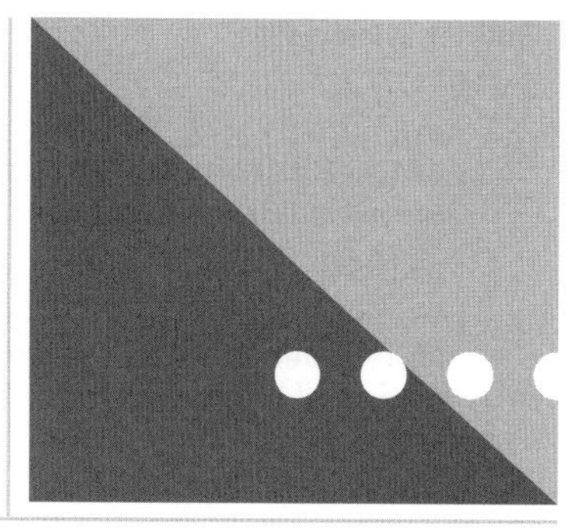

Aiello, L. C. and Wheeler, P. (1995) 'The expensive-tissue hypothesis: the brain and the digestive system in human primate evolution.' *Current Anthropology* **36**: 199–221.

Aiello, L. C., Bates, N. and Joffe, T. (2001) In defence of the expensive tissue hypothesis, in D. Falk and K. R. Gibson (eds) *Evolutionary Anatomy of the Primate Cerebral Cortex*. Cambridge, Cambridge University Press.

Alexander, R. D. (1979) *Darwinism and Human Affairs*. Seattle, University of Washington Press.

Alexander, R. D. (1987) *The Biology of Moral Systems*. New York, Aldine de Gruyter.

Alexander, R. D. (1990) *How Did Humans Evolve? Reflections on the Uniquely Unique Species*. Ann Arbor, MI, University of Michigan.

Alexander, R. D. and Noonan, K. (1979) Concealment of ovulation, parental care and human social evolution, in N. I. A. Chagon and W. Irons (eds) *Evolutionary Biology and Human Social Behaviour: An Anthropological Perspective*. North Scituate, MA, Duxbury Press.

Alexander, R. D., Hoogland, J. H. et al. (1979) Sexual dimorphisms and breeding systems in pinnipeds, ungulates, primates and humans, in N. I. A. Chagon and W. Irons (eds) *Evolutionary Biology and Human Social Behaviour: An Anthropological Perspective*. North Scituate, MA, Duxbury Press.

Allen, E., Alper, J. et al. (1975) Letter to the editor. *New York Review of Books* **22**: 43–4.

Allen, J. S. and Sarich, V. M. (1988) 'Schizophrenia in evolutionary perspective.' *Perspectives in Biology and Medicine* **32**: 132–53.

Alonzo, S. H. and Sinervo, B. (2001) 'Mate choice games, context-dependent good genes, and genetic cycles in the side-blotched lizard, *Uta stansburiana*.' *Behavioral Ecology and Sociobiology* **49**: 176–86.

American Psychiatric Association (1994) *Diagnostic and Statistical Manual of Mental Disorders* (DSM) (4th edn). Arlington, VA, American Psychiatric Association.

American Society of Plastic Surgeons (2007) Breast Surgery Statistics 1996. Retrieved August 2007 from http://www.plasticsurgery.org/media/statistics/1996statistics.cfm.

Andersson, M. (1982) 'Female choice selects for extreme tail length in a widowbird.' *Nature* **299**: 818–20.

Andersson, M. (1994) *Sexual Selection*. Princeton, Princeton University Press.

Andreasen, N. C. (1987) 'Creativity and mental illness prevalence rates in writers and their 1st degree relatives.' *American Journal of Psychiatry* **144**: 1288–92.

Apicella, C. L. and Marlowe, F. W. (2004) 'Perceived mate fidelity and paternal investment resemblance predicts men's investment in children.' *Evolution and Human Behaviour* **25**(6): 371–9.

Arbib, M. A. (2005) 'From monkey-like action recognition to human language: an evolutionary framework for neurolinguistics.' *Behavioral and Brain Sciences* **28**: 105–24.

Archer, J. (1992) *Ethology and Human Development*. Hemel Hempstead, Harvester Wheatsheaf.

Arechiga, J., Prado, C. et al. (2001) 'Women in transition: menopause and body composition in different populations.' *Collective Anthropology* **25**: 443–8.

Arnhart, L. (1998) *Darwinian Natural Right: The Biological Ethics of Human Nature*. New York, State University of New York Press.

Arnhart, L. (2005) Incest taboo as Darwinian natural right, in A. P. Wolf and W. H. Durham (eds) *Inbreeding, Incest and the Incest Taboo*. Stanford, CA, Stanford University Press.

Ashworth, T. (1980) *Trench Warfare, 1914–1918: The Live and Let Live System*. New York, Holmes & Meier.

Axelrod, R. (1984) *The Evolution of Cooperation*. New York, Basic Books.

Axelrod, R. and Hamilton, W. D. (1981) 'The evolution of cooperation.' *Science* **211**: 1390–6.

Badcock, C. (1991) *Evolution and Individual Behaviour: An Introduction to Human Sociobiology*. Oxford, Blackwell.

Badcock, C. (2000) *Evolutionary Psychology: A Critical Introduction*. Cambridge, Polity Press.

Badcock, C. (2004) Mentalism and mechanism: the twin modes of human cognition, in C. Crawford and C. Salmon (eds) *Evolutionary Psychology, Public Policy and Personal Decisions*. Mahwah, NJ, Lawrence Erlbaum.

Badcock, C. and Crepsi, B. (2006) 'Imbalanced genomic imprinting in brain development: an evolutionary basis for the aetiology of autism.' *Journal of Evolutionary Biology* **19**(4): 1007–32.

Bailey, A., LeCouteur, A., et al. (1995) 'Autism as a strongly genetic disorder: evidence from a British twin study.' *Psychological Medicine* **25**: 63–77.

Bailey, J. M. (1998) Can behaviour genetics contribute to evolutionary behavioural science?, in C. Crawford and D. L. Krebs (eds) *Handbook of Evolutionary Psychology*. Mahwah, NJ, Lawrence Erlbaum.

Baillargeon, R. (1987) 'Object permanence in 3½- and 4½-month old infants.' *Developmental Psychology* **23**: 655–64.

Baillargeon, R. and DeVos, J. (1991) 'Object permanence in young infants: further evidence.' *Development* **62**: 1227–46.

Baillargeon, R., Spelke, E. et al. (1985) 'Object permanence in five-month-old infants.' *Cognition* **20**: 191–208.

Baker, R. R. (1996) *Sperm Wars*. London, Fourth Estate.

Baker, R. R. and Bellis, M. A. (1989) 'Number of sperm in human ejaculates varies in accordance with sperm competition theory.' *Animal Behaviour* **37**: 867–9.

Baker, R. R. and Bellis, M. A. (1995) *Human Sperm Competition*. London, Chapman & Hall.

Barash, D. (1982) *Sociobiology and Behaviour*. New York, Elsevier.

Barash, D. (2003) *The Survival Game*. New York, Times Books.

Barber, N. (1995) 'The evolutionary psychology of physical attractiveness: sexual selection and human morphology.' *Ethology and Sociobiology* **16**: 395–424.

Bard, K. A. (2005) Emotions in chimpanzee infants: the value of a comparative developmental approach to understand the evolutionary bases of emotion, in J. Nadd and D. Muir (eds) *Emotional Development: Recent Research Advances*. Oxford, Oxford University Press.

Barkow, J. H. (1989) *Darwin, Sex, and Status*. Toronto, University of Toronto Press.

Barkow, J. H. (1992) Beneath new culture is old psychology: gossip and social stratification, in J. H. Barkow, L. Cosmides and J. Tooby (eds) *The Adapted Mind*. Oxford, Oxford University Press.

Barkow, J. H. and Burley, N. (1980) 'Human fertility, evolutionary biology, and the demographic transition.' *Ethology and Sociobiology* **1**: 163–80.

Barkow, J. H., Cosmides, L. and Tooby, J. (eds) (1992) *The Adapted Mind: Evolutionary Psychology and the Generation of Culture*. Oxford, Oxford University Press.

Baron-Cohen, S. (1997) How to build a baby that can read minds: cognitive mechanisms in mindreading, in S. Baron-Cohen (ed.) *The Maladapted Mind*. Hove, Psychology Press.

Baron-Cohen, S. (1999) The evolution of a theory of mind, in M. C. Corballis and S. E. G. Lea (eds) *The Descent of Mind*. Oxford, Oxford University Press.

Baron-Cohen, S. (2006) 'The hyper-systematizing, assortative mating theory of autism.' *Progress in Neuro-Psychopharmacology and Biological Psychiatry* **30**: 865–72.

Baron-Cohen, S., Leslie, A. M and Frith, U. (1985) 'Does the autistic child have a theory of mind?' *Cognition* **21**: 37–46.

Barrett, L. and Dunbar, R. I. M. (1994) 'Not now dear, I'm busy.' *New Scientist* **142**: 30–4

Bateson, P. (1980) 'Optimal outbreeding and the development of sexual preferences in Japanese quail.' *Zeitschrift fur Tierpsychologie* **53**: 321–49.

Bateson, P. (1982) 'Preferences for cousins in Japanese quail.' *Nature* **295**: 236–7.

Beach, F. A. (1950) 'The snark was a boojum.' *American Psychologist* **5**: 115–24.

Beck, A. T. (1976) *Cognitive Therapy and the Emotional Disorders*. New York, New American Library.

Beck, A. T. and Clark, D. A. (1988) 'Anxiety and depression: an information processing perspective.' *Anxiety Research* **1**: 23–36.

Beer, J. M. and Horn, J. M. (2000) 'The influence of rearing order on personality development within two adoption cohorts.' *Journal of Personality* **68**: 769–819.

Bell, G. (1982) *The Masterpiece of Nature*. London, Croom Helm.

Belsky, J., Steinberg, L. et al. (1991) 'Childhood experience, interpersonal development, and reproductive strategy: an evolutionary theory of socialization.' *Child Development* **62**: 647–70.

Benshoof, L. and Thornhill, R. (1979) 'The evolution

of monogamy and concealed ovulation in humans.' *Journal of Social and Biological Structures* **2**: 95–106.

Bernstein, H., Hopf, F. A. et al. (1989) The evolution of sex: DNA repair hypothesis, in C. Vogel, A. Rasa, E. and E. Voland (eds) *The Sociobiology of Sexual and Reproductive Strategies*. London, Chapman & Hall.

Bertelsen, A. Harvald, B. and Hauge, M. (1977) 'A Danish twin study of manic-depressive disorders.' *British Journal of Psychiatry* **130**: 330–51.

Betzig, L. (1982) 'Despotism and differential reproduction.' *Ethology and Sociobiology* **3**: 209–21.

Betzig, L. (1986) *Despotism and Differential Reproduction: A Darwinian View of History*. Hawthorne, NY, Aldine.

Betzig, L. (1992) 'Roman polygyny.' *Ethology and Sociobiology* **13**: 309–49.

Betzig, L. (1998) Not whether to count babies, but which, in C. Crawford and D. L. Krebs (eds) *Handbook of Evolutionary Psychology*. Mahwah, NJ, Lawrence Erlbaum.

Bevc, I. and Silverman, I. (1993) 'Early proximity and intimacy between siblings and incestuous behaviour: a test of the Westermarck hypothesis.' *Ethology and Sociobiology* **14**: 171–81.

Bevc, I. and Silverman, I. (2000) 'Early separation and sibling incest: a test of the revised Westermarck theory.' *Evolution and Human Behaviour* **21**(3): 163–85.

Birkhead, T. R. and Moller, A. P. (1992) *Sperm Competition in Birds: Evolutionary Causes and Consequences*. London, Academic Press.

Bittles, A. H. (1995) The influence of consanguineous marriage on reproductive behaviour in India and Pakistan, in C. G. N. Mascie-Taylor and A. J. Boyce (eds) *Mating Patterns*. Cambridge, Cambridge University Press.

Bittles, A. H. (2004) Genetic aspects of inbreeding and incest, in A. P. Wolf and W. H. Durham (eds) *Inbreeding, Incest and the Incest Taboo*. Stanford, CA, Stanford University Press.

Bittles, A. H. and Makov, U. (1988) Inbreeding in human populations: assessment of the costs, in C. G. N. Mascie-Taylor and A. J. Boyce (eds) *Mating Patterns*. Cambridge, Cambridge University Press.

Bittles, A. H. and Neel, J. V. (1994) 'The costs of human inbreeding and their implications for variation at the DNA level.' *Nature Genetics* **8**: 117–21.

Blackmore, S. (1999) *The Meme Machine*. Oxford, Oxford University Press.

Bloom, G. and Sherman, P. W. (2005) 'Dairying barriers affect the distribution of lactose malabsorption.' *Evolution and Human Behaviour* **26**(4): 301–13.

Blurton Jones, N. and Sibly, R. M. (1978) Testing adaptiveness of culturally determined behaviour: do bushmen women maximize their reproductive success in spacing births widely and foraging seldom?, in N. Blurton Jones and V. Reynolds (eds) *Human Behaviour and Adaptation*. London, Taylor and Francis.

Blurton Jones, N., Hawkes, K. et al. (1999) Some current ideas about the evolution of the human life history, in P. C. Lee (ed.) *Comparative Primate Socioecology*. Cambridge, Cambridge University Press.

Boas, F. (1911) *The Mind of Primitive Man*. New York, Macmillan.

Boaz, N. T. and Almquist, A. J. (1997) *Biological Anthropology*. Englewood Cliffs, NJ, Prentice Hall.

Bogin, B. (1999) *Patterns of Human Growth*. Cambridge, Cambridge University Press.

Bonner, J. T. (1988) *The Evolution of Culture in Animals*. Princeton, NJ, Princeton University Press.

Bowlby, J. (1969) *Attachment Theory, Separation Anxiety and Mourning*. New York, Basic Books.

Bowlby, J. (1980) *Attachment and Loss*, vol. 3, *Loss, Sadness and Depression*. New York, Basic Books.

Bowler, P. (1982) *Evolution: The History of an Idea*. Berkeley, University of California Press.

Boyd, R. and Richerson, P. J. (1995) 'Why does culture increase human adaptability?' *Ethology and Sociobiology* **16**: 125–43.

Boyer, P. (1994) *The Naturalness of Religious Ideas: A Cognitive Theory of Religion*. Berkeley, University of California Press.

Breland, K. and Breland, M. (1961) 'The misbehaviour of organisms.' *American Psychologist* **16**: 681–4.

Brown, A. (1999) *Darwin Wars: How Stupid Genes became Selfish Gods*. London, Simon & Schuster.

Brown, P., Sutikna, T. et al. (2004) 'A new small-bodied hominin from the Late Pleistocene of Flores, Indonesia.' *Nature* **431**: 1055–61.

BSSRS (1976) 'The new synthesis is an old story.' *New Scientist* 13 May: 346–8.

Buckle, L., Gallup, G. G. et al. (1996) 'Marriage as a reproductive contract: patterns of marriage, divorce, and remarriage.' *Ethology and Sociobiology* **17**: 363–77.

Bulik, C. M., Sullivan, P. F. et al. (1999) 'Fertility and reproduction in women with anorexia nervosa: a controlled study.' *Journal of Clinical Psychiatry* **60**: 130–5.

Burch, R. L. and Gallup, J. G. G. (2000) 'Perceptions of paternal resemblance predict family violence.' *Evolution and Human Behaviour* **21**(6): 429–37.

Burkhardt, R. W. (1983) The development of an evolutionary ethology, in D. S. Bendall (ed.) *Evolution from Molecules to Men*. Cambridge, Cambridge University Press.

Burley, N. (1979) 'The evolution of concealed ovulation.' *American Naturalist* **114**: 835–58.

Burton, R. (1973) 'Folk theory and the incest taboo.' *Ethos* **1**: 504–16.

Buss, D. and Barnes, M. (1986) 'Preferences in human mate selection.' *Journal of Personality and Social Psychology* **50**: 559–70.

Buss, D. M. (1989) 'Sex differences in human mate preferences: evolutionary hypotheses tested in 37 cultures.' *Behavioural and Brain Sciences* **12**: 1–49.

Buss, D. M. (1994) *The Evolution of Desire*. New York, HarperCollins.

Buss, D. M. (1999) *Evolutionary Psychology*. Needham Heights, Allyn & Bacon.

Buss, D. M., Larsen, R. J. et al. (1992) 'Sex differences in jealousy: evolution, physiology and psychology.' *Psychological Science* 3(4): 251–5.

Byne, W., Kemether, E., Jones, L. et al. (1999) The neurochemistry of schizophrenia, in D. S. Charney, E. J. Nestler and B. S. Bunney (eds) *The Neurobiology of Mental Illness*. Oxford, Oxford University Press.

Byrne, R. W. (1995) *The Thinking Ape*. Oxford, Oxford University Press.

Byrne, R. W. and Whiten A. (1988) *Machiavellian Intelligence: Social Expertise and the Evolution of Intellect in Monkeys, Apes and Humans*. Oxford, Clarendon Press.

Cadoret, R. J., Troughton, E. et al. (1990) 'Genetic and environmental factors in adoptee antisocial personality.' *European Archives of Psychiatry and Neurological Sciences* **239**: 231–40.

Call, J. and Tomasello, M. (1999) 'A nonverbal false belief task: the performance of children and great apes.' *Child Development* **70**: 381–95.

Calvin, W. H. (1982) *The Throwing Madonna: Essays on the Brain*. New York, McGraw-Hill.

Calvin, W. H. (1993) The unitary hypothesis: a common neural circuitry for novel manipulations, language, plan-ahead, and throwing?, in K. R. Gibson and T. Ingold (eds) *Tools, Language and Cognition in Human Evolution*. Cambridge, Cambridge University Press.

Campbell, B. G., Loy, J. D. and Cruz-Uribe, K. (2006) *Humankind Emerging* (9th edn). Boston, MA, Allyn & Bacon.

Campbell, D. (1974) Evolutionary epistemology, in P. A. Schilpp (ed.) *The Philosophy of Karl Popper*. LaSalle, IL, Open Court.

Cann, R. L., Stoneking, M. et al. (1987) 'Mitochondrial DNA and human evolution.' *Nature* **325**: 31–6.

Cartmill, M. (1974) 'Rethinking primate origins.' *Science* **184**: 436–42.

Cartwright, J. H. (2001) *Evolutionary Explanations of Human Behaviour*. Hove, Routledge.

Cartwright, J. H. and Baker, B. (2005) *Literature and Science*. Santa Barbara, ABC-Clio.

Cashden, E. (1989) Hunters and gatherers: economic behaviour in bands, in S. Plattner (ed.) *Economic Anthropology*. Stanford, CA, Stanford University Press.

Casscells, W., Schoenberger, A. et al. (1978) 'Interpretation by physicians of clinical laboratory results.' *New England Journal of Medicine* **299**: 999–1000.

Chagon, N. I. A. and Irons, W. (eds) *Evolutionary Biology and Human Social Behaviour: An Anthropological Perspective*. North Scituate, MA, Duxbury Press.

Chapman, T. (1997) The epidemiology of fears and phobias, in G. C. L. Davey (ed.) *Phobias: A Handbook of Theory, Research and Treatment*. Chichester, John Wiley & Sons.

Chisholm, J. S. (1996) 'The evolutionary ecology of attachment organization.' *Human Nature* **7**: 1–38.

Chomsky, N. (1959) 'A review of B. F. Skinner's verbal behavior.' *Language* **35**(1): 26–58.

Clutton-Brock, T. H. (1994) The costs of sex, in R. V. Short and E. Balaban (eds) *The Differences Between the Sexes*. Cambridge, Cambridge University Press.

Clutton-Brock, T. H. and Vincent, A. C. J. (1991) 'Sexual selection and the potential reproductive rates of males and females.' *Nature* **351**: 58–60.

Comer, R. (2001) *Abnormal Psychology* (4th edn). New York, Worth.

Conover, M. R. (1990) 'Evolution of a balanced sex ratio by frequency-dependent selection in fishes.' *Science* **250**: 1556–8.

Cook, E. W., Hodes, R. L. et al. (1986) 'Preparedness and phobia: effects of stimulus content on human visceral conditioning.' *Abnormal Psychology* **95**: 195–207.

Cook, M. and Mineka, S. (1987) 'Second order conditioning and overshadowing in the observational conditioning of fear in monkeys.' *Behaviour Research and Therapy* **25**: 349–64.

Cornell, R. E., Palmer, C. et al. (2005) 'Introductory psychology texts as a view of sociobiology/ evolutionary psychology's role in psychology.' *Human Nature* **3**: 355–74.

Cosmides, L. and Tooby, J. (1992) Cognitive adaptations for social exchange, in J. H. Barkow, L. Cosmides and J. Tooby (eds) *The Adapted Mind*. Oxford, Oxford University Press.

Cosmides, L. and Tooby, J. (1996) 'Are humans good intuitive statisticians after all? Rethinking some conclusions from the literature on judgement under uncertainty.' *Cognition* **58**: 1–73.

Cosmides, L. Tooby, J. and Barkow, J. (1992) Introduction: evolutionary psychology and conceptual integration, in J. H. Barkow, L. Cosmides and J. Tooby (eds) *The Adapted Mind*. Oxford, Oxford University Press.

Craddock, N. and Jones, I. (1999) 'Genetics of bipolar disorder.' *Journal of Medical Genetics* **36**: 585–94.

Crawford, C. (1993) 'The future of socio-biology: counting babies or studying proximity mechanisms.' *Trends in Evolution and Ecology* **8**: 184–7.

Crawford, C. (1998a) The theory of evolution in the study of human behaviour, in C. Crawford and D. L. Krebs (eds) *Handbook of Evolutionary Psychology*. Mahwah, NJ, Lawrence Erlbaum.

Crawford, C. (1998b) Environments and adaptations: then and now, in C. Crawford and D. L. Krebs (eds) *Handbook of Evolutionary Psychology*. Mahwah, NJ, Lawrence Erlbaum.

Creel, S. and Creel, N. M. (1995) 'Communal hunting and pack size in African wild dogs *Lycaon pictus*.' *Animal Behaviour* **50**: 1325–39.

Cronin, H. (1991) *The Ant and the Peacock*. Cambridge, Cambridge University Press.

Crow, J. F. (1997) 'The high spontaneous mutation rate. Is it a health risk?' *Proceedings of the National Academy of Sciences* **94**: 8380–6.

Crow, T. J. (2000) 'Schizophrenia as the price that *Homo sapiens* pays for language: a resolution of the central paradox in the origin of the species.' *Brain Research Reviews* **34**: 118–29.

Crusio, W. E. (1995) 'The sociopathy of sociobiology.' *Behaviour and Brain Sciences* **18**(3): 523.

Curry, O. (2005) Morality as natural history: an adaptationist account of ethics. PhD thesis. London School of Economics.

Cziko, G. (1995) *Without Miracles*. Cambridge, MA, MIT Press.

Daly, M. and Wilson, M. I. (1988a) *Homicide*. New York, Aldine De Gruyter.

Daly, M. and Wilson, M. I. (1988b) 'Evolutionary social psychology and family homicide.' *Science* **242**: 519–24.

Daly, M. and Wilson, M. I. (1999) 'Human evolutionary psychology and animal behaviour.' *Animal Behaviour* **57**: 509–19.

Daly, M., Wilson, M. I. et al. (1982) 'Male sexual jealousy.' *Ethology and Sociobiology* **3**: 11–27.

Damasio, A. (2000) *The Feeling of What Happens*. London, Vintage.

Darwin, C. ([1871/4] 1981) *The Descent of Man and Selection in Relation to Sex*. Princeton, NJ, Princeton University Press.

Darwin, C. (1858) Letter to Charles Lyell 18th June 1858, in F. Burkhardt and S. Smith (eds) (1991) *The Correspondence of Charles Darwin*, vol. 7. Cambridge, Cambridge University Press.

Darwin, C. (1859a) Letter to Alfred Russel Wallace, 13 November 1859, in F. Burkhardt and S. Smith (eds) (1991) *The Correspondence of Charles Darwin* vol. 7. Cambridge, Cambridge University Press.

Darwin, C. (1859b) *On the Origin of Species by Natural Selection*. London, John Murray.

Darwin, C. (1860) Letter to Charles Lyell, 25/2/1860, in F. Burkhardt and S. Smith (1991) *The Correspondence of Charles Darwin*, vol. 8. Cambridge, Cambridge University Press.

Darwin, C. (1872) *The Expression of the Emotions in Man and Animals*. London, John Murray.

Darwin, C. (1881) Letter to W. Graham, in F. Darwin. *Autobiography of Charles Darwin*. London, Watts, pp. 152–4.

Darwin, C. (1883) *The Descent of Man and Selection in Relation to Sex*. New York, Appleton.

Darwin, F. (1902) *The Life of Charles Darwin*. London, John Murray.

Darwin, F. (ed.) (1887) *The Life and Letters of Charles Darwin*, 3 vols. London, John Murray.

Davies, N. B. (1992) *Dunnock Behaviour and Social Evolution*. Oxford, Oxford University Press.

Davies, P., Fetzer, H. et al. (1995) 'Logical reasoning and domain specificity: a critique of the social exchange theory of reasoning.' *Biology and Philosophy* **10**(1): 1–37.

Davies, W., Isles, A. R. et al. (2004) 'Imprinted gene expression in the brain.' *Neuroscience and Behavioural Reviews* **20**: 1–10.

Davis, J. N. (1997) 'Birth order, sibship size, and status in modern Canada.' *Human Nature* **8**: 205–30.

Dawkins, M. S. (1986) *Unravelling Animal Behaviour*. Harlow, Longman.

Dawkins, R. ([1976] 1989) *The Selfish Gene*. Oxford, Oxford University Press.

Dawkins, R. (1982) *The Extended Phenotype: The Long Reach of the Gene*. Oxford, W. H. Freeman.

Dawkins, R. (1986) *The Blind Watchmaker*. London, Longman.

Dawkins, R. (1995) *River out of Eden: A Darwinian View of Life*. London, HarperCollins.

Dawkins, R. (1998) *Unweaving the Rainbow*. London, Penguin.

Dawkins, R. (2006) *The God Delusion*. London, Bantam.

De Sousa Campos, L., Otta, E. et al. (2002) 'Sex differences in mate selection strategies: content analyses and responses to personal advertisements in Brazil.' *Evolution and Human Behaviour* **23**(5): 395–406.

De Waal, F. (1997) 'The chimpanzee's service economy: food for grooming.' *Evolution and Human Behaviour* **18**: 375–86.

Deacon, T. (1997) *The Symbolic Species*. London, Penguin.

Deacon, T. W. (1992) The human brain, in S. Jones, R. Martin and D. Pilbeam (eds) *The Cambridge Encyclopedia of Human Evolution*. Cambridge, Cambridge University Press.

Degler, C. N. (1991) *In Search of Human Nature: The Decline and Revival of Darwinism in American Social Thought*. Oxford, Oxford University Press.

DeKay, W. T. (1995) Grandparental investment and the uncertainty of kinship. Seventh annual meeting of the Human Behavior and Evolution Society, Santa Barbara, CA.

Delprato, D. J. (1989) 'Hereditary determinants of fears and phobias: a critical review.' *Behaviour Therapy* **11**: 79–103.

Dennett, D. C. (1995) *Darwin's Dangerous Idea*. New York, Simon & Schuster.

Desmond, A. and Moore, J. (1991) *Darwin*. London, Michael Joseph.

DeSouza, M. J. and Metzger, D. A. (1991) 'Reproductive dysfunction in amenorrheic athletes and anorexic patients.' *Medicine and Science in Sports and Exercise* **56**: 20–7.

Dewsbury, D. A. (1984) *Comparative Psychology in the Twentieth Century*. Stroudsburg, PA, Hutchinson Ross.

Dewsbury, D. A. (1990) *Contemporary Issues in Comparative Psychology*. Sunderland, MA, Sinauer.

Diamond, J. (1991) *The Rise and Fall of the Third Chimpanzee*. London, Vintage.

Doolittle, W. and Sapienza, C. (1980) 'Selfish genes, the phenotype paradigm and genome evolution.' *Nature* **284**: 601–3.

Dugatin, L. A. (1996) 'Interface between culturally based preferences and genetic preferences: female choice in *Poecilla reticulata*.' *Proceedings of the National Academy of Sciences* **93**(7): 2770–3.

Dunbar, R. (1980) 'Determinants and evolutionary consequences of dominance among female Gelada baboons.' *Behavioural Ecology and Sociobiology* **7**: 253–65.

Dunbar, R. I. M. (1993) 'Coeveolution of neocortical size, group size and language in humans.' *Behavioural and Brain Sciences* **16**: 681–735.

Dunbar, R. I. M. (1995a) 'Are you lonesome tonight.' *New Scientist* **145**(1964): 12–16.

Dunbar, R. I. M. (1995b) 'Neocortext size and group size in primates: a test of the hypothesis.' *Journal of Human Evolution* **28**: 287–96.

Dunbar, R. I. M. (1996a) *Grooming, Gossip and the Evolution of Language*. London, Faber and Faber.

Dunbar, R. I. M. (1996b) Determinants of group size in primates: a general model, in W. G. Runciman, J. Maynard Smith and R. I. M. Dunbar (eds) *Evolution of Social Behaviour in Primates and Man*. Oxford, Oxford University Press.

Dunbar, R. I. M. and Aiello, L. C. (1993) 'Neocortex size, group size, and the evolution of language.' *Current Anthropology* **34**(2): 184–93.

Dunbar, R. I. M., Duncan, N. D. C. et al. (1994) 'Size and structure of freely forming conversational groups.' *Human Nature* **6**(1): 67–78.

Dunham, C., Myers, F. et al. (1991) *Mamatoto: A Celebration of Birth*. London, Virago.

Durant, J. (1981) Innate character in animals and man: a perspective on the origins of ethology, in C. Webster (ed.) *Biology, Medicine and Society 1840–1940*. Cambridge, University of Cambridge Press.

Durant, J. R. (1986) 'The making of ethology: the

association for the study of animal behaviour, 1936–1986.' *Animal Behaviour* **34**: 1601–16.

Durham, W. (1991) *Coevolution: Genes, Culture and Human Diversity*. Stanford, Stanford University Press.

Easton, J. A., Schipper, L. D. et al. (2006) 'Why the adaptationist perspective must be considered: the example of morbid jealousy.' *Behavioral and Brain Sciences* **26**: 411–12.

Eaton, B. S., Eaton, S. B. III et al. (1999) Paleolithic nutrition revisited, in E. Trevathan and J. J. McKenna (eds) *Evolutionary Medicine*. New York, Oxford University Press.

Eaton, S. B. and Cordain, L. (1997) Evolutionary aspects of diet: old genes, new fuels, in A. P. Simopoulos (ed.) *Nutrition and Fitness: Evolutionary Aspects, Children's Health, Programs and Policies*. Washington, DC, S. Karger.

Eibl-Eibesfeldt, I. (1970) *Ethology: The Biology of Behaviour*. New York, Holt, Rinehart & Winston.

Eibl-Eibesfeldt, I. (1989) *Human Ethology*. New York, Aldine de Gruyter.

Einon, D. (1998) 'How many children can one man have?' *Evolution and Human Behaviour* **19**: 413–26.

Ekman, P. (1973) Cross cultural studies of facial expressions, in P. Ekman (ed.) *Darwin and Facial Expression*. London, Academic Press.

Ekman, P., Friesen, W. V. et al. (1987) 'Universals and cultural differences in the judgments of facial expressions of emotion.' *Journal of Personality and Social Psychology* **53**(4): 712–17.

Elliot-Smith, G. (1927) *The Evolution of Man*. London, Oxford University Press.

Ellis, A. (1962) *Reason and Emotion in Psychotherapy*. Secaucus, NJ, Prentice Hall.

Ellis, B. J. and Graber, J. (2000) 'Psychological antecedents of variation in girls' pubertal timing: maternal depression, stepfather presence, and marital and family stress.' *Child Development* **71**: 485–501.

Emlen, S. T. (1995) 'An evolutionary theory of the family.' *Proceedings of the National Academy of Sciences* **92**: 8092–9.

Emlen, S. T. and Wrege, P. H. (1988) 'The role of kinship in helping decisions among white-fronted bee-eaters.' *Behavioural Ecology and Sociobiology* **23**: 305–15.

Engel, G. and Schmale, A. (1972) Conservation-withdrawal: a primary regulatory process for organismic homeostasis, in R. Porter and J. Night (eds) *Physiology, Emotion and Psychiatric Illness*. Amsterdam, Associated Scientific Publishers.

Erickson, C. J. and Zenone, P. G. (1976) 'Courtship differences in male ring doves: avoidance of cuckoldry?' *Science* **192**: 1353–4.

Euler, H. A. and Weitzel, B. (1996) 'Discriminating grandparental solicitude as reproductive strategy.' *Human Nature* **7**: 39–59.

Eyre-Walker, A. and Keightley, P. D. (1999) 'High genomic deleterious mutation rates in hominids.' *Nature* **397**: 344–7.

Eysenk, M. (2004) *Psychology: An International Perspective*. Hove, Psychology Press.

Falk, D. (1983) 'Cerebral cortices of East African early hominids.' *Science* **221**: 1072–74.

Fehr, E. and Gaechter, S. (2002) 'Altruistic punishment in humans.' *Nature* **10**: 137–40.

Feinman, S. and Gill, G. W. (1977) 'Females' response to male beardedness.' *Perceptual and Motor Skills* **94**: 533–4.

Feldman, M. W. and Laland, K. N. (1996) 'Gene–culture coevolutionary theory.' *Trends in Evolution and Ecology* **11**: 453–7.

Fendt, M. and Fanselow, M. S. (1999) 'The neuroanatomical and neurochemical basis of conditioned fear.' *Neuroscience and Biobehavioural Reviews* **23**: 743–60.

Fessler, D. M. T. and Navarrete, C. D. (2004) 'Third-party attitudes towards sibling incest: evidence for Westermarck's hypotheses.' *Evolution and Human Behaviour* **25**: 277–94.

Field, T., Sandburg, S., Garcia, R. et al. (1985) 'Pregnancy problems, postpartum depression, and early mother–infant interactions.' *Developmental Psychology* **21**(1): 1152–6.

Finch, C. E. and Sapolsky, R. M. (1999) 'The evolution of Alzheimer disease, the reproductive schedule, and apoE isoforms.' *Neurobiology and Aging* **20**: 407–28.

Finkielkraut, A. (1988) *The Undoing of Thought*. London, Claridge Press.

Fisher, R. A. (1930) *The Genetical Theory of Natural Selection*. Oxford, Clarendon Press.

Flaxman, S. M. and Sherman, P. W. (2000) 'Morning sickness: a mechanism for protecting mother and embryo.' *Quarterly Review of Biology* **75**: 113–48.

Flinn, M. (1988) 'Step-parent/step-offspring interactions in a Caribbean village.' *Ethology and Sociobiology* **9**: 335–69.

Fodor, J. (1983) *The Modularity of Mind*. Cambridge, MA, MIT Press.

Foley, R. A. (1987) *Another Unique Species*. Harlow, Longman.

Foley, R. A. (1989) The evolution of hominid social behaviour, in V. Standen and R. A. Foley (eds) *Comparative Socioecology*. Oxford, Blackwell Scientific.

Foley, R. A. (1992) Studying human evolution by analogy, in S. Jones, R. Martin and D. Pilbeam (eds) *The Cambridge Encyclopedia of Human Evolution*. Cambridge, Cambridge University Press.

Foley, R. A. (1996) An evolutionary and chronological framework for human social behaviour, in W. G. Runciman, J. Maynard Smith and R. Dunbar (eds) *Evolution of Social Behaviour Patterns in Primates and Man*. Oxford, Oxford University Press.

Foley, R. A. (1997) *Humans before Humanity*. Oxford, Blackwell.

Foley, R. A. (2002) Hominid evolution, in M. Pagel (ed.) *The Encyclopaedia of Evolution*. Oxford, Oxford University Press.

Folley, B. S. and Park, S. (2005) 'Verbal creativity and schizotypal personality in relation to prefrontal hemispheric laterality: a behavioural and near-infrared optical imaging study.' *Schizophrenia Research* **80**(2–3): 271–82.

Foster, K. R., Wenseleers, T. et al. (2001) 'Spite: Hamilton's unproven theory.' *Annales Zoologici Fennici* **38**: 229–38.

Frank, R. (1988) *Passions within Reason: The Strategic Role of the Emotions*. New York, W. W. Norton.

Frazer, J. G. (1910) *Totemism and Exogamy: A Treatise on Certain Early Forms of Superstition and Society*. London, Macmillan – now Palgrave Macmillan.

Freeman, D. (1996) *Margaret Mead and the Heretic*. London, Penguin.

Freud, S. ([1913] 1950) *Totem and Taboo*. New York, W.W. Norton.

Friday, A. E. (1992) Human evolution: the evidence from DNA sequencing, in S. Jones, R. Martin and D. Pilbeam (eds) *The Cambridge Encyclopedia of Human Evolution*. Cambridge, Cambridge University Press.

Gadgil, M. and Bossert, W. H. (1970) 'Life historical consequences of natural selection.' *American Naturalist* **104**: 1–24.

Gallese, V., Fadiga, L. et al. (1996) 'Action recognition in the premotor cortex.' *Brain* **119**: 593–609.

Gallese, V., Keysers, C. and Rizzolatti, G. (2004) 'A unifying view of the basis of social cognition.' *Trends in Cognitive Sciences* **8**(9): 396–403.

Gallup, G. G. (1970) 'Chimpanzees: self-recognition.' *Science* **167**: 86–7.

Gangstead, S. W. and Buss, D. M. (1993) 'Pathogen prevalence and human mate preference.' *Ethology and Sociobiology* **14**: 89–96.

Gangestad, S. W. and Simpson, J. A. (2000) 'The evolution of human mating: trade-offs and strategic pluralism.' *Behavioural and Brain Sciences* **23**: 573–87.

Gangestad, S. W. and Thornhill, R. (1994) 'Facial attractiveness, developmental stability and fluctuating asymmetry.' *Ethology and Sociobiology* **15**: 73–85.

Garber, P. A. (1989) 'Role of spatial memory in primate foraging patterns: *Saguinus mystax* and *Saguinus fuscicollis*.' *American Journal of Primatology* **19**: 203–16.

Garcia, J. and Koelling, R. (1966) 'Relation cue to consequences in avoidance learning.' *Psychonomic Science* **4**: 123–4.

Garcia, J., Ervin, F. and Koelling, R. (1966) 'Learning with prolonged delay of reinforcement.' *Psychonomic Science* **5**: 121–2.

Gaulin, S. J. C. and Hoffman, H. A. (1988) Evolution and development of sex differences in spatial ability, in L. Betzig, M. Borgerhoff Mulder and P. Turke (eds) *Human Reproductive Behaviour: A Darwinian Perspective*. Cambridge, Cambridge University Press.

Geary, D. C. (1998) *Male, Female: The Evolution of Human Sex Differences*. Washington DC, American Psychological Association.

Gentilucci, M. (2003) 'Grasp observation influences speech production.' *European Journal of Neuroscience* **17**: 179–84.

Ghiselin, M. T. (1974) *The Economy of Nature and the Evolution of Sex*. Berkeley, University of California Press.

Gigerenzer, G. (2000) *Adaptive Thinking*. New York, Oxford University Press.

Gigerenzer, G. (2001) The adaptive toolbox, in G. Gigerenzer and R. Selten (eds) *Bounded Rationality: The Adaptive Toolbox*. Cambridge, MA, MIT Press.

Gigerenzer, G. and Hug, K. (1992) 'Domain-specific reasoning: social contracts, cheating and perspective change.' *Cognition* **43**: 127–71.

Gigerenzer, G. and Selten, R. (eds) (2001) *Bounded Rationality: The Adaptive Toolbox*. Cambridge, MA, MIT Press.

Gigerenzer, G., Hoofrage, U. et al. (1991) 'Probabilistic mental models: a Brunswikian theory of confidence.' *Psychological Review* **98**: 506–28.

Gigerenzer, G., Todd, P. M. and ABC Research Group (eds) (1999) *Simple Heuristics that Make us Smart.* New York, Oxford University Press.

Gil-da-Costa, R., Martin, A. et al. (2006) 'Species-specific calls activate homologs of Broca's and Wernicke's areas in the macaque.' *Nature Neuroscience* **9**: 1064–70.

Goldman, L. S., Genel, M., et al. (1998) 'Diagnosis and treatment of attention-deficit/hyperactivity disorder in children and adolescents.' *Journal of the American Medical Association* **279**: 1100–7.

Goldstein, D. G. and Gigerenzer, G. (1999) The recognition heuristic, in G. Gigerenzer, P. M. Todd and ABC Research Group (eds) *Simple Heuristics that Make us Smart.* New York, Oxford University Press.

Gonzaga, G. C., Haselton, M. G., Smurda, J., Davies, M. and Poore, J. (2007) 'The role of love and desire in suppressing the thought of attractive romantic alternatives.' Manuscript under review at *Evolution and Human Behavior.*

Gopnik, M., Dalalakis, J., Fukuda, S. E., Fukuda, S. and Kehayia, E. (1996) 'Genetic language impairment: unruly grammars.' *Proceedings of the British Academy* **88**: 223–49.

Gottesman, I. I. (1991) *Schizophrenia Genesis: The Origins of Madness.* New York, W. H. Freeman.

Gottlieb, G. (1971) *Development of Species Identification in Birds.* Chicago, University of Chicago Press.

Gould, R. G. (2000) 'How many children could Moulay Ismail have had?' *Evolution and Human Behaviour* **21**(4): 295.

Gould, S. J. (1978) *Ever Since Darwin.* London, Burnett.

Gould, S. J. (1981) *The Mismeasure of Man.* London, Penguin.

Gould, S. J. (1998) 'Let's leave Darwin out of it.' *New York Times* 29 May.

Gould, S. J. and Lewontin, R. C. (1979) 'The spandrels of San Marco and the Panglossian paradigm: a critique of the adaptionist programme.' *Proceedings of the Royal Society of London* **205**: 581–98.

Grammer, K. (1992) 'Variations on a theme: age dependent mate selection in humans.' *Behaviour and Brain Sciences* **15**: 100–2.

Grammer, K. and Thornhill, R. (1994) 'Human facial attractiveness and sexual selection: the roles of averageness and symmetry.' *Journal of Comparative Psychology* **108**: 233–42.

Gravlee, C. C., Bernard, H. R. et al. (2003) 'Heredity, environment, and cranial form: a reanalysis of Boas's immigrant data.' *American Anthropologist* **105**(1): 123–36.

Gray, P. B., Kahlenberg, S. M. et al. (2002) 'Marriage and fatherhood are associated with lower testosterone in males.' *Evolution and Human Behaviour* **23**: 193–201.

Greenlees, I. A. and McGrew, W. C. (1994) 'Sex and age differences in preferences and tactics of mate attraction: analysis of published advertisements.' *Ethology and Sociobiology* **15**: 59–72.

Griffin, R. and Baron-Cohen, S. (2002) The intentional stance: developmental and neurocognitive perspectives, in A. Brook and D. Ross (eds) *Daniel Dennett.* Cambridge, Cambridge University Press.

Griffiths, P. E. (2001) From adaptive heuristic to phylogenetic perspective: some lessons from the evolutionary psychology of emotion, in H. R. Holcolm III (ed.) *Conceptual Challenges in Evolutionary Psychology: Innovative Research Strategies.* Norwell, MA, Kluwer Academic.

Grosberg, R. K. and Quinn, J. F. (1986) 'The genetic control and consequences of kin recognition by the larvae of a colonial marine invertebrate.' *Nature* **322**: 456–9.

Gruber, H. E. (1974) *Darwin on Man: A Psychological Study of Scientific Creativity*, together with Darwin's early and unpublished notebooks transcribed and annotated by Paul H. Barrett. London, Wildwood House.

Guttentag, M. and Secord, P. (1983) *Too Many Women?* Beverly Hills, CA, Sage.

Hagen, E. H. (1999) 'The functions of post-partum depression.' *Evolution and Human Behaviour* **20**: 325–59.

Hagen, E. H. (2002) 'Depression as bargaining: the case post-partum.' *Evolution and Human Behaviour* **23**: 323–36.

Hagen, E. H. (2003) The bargaining model of depression, in P. Hammerstein (ed.) *Genetic and Cultural Evolution of Cooperation.* Cambridge, MA, MIT Press.

Haig, D. (1993) 'Genetic conflicts in human pregnancy.' *Quarterly Review of Biology* **68**(4): 495–532.

Haig, D. (1997) The social gene, in J. R. Krebs and N. B. Davies (eds) *Behavioural Ecology.* Oxford, Blackwell Science.

Haig, D. (2000) 'The kinship theory of genomic imprinting.' *Annual Review of Ecology and Systematics* **31**: 9–32.

Halberstadt, J. and Rhodes, G. (2000) 'The attractiveness of nonface averages: implications for an evolutionary explanation of the attractiveness of average faces.' *Psychological Science* **11**(4): 285–9.

Hamilton, W. D. (1964) 'The genetical evolution of social behaviour, 1.' *Journal of Theoretical Biology* **7**: 1–16.

Hamilton, W. D. (1970) 'Selfish and spiteful behaviour in an evolutionary model.' *Nature* **228**: 1218-20.

Harcourt, A. H. (1991) 'Sperm competition and the evolution of non-fertilizing sperm in mammals.' *Evolution* **45**(2): 314–28.

Harcourt, A. H., Harvey, P. H. et al. (1981) 'Testis weight, body weight and breeding system in primates.' *Nature* **293**: 55–7.

Hardin, G. (1968) 'The tragedy of the commons.' *Science* **162**: 1243–8.

Hare, R. D. (1993) *Without Conscience: The Disturbing World of the Psychopaths Among Us*. New York, Simon & Schuster.

Harpending, H. and Sobus, J. (1987) 'Sociopathy as an adaptation.' *Ethology and Sociobiology* **8**: 63–72.

Harris, C. R. (2003) 'A review of sex differences in sexual jealousy, including self-report data, psychophysiological responses, interpersonal violence, and morbid jealousy.' *Personality and Social Psychology Review* **7**: 102–28.

Harris, G. T., Zoe Hilton, N. et al. (2007) 'Children killed by genetic parents versus stepparents.' *Evolution and Human Behaviour* **28**(2): 85–95.

Hartung, J. (1976) 'On natural selection and the inheritance of wealth.' *Current Anthropology* **17**: 607–22.

Hartung, J. (1982) 'Polygyny and the inheritance of wealth.' *Current Anthropology* **23**: 1–12.

Hartung, J. (1985a) 'Review of Shepher's incest: a biosocial view.' *American Journal of Physical Anthropology* **67**(2): 169–71.

Hartung, J. (1985b) 'Matrilineal inheritance: new theory and analysis.' *Behavioral and Brain Science* **8**: 661–8.

Harvey, P. H. and Bradbury, J. W. (1991) Sexual selection, in J. R. Krebs and N. B. Davies (eds) *Behavioural Ecology*. Oxford, Blackwell Scientific.

Harwood, J. (1976) 'The race–intelligence controversy: a sociological approach – professional factors.' *Social Studies of Science* **6**: 369–94.

Harwood, J. (1977) 'The race–intelligence controversy: a sociological approach II – external factors.' *Social Studies of Science* **7**: 1–30.

Haselton, M. G. and Buss, D. M. (2000) 'Error management theory: a new perspective on biases in cross-sex mind reading.' *Journal of Personality and Social Psychology* **78**: 81–91.

Haselton, M. G. and Ketelaar, T. (2006) Irrational emotions or emotional wisdom? The evolutionary psychology of emotions and behaviour, in J. P. Forgas (ed.) *Hearts and Minds: Affective Influences on Social Cognition and Behaviour*. New York, Psychology Press.

Haselton, M. G., Buss, D. M. et al. (2005) 'Sex, lies and strategic interference: the psychology of deception between the sexes.' *Personality and Social Psychology Bulletin* **31**: 3–23.

Hatfield, E. and Sprecher, S. (1986) *Mirror, Mirror: The Importance of Looks in Everyday Life*. Albany, NY, University of New York Press.

Healey, M. D. and Ellis, B. J. (2007) 'Birth order, conscientiousness, and openness to experience: tests of the family-niche model of personality using a within-family methodology.' *Evolution and Human Behaviour* **28**(1): 55–9.

Hebb, D. O. (1949) *The Organization of Behavior*. New York, Wiley.

Heinrich, J. and Boyd, R. (1998) 'The evolution of conformist transmission and the emergence of between-group differences.' *Evolution and Human Behaviour* **19**: 215–42.

Heinrich, J., Albers W., Boyd R. et al. (1999) Group report: what is the role of culture in bounded rationality, in G. Gigerenzer and R. Selten (eds) *Bounded Rationality*. Cambridge, MA, MIT.

Henriques, G. R. (2002) 'The harmful dysfunction analysis and the differentiation between mental disorder and disease.' *Scientific Review of Mental Health Practice* **1**: 157–73.

Herrnstein, R. and Murray, C. (1994) *The Bell Curve: Intelligence and Class Structure in American Life*. New York, Free Press.

Hertwig, R. and Gigerenzer, G. (1999) 'The "conjunction fallacy revisited": how intelligent inferences look like reasoning errors.' *Journal of Behavioural Decision Making* **12**: 275–305.

Heyes, C. (2006) Beast machines: questions of animal consciousness, in M. S. Davies and L. Weiskrantz (eds) *Frontiers of Consciousness*. Oxford, Oxford University Press.

Heyes, C. M. (1998) 'Theory of mind in nonhuman primates.' *Behavioral and Brain Sciences* **21**(1): 101–34.

Hill, K. (1982) 'Hunting and human evolution.' *Journal of Human Evolution* **11**: 521–44.

Hill, K. (1993) 'Life history theory and evolutionary anthropology.' *Evolutionary Anthropology* **2**: 78–88.

Hill, K. and Hurtado, A. M. (1997) The evolution of premature reproductive senescence and menopause in human females: an evaluation of the grandmother hypothesis, in L. Betzig (ed.) *Human Nature: A Critical Reader*. Oxford, Oxford University Press.

Hill, K. and Hurtado, A. M. (1996) *Demographic/Life History of Ache Foragers*. Hawthorne, NY, Aldine de Gruyter.

Hill, K. and Kaplan, H. (1988) Tradeoffs in male and female reproductive strategies among the Ache, in L. Betzig, M. Borgerhoff Mulder and P. Turke (eds) *Human Reproductive Behaviour*. Cambridge, Cambridge University Press.

Hinde, R. A. (1982) *Ethology*. Oxford, Oxford University Press.

Holloway, R. L. (1975) Early hominid endocasts: volumes, morphology, and significance for hominid evolution, in R. Tuttle (ed.) *Primate Functional Morphology and Evolution*. The Hague, Mouton.

Holloway, R. L. (1983) 'Human paleontological evidence relevant to language behaviour.' *Human Neurobiology* **2**: 105–14.

Holloway, R. L. (1996) Evolution of the human brain, in A. Lock and C. R. Peters (eds) *Handbook of Symbolic Evolution*. Oxford, Oxford University Press.

Holmes, W. G. and Sherman, P. W. (1982) 'The ontogeny of kin recognition in two species of ground squirrels.' *American Zoologist* **22**: 491–517.

Hoogland, J. L. (1983) 'Nepotism and alarm calls in the black-tailed prairie dogs (*Cynomys ludovicianus*).' *Animal Behaviour* **31**: 472–9.

Hosken, F. P. (1979) *The Hosken Report: Genital and Sexual Mutilation of Females*. Lexicon, MA, Women's International Network News.

Howell, N. (1979) *The Demography of the Dobe !Kung*. New York, Academic Press.

Hrdy, S. B. (1979) 'Infanticide among animals: a review, classification and examination of the implications for the reproductive strategies of females.' *Ethology and Sociobiology* **1**: 13–40.

Huang, Z., Willet, W. C. et al. (1999) 'Waist circumference, waist : hip ratio, and risk of breast cancer in the nurses' Helath study.' *American Journal of Epidemiology* **150**: 1316–24.

Hugdahl, K. and Karker, A. C. (1981) 'Biological vs. experiential factors in phobic conditioning.' *Behavioural Research and Therapy* **19**: 109–15.

Hull, D. L. (1981) Units of evolution: a metaphysical essay, in V. J. Jensen and R. Harre (eds) *The Philosophy of Evolution*. Brighton, Harvester.

Hume, D. ([1739] 1985) *A Treatise of Human Nature*. London, Penguin Classics.

Humphreys, L.G. (1939) 'Acquisition and extinction of verbal expectations in a situation analogous to conditioning.' *Journal of Experimental Psychology* **25**: 294–301.

Huxley, J., Mayr, E. et al. (1964) 'Schizophrenia as a genetic morphism.' *Nature Genetics* **204**: 220–1.

Hyland, M. E. (1991) 'What are the incest rules of the Upper Paleolithic people? Putting evolution into an evolutionary analysis.' *Behavioural and Brain Sciences* **14**: 247–93.

Iacoboni, M., Molnar-Szakacs, I. et al. (2005) 'Grasping the intentions of others with one's own mirror neuron system.' *PLoS Biology* **3**(3): e79.

Ingman, M., Kaessmann, H. et al. (2000) 'Mitochondrial genome variation and the origin of modern humans.' *Nature* **408**: 708–13.

Irons, W. (1979) Natural selection, adaptation and human social behaviour, in N. I. A. Chagnon and W. Irons (eds) *Evolutionary Biology and Human Social Behaviour: An Anthropological Perspective*. North Scituate, MA, Duxbury.

Irons, W. (1983) Human female reproductive strategies, in S. K. Wassler (ed.) *Social Behaviour of Female Vertebrates*. New York, Academic Press.

Isbell, L. A. and Young, T. P. (1996) 'The evolution of bipedalism in hominids and reduced group size in primates.' *Journal of Human Evolution* **26**: 183–202.

James, W. (1890) *Principles of Psychology*. New York, Henry Holt.

Jamison, K. R. (1994) *Touched With Fire: Manic Depressive Illness and the Artistic Temperament*. New York, Free Press.

Jankowiak, W. and Diderich, M. (2000) 'Sibling solidarity in a polygamous community in the USA: unpacking inclusive fitness.' *Evolution and Human Behaviour* **21**: 125–39.

Jefferson, T., Herbst, J. H. et al. (1998) 'Associations between birth order and personality traits: evidence from self-reports and observer ratings.' *Journal of Research in Personality* **32**: 498–509.

Jerison, H. J. (1973) *Evolution of the Brain and Intelligence*. New York, Academic Press.

Johnson-Laird, P. N. and Oatley K. (1992) 'Basic

emotions, rationality, and folk theory.' *Cognition and Emotion* 6(3/4): 201–23.

Johnston, V. S., Hagel, R. et al. (2001) 'Male facial attractiveness: evidence for hormone-mediated adaptive design.' *Evolution and Human Behaviour* **22**: 251–69.

Kahneman, D., Slovic, P. et al. (eds) (1982) *Judgement under Uncertainty*. Cambridge, Cambridge University Press.

Kalikow, T. (1983) 'Konrad Lorenz's ethological theory: explanation and ideology 1938–1943.' *Journal of the History of Biology* **16**: 39–73.

Kanner, L. (1943) 'Autistic disturbances of affective contact.' *Nervous Child* **2**: 217–50.

Kant, I. (1785) First section transition from the common rational knowledge of morality to the philosophical, in I. Kant *Fundamental Principles of the Metaphysic of Morals*, trans. T. Kingsmill Abbot.

Kaplan, H. S. and Gangestad, S. W. (2005) Life history theory and evolutionary psychology, in D. Buss (ed.) *The Handbook of Evolutionary Psychology*. Hoboken, NJ, John Wiley & Sons.

Kaplan, H. S. and Lancaster, J. B. (1999) Skills-based competitive labour markets, the demographic transition, and the interaction of fertility and parental human capital in the determination of child outcomes, in L. Cronk, W. Irons and N. Chagnon (eds) *Human Behaviour and Adaptation: An Anthropological Perspective*. New York, Aldine de Gruyter.

Kaplan, H. S., Hill, K. et al. (2000) 'A theory of human life history evolution: diet, intelligence, and longevity.' *Evolutionary Anthropology* **9**: 156–85.

Kaplan, H. S., Lancaster, J. B. et al. (1995) 'Does observed fertility maximize fitness among New Mexican men? A test of an optimality model and a new theory of parental investment in the embodied capital of offspring.' *Human Nature* **6**: 325–60.

Karlsson, J. L. (1974) 'Inheritance of schizophrenia.' *Acta Psychiatrica* **247**: 77–88.

Keller, L. and Ross, K. G. (1998) 'Selfish genes: a green beard effect.' *Nature* **394**: 573–5.

Kenny, A. (1986) *Rationalism, Empiricism and Idealism*. Oxford, Oxford University Press.

Kenny, C. T. and Fletche, D. (1973) 'Effects of beardedness on person perception.' *Perceptual and Motor Skills* **37**: 413–14.

Kenrick, D. T., Sadalla, E. et al. (1990) 'Evolution, traits, and the stages of human courtship: qualifying the parental investment model.' *Journal of Personality* **58**: 97–116.

Ketelaar, T. and Au, W. T. (2003) 'The effects of guilty feelings on the behaviour of uncooperative individuals in repeated social bargaining games: an affect-as-information interpretation of the role of emotion in social interaction.' *Cognition and Emotion* **17**: 429–53.

Ketelaar, T. and Goodie, A. S. (1998) 'The satisficing role of emotions in decision making.' *Psykhe: Revista de la Escuela de Psicologia* **7**: 63–77.

Keverne, E. B., Fundele, R. et al. (1996) 'Genomic imprinting and the differential roles of parental genomes in brain development.' *Developmental Brain Research* **92**: 91–100.

Kipling, R. (1967) *Just So Stories*. London, Macmillan – now Palgrave Macmillan.

Kirkpatrick, D. R. (1984) 'Age, gender, and patterns of common intense fears among adults.' *Behaviour Research and Therapy* **22**: 141–50.

Kirkpatrick, M. and Ryan, M. (1991) 'The evolution of mating preferences and the paradox of the lek.' *Nature* **350**: 33–8.

Kitcher, P. (1985) *Vaulting Ambition*. Cambridge, MA, MIT Press.

Klineberg, O. (1940) *Social Psychology*. New York, Holt.

Klinnert, M. D., Campos, J., Sorce, J. et al. (1982) The development of social referencing in infancy, in R. Plutchik and H. Kellerman (eds) *Emotion: Theory, Research and Experience*, vol. 2, *Emotion in Early Development*. New York, Academic Press.

Kloss, R. J. and Nesse, R. M. (1992) 'Trisomy: chromosome competition or maternal strategy.' *Ethology and Sociobiology* **13**: 283–7.

Kohlberg, L. (1984) *Essays in Moral Development*, vol. 2, *The Psychology of Moral Development*. New York, Harper & Row.

Kohn, M. and Mithen, S. (1999) 'Handaxes: products of sexual selection.' *Antiquity* **73**: 518–26.

Krebs, D. L. (2000) The evolution of moral dispositions in the human species, in D. Lecroy and P. Moller (eds) *Evolutionary Perspectives on Human Reproductive Behaviour*. New York Academy of Sciences.

Krebs, D. L. (2005) The evolution of morality, in D. Buss (ed.) *The Handbook of Evolutionary Psychology*. Hoboken, NJ, John Wiley & Sons.

Kuch, K., Cox, B. J. et al. (1994) 'Phobias, panic and pain in 55 survivors of road vehicle accidents.' *Journal of Anxiety Disorders* **8**: 181–7.

Kvarnemo, C. and Ahnesjo, I. (1996) 'The dynamics

of operational sex ratios and competition for mates.' *Trends in Evolution and Ecology* **11**: 4–7.

Lack, D. (1943) *The Life of the Robin*. London, Penguin.

Laitman, J. T. (1984) 'The anatomy of human speech.' *Natural History* August: 20–27.

Laland, K. N. and Brown, G. R. (2002) *Sense and Nonsense: Evolutionary Perspectives on Human Behaviour*. Oxford, Oxford University Press.

Laland, K. N., Richerson, P. J. et al. (1996) Developing a theory of animal social learning, in C. M. Heyes and B. G. Galef (eds) *Social Learning in Animals: The Roots of Culture*. New York, Academic Press.

Lalumière, M. L. and Seto, M. C. (1998) 'What's wrong with psychopaths?' *Psychiatry Rounds* **2**: 6.

Lalumière, M. L., Harris, G. T. et al. (2001) 'Psychopathy and developmental instability.' *Evolution and Human Behaviour* **22**: 75–92.

Lancaster, J. B. (1997) An evolutionary history of human reproductive strategies and the status of women in relation to population growth and social stratification, in P. A. Gowaty (ed.) *Evolution and Feminism*. New York, Chapman & Hall.

Lancaster, J. B. and King, B. J. (1995) An evolutionary perspective on menopause, in V. Kerns and J. K. Brown (eds) *In Her Prime*. Boston, MA, Bergin & Garvey.

Lancaster, J. B. and Lancaster, C. S. (1983) Parental investment: the hominid adaptation, in D. J. Ortner (ed.) *How Humans Adapt: A Biocultural Odyssey*. Washington, DC, Smithsonian Institution Press.

Langlois, J. H. (1990) 'Infants differential social responses to attractive and unattractive faces.' *Developmental Psychology* **26**: 153–9.

Langlois, J. H. and Roggman, L. A. (1990) 'Attractive faces are only average.' *Psychological Science* **1**: 115–21.

Langlois, J. H., Kalakanis, L. E. et al. (2000) 'Maxims or myths of beauty: a meta-analytic and theoretical review.' *Psychological Bulletin* **126**: 390–423.

Lanska, J. D., Lanska, M. J. et al. (1985) 'Factors influencing anatomical location of fat tissue in 52,953 women.' *International Journal of Obesity* **9**: 29–38.

Leakey, R. (1994) *The Origin of Humankind*. London, Weidenfeld & Nicolson.

Leavitt, G. (1990) 'Sociobiological explanations of incest avoidance: a critical review of evidential claims.' *American Anthropologist* **92**: 971–93.

LeDoux, J. (1997) *The Emotional Brain: The Mysterious Underpinnings of Emotional Life*. New York, Simon & Schuster.

Lee, R. B. (1979) *The !Kung San: Men, Women, and Work in a Foraging Society*. Cambridge, Cambridge University Press.

Lehrman, D. S. (1953) 'A critique of Konrad Lorenz's theory of instinctive behaviour.' *Quarterly Review of Biology* **28**: 337–63.

Leslie, A. M. (1984) 'Spatiotemporal continuity and the perception of causality in infants.' *Perception* **13**: 287–305.

Leslie, A. M. (1987) 'Pretense and representation: the origins of "theory of mind".' *Psychological Review* **94**: 412–26.

Lévi-Strauss, C. (1956) The family, in H. L. Shapiro (ed.) *Man, Culture and Society*. London, Oxford University Press.

Lewin, R. (2005) *Human Evolution: An Illustrated Introduction*. Oxford, Blackwell.

Lieberman, D., Tooby, J. and Cosmides, L. (2003) 'Does morality have a biological basis? An empirical test of the factors governing moral sentiments relating to incest.' *Proceedings of the Royal Society of London*. Series B **270**: 819–26.

Lilienfield, S. O. and Marino, L. (1999) 'Essentialism revisited: evolutionary theory and the concept of mental disorder.' *Journal of Abnormal Psychology* **108**: 400–11.

Little, A. C., Penton-Voak, I. S. et al. (2000) Evolution and individual differences in the perception of attractiveness: how cyclic hormonal changes and self-perceived attractiveness influence female preferences for male faces, in G. Rhodes and L. Zebrowitz (eds) *Facial Attractiveness*. Westport, Ablex.

Lopez, A. D. et al. (2006) *The Global Burden of Disease and Risk Factors*, DCPP Publications.

Lorenz, K. (1953) *King Solomon's Ring*. London, Reprint Society.

Low, B. (1989) 'Cross-cultural patterns in the training of children: an evolutionary perspective.' *Journal of Comparative Psychology* **103**(4): 311–19.

Low, B., Alexander, R. D. et al. (1987) 'Human hips, breasts and buttocks: Is fat deceptive?' *Ethology and Sociobiology* **8**: 249–57.

Lown, B. A. (1975) 'Comparative psychology twenty five years after.' *American Psychologist* **30**: 858–9.

Lumsden, C. J. and Wilson, E. O. (1981) *Genes, Mind and Culture*. Cambridge, MA, Harvard University Press.

Lumsden, L. J. and Wilson, E. O. (1982) 'Precis of genes, mind and culture.' *Behavioural and Brain Sciences* **5**: 1–7.

Mace, R. and Cowlishaw, G. (1996) 'Cross-cultural patterns of marriage and inheritance: a phylogenetic approach.' *Ethology and Sociobiology* **17**: 87–97.

McGaugh, J. L., Roozendaal, B. et al. (1999) Modulation of memory storage by stress hormones and the amygdaloid complex, in M. Gazzaniga (ed.) *Cognitive Neuroscience* (2nd edn). Cambridge, MA, MIT Press.

McGill, T. E. (1965) *Readings in Animal Behaviour.* New York, Rinehart & Winston.

McGuffin, P., Katz, R. et al. (1996) 'A hospital-based twin register of the hereditability of DSM-IV unipolar depression.' *Archives of General Psychiatry* **53**: 129–36.

McGuire, M. T. and Troisi, A. (1998) *Darwinian Psychiatry.* Oxford, Oxford University Press.

McGuire, M. T., Troisi, A. and Raleigh, M. M. (1997) Depression in evolutionary context, in Baron-Cohen S. (ed.) *The Maladapted Mind.* Hillsdale, NJ, Lawrence Erlbaum.

MacKenzie, D. (1976) 'Eugenics in Britain.' *Social Studies of Science* **6**: 499–532.

McKusick, V. A. (2000) 'Ellis-van Creveld syndrome and the Amish.' *Nature Genetics* **24**: 203–4.

McKusick, V. A., Egeland, J. A. et al. (1964) 'Dwarfism in the Amish: the Ellis-van Creveld syndrome.' *Bulletin Johns Hopkins Hospital* **115**: 306–36.

MacLean, P. D. (1972) 'Cerebral evolution and emotional processes: new findings on the striatal complex.' *Annals of the New York Academy of Sciences* **193**: 137–49.

McLellan, B. and McKelvie, S. J. (1993) 'Effects of age and gender on perceived facial attractiveness.' *Canadian Journal of Behavioural Science* **25**: 135–42.

Manning, J. and Pickup, L. J. (1998) 'Symmetry and performance in middle distance runners.' *International Journal of Sports and Medicine* **19**: 205–9.

Manning, J. T. (1995) 'Fluctuating asymmetry and body weight in men and women: implications for sexual selection.' *Ethology and Sociobiology* **16**: 145–53.

Manning, J. T., Koukourakis, K. et al. (1997) 'Fluctuating asymmetry, metabolic rate and sexual selection in human males.' *Evolution and Human Behaviour* **18**: 15–21.

Manning, J. T., Scutt, D. et al. (1996) 'Asymmetry and the menstrual cycle in women.' *Ethology and Sociobiology* **17**: 129–43.

Marks, I. M. and Nesse, R. M. (1994) 'Fear and fitness: an evolutionary analysis of anxiety disorders.' *Ethology and Sociobiology* **15**: 247–61.

Marks, I. M. and Tobena, A. (1990) 'Learning and unlearning fear: a clinical and evolutionary perspective.' *Neuroscience and Biobehavioural Reviews* **14**: 365–84.

Marlowe, F. W. (1998) 'The nubility hypothesis: the human breast as an honest signal of residual reproductive value.' *Human Nature* **9**: 263–71.

Marlowe, F. W. and Wetsman, A. (2001) 'Preferred waist-to-hip ratio and ecology.' *Personality and Individual Differences* **30**: 481–9.

Marlowe, F. W., Apicella, C. L. et al. (2005) 'Men's preferences for women's profile waist-to-hip ratio in two societies.' *Evolution and Human Behaviour* **26**(6): 458–69.

Marr, D. (1982) *Vision.* New York, W. H. Freeman.

Martin, R. D. (1981) 'Relative brain sizes and basal metabolic rate in terrestrial vertebrates.' *Nature* **293**: 57–60.

Martin, R. D. (1990) *Primate Origins and Evolution.* Princeton, Princeton University Press.

Maynard Smith, J. (1974) 'The theory of games and the evolution of animal conflicts.' *Journal of Theoretical Biology* **47**: 209–21.

Maynard Smith, J. (1989) *Evolutionary Genetics.* Oxford, Oxford University Press.

Maynard Smith, J. (ed.) (1982) *Evolution Now: A Century after Darwin.* London, Macmillan – now Palgrave Macmillan.

Mayr, E. (1976) *Evolution and the Diversity of Life: Selected Essays.* Cambridge, MA, Harvard University Press.

Mead, M. (1928) *Coming of Age in Samoa: A Psychological Study of Primitive Youth for Western Civilization.* New York, William Morrow.

Mealey, L. (1997) The sociobiology of sociopathy: an integrated evolutionary model, in S. Baron-Cohen (ed.) *The Maladapted Mind.* Hove, Psychology Press.

Mealey, L., Bridgestock, R. et al. (1999) 'Symmetry and perceived facial attractiveness.' *Journal of Personality and Social Psychology* **76**: 151–8.

Mealey, L., Daood, C. et al. (1996) 'Enhanced memory for faces of cheaters.' *Ethology and Sociobiology* **17**: 119–28.

Meister, I. G., Boroojerdi, B. et al. (2003) 'Motor

cortex hand area and speech: implications for the development of language.' *Neuropsychologia* **41**: 401–6.

Metcalf, R. A. and Whitt, G. S. (1977) 'Relative inclusive fitness in the social wasp *Polistes metricus*.' *Behavioural Ecology and Sociobiology* **2**: 353–60.

Meyer, R. G. and Deitsch, S. E. (1996) *The Clinician's Handbook: Integrated Diagnostics, Assessment, and Intervention in Adult and Adolescent Psychopathology*. Boston, Allyn & Bacon.

Miller, G. A. (2003) 'The cognitive revolution: a historical perspective.' *Trends in Cognitive Science* **7**(3): 141–5.

Miller, G. F. (1998) How mate choice shaped human nature: a review of sexual selection and human evolution, in C. Crawford and D. L. Krebs (eds) *Handbook of Evolutionary Psychology.* Mahwah, NJ, Lawrence Erlbaum.

Miller, G. F. (1999a) Sexual selection for cultural displays, in R. Dunbar, C. Knight and C. Power (eds) *The Evolution of Culture*. Edinburgh, Edinburgh University Press.

Miller, G. F. (1999b) 'Waste is good'. *Prospect* **38**: February.

Miller, G. F. (2000) *The Mating Mind: How Sexual Choice Shaped the Evolution of Human Nature.* London, Heinemann/Doubleday.

Milton, K. (1988) Foraging behaviour and the evolution of primate intelligence, in R. W. Byrne and A. Whiten (eds) *Machiavellian Intelligence*. Oxford, Oxford University Press.

Misra, A. and Vikram, N. (2003) 'Clinical and pathophysiological consequences of abdomininal adiposity and abdominial adipose tissue deposits.' *Nutrition* **19**: 456–7.

Mithen, S. (1996) *The Prehistory of the Mind: The Cognitive Origins of Art, Religion and Science*. London, Thames & Hudson.

Mock, D. W. and Parker, G. A. (1997) *The Evolution of Sibling Rivalry.* Oxford, Oxford University Press.

Moffitt, T. E. (1987) 'Parental mental disorder and offspring criminal behaviour: an adoption study.' *Psychiatry* **50**: 346–60.

Moller, A. P. (1987) 'Behavioural aspects of sperm competition in swallows (*Hirundo rustica*).' *Behaviour* **100**: 92–104.

Moore, H.D.M., Martin, M. and Birkhead, T.R. (1999) 'No evidence for killer sperm or other selective interactions between human spermatozoa in ejaculates of different males in vitro.' *Proceedings of the Royal Society of London*. Series B Biological Sciences **266**(1436): 2343–50.

Morgan, C. L. (1894) *An Introduction to Comparative Psychology*. London, Walter Scott.

Murdock, G. P. and White, D. R. (1969) 'Standard cross-cultural sample.' *Ethnology* **9**: 329–69.

Murphy, D. and Stich, S. (2000) Darwin in the madhouse, in P. Caruthers and A. Chamberlain (eds) *Evolution and the Human Mind.* Cambridge, Cambridge University Press.

Myers, D. G. (1993) *Social Psychology*. New York, McGraw-Hill.

National Heart, Lung and Blood Institute (NHLBI) (1998) *Clinical Guidelines on the Identification, Evaluation and Treatment of Overweight and Obesity in Adults: The Evidence Report.* Bethesda, MD, National Institute of Health.

Nesse, R. M. (1999) Testing evolutionary hypotheses about mental disorders, in S. C. Stearns (ed.) *Evolution in Health and Disease.* Oxford, Oxford University Press.

Nesse, R. M. (2001) *Evolution and the Capacity for Commitment.* New York, Russell Sage.

Nesse, R. M. (2004) 'Cliff-edged fitness functions and the persistence of schizophrenia.' *Behavioural and Brain Sciences* **27**(6): 862–3.

Nesse, R. M. (2005) 'Natural selection and the regulation of defenses: a signal detection analysis of the smoke detector principle.' *Evolution and Human Behaviour* **26**: 88–105.

Nesse, R. M. and Williams, G. C. (1995) *Evolution and Healing: The New Science of Darwinian Medicine.* London, Weidenfield & Nicolson.

Nettle, D. (2004) 'Evolutionary origins of depression: a review and reformulation.' *Journal of Affective Disorders* **81**: 91–102.

Nettle, D. and Clegg, H. (2006) 'Schizotypy, creativity and mating success in humans.' *Proceedings of Royal Society London*, B-Biological Sciences **273**(1586): 611–15.

Neuhauser, M. and Krackow, S. (2006) 'Adaptive-filtering of trisomy 21: risk of Down syndrome depends on family size and age of previous child.' *Naturwissenschaften* DOI 10.1007/s00114-006-0165-3.

Nowak, M. and Sigmund, K. (1998) 'Evolution of indirect reciprocity by image scoring.' *Nature* **393**: 573–6.

Nunn, C. L., Gittleman, J. L. and Antonovics, J. (2000) 'Promiscuity and the primate immune system.' *Science* **290**(5494): 1168–70.

O'Neill, P. and Petrinovich, L. (1998) 'A preliminary cross-cultural study of moral intuitions.' *Evolution and Human Behaviour* **19**(6): 349–67.

Oaksford, M. and Chater, N. (1994) 'A rational analysis of the selection task as optimal data selection.' *Psychological Review* **101**: 608–31.

Ohman, A. (1979) Fear relevance, autonomic conditioning, and phobias: a laboratory model, in P. O. Sjoden, S. Bates and W. S. Dockens III (eds) *Trends in Behaviour Therapy*. New York, Academic Press.

Ohman, A. and Mineka, S. (2001) 'Fears, phobias, and preparedness: toward an evolved module of fear and fear learning.' *Psychological Review* **108**(3): 483–522.

Ohman, A. and Soares, J. J. F. (1994) 'Unconscious anxiety: phobic responses to masked stimuli.' *Journal of Abnormal Psychology* **103**: 231–40.

Oldroyd, D. R. (1980) *Darwinian Impacts: An Introduction to the Darwinian Revolution*. Buckingham, Open University Press.

Ottenheimer, M. (1996) *Forbidden Relatives*. Urbana, IL, University of Illinois Press.

Pagel, M. (1998) 'Mother and father in surprise genetic agreement.' *Nature* **397**: 19–20.

Panksepp, J. (2006) 'Emotional endophenotypes in evolutionary psychiatry.' *Progress in Neuro-Psychopharmacology and Biological Psychiatry* **30**: 774–84.

Parker, G. A. (1970) 'Sperm competition and its evolutionary consequences in the insects.' *Biological Review* **45**: 525–67.

Parker, G. A., Baker, R. R. et al. (1972) 'The origin and evolution of gamete dimorphism and the male–female phenomenon.' *Journal of Theoretical Biology* **36**: 529–53.

Pashos, A. (2000) 'Does paternal uncertainty explain discriminative grandparental solicitude.' *Evolution and Human Behaviour* **21**: 97–111.

Pasquali, R., Gambineri, A. et al. (1999) 'The natural history of the metabolic syndrome in young women with the polycystic ovary syndrome and the effect on long-term oestrogen-progestagen treatment.' *Clinical Endocrinology* **50**: 517–27.

Passingham, R. (1988) *The Human Primate*. Oxford, W. H. Freeman.

Paul, R. A. (1991) 'Psychoanalytic theory and incest avoidance rules.' *Behavioural and Brain Sciences* **14**: 276–7.

Pearce, D., Markandya, A. and Barbier, E. B. (1989) *Blueprint for a Green Economy*. London, Earthscan.

Peccei, J. S. (1995) 'The origin and evolution of menopause: the altriciality-lifespan hypothesis.' *Ethology and Sociobiology* **16**: 425–49.

Pedersen, F. A. (1991) 'Secular trends in human sex ratios: their influence on individual and family behaviour.' *Human Nature*: 271–91.

Pellegrini, R. J. (1973) 'Impressions of male personality as a function of beardedness.' *Psychology* **10**: 29–33.

Penfield, W. and. Faulk, M. E (1955) 'The insula: further observations on its function.' *Brain* **78**: 445–70.

Pennington, R. (1992) 'Did food increase fertility? An evaluation of !Kung and Herero history.' *Human Biology* **64**: 497–521.

Pennisi, E. (2001) 'Tracking the sexes by their genes.' *Science* **291**: 1733–4.

Penton-Voak, I. S. and Chen, J. (2004) 'High salivary testosterone is linked to masculine male facial appearance in humans.' *Evolution and Human Behaviour* **25**(4): 229–42.

Penton-Voak, I. S. and Perrett, D. I. (2000) 'Female preference for male faces changes cyclically: further evidence.' *Evolution and Human Behaviour* **21**(1): 39–49.

Perrett, D. I., Burt, D. M. et al. (1999) 'Symmetry and human facial attractiveness.' *Evolution and Human Behaviour* **20**: 295–307.

Perrett, D. I., Lee, K. J. et al. (1998) 'Effects of sexual dimorphism on facial attractiveness.' *Nature* **394**: 884–7.

Perrett, D. I., May, K. A. et al. (1994) 'Facial shape and judgements of female attractiveness.' *Nature* **368**: 239–42.

Perusse, D. (1993) 'Cultural and reproductive success in industrial societies: testing the relationship at the proximate and ultimate levels.' *Behavioural and Brain Sciences* **16**: 267–323.

Petrinovich, L., O'Neill, P. et al. (1993) 'An empirical study of moral intuitions: towards an evolutionary ethics.' *Journal of Personality and Social Psychology* **64**: 467–78.

Pettijohn, T. F. I. and Tesser, A. (1999) 'An investigation of popularity in environmental context: facial feature assessment of American movie actresses.' *Media Psychology* **1**: 229–47.

Phillips, M. L., Young, A. W. et al. (1997) 'A specific neural substrate for perceiving facial expressions of disgust.' *Nature* **389**: 495–8.

Piaget, J. (1932) *The Moral Development of the Child.* London, Routledge Kegan & Paul.

Pietrzak, R. H., Laird, J. D. et al. (2002) 'Sex differences in human jealousy: a coordinated study of forced-choice, continuous rating-scale, and physiological responses on the same subjects.' *Evolution and Human Behaviour* **23**(2): 83–95.

Pinker, S. (1994) *The Language Instinct.* London, Penguin.

Pinker, S. (1997) *How the Mind Works.* New York, Norton.

Pinker, S. and Bloom, P. (1990) 'Natural language and natural selection.' *Behavioural and Brain Sciences* **13**: 707–84.

Pitchford, I. (2001) 'The origins of violence: is psychopathy an adaptation?' *Human Nature Review* **1**: 28–36.

Platek, S. M., Burch, R. L. et al. (2003) 'Reactions to children's faces: resemblance affects males more than females.' *Evolution and Human Behaviour* **23**: 159–166.

Platek, S. M., Raines, D. M. et al. (2004) 'Reactions to children's faces: males are more affected by resemblance than females are, and so are their brains.' *Evolution and Human Behaviour* **25**(6): 394–406.

Plomin, R. (1990) *Behavioural Genetics: A Primer.* New York, Freeman.

Plotkin, H. (2004) *Evolutionary Thought in Psychology: A Brief History.* Oxford, Blackwell.

Plotnik, J. M., de Waal, F. B. et al. (2006) 'Self-recognition in an Asian elephant.' *Proceedings of the National Academy of Sciences of America* **103**(45): 17053–7.

Polimeni, J. and Reiss, J. P. (2003) 'Evolutionary perspectives on schizophrenia.' *Canadian Journal of Psychiatry* **48**(1): 34–9.

Popper, K. (1959) *The Logic of Scientific Discovery.* London, Hutchinson.

Popper, K. (1963) *Conjectures and Refutations: The Growth of Scientific Knowledge.* London, Routledge & Kegan Paul.

Potts, B. (1995) 'Queen Victoria's gene.' *Biological Sciences Review* November: 18–22.

Premack, D. and Woodruff, G. (1978) 'Does the chimpanzee have a theory of mind?' *Behavioural and Brain Sciences* **1**: 515–26.

Preuschoft, S. (1992) 'Laughter and smile in Barbary macaques.' *Ethology* **91**: 220–36.

Price, J., Sloman, L. et al. (1997) The social competition hypothesis of depression, in S. Baron-Cohen (ed.) *The Maladapted Mind.* Hove, Psychology Press.

Price, L. H. (1968) The genetics of depressive behaviour, in A. Coppen and S. Walk (eds) Recent developments in affective disorders. *British Journal of Psychiatry* Special Publication No 2.

Prochaska, J. O. and Norcross, J. C. (1994) *Systems of Psychotherapy: A Transtheoretical Analysis* (3rd edn). Pacific Grove, CA, Brooks/Cole.

Profet, M. (1992) Pregnancy sickness as adaptation: a deterrent to maternal ingestion of teratogens, in J. H. Barkow, L. Cosmides and J. Tooby (eds) *The Adapted Mind: Evolutionary Psychology and the Generation of Culture.* New York, Oxford University Press.

Profet, M. (1993) 'Menstruation as a defense against pathogens transported by sperm.' *Quarterly Review of Biology* **68**: 335–86.

Pury, C. and Mineka, S. (1997) 'Fear-relevant covariation bias for blood-injury-relevant stimuli and aversive outcomes.' *Behavioural Research and Therapy* **35**: 35–47.

Pusey, A. (2004) Inbreeding avoidance in primates, in A. P. Wolf and W. H. Durham (eds) *Inbreeding, Incest and the Incest Taboo.* Stanford, CA, Stanford University Press.

Ramachandran V (2000) http://www.edge.org/3rd_culture/ramachandran/ramachandran_p1.html, accessed 2 March 2007.

Ramachandran, V. and Oberman, L. M. (2006) 'Broken mirrors.' *Scientific American* **295**(5): 39–45.

Rapoport, A. and Chammah, A. M. (1965) *Prisoners' Dilemma: A Study in Conflict and Cooperation.* Ann Arbor, MI, University of Michigan.

Rest, J. (1975) 'Longitudinal study of the defining issues test: a strategy for analysing developmental change.' *Developmental Psychology* **11**: 738–48.

Reynolds, J. D. and Harvey, P. H. (1994) Sexual selection and the evolution of sex differences, in R. V. Short and E. Balban (eds) *The Differences Between the Sexes.* Cambridge, Cambridge University Press.

Rhodes, G. (2006) 'The evolutionary psychology of facial beauty.' *Annual Review of Psychology* **57**: 199–226.

Richards, J. R. (1987) *Darwin and the Emergence of Evolutionary Theories of Mind and Behaviour.* Chicago, University of Chicago Press.

Richards, J. R. (2000) *Human Nature after Darwin.* London, Routledge.

Richards, R., Kinney, D., Lundy, I. and Benet, M.

(1988) 'Creativity and manic-depresives, cyclothymes, and their normal first-degree relatives.' *Journal of Abnormal Psychology* **97**: 281–8.

Richerson, P. J. and Boyd, R. (2005) *Not by Genes Alone*. Chicago, University of Chicago Press.

Ridley, M. (1993) *The Red Queen*. London, Viking.

Ridley, M. (1996) *The Origins of Virtue*. London, Viking.

Ridley, M. (2003) *Nature via Nurture*. London, Fourth Estate.

Rizzolatti, G. and Arbib, M. A. (1998) 'Language within our grasp.' *Trends in Neuroscience* **21**(5): 188–94.

Rizzolatti, G., Fogassi, L. and Gallese, V. (2006) 'Mirrors in the mind.' *Scientific American* **295**(5): 30–7.

Roberts, J. M. and Lowe, C. R. (1975) 'Where have all the conceptions gone?' *Lancet* **1**: 498–9.

Rodman, P. S. and McHenry, H. M. (1980) 'Bioenergetics of hominid bipedalism.' *Journal of Physical Anthropology* **52**: 103–6.

Rogers, A. R. (1989) 'Does biology constrain culture.' *American Anthropologist* **90**: 819–31.

Rolls, E. T. (1996) 'The orbitofrontal cortex.' *Philosophical Transactions of the Royal Society*, B **351**: 1433–44.

Romanes, G. J. (1887) Mental differences between men and women, in C. L. Morgan (ed.) *Essays by George John Romanes*. London, Longmans, Green.

Romanes, G. J. (1888) *Mental Evolution in Man*. New York, Appleton.

Rose, S. (1978) 'Pre-Copernican sociobiology.' *New Scientist* **80**: 45–6.

Rose, S., Kamin, L.J. and Lewontin, R.C. (1985) *Not in Our Genes*. London, Penguin.

Rosenham, D. L. and Seligman, M. E. P. (1989) *Abnormal Psychology*. New York, Norton.

Roth, G. and Dicke, U. (2005) 'Evolution of the brain and intelligence.' *Trends in Cognitive Science* **9**(5): 250–7.

Rush, A. J., Kain, J. W. et al. (1991) Neurological bases for psychiatric disorders, in R. N. Rosenberg (ed.) *Comprehensive Neurology*. London, Raven Press.

Ruvolo, M., Pan, D. et al. (1994) 'Gene trees and hominid phylogeny.' *Proceedings of the National Academy of Science* **91**: 8900–4.

Saino, N., Bolzern, A. M. et al. (1997) 'Immunocompetence, ornamentation, and viability of male barn swallows (*Hirundo rustica*).' *Proceedings of the National Academy of Science* **94**: 549–52.

Salmon, C. A. (1999) 'On the impact of sex and birth order on contact with kin.' *Human Nature* **10**: 183–97.

Samuels, R. (1998) 'Evolutionary psychology and the massive modularity hypothesis.' *British Journal for the Philosophy of Science* **49**: 575–602.

Saroglou, V. and Fiasse, L. (2003) 'Birth order, personality, and religion: a study among young adults from a three-sibling family.' *Personality and Individual Differences* **35**: 19–29.

Scheib, J. E., Gangestad, S. W. et al. (1999) 'Facial attractiveness, symmetry and cues to good genes.' *Proceedings of the Royal Society of London*, Series B **266**: 1913–17.

Schiff, B. B. and Lamon, M. (1994) 'Inducing emotion by unilateral contraction of face muscles.' *Cortex* **30**: 247–54.

Schmitt, D. (2003) 'Universal sex differences in the desire for sexual variety: tests from 52 nations, 6 continents and 13 islands.' *Journal of Personality and Social Psychology* **85**(1): 85–104.

Schmitt, D. (2004) Fundamentals of human mating strategies, in D. Buss (ed.) *The Handbook of Evolutionary Psychology*. Hoboken, NJ, John Wiley & Sons.

Schmitt, D. (2005) 'Sociosexuality from Argentina to Zimbabwe: a 48-nation study of sex, culture, and strategies of human mating.' *Behavioural and Brain Sciences* **28**: 247–311.

Schneider, M. A. and Hendrix, L. (2000) 'Olfactory sexual inhibition and the Westermarck effect.' *Human Nature* **11**(1): 65–91.

Scott, J. P. (1973) 'The organisation of comparative psychology.' *Annals of the New York Academy* **223**: 7–40.

Scourfield, J., McGuffin, P. et al. (1997) 'Genes and social skills.' *BioEssays* **19**: 1125–7.

Sear, R., Mace, R. et al. (2000) 'Maternal grandmothers improve nutritional status and survival of children in rural Gambia.' *Proceedings of the Royal Society* Series B **267**: 1641–7.

Seligman, M. E. P. (1971) 'Phobias and preparedness.' *Behaviour Therapy* **2**: 307–20.

Shackelford, T. K. and Larsen, R. J. (1999) 'Facial attractiveness and physical health.' *Evolution and Human Behaviour* **20**: 71–6.

Shapiro, L. and Epstein, W. (1998) 'Evolutionary theory meets cognitive psychology: a more selective perspective.' *Mind and Language* **13**(2): 171–94.

Shepard, R. and Metzler, M. J. (1971) 'Mental rotation of three dimensional objects.' *Science* **171**(972): 701–3.

Shepher, J. (1971) 'Mate selection among second generation kibbutz adolescents and adults: incest avoidance and negative imprinting.' *Archives of Sexual Behaviour* **1**(4): 293–307.

Sherman, P. W. and Billing, J. (1999) 'Darwinian gastronomy: why we use spices.' *Bioscience* **49**: 453–63.

Sherman, P. W. and Hash, G. A. (2001) 'Why vegetable recipes are not very spicy.' *Evolution and Human Behavior* **22**(3): 147–64.

Sherman, P. W. and Reeve H. K. (1997) Forward and backward: alternative approaches to studying human behaviour, in L. Betzig (ed.) *Human Nature.* Oxford, Oxford University Press.

Short, R. V. (1994) Why sex, in R. V. Short and E. Balaban (eds) *The Differences Between the Sexes. Cambridge.* Cambridge, Cambridge University Press.

Short, R. V. and Balban, E. (eds) (1994) *The Differences Between the Sexes.* Cambridge, Cambridge University Press.

Sigmund, K. (1993) *Games of Life: Explorations in Ecology, Evolution and Behaviour.* Oxford, Oxford University Press.

Sillen-Tullberg, B. and Moller, A. (1993) 'The relationship between concealed ovulation and mating systems in anthropoid primates: a phylogenetic analysis.' *American Naturalist* **141**: 1–25.

Silverman, I. (1987) Race, race differences, and race relations: perspectives from psychology and sociobiology, in C. Crawford, M. Smith and D. Krebs (eds) *Sociobiology and Psychology: Ideas, Issues and Applications.* Mahwah, NJ, Lawrence Erlbaum.

Silverman, I. and M. Eals (1992) Sex differences in spatial abilities: evolutionary theory and data, in J. H. Barkow, L. Cosmides and J. Tooby (eds) *The Adapted Mind.* Oxford, Oxford University Press.

Silverman, I., Choi, J. et al. (2000) 'Evolved mechanisms underlying wayfinding: further studies on the hunter-gatherer theory of spatial sex differences.' *Evolution and Human Behaviour* **21**(3): 201–15.

Simon, H. (1956) 'Rational choice and the structure of environments.' *Psychology Review* **63**: 129–38.

Simon, H. (1990) 'Invariants of human behaviour.' *Annual Review of Psychology* **41**: 1–19.

Simoons, F. J. (2001) 'Persistence of lactase activity among northern Europeans: a weighing of the evidence for the calcium absorption hypothesis.' *Ecology of Food and Nutrition* **40**: 397–469.

Simpson, J. A. and Gangestad, S. W. (1991) 'Individual differences in sociosexuality: evidence for convergent and discriminant validity.' *Journal of Personality and Social Psychology* **60**: 870–83.

Singer, P. (1998) Evolutionary workers' party. *The Times Higher Educational Supplement* 15 May: 15–17.

Singh, D. (1993) 'Adaptive significance of female attractiveness.' *Journal of Personality and Social Psychology* **65**: 293–307.

Singh, D. (1995) 'Female judgement of male attractiveness and desirability for relationships: role of waist-to-hip ratios and financial status.' *Journal of Personality and Social Psychology* **69**(6): 1089–101.

Singh, D. and Luis, S. (1995) 'Ethnic and gender consensus for the effect of waist-to-hip ratio on judgement of women's attractiveness.' *Human Nature* **6**(1): 51–65.

Skinner, B. F. (1957) *Verbal Behaviour.* New York, Appleton-Century-Crofts.

Skinner, B. F. (1976) *Walden Two.* Englewood Cliffs, NJ, Prentice Hall.

Skuse, D. H., James, R. S. et al. (1997) 'Evidence from Turner's syndrome of an imprinted X-linked locus affecting cognitive function.' *Nature Genetics* **387**: 705–8.

Smith, E. A. and Bliege Bird, R. L. (2000) 'Turtle hunting and tombstone opening: public generosity as costly signalling.' *Evolution and Human Behaviour* **21**: 245–61.

Smith, M., Kish, B. J. and Crawford, C. B. (1987) 'Inheritance of wealth as human kin investment.' *Ethology and Sociobiology* **8**: 171–82.

Smith, R. (1997) *The Fontana History of the Human Sciences.* London, Fontana.

Smith, R. L. (1984) Human sperm competition, in R. L. Smith (ed.) *Sperm Competition and the Evolution of Animal Mating Systems.* London, Academic Press.

Smoller, J. W. and Tsuang, M. T. (1998) 'Panic and phobic anxiety: defining phenotypes for genetic studies.' *American Journal of Psychiatry* **155**(9): 1152–62.

Spain, D. H. (1991) 'Muddled theory and misinterpreted data.' *Behavioural and Brain Sciences* **14**: 278–9.

Sparks, C. S. and Jantz, R. L. (2002) 'A reassessment of human cranial plasticity: Boas revisited.' *Proceedings of the National Academy of Science* **99**(23): 14636–9.

Spence, S., Shapiro, D. et al. (1996) 'The role of the right hemisphere in the physiological and cognitive processing of emotional stimuli.' *Psychophysiology* **33**: 112–22.

Spencer, H. (1855) *Principles of Psychology*. London, Longman, Brown, Green and Longmans.

Springer, S. P. and Deutch, G. (1998) *Left Brain, Right Brain: Perspectives from Cognitive Neurosciences*. New York, Freeman.

Stanford, C. B. (1998) 'The social behaviour of chimpanzees and bonobos.' *Current Anthropology* **39**(4): 399–419.

Steen, R. G. (1996) *DNA and Destiny: Nature and Nurture in Human Behaviour*. New York, Plenum.

Steffenburg, S., Gillberg, C. et al. (1989) 'A twin study of autism in Denmark, Finland, Iceland, Norway and Sweden.' *Journal of Child Psychology and Psychiatry* **30**: 405–16.

Sternglanz, S. H., Gray, J. L. and Murakami, M. (1977) 'Adult preferences for infantile facial features: an ethological approach.' *Animal Behaviour* **25**: 108–15.

Stevens, A. and Price, J. (1996) *Evolutionary Psychiatry*. London, Routledge.

Strachan, T. and Read, A. P. (1996) *Human Molecular Genetics*. Oxford, Bios Scientific.

Streeter, S. A. and McBurney, D. H. (2003) 'Waist-hip ratio and attractiveness: new evidence and a critique of "a critical test".' *Evolution and Human Behaviour* **24**(2): 88–99.

Stringer, C. B. and Andrews, P. (1988) 'Genetic and fossil evidence for the origin of modern humans.' *Science* **239**: 1263–8.

Sudbury, P. (1998) *Human Molecular Genetics*. London, Addison-Wesley.

Sugiyama, L. (2004) 'Is beauty in the context-sensitive adaptations of the beholder? Shiwiar use of waist-to-hip ratio in assessments of female mate value.' *Evolution and Human Behaviour* **25**(1): 51–62.

Sullivan, P. F., Neale, M. C. and Kendler, K. S. (2000) 'Genetic epidemiology of major depression: review and meta-analysis.' *American Journal of Psychiatry* **157**: 1552–62.

Sulloway, F. J. (1996) *Born to Rebel: Birth Order, Family Dynamics and Creative Lies*. New York, Pantheon.

Surbey, M. K. (1990) Family composition, stress, and the timing of human menarche, in T. E. Ziegler and F. B. Bercovitvch (eds) *Socioendocrinology of Primate Reproduction*. New York, Wiley-Liss.

Swami, V. and Furnham, A. (2007) *The Psychology of Physical Attraction*. London, Psychology Press.

Swami, V. and Tovee, M. J. (2005a) 'Male physical attractiveness in Britain and Malaysia: a cross-cultural study.' *Body Image* **2**: 383–93.

Swami, V. and Tovee, M. J. (2005b) 'Female physical attractiveness in Britain and Malaysia: a cross-cultural study.' *Body Image* **2**: 115–28.

Swami, V. and Tovee, M. J. (2006) 'Does hunger influence judgements of female physical attactiveness?' *British Journal of Psychology* **97**: 353–63.

Symons, D. (1979) *The Evolution of Human Sexuality*. Oxford, Oxford University Press.

Symons, D. (1992) On the use and misuse of Darwinism, in J. H. Barkow, L. Cosmides and J. Tooby (eds) *The Adapted Mind*. Oxford, Oxford University Press.

Symons, D. and Ellis, B. (1989) Human male–female differences in sexual desire, in A. E. Rasa, C. Vogel and E. Voland (eds) *The Sociobiology of Sexual and Reproductive Strategies*. London, Chapman & Hall.

Szasz, T. S. (1974) *The Age of Madness: The History of Involuntary Hospitalisation*. New York, Jason Aronson.

Thompson, P. (2000) *The Brain: A Neuroscience Primer*. New York, Worth.

Thornhill, N. W. (1991) 'An evolutionary analysis of rules regulating human inbreeding and marriage.' *Behavioural and Brain Sciences* **14**: 247–93.

Thornhill, R. and Gangestad, S. W. (1993) 'Human facial beauty: averageness, symmetry and parasite resistance.' *Human Nature* **4**: 237–69.

Thornhill, R. and. Gangestad, S. W. (1994) 'Human fluctuating asymmetry and sexual behaviour.' *Psychological Science* **5**: 297–302.

Thornhill, R., Gangestad, S. W. et al. (1996) 'Human female orgasm and male fluctuating asymmetry.' *Animal Behaviour* **50**: 1601–15.

Thorpe, W. H. (1961) *Bird Song*. London, Cambridge University Press.

Tienari, P. (1991) 'Interaction between genetic vulnerability and family environment: the Finnish adoptive family study of schizophrenia.' *Acta Psychiatrica* **84**: 460–5.

Tierson, F. D., Olsen, C. L. et al. (1985) 'Influence of cravings and aversion on diet in pregnancy.' *Ecology of Food and Nutrition* **17**: 117–29.

Tinbergen, N. (1952) 'The curious behaviour of the stickleback.' *Scientific American* **187**(Dec): 22–6.

Tinbergen, N. (1963) 'On the aims and methods of ethology.' *Zeitschrift fur Tierpsychologie* **20**: 410–33.

Todd, P. and Miller, G. (1999) From pride and prejudice to persuasion, in G. Gigerenzer, P. Todd and ABC Research Group (eds) *Simple Heuristics that Make Us Smart*. Oxford, Oxford University Press.

Tomarken, A. J., Mineka, S. et al. (1989) 'Fear-relevant selective associations and covariation bias.' *Journal of Abnormal Psychology* **98**: 381–94.

Tomasello, M. and Call, J. (1997) *Primate Cognition*. Oxford, Oxford University Press.

Tooby, J. and Cosmides, L. (1990) 'The past explains the present: adaptations and the structure of ancestral environments.' *Ethology and Sociobiology* **11**: 375–424.

Tooby, J. and Cosmides, L. (1992) The psychological foundations of culture, in J. H. Barkow, L. Cosmides and J. Tooby (eds) *The Adapted Mind*. Oxford, Oxford University Press.

Tooby, J. and Cosmides, L. (2005) Conceptual foundations of evolutionary psychology, in D. Buss (ed.) *The Handbook of Evolutionary Psychology*. Hoboken, NJ, John Wiley & Sons.

Tovee, M. J. and Cornelissen, P. L. (1999) 'The mystery of human beauty.' *Nature* **339**: 215–16.

Tovee, M. J. and Cornelissen, P. L. (2001) 'Female and male perceptions of female attractiveness in front-view and profile.' *British Journal of Psychology* **92**: 391–402.

Tovee, M. J., Swami, V. et al. (2006) 'Changing perceptions of attractiveness as observers are exposed to different culture.' *Evolution and Human Behaviour* **27**(6): 443–57.

Tovee, M. J., Tasker, K. et al. (2000) 'Is symmetry a visual clue to attractiveness in the human female body.' *Evolution and Human Behaviour* **21**(3): 191–201.

Trivers, R. L. (1971) 'The evolution of reciprocal altruism.' *Quarterly Review of Biology* **46**: 35–57.

Trivers, R. L. (1972) Parental investment and sexual selection, in B. Campbell (ed.) *Sexual Selection and the Descent of Man*. Chicago, Aldine.

Trivers, R. L. (1974) 'Parent–offspring conflict.' *American Zoologist* **14**: 249–64.

Trivers, R. L. (1981) Sociobiology and politics, in E. White (ed.) *Sociobiology and Human Politics*. Lexington, MA, Lexington Books.

Trivers, R. L. (1985) *Social Evolution*. Menlo Park, CA, Benjamin/Cummings.

Trivers, R. L. and Willard, D. E. (1973) 'Natural selection of parental ability to vary the sex ratio of offspring.' *Science* **179**: 90–2.

Tryon, R. C. (1940) 'Genetic differences in maze-learning ability in rats.' *Yearbook of the National Society for the Study of Education* **39**: 111–19.

Tsuang, M. T., Stone, W. S., et al. (2001) 'Genes, environment and schizophrenia.' *British Journal of Psychiatry* **178**: s18–24.

Tullberg, B. S. and Lummaa, V. (2001) 'Induced abortion ratio in Sweden falls with age, but rises again before menopause.' *Evolution and Human Behaviour* **22**: 1–10.

Turke, P. W. (1989) 'Evolution and the demand for children.' *Population and Development Review* **15**: 61–90.

Tutin, C. (1979) 'Responses of chimpanzees to copulation, with special reference to interference by immature individuals.' *Animal Behaviour* **27**: 845–4.

Tversky, A. and Kahneman, D. (1983) 'Extensional versus intuitive reasoning: the conjunction fallacy in probability judgment.' *Psychological Review* **90**: 293–315.

Underhill, P. A., Shen, P. et al. (2000) 'Y chromosome sequence variation and the history of human populations.' *Nature Genetics* **26**: 358–61.

Valenstein, E. S. (1998) *Blaming the Brain: The Real Truth about Drugs and Mental Health*. New York, Free Press.

Van Dongen, P. A. M. (1998) Brain size in vertebrates, in R. Nieuwenhuys (ed.) *The Central Nervous System of Vertebrates*, vol. 3. Berlin, Springer.

Van Hoof, J. (1972) A comparative approach to the phylogeny of laughter and smiling, in R. A. Hinde (ed.) *Non-verbal Communication*. Cambridge, Cambridge University Press.

Van Hoof, M. H., Voorhorst, F. J. et al. (2000) 'Insulin, androgen and gonadotrophin concentration, body mass index, and waist-to-hip ratio in the first years after menarche in girls with regular menstrual cycle, or oligomenorrhea.' *Journal of Clinical Endocrinology and Metabolism* **85**: 1394–1400.

Van Valen, L. (1973) 'A new evolutionary law.' *Evolutionary Theory* **1**: 1–30.

Voland, E. (1990) 'Differential reproductive success within the Krummhorn population.' *Behavioural Ecology and Sociobiology* **26**: 65–72.

Voland, E. and Engel, C. (1989) Women's reproduction and longevity in a premodern population, in E. Rasa, C. Vogel and E. Voland (eds) *The Sociobiology of*

Sexual and Reproductive Strategies. London, Chapman & Hall.

Von Neumann, J. and Morgenstern, O. (1944) *Theory of Games and Economic Behaviour*. Princeton University Press.

Vrba, E. S. (1999) Habitat theory in relation to the evolution in African neogene and hominids, in T. G. Bromage and F. Schrenk (eds) *African Biogeography, Climate Change, and Early Hominid Evolution*. New York, Oxford University Press.

Wagstaff, J., Knoll, J. H. et al. (1992) 'Maternal but not paternal transmission of 15q11-q13 linked nondeletion Angelman syndrome leads to phenotypic expression.' *Nature Genetics* **1**: 291–4.

Wakefield, J. C. (1992) 'The concept of mental disorder: on the boundary between biological facts and social values.' *American Psychologist* **47**: 373–88.

Wallace, A. R. (1889) *Darwinism*. London, Macmillan and Son – now Palgrave Macmillan.

Wallace, A. R. (1905) *My Life: A Record of Events and Opinions*. London, Chapman & Hall.

Waller, B. and Dunbar, R. I. M. (2005) 'Differential behavioural effects of silent bared teeth display and relaxed open mouth display in chimpanzees (*Pan Troglodytes*).' *Ethology* **111**: 129–42.

Walter, A. (2006) 'The anti-naturalistic fallacy: evolutionary moral psychology and the insistence of brute facts.' *Evolutionary Psychology* **4**: 33–48.

Warner, H., Martin, D. E. et al. (1974) 'Electro-ejaculation of the great apes.' *Annals of Biomedical Engineering* **2**: 419–32.

Wason, P. (1966) Reasoning, in B. M. Foss (ed.) *New Horizons in Psychology*. Harmondsworth, Penguin.

Watson, J. B. (1930) *Behaviourism*. New York, Norton.

Watson, J. P. and Crick, F. H. C. (1953) 'A structure for deoxyribonucleic acid.' *Nature* **171**: 737–8.

Watson, P. J. and Andrews P. W. (2002) 'Towards a revised evolutionary adaptationist analysis of depression: the social navigation hypothesis.' *Journal of Affective Disorders* **72**: 1–14.

Wedekind, C. and Furi, S. (1997) 'Body odour preferences in men and women: do they aim for specific MHC combinations or simply heterozygosity?' *Proceedings of the Royal Society of London*, Series B **264**: 1471–9.

Wedekind, C., Seebeck, T., Bettens, F. et al. (1995) 'MHC-dependent mate preferences in humans.' *Proceedings of the Royal Society in London*, B **260**: 245–9.

Weeden, J. and Sabini, J. (2005) 'Physical attractiveness and health in Western societies: a review.' *Psychological Bulletin* **131**: 635–53.

Weiskrantz, L. (1956) 'Behavioural changes associated with ablation of the amygdaloid complex in monkeys.' *Journal of Comparative and Physiological Psychology* **49**: 381–91.

West-Eberhard, M. J. (1975) 'The evolution of social behaviour by kin selection.' *Quarterly Review of Biology* **50**: 1–33.

Westendorp, R. G. J. and Kirkwood, T. B. L. (1998) 'Human longevity at the cost of reproductive success.' *Nature* **396**: 743–6.

Westermarck, E. A. (1891) *The History of Human Marriage*. New York, Macmillan.

Wheeler, P. E. (1991) 'The thermoregulatory advantages of hominid bipedalism in open equatorial environments: the contribution of increased convective heat loss and cutaneous evaporative cooling.' *Journal of Human Evolution* **21**: 107–15.

White, L. (1948) 'The definition and prohibition of incest.' *American Anthropologist* **50**(part 1): 417.

Wicker, B., Keysers, C. et al. (2003) 'Both of us disgusted in my insula: the common neural basis of seeing and feeling disgust.' *Neuron* **40**(3): 655–64.

Wierson, M., Long, P. J. et al. (1993) 'Toward a new understanding of early menarche: the role of environmental stress in pubertal timing.' *Adolescence* **23**: 913–24.

Wilkinson, G. S. (1984) 'Reciprocal food sharing in vampire bats.' *Nature* **308**: 181–4.

Wilkinson, G. S. (1990) 'Food sharing in vampire bats.' *Scientific American* **262**: 76–82.

Williams, G. C. (1957) 'Pleitropy, natural selection and the evolution of senescence.' *Evolution* **11**: 398–411.

Williams, G. C. (1966) *Adaptation and Natural Selection*. Princeton, Princeton University Press.

Williams, G. C. (1975) *Sex and Evolution. Monographs in Population Biology*. Princeton, Princeton University Press.

Williams, G. C. and Nesse, R. M. (1991) 'The dawn of Darwinian medicine.' *Quarterly Review of Biology* **66**: 1–22.

Williams, J. H. G., Whitten A., Suddendorf, T. and Perrett, D. I. (2001) 'Imitation, mirror neurons and autism.' *Neuroscience and Biobehavioral Reviews* **25**: 287–95.

Wilson, D. S. (1992) Group selection, in E. F. Keller

and E. A. Lloyd (eds) *Keywords in Evolutionary Biology.* Cambridge, MA, Harvard University Press.

Wilson, D. S. (1994) 'Adaptive genetic variation and human evolutionary psychology.' *Ethology and Sociobiology* **15**: 219–35.

Wilson, D. S. and Sober, E. (1994) 'Re-introducing group selection to the human behavioural sciences.' *Behavioural and Brain Sciences* **17**: 585–654.

Wilson, E. O. (1975) *Sociobiology: The New Synthesis.* Cambridge, MA, Harvard University Press.

Wilson, E. O. (1998) *Consilience: The Unity of Knowledge.* New York, Knopf.

Wilson, E. O. and Ruse, M. (1985) 'The evolution of ethics.' *New Scientist* 17 October: 50–2.

Wilson, M. and Daly, M. (1992) The man who mistook his wife for a chattel, in J. H. Barkow, J. Tooby and L. Cosmides (eds) *The Adapted Mind.* Oxford, Oxford University Press.

Wimmer, H. and Perner, J. (1983) 'Beliefs about beliefs: representation and the constraining function of wrong beliefs in young children's understanding of deception.' *Cognition* **13**: 103–28.

Wirtz, P. (1997) 'Sperm selection by females.' *Trends in Ecology and Evolution* **12**(5): 172–3.

Wispe, L. G. and Thompson, J. W. (1976) 'The war between worlds: biological versus social evolution and some related issues.' *American Psychologist* **31**: 346.

Wolf, A. P. (1993) 'Westermarck redivivus.' *Annual Review of Anthropology* **22**: 157–75.

Wolf, A. P. (2004) Introduction, in A. P. Wolf and W. H. Durham (eds) *Inbreeding, Incest and the Incest Taboo.* Stanford, CA, Stanford University Press.

Wolpoff, M. H., Wu, X. et al. (1984) Modern *Homo sapiens* origins: a general theory of hominid evolution involving the fossil evidence from East Asia, in F. Smith and S. F. (eds) *The Origins o Humans: A World Survey of the Fossil Ev* York, Alan R Liss.

Workman, L. and Reader, W. (2004) *Ev Psychology*. Cambridge, Cambridge Un

Workman, L., Peters, S. et al. (2000) 'L of perceptual processing of pro- and a emotions displayed in chimeric faces.' 237–49.

Wright, R. (1994) *The Moral Animal: E Psychology and Everyday Life*. London,

Wyles, J. S., Kunkei, J. G. et al. (1983) iour and anatomical evolution.' *Proce National Academy of Sciences* **80**: 4394

Wynn, T. (1988) Tools and the evolut intelligence, in R. W. Byrne and A. W *Machiavellian Intelligence*. Oxford, Ox Press.

Wynne-Edwards, V. C. (1962) *Animal Relation to Social Behaviour*. Edinburg Boyd.

Yamazaki, K., Beauchamp, G. K. et al ial imprinting determines H–2 select erences.' *Science* **240**: 1331–2.

Young, J. Z. (1981) *The Life of Vertebr* Oxford University Press.

Yu, D. W. and Shepard, G. H. (1998) the eye of the beholder?' *Nature* **396**

Zaadstra, B. M., Seidell, J. C. et al. (1 female fecundity: prospective study fat distribution on conception rates *Journal* **306**: 484–7.

Zahavi, A. (1975) 'Mate selection: a handicap.' *Journal of Theoretical Biol*

Zihlman, A. (2000) *The Human Evol Book*. New York, HarperCollins.

Index

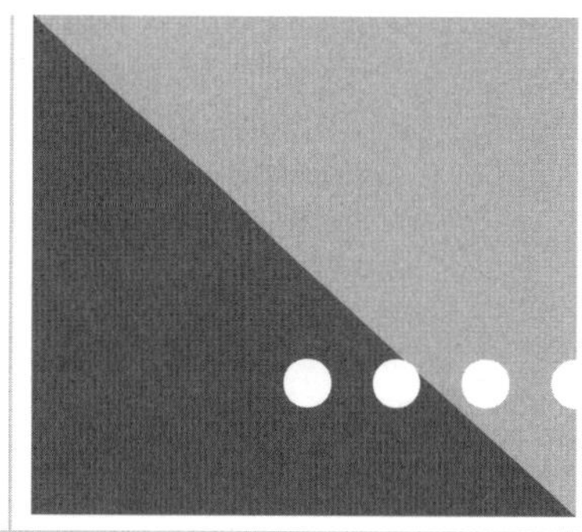

Scientific names for species and titles of books are given in *italics*.
Page numbers of glossary definitions are given in **bold**.

S

T

U

V

W

X

Y

Z